The LIVING EARTH

THOMAS Y. CROWELL COMPANY
New York tyc Established 1834

The
LIVING

ULDIS ROZE
*Queens College
of the City University of New York*

An introduction to biology

EARTH

Copyright © 1976 by Thomas Y. Crowell Company, Inc.

All Rights Reserved

Except for use in a review, the reproduction or utilization of this work in any form or by any electronic, mechanical, or other means, now known or hereafter invented, including photocopying and recording, and in any information storage and retrieval system is forbidden without the written permission of the publisher. Published simultaneously in Canada by Fitzhenry & Whiteside, Ltd., Toronto

The Living Earth: An Introduction to Biology is published in cooperation with Landsberger Publishing Corporation.

Cover Design: Richard Ference

Design by Angela Foote

Illustrations by J. Skorpil

Library of Congress Cataloging in Publication Data
Roze, Uldis, 1938–
   The living earth.
   Bibliographies
   Includes index.
   1. Biology. I. Title.
QH308.2.R69   1976   574   75-33889
ISBN 0-690-00856-2

Thomas Y. Crowell Company
666 Fifth Avenue
New York, New York 10019

Manufactured in the United States of America

For Rachel

Preface

In writing *The Living Earth,* I have held to one basic belief: Biology is an exciting, vibrant, important study, and anyone writing on biology who does not convey this excitement should try again. A text should be a pleasure to read, not a struggle. In this book I have not watered down the intellectual content of the subject matter, but have tried to look at each subject freshly, to avoid jargon, and to relate biology to everyday life. I have been greatly helped by my students, who have not hesitated to reject staleness, ambiguity, and appeal to authority, and who have demanded a biology that makes sense.

The organization of the book is straightforward and is one used by many biology texts: a progression from simpler to more complex assemblies of life—1. Cells, 2. Organisms, 3. Populations, 4. Communities. I have tried to keep each part relatively independent of the others so that they may be presented, for example, in the order 3-4-2-1. This flexibility is useful because, in the fall semester, the field biology (Part 4) must be presented early, before the snows. In the spring it must be presented late, after the flowers have opened. Selection of topics has been guided by the principle of range with depth. Biology is a vast topic: a single year's output of *Biological Abstracts* now fills an entire library shelf. This profusion of information leads to two common pitfalls. In one case, an attempt is made to cover everything, with the result that no time remains to say anything of substance. In the other case, the author may concentrate only on his or her own area of interest, to produce a specialized book. I have tried to steer between these extremes by covering a wide range of subject matter. However, at each level I have selected a few points to explore at greater length. Thus, I do not discuss all known biome types but do spend some time on the deciduous forest and the salt marsh. I do not discuss all the human organs but instead take a more leisurely look at the heart, the brain, the reproductive system. Some related systems are referred to in brief appendices in the interest of instructional flexibility. Similarly, energy metabolism and weak chemical bonds are discussed in appendices, to be used according to individual preference.

The general emphasis is on a biology that is accessible. In the first place, this means a strong emphasis on field biology in addition to the normal emphasis on laboratory discoveries. In this respect the book is unique. Very few students have access to laboratories or will read scientific papers, but all will spend time in the woods, ponds, marshes, and city lots. These I discuss. Likewise, the wild animals described are largely local insects and mammals, and the physiological systems are those of the human body. The student is encouraged to experience and observe life at first hand. At the same time, biology is also a cooperative human activity, and to emphasize this I have used judicious reference to living and dead scientists. I have also included reading lists at the ends of chapters. These books are selected for readability and literary value. Occasional brief excerpts are quoted as writing samples to encourage further reading.

Terminology is always heavy in an introductory text on biology, but an attempt has been made to keep it to a minimum (for example, intermediates of the Krebs cycle are not named, nor are most of the steps of mitosis and meiosis). Where a shorter or simpler term is available, it is used (e.g. sperm duct *v*. vas deferens). As a further helpful device, most new terms are supplied with etymologies as an aid to understanding.

While the book would, I feel, be useful to all biology students, it has been written with the needs of the nonmajor in mind: the intelligent student without an extensive chemical background. A number of excellent texts are available for the student with a strong chemical background. The student without such a background has been too often overlooked.

If this book has an underlying philosophy, it might be summarized as a caring for life. Although the book has a wide range, its guiding spirits are people like Rachel Carson, Edwin Way Teale, Jane Goodall, Aldo Leopold. Though sympathetic to the aims of groups such as the Sierra Club and the Environmental Defense Fund, it does not approach the problems of life from an adversary point of view. Rather, it tries to show that life at every level is complex, surprising, and worthy of respect and nurture.

Biology is very much a cooperative enterprise, and this book has not been an exception. I want to thank my friends and colleagues at Queens College who have assisted by critically reading portions of the manuscript: David Ewert, Andrew Greller, Carl Hirsch, Brian Malone, Leslie Marcus, Milton Nathanson, and Marvin Wasserman. In addition many others at Queens College and outside have opened their photograph collections and have helped with discussions of specific points. Their names are too numerous to mention, but they know who they are. Any errors which remain I accept as my own. Finally, a special thanks is due to Susan Munger, Biology Editor at Thomas Y. Crowell Company.

Uldis Roze
October, 1975

Contents

I. The Cell

Chapter 1 / Architecture of the Cell   2

    What Is a Cell? 5   The Sizes of Cells 5   Small Size Does Not Mean the Absence of Complexity 6   The Cell Wall 7   The Cell Membrane 7   Transport across the Cell Membrane 10   The Cell Membrane and Flow of Water 13   Internal Membranes 18   Lysosomes 19   The Nuclear Membrane 21   The Mitochondrion 23   The Chloroplast 25   Microtubules and Microfilaments 25

Chapter 2 / Cell Metabolism   32

    Macromolecules 36   Construction of Macromolecules 38   Metabolic Pathways 39   Enzymes as

Catalysts 40   Enzymes—Specificity 42   Enzymes as Proteins 44   The Advantages of Instability 46   Enzymes as Control Valves 48   Birth Defects 49

Appendix A: Energy Metabolism 50   Respiratory Energy Metabolism: Glycolysis, the Krebs Cycle, and Electron Transport 50   Photosynthesis 56
Appendix B: Weak Chemical Bonds 62

Chapter 3 / DNA and Protein Synthesis   68

DNA and Information 70   From DNA to Cell 71   The Replication of DNA 72   Proof for the Watson-Crick Mechanism 75   DNA and Protein Synthesis 78   Participation of RNA in Protein Synthesis: Messenger RNA 79   The Genetic Code 81   RNA and Protein Synthesis: Transfer RNA 82   RNA and Protein Synthesis: Ribosomal RNA 83   Control of DNA Function 86   The Control of Lactose Metabolism in *E. Coli* 87

II. The Organism

Chapter 4 / The Circulatory System   96

The Blood 100   *Plasma* 100   *The Red Blood Cells* 105   *The White Blood Cells* 108   The Heart and Blood Vessels 112   *The Heart* 114   *Blood Vessels* 115   Regulatory Mechanisms in the Circulatory System 118   *Intrinsic Mechanisms* 119   *Extrinsic Mechanisms* 120

Appendix A: The Digestive System 121
Appendix B: The Kidney 125   The Artificial Kidney 128

Chapter 5 / The Nervous System   132

   The Neuron 135   *The Nerve Impulse* 135   *The Synapse* 140   Fast Neuron Transport 144   The Peripheral Nervous System 145   *Voluntary Nerves and the Nerve Reflex* 147   The Autonomic Nervous System 147   The Central Nervous System—Vegetative Functions 149   *The Hindbrain* 151   *The Midbrain* 153   *The Forebrain* 153   The Central Nervous System—The Cerebral Cortex 160   *Motor and Sensory Cortex* 161   Memory and Learning 164

   Appendix A: The Sense Organs 167
      The Eye 167   The Ear 169   Taste and Smell 171   Other Body Senses 172
   Appendix B: Muscle 174
   Appendix C: The Hormonal System 178

Chapter 6 / Reproduction   188

   Sex on the Cellular Level 190   Production of Gametes and Sex Hormones 194   *Testes* 194   *The Ovaries* 200   Fertilization 207   *Role of Male Sex Organs* 207   *Female Genitalia* 209   *The Path of Sperm* 210   Pregnancy 211   *Role of Progesterone* 211   *Embryonic Development* 213   Birth 223   Contraception 225   *Future Methods of Birth Control* 230

III. Populations

Chapter 7 / Genetics—The Science of Inheritance   234

   Gene Expression 238   Genes and Chromosomes 238   *The Diploid State* 239   *Genes and the Environment* 242   *Polygenic and Pleiotropic Effects* 245   Meiosis and Gene Transmission 247   *Maternal and*

Paternal Contributions 247   *Crossing Over* 252
Independent Assortment and the Laws of Chance 253
Population Gentics 257   *Gene Frequencies* 257
Human Races 260   *Changes in Gene Frequencies* 262

Chapter 8 / Evolution   266

The Process of Evolution 269   *The Species* 269
Evolutionary Changes in the Species 273   *The Formation of New Species* 279   *Faunal Regions and Continental Drift* 283   *Adaptive Radiation* 289
Construction of Evolutionary Trees 296   *Evolution at the Molecular Level* 300   *Unanswered Questions in Evolution* 309

Chapter 9 / The Variety of Life   314

The Monerans 320   *The Bacteria* 320   *The Rickettsiae* 326   *The Blue-Green Algae* 326   *The Viruses* 328   *The Protistans* 331   *The Fungi* 333

Chapter 10 / The Plants   340

The Algae 342   *The Bryophytes* 345   *The Tracheophytes* 347   *The Flower* 349   *The Leaf* 358   *Xylem and Water Transport* 360   *The Root and Water Transport* 361   *The Phloem and Carbohydrate Transport* 363   *Anchoring and Support Against Gravity* 366   *The Hormonal System of Plants* 367

Chapter 11 / The Animals   374

The Coelenterate Phylum 376   *The Flatworm Phylum* 378   *The Annelid Phylum* 380   *The Arthropod Phylum* 383   *The Life of the Insects* 388

The Chordate Phylum 393   *The Amphibians* 396
The Reptiles 398   *The Birds* 402   *The Mammals* 405   *The Primates* 408   Homo Sapiens 411   Human Evolution 416   *The Future Evolution of Homo Sapiens* 420

IV. Communities

Chapter 12 / Field Biology   426

The Soil 429   *Major Soil Animals* 430   *Minor Soil Organisms* 432   *Feeding Relationships* 432   The Salt Marsh 433   *Salt Marsh Animals* 436   *Feeding Relationships* 440   The Freshwater Pond 441   *The Deep Pond* 442   *The Shallow Pond* 443   *The Edge of the Pond* 443   *Pond Insects* 444   *Pond Life and Pond Stability* 447   The Forest 448   *The Forest and the Seasons* 449   *Structure of the Forest* 455   Nature in the City 458   *Parks* 459   *The Vacant Lot* 460

Chapter 13 / The Growth and Regulation of Population 468

The Population Growth Curve 470   *Wildlife Management* 472   *Carrying Capacity for Homo Sapiens* 475   Population Control Factors 477   *Predation* 478   *Parasitism* 488   *Competition* 494   *Endocrine Control of Birthrate* 498   *Emigration* 499

Chapter 14 / Energy Flow and Materials Cycling   504

Energy Flow 506   Energy Flow Through Communities 507   *Community Productivities* 509   *Energy Budgets and Community Structure* 513   Use of

Energy by Consumers 517   *Energy Budget of a Community* 519   *Summary of Energy Flow* 526   The Cycling of Materials 527   *Carbon Cycle* 528   *The Nitrogen Cycle* 531

The Classification of Living Organisms 537

Glossary 543

Photograph Acknowledgments 565

Index 571

Introduction

Some years ago, soon after our marriage, my wife and I went to England to visit with her family and see the green countryside. I particularly remember a visit with an old farmer who lived alone in a stone cabin outside a small village in Wiltshire. He had grown up in the house, had quarried slate which covered the local roofs, had planted trees, herded cattle, and raised a family. He still fed his own chickens every morning, grew his own gooseberries and vegetables, and played cricket with his grandchildren when they came for visits. But he also read the London paper and had recently acquired a television, and I remember his eager questions about student rioting at American universities and about the chances of Muhammed Ali. He was a man with firm roots in his living countryside, but also with a thirst for the ideas and aspirations of his time. One could do worse than try to emulate this world view in a biology text.

The study of biology has always been pulled in two directions. The field biologist might spend years observing the digger wasps on a sand dune, or living with mountain gorillas in the rain forest, or tracking down spiderworts across the American southwest. Under the eyes of the field biologist, the commonplace events of a roadside, a back yard, or a small pond assume a new complexity and magic. Field biology is a low-budget activity accessible to everyone, even the streetwise inner city dweller or the car-bound suburbanite. It is the perception of the field biologist that guides the ethos of the environmental movement. An environmental concern not grounded in such first-hand experience will not survive the bitter political and economic winds that appear to be in store for it.

The other kind of biologist may study the same phenomena as the biologist in the field but does so from the laboratory. The past quarter century has been a kind of golden age for laboratory biologists. Immense vistas have opened up at every level. The secrets of the cell nucleus have been breached. The genetic code has been cracked, self-replicating viruses have been synthesized from inert chemicals, cells of rats and humans may be fused into a single cell, and human genes may be inserted into the bodies of bacteria. At every level there has been a cross-fertilization from other sciences. The geological theory of continental drift has given a rational explanation for observations in zoogeography. Chemical techniques have freed evolutionists from their close dependence on the fossil record. And long before this, the use of radioactive isotopes as tracers unravelled the intricacies of cellular metabolism. Unfortunately, the world of the laboratory is not a low-budget world and is not accessible to more than a very few. But all who have ever taken part in this world have felt an excitement that is a mind-altering experience. And there is no question that the discoveries of the laboratory will not remain buried inside libraries but will emerge to affect your life and mine. Not to know of these discoveries is like not knowing of Columbus in 1492 or of Galileo in 1500.

There is perhaps a third kind of biology which illuminates both of the preceding. This is armchair biology—the biology of books and magazines. Students of biology are lucky in possessing a literature that is often a joy to read. The proof comes from the fact that works of many of the greatest of biologists have graced the lists of national best-sellers. I have suggested books for further reading in end-of-chapter bibliographies and as illustration have included random paragraphs from several authors. Harder to include are examples from visual atlases, such as the one by R. G. Kessel and C. Y. Shih on scanning electron microscopy, Andreas Feininger on shells, Stephen Dalton on insects in flight. Impossible to include is the flavor of beautiful and well-written magazines, such as *Natural History, Audubon, Scientific American, Smithsonian*. But these are accessible to everyone.

In this book I have tried to strike a balance between the biology of the field and the biology of the laboratory because each adds its own dimension to the fuller understanding of life. Finally, I hope that students who have read this book will want to read further. No true student of biology can study without reading beyond textbooks.

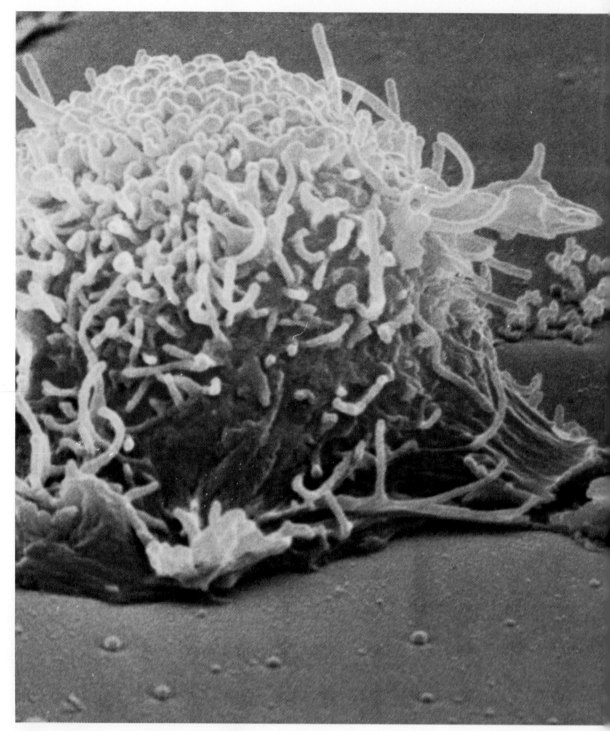

Part I

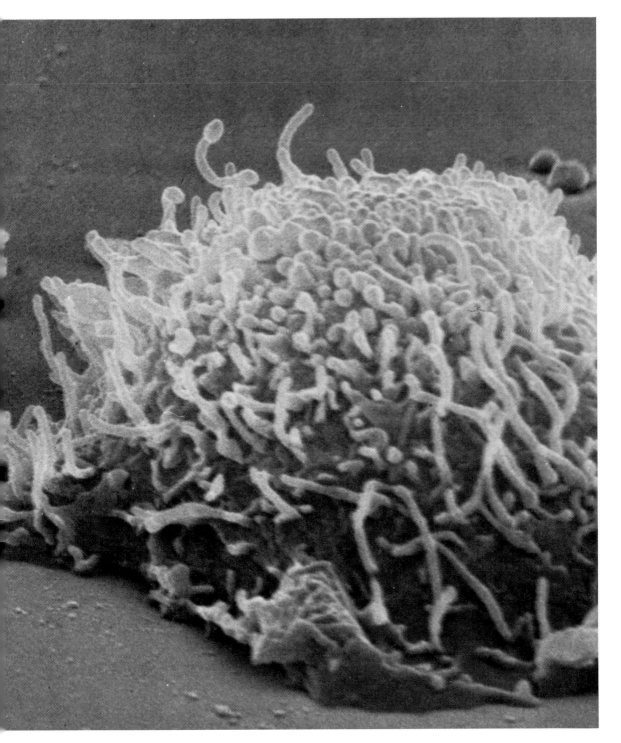

The Cell

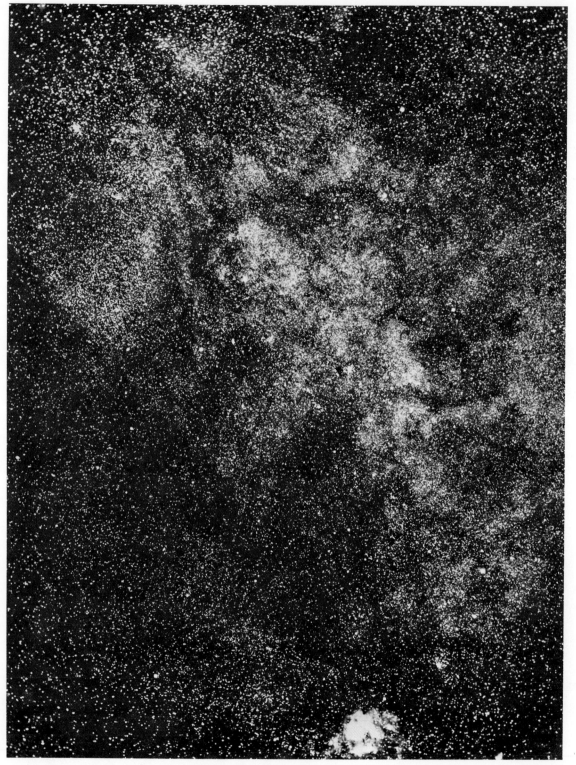

Figure 1-1 THE SUMMER SKY. A portion of the Milky Way in the region of Sagittarius.

chapter 1

Architecture of the Cell

Who, while gazing up at the glittering wash of stars on a dark summer night, has not felt a distance and loneliness in the presence of the universe pulsing? Long ago, the Greeks made efforts to humanize the stars, to order them into constellations more manageable to the human psyche. But these efforts to bring the stars into the scale of human experience were unavailing: the universe has remained a mysterious place. If anything, our twentieth century astronomers have made it more strange while adding encyclopedias filled with facts to the libraries.

Consider the death of a star. For billions of years, a star may radiate away its substance into space until finally comes the time of reckoning. With its fuel gone, the star begins a catastrophic collapse, and its ultimate fate depends on its initial mass. A star the size of the sun contracts until it collapses into a ball about the size of the earth. The star is now called a white dwarf, and it slowly radiates away its remaining heat, becoming a black cinder floating in space. If the initial mass of the star is somewhat larger (about 1.4 times the mass of the sun), the contraction process continues beyond the white dwarf stage and the star collapses into a neutron star. An enormous explosion accompanies the process, and for a moment the dying star is brighter than an entire galaxy made up of billions of stars. The remaining mass of the star is now compressed into a ball about 6 miles in diameter, with a density equal to that found in the central nucleus of an atom. A tablespoon of material from the neutron star weighs as much as a small planet 100 miles in diameter. The neutron star stops contracting because the enormous forces of gravitational collapse are counterbalanced by the even more massive force of nuclear repulsion, the strongest force now known in the universe. Direct proof for the existence of neutron stars has come only within the past few years, as astronomers discovered a series of small, periodically flashing bodies called pulsars. But what if the parent star initially had an even greater mass, say, two or three times the mass of the sun? (Many such stars are known.) In that case the gravitational forces generated during its collapse would be greater than even the nuclear force could withstand, and the star would reach the neutron star stage but keep right on collapsing. Its further lifetime would now be measured in seconds, not eons. It would shrink to the size of a helium balloon, then to the size of a grape, then to the size of a bacterium, and so on to infinity. Even before it reached these dimensions, its density would be so great that no light could escape from its surface—it would exist as a black hole in space. Do huge black holes inhabit the centers of galaxies, holding them together gravitationally? Do the black holes reach a new equilibrium dimension, stabilized by forces not yet discovered in the universe? Do they continue their contraction to emerge as antimatter in some antiuniverse? These are questions that cannot be answered at the present time.

How does all of this relate to the cell? The point is that the cell is also strange, in some ways as strange as the neutron star or the black hole. Our intuitive universe is made up of objects of a size that we can relate to—from insects to elephants, from pollen grains to redwood trees. Even the mountains and rivers and coastlines we know are no larger than our visual fields can encompass. But with stars or with cells, our senses fail. We must reinforce them with telescopes, microscopes, and a whole series of other sensory crutches. The result is a reality that flies in the face of much that we have come to think of as normal. And yet the gravity that makes an apple fall is the same gravity that collapses a neutron star. The natural laws we observe daily in our lives extend to the very large and to the very small and will act as our guides in these unfamiliar worlds. Let us now enter the unfamiliar world of the cell.

What Is a Cell?

We will say now, rather circumspectly, that a cell from an animal is a small, complete animal, and that a cell from a plant is a small, complete plant. Not complete of course in the sense that they have antlers or tails or flowers or roots, but complete in the sense that the animal cell can lead an independent, animallike existence, and the plant cell an independent, plantlike existence. In fact, our animalness is simply a reflection of the animal way of life of our cells, and it can be said with some justification that the purpose of all of our physical activities is the satisfaction of the needs of the little animals that make up our bodies. Let us begin by looking at the physical appearance of the cell.

The Sizes of Cells

Certainly the first characteristic that strikes the observer is the extremely small size of most cells. Our bodies contain around 100 trillion of these tiny animals, and this means sizes in the micron range. A micron (μ) is one millionth of a meter (a meter equals 3.281 feet). For the sake of illustration, a bee is roughly 15,000 microns long, and if a micron were drawn to the scale of 1 inch, the same bee would have to be drawn to the dimensions of the World Trade Center, the world's second tallest building. Typical animal cells may be as small as 5μ in diameter, or as large as 50μ in diameter. Is there any advantage to being this small? Indeed there is, and it has to do with the overall rate at which cells live; small cells live more intensely than large ones, the reason for this being a geometric one. A living cell continually imports and exports materials necessary for life; oxygen and foods come in, and carbon dioxide and waste products go out. The bulkier the cell, the greater the traffic in these materials, and the difficulty is that, as a roughly spherical cell increases in volume, its surface area increases much more slowly. Since all traffic must pass in and out via the surface, the surface membrane quickly becomes a bottleneck limiting the overall flow and thus of necessity limiting the intensity of the life processes within.

Perhaps the most extreme example is the yolk of a hen's egg. The yolk is a single cell (in fact, the largest known cells in the world are the yolks of birds' eggs). Being an animal cell, it undergoes normal life processes; it takes in oxygen and emits carbon dioxide and wastes, it produces heat, and so on. But these activities are all so sluggish that hardly anyone notices that in a boiled egg these processes have come to a halt. The boiled egg has stopped breathing. In the egg, the mass of the yolk is so enormous compared to its surface area that normal diffusion processes responsible for moving things in and out of the cell have not been able to provide the cell with a "living standard" anywhere comparable to that of more typical body cells. A more mathematical explanation of these factors is given in Table 1. Thus, if materials flow in and out at equal rates per unit surface, the egg yolk cell can live only at about one-thousandth the rate of a normal cell.

TABLE 1
Surface/volume ratios in normal cell and egg yolk cell

NORMAL CELL	EGG YOLK CELL
Radius $= 10 \, \mu$	Radius $= 10^4 \, \mu$[a]
Volume $= \frac{4}{3} \pi r^3 \approx 4{,}000 \, \mu^3$	Volume $= \frac{4}{3} \pi r^3 \approx 4 \times 10^{12} \, \mu^3$
Surface area $= 4\pi r^2 \approx 1{,}200 \, \mu^2$	Surface area $= 4\pi r^2 \approx 12 \times 10^8 \, \mu^2$
Surface/volume $\approx \frac{1{,}200}{4{,}000} = 0.3$	Surface/volume $\approx 3 \times 10^{-4} = 0.0003$

[a] 10^4 is a mathematical notation for 10,000, or 1 followed by four zeros.

Small Size Does Not Mean the Absence of Complexity

Having made the point that well over 99 percent of all cells are minute in size, one should not assume that they are simply disorganized blobs of protoplasmic substance. Tiny though they are, cells have an intricate architecture, and it would be no exaggeration to say that each living cell contains more precision parts than the family automobile or television set. Only a precision instrument would be capable of the extraordinary behavior that characterizes all cells. If say, GM or IBM tried to mass-produce cells and sell them on the open market, it is certain that they would cost thousands and possibly tens of thousands of dollars apiece, if they could be produced at all.

As we begin to look at the details of cellular architecture, let us keep in mind that the materials used in cell construction are not the familiar rigid ones of everyday experience. The world of the cell is a "soft" world, where everything bends and changes shape from moment to moment and moves around from here to there; but this softness and changeability should never be confused with formlessness. The instructions for making a living cell out of its normal "soft" components are as rigidly specific as would be the instructions for making a cell out of glass or stainless steel.

The Cell Wall

Essentially all plant and bacterial cells, and many one-celled protistans (Chapter 9) are surrounded by a thick, semirigid *cell wall*. In fact, the rigidity of lumber results from the presence of innumerable tiny cell walls, still locked together in the same arrangement as when the cells were living. In green plants, these cell walls are made of cellulose, the same material that constitutes this sheet of paper. This cell wall is not living, but should be considered a secretion product of the cell, much like hormones, waste products, and other materials. Cell walls serve the plant in many far-reaching ways. They are the analog of the animal skeletal system, and incidentally are much better at their job than any animal skeleton. The tallest living trees, the California redwoods, are also the largest and heaviest things that have ever lived; they are far larger than whales or ancient dinosaurs. Their size is made possible entirely by the properties of their cell walls—any known animal of this size would be immediately crushed to death by its own weight. Plants also use their cell walls for the purpose of transporting foods and fluids, for protection against osmotic stress (see page 15), and for a variety of special jobs such as controlling water loss from leaves.

The cell walls of bacteria are made of materials other than cellulose. The importance of the cell wall to bacterial survival can be appreciated from the success of penicillin as an antibacterial agent. Penicillin does not kill bacteria outright, but it prevents a bacterial cell from making a new cell wall. As long as the bacterium rests quietly, it is in no danger. But when it tries to divide, it cannot wrap itself in a new cell wall and either explodes because of high internal osmotic pressure or is quickly dispatched by cellular defense mechanisms.

In contrast to plants, bacteria, and many protistans and fungi, most animal cells lack cell walls. This accounts for the mobile, fluid body form of most animals.

The Cell Membrane

Inside any cell wall, and enclosing the fluid components of the cell (the cytoplasm), is the *cell membrane*. Much thinner than the cell wall, the membrane of the cell is only a few molecules thick. In fact, this structure is so thin that for decades its very existence was questioned, until the perfection of the electron microscope settled the argument by allowing clear photographs to be made of a membrane enclosing all cells.

What are the properties of this cell membrane? First, let us note one point which we shall come back to again many times. All life arose in water, and in a sense it has never left this element. All cells are immersed in water. Water seeps around every living cell in the body and makes up the bulk of its internal contents as well. Organisms such as the cactus and the desert rat carry around a little puddle of water within themselves. For such organisms, much of their effort in terms of behavior and physical structure is devoted to conserving and protecting this fluid cargo.

Now, it happens that water is perhaps the world's best solvent; it dissolves an incredibly long list of other compounds. After more than a century of work, chemists have discovered only one or two exotic compounds that can approach common, everyday water in power and versatility as a solvent. This means that the materials used in constructing the cell membrane will have to be very special if they are to withstand the powerful solvent action of water.

Fortunately, there is one class of biological compounds that has this necessary resistance—the class of compounds known as the lipids. Some everyday examples of lipids are butter, candle wax, oil, and grease. Chemically, butter is very different from petroleum oil, but they share the property of water insolubility. It is not these common examples of lipids that are used in cell membranes, but a special class of lipids called *phospholipids*. A functional (though not structural) equivalent of a phospholipid is soap.

What is the secret of the great cleansing power of soap? Soap happens to be a very strange molecule—it has two contradictory sets of properties. One end behaves like a good lipid should—it abhors water and associates only with others of its kind. This is called the *hydrophobic* ("water-fearing") end. The other end of the soap molecule loves water so much that it goes into solution and drags the hydrophobic tail along with it. This end is called the *hydrophilic* ("water-loving") end. When soap is used in cleaning, the hydrophobic end clamps on to grease and fats. The hydrophilic end then carries the whole complex into water solution (see Fig. 1-2).

How is all this related to phospholipids and cell membranes? Phospholipids, like natural soaps, are bifunctional molecules and, if it were not for their scarcity and expense, would make excellent soaps. The interesting thing is that scientists can use either of these substances to create artificial membranes only one molecular layer thick. It is very simple. Drop a tiny pinch of soap or phospholipid on top

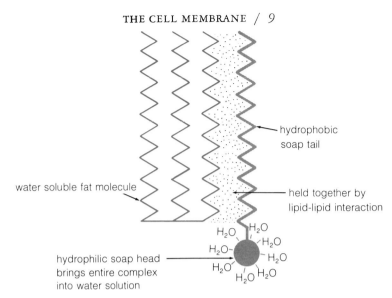

Figure 1-2 INTERACTION BETWEEN GREASE AND SOAP. The hydrophobic end of the soap molecule interacts with the large, almost totally hydrophobic grease molecule. The hydrophilic end of the soap molecule drags the entire complex into water solution.

of a tray of water. The hydrophilic heads of the molecules will bury themselves in the water, while the hydrophobic tails will resist getting wet and will stick up into the air. If we're using phospholipids, the molecules will be arranged as shown in Figure 1-3. The molecules can now be pushed close together by means of a thin, floating platinum wire. The result is a membrane. This artificial membrane is quite stable and will maintain itself by natural cohesive forces even if the floating wire barrier is taken away.

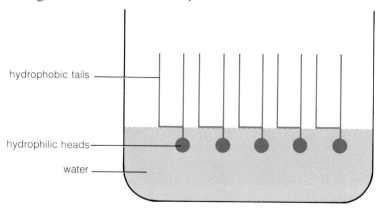

Figure 1-3 PHOSPHOLIPIDS AT AN AIR-WATER INTERFACE. Hydrophilic ends of the phospholipid molecules are buried in water. The hydrophobic ends are excluded from water and remain sticking up in the air. They interact with each other to form an elastic membrane.

In the case of the cell membrane, there is a double array of such phospholipid molecules. Such a membrane is a model of elegance and economy. It is vanishingly thin, yet strong enough to remain intact while the cell twists and moves about. The inner and outer surfaces of the membrane are wetted by water, while the middle portion remains tight and dry.

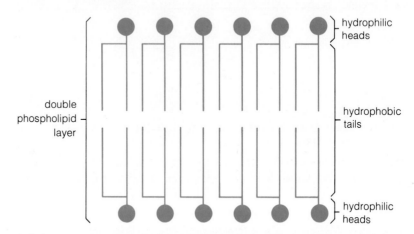

Figure 1-4 ARRANGEMENT OF PHOSPHOLIPIDS IN A NATURAL MEMBRANE. Phospholipids are joined in a double array with hydrophobic regions arranged towards the middle and hydrophilic heads to the outside.

Transport across the Cell Membrane

Unfortunately, this model is a bit too simple to account for both the behavior and components of the real cell membrane. Any molecular traffic through the model membrane would be based on two factors: the size of the molecule[1] (small molecules squeeze through more quickly) and lipid solubility (lipid-soluble molecules such as alcohol pass through more quickly). It so happens that experimentation fully confirms these predictions. Small molecules penetrate more quickly than large ones with the same chemical personality, and alcohol, being lipid-soluble, passes through with remarkable ease. In fact, one can get mildly drunk simply by pouring distilled alcohol into one's galoshes! Had the experimenters stopped there, we might have had a simple and satisfying theory of cell membrane structure. Unfortunately, they did not stop, but turned up a number of additional facts

---

1. Cells contain a number of chemicals, including molecules and free ions. Molecules are the smallest units of a chemical compound that still preserve the properties of that compound. Molecules consist of atoms held together by covalent bonds.

completely incompatible with the proposed model. For instance, it was discovered that a number of rather large molecules with no lipid solubility were sprinting in and out of cells much more quickly than their smaller cousins. An example of such a molecule is the common sugar glucose. As if that were not puzzling enough, glucose could enter a cell against a concentration gradient. That is to say, even if the inside of a cell contained a higher concentration of glucose than the outside, glucose would still continue to enter. This is very strange behavior—a bit like a faint breath of perfume in a room trying to squeeze back into the bottle. As might be expected, this is not something the glucose molecule does of its own volition—it must be actively pumped into the cell by some mechanism located in the cell membrane and powered by the cell's energy supply. As soon as there is interference with this energy supply, for example, by a poison such as cyanide, the pumping action stops. Such energy-requiring molecular pumps are encountered in every cell in the body, though different cells may be engaged in pumping different molecules. The process is called *active transport*.

Thus the cell membrane performs two functions. First, it must be able to *recognize* the glucose molecule as being different from others of approximately the same size and chemical properties. The fine degree of discrimination required is something a lipid surface is not capable of. This is because essentially all lipid molecules making up a membrane are alike, somewhat like bricks in a brick wall.

Such recognitive faculties in the biological world are relegated largely to a single class of molecules—the *proteins*. We will hear more about proteins later, but for now let us remember that proteins are another class of molecules found in the cell membrane. They are very large molecules possessed of extraordinary surface complexity. These richly varied surfaces can easily distinguish between a glucose molecule and, say, a galactose molecule which has the same size and shape and differs only in one insignificant detail (Fig. 1-5).

The second function the cell membrane performs is *pumping*. Again, a biologist might have predicted that the rather complex machinery required is beyond the capabilities of an undifferentiated lipid surface, and that protein molecules must be called upon for the special talents required. Careful analysis of membrane constituents shows that both lipids and proteins are present, and both contribute their special properties to formation of the total membrane. Lipids provide a tough, yet thin and elastic, waterproof sheet, while proteins are active in recognition and pumping functions. Cell biologists used to think that proteins were smeared over the lipid surface like peanut butter over a sandwich, but newer evidence suggests a more discontinuous distribution, as shown in Figure 1-6. A number of proteins may be

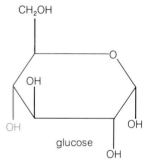

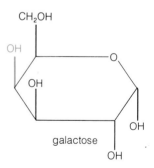

Figure 1-5
CHEMICAL DIFFERENCE BETWEEN GLUCOSE AND GALACTOSE. Carbon atoms with linked hydrogens are represented by a linear skeleton. The two molecules differ in the orientation of a single —OH group.

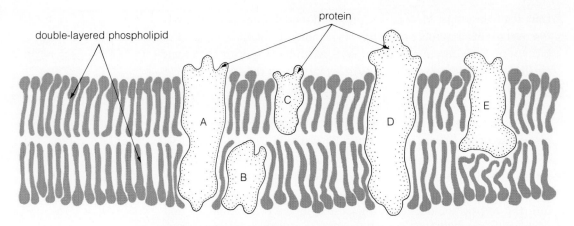

Figure 1-6 CROSS SECTION OF NATURAL MEMBRANE. The double layer of phospholipids is partially or completely pierced by molecules of protein. These act to control traffic across the cell membrane and to confer individuality on the cell.

associated with a single functional complex, as A–B–C. The A–B–C complex and D function both as recognition sites and as active transport pumping sites.

A large molecule such as galactose or fructose can enter a cell by a second mechanism called *facilitated diffusion*. In this case a specific membrane protein, called a *permease,* combines loosely with the molecule to be transported and then diffuses through the lipid layer to the inner surface where the loose bond is broken. The transported mole-

TABLE 2
Cellular transport processes

	ACTIVE TRANSPORT	FACILITATED DIFFUSION	PASSIVE DIFFUSION
Energy required?	Yes	No	No
Specific protein required?	Yes	Yes	No
Dependent on concentration gradient?	Independent	Dependent	Dependent
Examples of molecules and ions transported	Glucose, Na^+	Galactose	Water and other small molecules; alcohol, ether, and other lipid-soluble molecules

cule then remains inside the cell, while the permease swims back to the surface where it may pick up another suitable molecule and repeat the process. Facilitated diffusion requires no energy input from the cell and works only with a concentration gradient, never against one. Thus, if more galactose is present inside the cell than outside, the net flow will be directed outward. Protein molecule E in Figure 1-6 may act as a permease. A summary of cellular transport processes is presented in Table 2.

The Cell Membrane and Flow of Water

We have considered several properties of the cell membrane—its water stability and its recognition and transport functions. There is one additional bit of curious behavior we should examine: *osmotic pressure* responses. The concept of osmotic pressure may seem esoteric, but it is as common in everyday life as wind and rain. A man waters his lawn, and slowly the drooping grasses and flowers revive. How did the water get from the ground into the roots and stems? As a result of osmotic pressure differences. Or consider a woman playing tennis on a hot summer day. After an exhausting game, she has an overwhelming craving for iced tea. Why is she thirsty? Because the osmotic pressure of her blood has increased slightly and this increase is monitored by a small region in the hypothalamus. Or consider a road crew spreading salt on a snowy street. The following spring, the grass and shrubs along the roadside die. What killed them? Excessive osmotic pressure in the groundwater.

What is meant by osmotic pressure? Briefly, it is a measure of solute concentration in water. Pure water has an osmotic pressure of zero, while any addition of salts, sugars, urea, or any dissolved substance increases the osmotic pressure. The more dissolved molecules per unit volume, the higher the osmotic pressure rises. But osmotic pressure is a little bit like radio waves—they are around us all the time, but we cannot demonstrate their presence without a radio receiver. The "demonstrating instrument" for osmotic pressure is a semipermeable membrane. All membranes of living cells are semipermeable. That is, they permit some molecules such as water to slip through easily, while holding back or completely barring others (such as salts, large sugars, and proteins). The easiest way to understand this behavior is to imagine the membrane sprinkled with tiny pores, large enough to let water molecules through but not large enough for bulkier molecules such as sugars and proteins. Such pores may be actual physical structures, or they may simply consist of loose spaces between the tightly packed components of the membrane.

If we now place the membrane between pure water on the right and water plus sugar on the left, as shown in Figure 1-7, a rush of water molecules through the membrane in both directions will occur. If the water molecules on both sides of the membrane were equally active, the number entering either compartment would equal the number leaving, and the fluid levels on both sides would remain the same. But the water molecules on the two sides are not equally free to move. The activity of the water molecules on the left has been reduced by the addition of sugar. The sugar molecules are restricting water movement by exerting an attraction on the water around them and by interrupting the free movement of water molecules. The result is a rise in fluid level in the left-hand compartment and a drop in fluid level on the right.

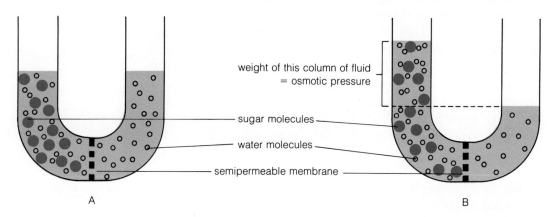

Figure 1-7 SEMIPERMEABLE MEMBRANE AND OSMOTIC PRESSURE. (A) Beginning of experiment. Equal volumes of fluid are contained in both arms of the apparatus, which are separated by a semipermeable membrane. Fluid on right is pure water, while fluid on left contains water plus larger sugar molecules. (B) End of experiment. A net movement of water has occurred from right to left. The difference in fluid levels between the two arms creates a hydrostatic pressure that is equal to the osmotic pressure of the sugar solution.

After a period of some hours, the two levels stop moving, and an equilibrium is reached in which the number of water molecules going from left to right equals the number going from right to left. Why did the fluid level on the left stop rising? Remember, the sugar molecules are too large to pass through the small pores of the membrane, and they are still all in the left-hand compartment and are still reducing the activity of the water on that side. But a new force now exists that is compensating for this reduction in activity—the hydrostatic pressure of the head of water on the left. This hydrostatic pressure (which equals the osmotic pressure) for most living cells is in the neighborhood of 0.2 atmospheres, or 3 feet of water. That is to say, if living cells were immersed in distilled water and a glass capillary were pushed through the cell membrane, water would enter the cell through the cell membrane and expand outward through the capillary to a height of perhaps 3 feet.

This experiment can be performed only with a plant cell, where a firm, meshlike cell wall prevents the cell from bursting. An animal cell lacking such external support would explode in a fraction of a second because of the inrush of water. Those naked cells that must live in fresh water have evolved elaborate protective devices to prevent this. Single-celled organisms such as *Tokophrya* and *Amoeba* possess internal bailers, called *contractile vacuoles,* which pump the incoming water out as fast as it enters (Fig. 1-8). But in the animal world, the most important self-protective device is the avoidance of fresh water as an environment. Thus almost all the cells of the human body are bathed in a fluid having the same osmotic pressure as that inside the cell. The cells lining the outside of the skin are dead cells and repel water. However, this is not true of the cells inside your mouth and throat. What happens when you take a drink of fresh water?

Figure 1-8
CONTRACTILE VACUOLE (CV) IN TOKOPHRYA INFUSIONUM. Water is expelled through a permanent pore (p), basal bodies (bb) seen in longitudinal section.

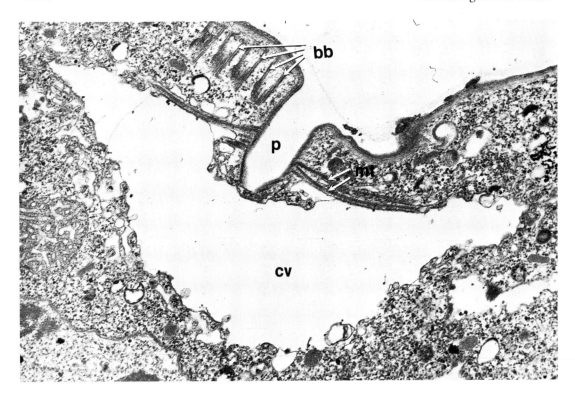

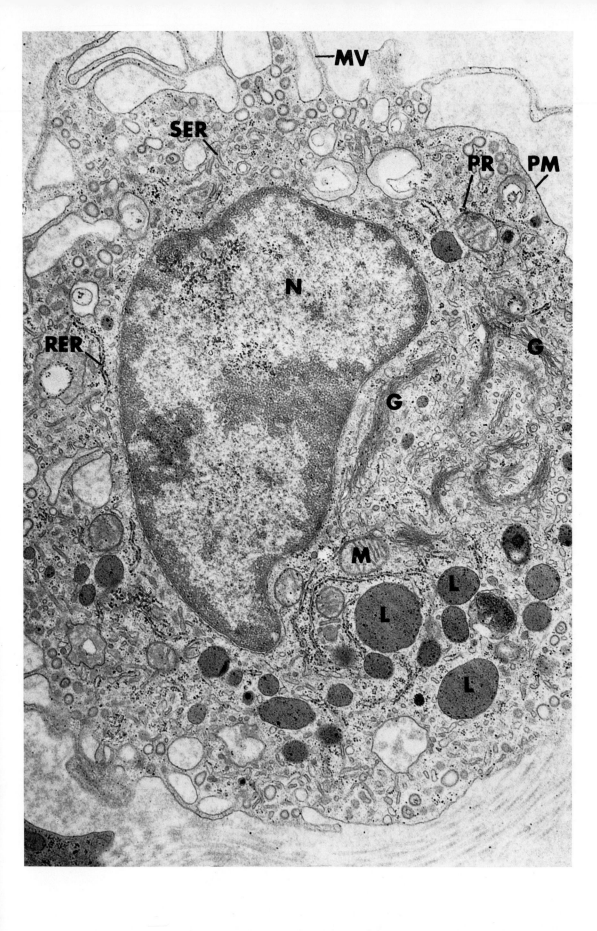

ARCHITECTURE OF THE CELL / 17

Figure 1-9 CELL ORGANELLES IN AN ANIMAL CELL. Visible are the nucleus (N), microvilli (MV), plasma membrane (PM), "rough" endoplasmic reticulum (RER) covered with ribosomes, as well as "smooth" endoplasmic reticulum without ribosomes (SER), cluster of ribosomes ("polysomes") engaged in protein synthesis (PR), mitochondria (M), Golgi body (G), and lysosomes (L).

Figure 1-10 GOLGI SECRETION MECHANISMS (ARROW) (BELOW). Protein is synthesized on the rough endoplasmic reticulum and moves along the inner compartment (cisterna) to the Golgi body. Golgi modifies the ER product and wraps a membrane around it. Two kinds of structures are formed: (A) lysosomes remain inside the cell; (B) secretory vesicles discharge the protein outside the cell.

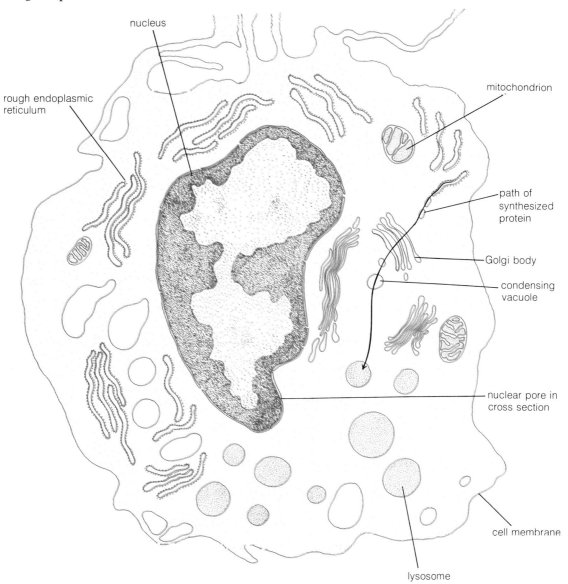

Internal Membranes

Let us now peer inside the cell and observe more of the strange landscape of these organisms. It seems that, once living things invented the membrane, they recognized a happy invention and kept using it inside the cell as well.[2] In fact, a typical cell contains more membrane material inside than on its surface. Most of these membranous structures are interconnected and, in order to travel from the surface to almost any point in the interior, one need never leave the surface of a membrane (Fig. 1-9).

The bulk of the membranous component is contained in the *endoplasmic reticulum* (literally, "lacelike network inside the plasma"). If we could observe a typical cell enlarged 100,000 times, the cell would appear human-sized, about 6 feet long. Snapping and rippling, the cell membrane would enclose a translucent assembly of seaweed-like material eddying back and forth inside the moving cell plasma. This is the endoplasmic reticulum, and much of it would be covered with matchhead-sized pellets called *ribosomes*. (Literally, this means "bodies containing ribose." Ribose is a sugar found in RNA, about which more will be said later.) This kind of "rough" endoplasmic reticulum acts as the factory area of the cell, manufacturing its most important components, the proteins.

If we search along the flat sheets of our blown-up endoplasmic reticulum, we will discover here and there (though not in every cell) an area where the membrane is tightly pressed together into a flattened artichokelike arrangement. This area is called a *Golgi body* (after Camillo Golgi, the great Italian physician who first described it in 1898). If the endoplasmic reticulum is a protein factory, then the Golgi body is its shipping department. Note that not every cell exports its proteins and, if the protein is used locally, no Golgi body can be found. However, cells like those of the pancreas or mammary gland produce large amounts of protein for use in other parts of the body or even outside the body. This process is called *secretion,* and secreting cells have very active and prominent Golgi bodies. Protein is conveyed from its site of manufacture on the endoplasmic reticulum to the Golgi body. Here it is concentrated, a membrane is wrapped around it, and the entire package, called a *vesicle,* then travels through the cell plasma to the exterior, where the packaging membrane merges with the cell membrane and releases the cargo to the outside (Fig. 1-10).

---

2. All of the internal cell structures we are about to describe are found only in cells of higher organisms or *eukaryotes* (Chap. 9). Such internal structures are not present in the form described in cells of bacteria or blue-green algae.

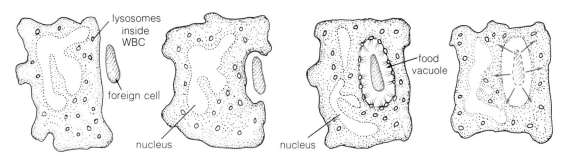

A. WBC bearing down on bacterium

B. Engulfment begins

C. Bacterium completely engulfed, lysosomes begin discharge of enzymes into food vacuole

D. Bacterium largely dissolved. Small molecules produced are absorbed into WBC

Lysosomes

There is a very interesting and rather recently discovered class of intracellular bodies called *lysosomes* (literally, "bodies that dissolve"). A lysosome is presumably produced by a Golgi body in the same way a secretory vesicle is produced. However, the cargo is an extremely dangerous one: a collection of very active and corrosive proteins capable of destroying an entire cell, including the cell that manufactured the lysosome in the first place. Like hand grenades, they remain inert until the pin is pulled. What constitutes "pulling the pin" for a lysosome? It seems to be an encounter of the host cell membrane with a foreign cell membrane, or perhaps only with a fragment of a foreign cell membrane. The sequence of events is illustrated in Figure 1-11. This diagram illustrates the attack of a white blood cell on a bacterium, but the same events occur in other lysosome interactions as well. In an uncanny way, the white blood cell detects the presence of a foreign body and moves it. As soon as the object of attack is reached, the white blood cell begins to ooze around it and encloses it in its own cell membrane. The infolding pinches off, migrates inward, and is now called a food vacuole. Even before the food vacuole is fully formed, lysosomes begin to discharge their corrosive proteins into it, and the foreign body dissolves. The white blood cell has literally eaten the foreign body. The process is called *phagocytosis* ("cellular feeding") and represents an evolutionarily ancient mode of feeding.

It is interesting that the white blood cell does not show any of this aggressive behavior toward other cells from its own body. For instance, a horse white blood cell can bump repeatedly into horse red blood cells without discharging a single lysosome. But now, for the sake of experiment, let us introduce some human red blood cells into the horse. Under the microscope, horse and human red blood cells are indistinguishable. But the horse white blood cell has no difficulty

Figure 1-11
A WHITE BLOOD CELL DISPATCHES A BACTERIUM.

20 / *The Cell: Architecture of the Cell*

in distinguishing between the two. It immediately attacks and destroys the human cells. How does the horse white blood cell know what is horse and what is nonhorse? Such knowledge depends on the membrane recognition sites mentioned earlier (page 12) in connection with active transport. The membrane proteins of every species are unique to that species and are recognized as such by the body's own cells. Otherwise intraorganismic cannibalism would result, with one cell eating another, and organisms would annihilate themselves.

Even so, self-destruction does occur, sometimes by plan and sometimes by biological sabotage. Let us consider planned demolition first. A frog starts its life as a tadpole, a long-tailed, plant-eating, fishlike organism. Gradually the tail of the tadpole disappears, limbs sprout and elongate, and the tadpole becomes an adult frog (Fig. 1-12). At the same time, changes are also taking place internally; the long, coiled intestinal tract of the tadpole is replaced by a short one in the frog, and food habits change from plant-eating to meat-eating. How does the tadpole rid itself of its long tail and intestinal tract?

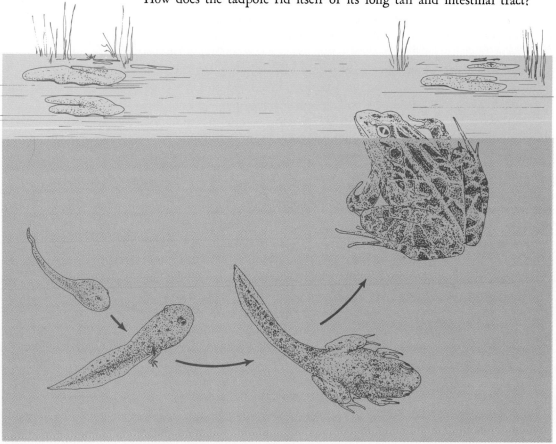

Figure 1-12 METAMORPHOSIS OF A TADPOLE INTO A FROG. Before the tail is completely resorbed, limbs begin to sprout. The frog does not leave the water until the process is complete.

This is an orderly demolition process in which lysosomes are programmed to destroy unwanted cells at the right time. The intimate details of this process are still not understood, though thyroid hormone levels are known to play a triggering role. Almost all animal cells contain lysosomes, and the lysosomal membrane, which is all that stands between the cell and its own destructive power, is sensitive to various stimuli such as oxygen deprivation and other treatments. The appropriate trigger thus causes the lysosome to destroy the cell from within.

Another example of such intraorganismic cell destruction occurs in our own bodies. Human red blood cells are made in the bone marrow, move out into the circulatory system, live there for 120 days, and are then destroyed in the spleen and liver. Your body contains billions of red blood cells that are 110 days old, but very few that are 130 days old. Even though a 130-day-old blood cell looks identical to a 110-day-old one under the microscope, the two are treated very differently by specialized white blood cells lurking in the liver and spleen. These executioner cells see the 130-day cell as a foreign body and immediately attack it as it tries to pass. Once more, it is presumed that a change has taken place in the identifying proteins on the cell membrane, costing the cell its life. In cellular society, old age is not respected.

Another case of self-destruction involves a kind of fifth-column attack. When a white blood cell attacks a bacterium, the white blood cell does not always win. Having no teeth, claws, or even means of escape, the bacterium fights back with chemical subterfuge. What triggers the white blood cell's discharge of lysosomes is contact with the identifying proteins on the cell wall of the bacterium. Some bacteria are capable of releasing soluble molecules into the bloodstream that have the same characteristics as the triggering molecules on their cell walls. These "pseudotriggers" penetrate the cell membrane of the white blood cell and cause a catastrophic discharge of the lysosomes into the white blood cell's own interior. The cell has thus been tricked into committing suicide and dies an agonizing death. However, even in death it performs one last service for its parent organism. The dying cell breaks down tissues leading to the skin surface, allowing the internal mass of bacteria and white blood cells to drain harmlessly to the outside. The draining material is called pus.

The Nuclear Membrane

If we now return to our 100,000-fold magnified cell, we will notice a large spherical object about the size of a medicine ball, enmeshed in the folds of the endoplasmic reticulum. This is the *nucleus* of the

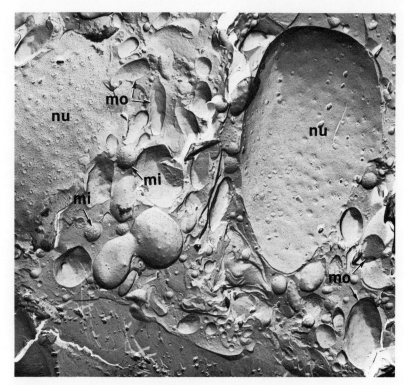

Figure 1-13 TWO FREEZE-FRACTURED CELLS SHOWING NUCLEAR PORES. Besides the two nuclear membranes, a number of mitochondrial membranes are also visible. These are seen both from the inside (mi) and outside (mo). This electron microscope photo differs from Figure 1-9 in two respects. First, the cells shown are not sliced into thin sections by a knife but are frozen and then cracked. The broken surface is uneven and tends to follow contours formed by internal membranes. Second, the electron beam of the microscope does not see the material as a transparent layer but scans its surface features.

cell, and it is enclosed in a double membrane, the *nuclear membrane,* which is continuous with the endoplasmic reticulum.

As even a nonbiologist knows, the nucleus of the cell is the final control and information center for the entire cell. Information stored here accurately describes the limits of a cell's capabilities. This topic is so important and actively developing that we will devote a separate chapter to it (Chapter 3).

The nuclear cargo is wrapped in a rather peculiar membrane. High-magnification electron micrographs show that the membrane has large pores, nevertheless it does not behave as a fishnet. Isolated nuclei are osmotically active, showing that the pores are plugged so as to prevent the passage of larger molecules. The function of the pores remains a matter of speculation. It is generally felt that they function in the exchange of materials between nucleus and cytoplasm.

Another peculiarity of the nuclear membrane is that it is doubled back on itself. This reflects the fact that it must periodically disappear. It must do this every time the cell divides, in order to make sufficient room for the nuclear material to sort itself out. At this time the double membrane is retracted and presumably becomes part of the endoplasmic reticulum. After the division process has been completed, the nuclear membrane re-forms itself around two newly formed nuclei.

The Mitochondrion

Swarming throughout the cytoplasm, and not attached to the great meshwork of the endoplasmic reticulum, are perhaps 800 to 1,000 small inclusions called *mitochondria* (singular, *mitochondrion*). The name derives from the Greek *mitos* ("thread") and *chondros* ("grain") and refers to the changeable appearance of these bodies with time. At times they may appear long and threadlike, but they may balloon out in bizarre bulges, break into smaller pieces, or meet and fuse. Perhaps this restless, agitated existence is a reflection of their function within the cell; they are power plants which provide the cell with the chemical energy it needs to survive. Any sabotage of these power plants leads to instant death of the cell, and it is interesting that some of the most potent poisons known, such as cyanide, work their effect in precisely this fashion.

Like the cell nucleus, the mitochondrion is bounded by a double membrane. The outer membrane is smooth, while the inner one is thrown into regular folds called *cristae* ("crests"). As a result, the inner membrane has a much higher surface area than the outer one—of the order of 10 times greater. Another difference between the inner and outer membranes is the fact that the inner membrane is coated with lollipoplike protrusions called *elementary particles*. The space inside the mitochondrion not filled by cristae and elementary particles is occupied by the *matrix substance,* made up of extremely fine granules. The matrix substance and the elementary particles constitute the working machinery of the mitochondrion. They convert the chemical energy of food molecules into a form of energy usable by the cell, just as a power plant converts the chemical energy of coal into electrical energy usable in light bulbs, motors, and appliances.

What is the function of the outer membrane? It appears to act as a control valve, determining the rate at which foodstuffs enter the mitochondrion for conversion into usable energy; and it does this in response to body levels of thyroid hormone. A person with high levels of thyroid hormone tends to be active, nervous, and quick in movement; he may eat heartily but remain thin and have a high body temperature. All this is a reflection of overactive mitochondria which "burn up" food molecules and convert them into phosphate-bond energy, the form of energy used by the body to perform work.

If mitochondria are independent bodies, how do they reproduce? When a cell divides, the division mechanism does not provide for the splitting in two of mitochondria. All that happens is that those mitochondria located in the right-hand side of a cell become part of the right-hand daughter cell, and those on the left become part of the left-hand daughter cell. If a cell originally carried 800 mitochondria, each daughter cell of the first generation will receive approximately 400, the second generation will receive 200, and so on quickly down to zero. Since the number of mitochondria per cell remains constant, obviously a replacement mechanism must be at work. It was observed that mitochondria replicate themselves independently of cell division. But what came as a surprise was the discovery that mitochondria carry their own DNA, or nuclear material. This mitochondrial DNA is arranged in a circular pattern, instead of the linear pattern found in the nucleus of higher cells (Fig. 1-14). There is another place where circular DNA is encountered: in the bacteria. In fact, a mitochondrion is approximately the same size as a bacterium. Does this mean that mitochondria represent degenerate bacteria that invaded higher cells hundreds of millions of years ago and then stayed on as welcome guests? The hypothesis sounds very attractive to most biologists.

Figure 1-14 CIRCULAR DNA FROM MITOCHONDRIA. The amount of DNA contained in these circles is far less than in even the simplest bacterium.

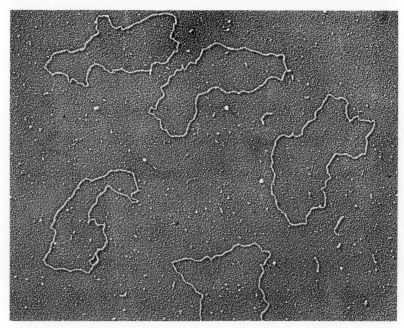

The Chloroplast

Plant cells carry another organelle, the *chloroplast,* that shows a certain structural resemblance to the mitochondrion. The chloroplast also has an outer membrane enclosing a richly folded assembly of inner membranes. These inner membranes are stacked in close arrays called *grana.* The membranes of the granum form a multilayer sandwich, and here the key ingredient is chlorophyll, the molecule a plant cell uses to capture sunlight energy and convert it into the chemical energy of sugar. Chlorophyll extracted from a chloroplast cannot convert light energy into chemical energy—it can do this only while accurately positioned inside the structure of the granum (see Fig. 2-19).

A typical plant cell contains 20 to 100 chloroplasts and, again, an independent mechanism of reproduction ensures that all daughter cells receive a sufficient number. Like the mitochondrion, a chloroplast carries its own circular DNA and is thought to represent an ancient microorganism that has assumed a symbiotic internal existence.

Microtubules and Microfilaments

A series of intracellular landmarks is constructed not of membranes but of extremely fine filaments. These filaments may be organized into microtubules such as the spindle fibers. Microtubules in turn may be aggregated into more complex assemblies such as basal bodies, cilia and flagella, and centrioles. Before describing these structures more closely, we should realize that the study of these cell organelles is one of the most active areas of current cellular research. Our understanding of microtubules and microfilaments is therefore in a state of flux.

Microfilaments are proteinaceous, threadlike structures so thin they are barely visible at the highest magnification of the electron microscope. Several classes of microfilaments may exist. One class appears to be involved with cellular movement and change of shape. This is perhaps not unexpected, since the contractile machinery of muscle also consists of thin filaments of proteins called actin and myosin. The two interact with each other to produce contraction (Fig. 1-25). However, even cells that are not part of muscle tissue may be able to move or contract. The advancing edge of an amoeba cell (which moves by a creeping motion) contains microfilaments. Microfilaments are also responsible for forming the cleavage furrow that constricts and then splits a dividing animal cell. Here the microfilaments are arranged in a concentric mass just under the cell membrane.

Another class of microfilaments apparently acts as an intracellular

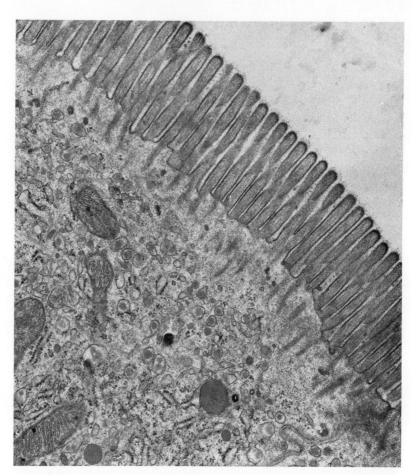

Figure 1-15
MICROVILLI OF AN INTESTINAL CELL. Each projecting microvillus is stiffened by a cluster of internal microfilaments.

skeleton. For example, a cell from the lining of the small intestine has an inner border which is thrown into tightly packed folds called *microvilli* (Fig. 1-15). The cell membrane by itself could never maintain this deeply furrowed appearance if the cytoplasm consisted of only a simple fluid. It is the microfilaments so abundantly scattered here that stiffen the cell and enable it to maintain this highly unlikely shape.

Microtubules are more easily observed in electron microscope photos, and appear to be present in all higher cells observed[3] (Fig. 1-16, 1-17). Microtubules are straight, unbranched structures that may run for many microns through a cell. Under extremely high magnification, microtubules may be resolved into a series of subunits (Fig. 1-18). The subunits appear to be a set of unstable protofilaments, typically 13 in number (Fig. 1-19). Each filament can dissociate into a train of globular protein molecules called *tubulin*. Microtubules may therefore appear and disappear in a cell as needed. The

---

3. That is, in the cells of eukaryotes: cells with a true nucleus and other organelles such as mitochondria, Golgi bodies, and so on.

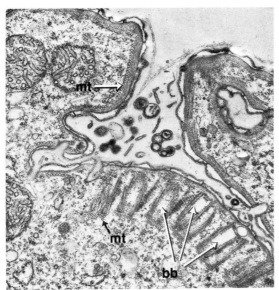

Figure 1-16 MICROTUBULES IN A PROTOZOAN. The microtubules (mt) are seen in cross section. Also visible are basal bodies (bb) in longitudinal section.

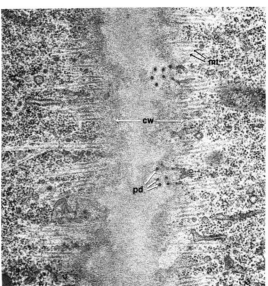

Figure 1-17 MICROTUBULES IN A PLANT CELL. Microtubules (mt) are seen in longitudinal section. Also visible is the cell wall (cw) separating two cells. The cell wall is pierced by small openings called plasmodesmata (pd) through which cytoplasm of adjacent cells can communicate.

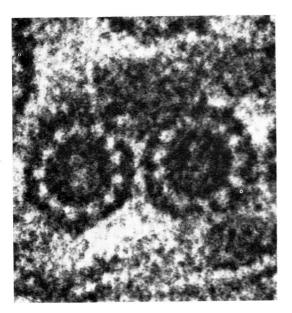

Figure 1-18 TWO MICROTUBULES IN CROSS SECTION. These microtubules from a brain cell in a goldfish are shown to consist of 13 subunits (tubulin molecules).

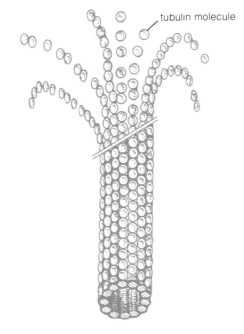

Figure 1-19 ASSEMBLY OF A MICROTUBULE FROM TUBULIN SUBUNITS. The microtubule can fray into unstable protofilaments. Each protofilament is assembled from a string of tubulin molecules.

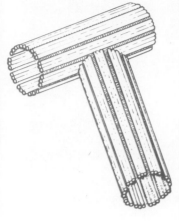

Figure 1-20
CENTRIOLES. Centrioles are always paired and consist of clusters of microtubules in a "3 x 9" arrangement.

spindle fibers present in a cell during mitosis are one example of such appearing and disappearing microtubules (Fig. 1-21).

As mentioned before, individual microtubules may in turn be aggregated into more complex assemblies. Among the strangest are the *centrioles* found in animal cells. Centrioles consist of tubular assemblies of microtubules arranged in a characteristic "3 x 9" format as shown in Figure 1-20. Centrioles come in pairs, with each centriole aligned perpendicular to the other. These organelles are self-reproducing, and reproduce in synchrony with the cell so as to maintain the normal number of two per cell. During cell division, the centrioles direct the assembly of spindle fibers. The spindle fibers then form attachments to the chromosomes and move them to their proper resting places in the future daughter cells (Fig. 1-21).

The *basal body* (Fig. 1-22) is a modified centriole associated with outgrowths called *cilia* (Fig. 1-23) (short) or *flagella* (long). The cilium or flagellum constantly makes smooth, whiplike motions which drive the cell forward (if it is free-moving), or set up external water currents (if the cell is stationary like those in the windpipe). The movements of the cilia and flagella are controlled by the basal bodies, and the degree of control is quite impressive. A "hairy" cell like *Paramecium* contains several hundred cilia, yet they all beat in synchrony to provide rapid travel, and *Paramecium* can exert instantaneous control over its movements, for instance, turning away from a noxious fluid. A similar synchronization of ciliar beating is observed in the cells of the human throat. All cilia beat in unison to remove particles of grime and dust from the throat (Fig. 1-24).

Centrioles, basal bodies, and cilia and flagella all contain microtubules in various arrangements. Whether organized in such complex assemblies or loosely dispersed, microtubules are associated with cellular movement. For example, the intracellular streaming that keeps the cytoplasm well-stirred (especially in plant cells) appears to depend on microtubules. So does the rapid movement of proteins and cell organelles down a long nerve cell (Chap. 5). The feeding tentacles of one-celled suctorians which appear to be "sucking" out the cytoplasm of their prey cells are well-laced with microtubules. In fact, the suctorians do not depend on suction for their food but on movement of prey cytoplasm by microtubules (Chap. 9).

How do microtubules bring about intracellular movement? Perhaps the most fruitful analogy is that of muscular movement (Fig. 1-25). Muscular contraction results when the two filamentous proteins actin and myosin interdigitate and slide past each other (Chapter 5, Appendix B). This sliding motion is thought to depend on a walk-like mechanism, analogous to a myosin mouse pushing an actin running wheel. It has been proposed that the tubulin of microtubules is a

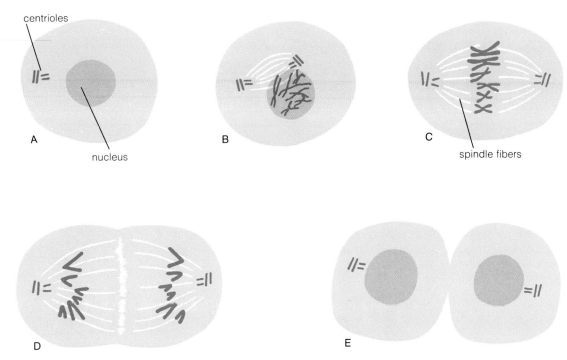

Figure 1-21 MITOSIS. This type of cell division yields chromosome counts in each daughter cell that are identical to that in the parent cell. In plant cells, the entire process is completed without the aid of centrioles. (A) Resting cell (interphase): nucleus without visible chromosomes, centrioles without spindle fibers. (B) Early mitosis (prophase): centrioles replicate, nuclear membrane disappears, chromosomes become visible, spindle fibers begin to arc across cell. Chromosomes have already doubled at this point, and daughter strands are held together at a single point (the centromere). (C) Middle mitosis (metaphase): nuclear membrane absent, chromosomes fully condensed and attached to spindle fibers via centromeres. (D) Late mitosis (anaphase): daughter chromosomes traveling to opposite poles of cell, guided by spindle fibers. (E) End of mitosis (new interphase): nuclear membranes have re-formed around chromosomes. Two daughter cells are now present with individual cell membranes and equal nuclear contents.

functional (though not structural) analogue of the muscle protein myosin. Thus a similar "walking" mechanism also produces relative movement, but here the microtubule forms a stationary stage against which other cytoplasmic constituents are moved.

The foregoing list of cellular structures has not been an exhaustive one. We have not mentioned organelle substructures found only in specialized cells, such as synaptic knobs, sieve plates, acrosomes, and mesosomes. We have not even mentioned some rather widely occurring organelles such as peroxysomes, pinocytic vesicles, storage

30 / *The Cell: Architecture of the Cell*

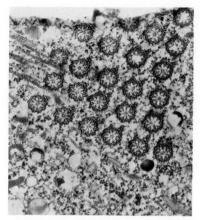

Figure 1-22 BASAL BODIES IN CROSS SECTION. Basal bodies are modified centrioles and have a "3 x 9" arrangement of microtubules.

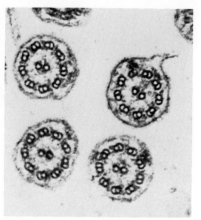

Figure 1-23 CILIA IN CROSS SECTION. The cilia have a "2 x 9 x 2" arrangement of microtubles.

Figure 1-24 CILIATED CELLS OF HUMAN TRACHEA (WINDPIPE). Small tufted structures are the tops of goblet cells, which secrete mucus. The mucus traps dust and other foreign objects. Rhythmic coordinated beating of the cilia moves the mass towards the throat and away from the lungs.

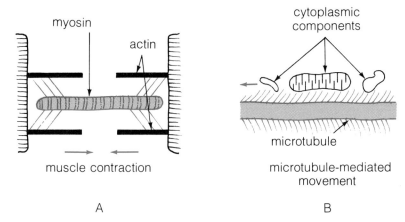

Figure 1-25 CELLULAR MOVEMENT MEDIATED BY MUSCLE PROTEINS AND MICROTUBULES. (A) Muscle contracts because of interaction between two proteins. Cross arms of myosin move filaments of actin inward to produce a sliding motion. (B) Proposed mechanism for microtubule-mediated movement. The microtubule remains stationary, while its cross arms move organelles, such as the mitochondrion and vesicles shown here, about the cell.

granules, and cell vacuoles. However, this brief look should make plain the fact that, down to the last micron of visible space, the living cell everywhere presents a picture of structure and organization. It is for this reason that living things are commonly called organisms—they are organized!

Bibliography

Lewis Thomas, *The Lives of a Cell*, Viking Press, 1974. This little collection of essays touches on a broad range of biological topics, from cell structure to anthropology. One of the threads of the author's argument is the importance of cooperation in biological phenomena. The book has been a national best-seller. GENERAL READING

W. Bloom and D. W. Fawcett, *A Textbook of Histology*, 10th ed., W. B. Saunders, 1975. An exhaustive discussion of all aspects of cell structure. TEXTBOOKS

A. B. Novikoff and E. Holtzman, *Cells and Organelles*, Holt, Rinehart, and Winston, 1970. A small, concise treatment of the eukaryote cell that is both readable and accurate.

A. Claude, "The Coming of Age of the Cell," *Science 189*, p. 433 (1975). The Nobel Prize address of a pioneer of modern cell studies. ARTICLES

D. Kennedy, *The Living Cell, Readings from Scientific American*, W. H. Freeman and Co., 1974.

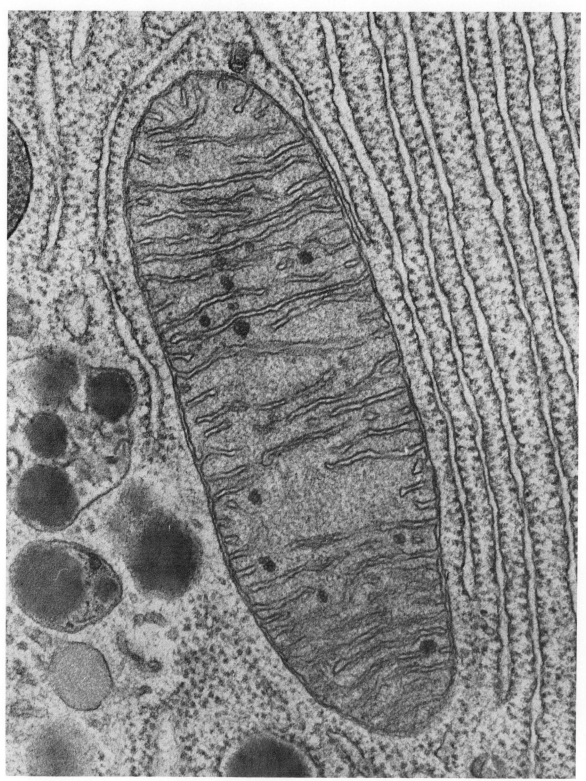

Figure 2-1 A MITOCHONDRION IN CROSS SECTION. The inner membrane is thrown into folds called *cristae*.

chapter 2

We have taken a short tour through the architecture of the cell. Is there any one cellular component that we can point to and say, "This is the one that is responsible for the phenomenon of life?"

Before answering this question, let us first see what we mean by the term "life." The question may not be as academic as it sounds. During the coming decade, space probes will be soft-landed on Mars and perhaps on other planets, and a search for extraterrestrial life

Cell Metabolism

will be made. How will we recognize life if it is present?[1] There is little reason to expect that life on other planets, if it exists, even remotely resembles the external forms we know on earth. Life is associated with an ability, not with any set of physical structures. The ability most importantly associated with life is the ability of an organized system to maintain and extend its organization. That is not as simple as it may sound.

Consider an obviously nonliving thing such as a stereo set. A stereo set is a highly organized, quality-controlled object. When the set is new, it produces a clear, rich tone and responds precisely to the tuning selector. But as the set becomes older, and perhaps without visible outward change, its performance deteriorates. There may be garbling of sound or strange background noises; the controls may stick. No one will be surprised, because this is what we expect of inanimate objects, no matter how finely made. Our watches, houses,

---

1. Recently, students in a biology class at an eastern university were asked this question on an exam. One student answered, "Ask the local inhabitants. Even a negative answer will be significant." The student received an A—.

Figure 2-2 THE AGING OF A LUXURY CAR. Manufacturer's ad 1959 (top). Final rest 1974 (below). This car has reached its final position not because of shoddy workmanship or neglect, but because of the imperceptible toll exacted by the Second Law of Thermodynamics.

cars, and personal possessions all deteriorate physically with age. The universal observation that new things break down and wear out is expressed in the form of a natural law, the *second law of thermodynamics*. The second law can be expressed in several equivalent forms. One formulation of the second law says that all spontaneous processes tend toward randomness and disorder. Randomness can be quantitatively measured in units physicists call entropy units. The converse of the second law would state that situations do *not* spontaneously progress from disorder to order. That is to say, crumpled fenders do not by themselves become smooth, spilled gravy does not crawl back into the gravy boat, and unraveled sweaters do not knit themselves up again. The fender *can* be straightened, the gravy mopped up, and the sweater mended, but only if someone does work to accomplish these tasks. While the damage could have occurred spontaneously, the repair processes can occur only with the expenditure of work.

This brings us back to our definition of life. Living organisms do not appear to deteriorate with age. A redwood tree, for example, may live for 4,000 years and could perhaps survive much longer if not brought down by some external cataclysm such as a chain saw or a devastating flood. Does this mean that living organisms do not obey the second law of thermodynamics? The great Austrian physicist Erwin Schrödinger (winner of the 1933 Nobel Prize in physics) made just such a proposal during the 1940s. The Schrödinger proposal sounded very plausible, but only so long as the living organism was observed in isolation from the universe around it. If we enlarge our area of observation to include the living organism together with its environment, we find that living organisms do in fact obey the second law of thermodynamics.

Figure 2-3 A HOUSE MOUSE AT HOME.

Though the proper frame of observation should include the entire universe, we can adequately demonstrate second-law effects in living organisms by choosing a much smaller frame of reference. For example, we might choose to observe a family of mice living in a house. If the mice prosper, their total number may increase. Mice are highly ordered, complex, intricate creatures. The maintenance and expansion of such intricately organized assemblies flies in the face of second-law prediction. But the mice are unwelcome guests in the house—they make nests in the furniture, chew holes in paper and clothing, destroy stores of food, and strip insulation from electric wiring. In total, the house is randomized and disordered much faster than if the mice were absent. We therefore draw the general conclusion that living things maintain and expand their own organization at the expense of the organization of their environment. In fact, it would be impossible for any living organism to remain alive outside its environmental matrix. We know this as an ecological law, but it is a law with a thermodynamic basis.

Macromolecules

Let us next explore the nature of the linkage between living organisms and their environment and, for the sake of illustration, let us use again the example of our family of mice. The mice may leave their mark on the house for a variety of personal reasons. They may build nests, construct runways, or simply chew on materials in order to sharpen their teeth. But most fundamentally, mice must eat to stay alive, and it is their feeding behavior that leaves the most extensive imprint on the house.

What happens when a mouse feeds itself? To trace out the reasons for the feeding, and to understand the basic processes involved, we must examine this process at the cellular and molecular levels. For the sake of illustration, let us suppose that the mouse has discovered a bag of dried kidney beans which it then proceeds to devour over a matter of days. Superficially we may observe that the beans disappear and are converted into mouse tissue. We might also observe that, no matter what the original food source is, it does not influence the final result; kidney beans as well as bread crumbs or scraps of meat are always converted into mouse tissue. To understand why this happens, we have to look inside the cells of the mouse and the kidney bean.

A kidney bean, like a mouse, is made up of microscopic cells and is a living organism. (This is true even of the dried kidney bean.) Because they are living organisms, the kidney bean as well as the mouse contain many very large, specialized molecules which are found nowhere outside the living world. These important molecules are called *macromolecules* ("large molecules").[2] Four distinct classes of macromolecules are known in the living world: proteins, polysaccharides, nucleic acids, and fats (Table 1).

TABLE 1
Biological macromolecules

CLASS OF MACRO-MOLECULE	TYPE OF SUBUNIT	FUNCTIONS IN CELL	EXAMPLES
Protein	Amino acid	Enzymes, antibodies, structural	Egg albumin, hemoglobin
Polysaccharide	Simple sugar	Short-term energy storage, structural	Starch, cellulose
Nucleic acid	Nucleotide	Information storage	DNA, RNA
Fat	Fatty acid plus glycerol	Long-term energy storage, structural	Olive oil, butter

In this group, the fats have by far the smallest molecular dimensions. The other classes of macromolecules achieve enormous sizes because they are *polymers* ("many parts"). That is, they are made up of hundreds, thousands, or sometimes millions of smaller repeating units, arranged like pearls in a necklace. For proteins, these repeating units are called amino acids, for polysaccharides, they are called simple sugars, and for nucleic acids, nucleotides. Because the nature of the repeating units is so different in the different classes of macromolecules, the final molecules wind up with dramatically different chemical properties and with different functions in the cell. The chemistry of living organisms is largely the chemistry of macromolecules.

For the time being, let us concern ourselves with only one class of macromolecules, the polysaccharides. Two of the most common polysaccharides known are starch (which accounts for most of the dry weight of a loaf of bread) and cellulose (which accounts for most of

---

2. A molecule is an assembly of atoms linked by covalent bonds. Covalent bonds are bonds that arise from the sharing of electrons between atoms. They are generally quite strong.

38 / *The Cell: Cell Metabolism*

the weight of this book). We will concern ourselves here with starch, because it makes up well over 90 percent of the dry weight of the kidney beans being consumed by our hypothetical mouse. What happens to this bean starch inside the body of the mouse? It is transformed into a type of starch characteristic of the mouse.

Construction of Macromolecules

Bean starch and mouse starch are alike in that the repeating unit in both is the simple sugar glucose. However, they differ in the manner in which the glucose is polymerized. In bean starch, the glucose units form long, straight chains, while in mouse starch a branching structure is developed (Fig. 2-4). How does the chemical machinery of the mouse transform a straight-chain molecule into a branching structure? It first dismantles the bean starch molecules into separate glucose units before reassembling them in the new way characteristic of the mouse. This is where the second law of thermodynamics comes in again: it is easy to disassemble, but it is hard to reassemble. A slight nick with the scissors will demolish a necklace, sending the beads scattering into every corner of the room. But to put them back together again requires patient work. What is the source of this work in the mouse? The source is energy from individual glucose molecules, torn apart still further and combined with oxygen to yield carbon dioxide, water, and a huge amount of energy. Details of this process (respiratory energy metabolism) are given in Appendix A. This energy is tapped by the cells of the mouse in order to reassemble the glucose molecules into their new configurations. The whole process may be written as in Figure 2-5.

An analogous picture could be drawn for the lipids, proteins, and, to a smaller extent, the nucleic acids. The macromolecules in the food source provide the mouse with two essentials: *raw materials* to construct mouse macromolecules, and an *energy source* to carry out the construction process. As soon as the new macromolecules have been synthesized, they start to behave like any inanimate object. That is to say, they begin to fall prey to second-law vagaries; they become damaged, fall apart, and stop working as predictably as any man-made appliance. The mouse therefore has no choice but to keep eating in order to replace or repair its crumbling chemical plant. Starvation leads to death, and this is true even if the mouse lies still and does not twitch a whisker. Its energy consumption at rest represents the effort necessary to forestall second-law catastrophe. A living system is therefore a steady-state system; materials continually flow in and waste materials flow out, yet the internal arrangement remains the

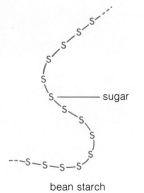
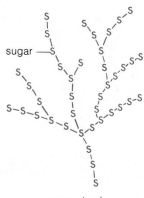

Figure 2-4
STRUCTURES OF PLANT AND ANIMAL STARCH. Both consist of long chains of sugar (glucose) units linked by chemical bonds. Note that animal starch has the more "treelike" structure.

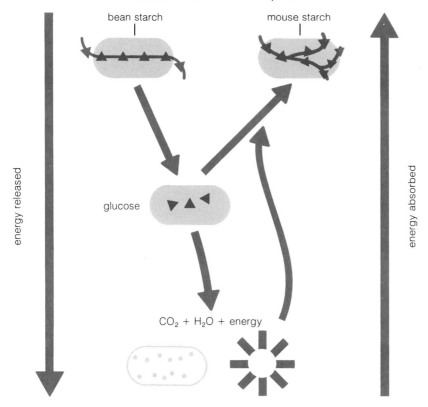

Figure 2-5 TRANSFORMATION OF MACROMOLECULES INSIDE A LIVING CELL.

same. This rate of throughput can be very high. A mouse eats its own body weight in food approximately every 24 hours. If we assume that half of this passes right through the digestive system without being absorbed, then, on the average, we have a "new" mouse every 48 hours. Humans have a slower rate of throughput; we eat our own body weight in approximately 2 months, hence with 50 percent absorption it would take 4 months to create a "new" human.[3]

Metabolic Pathways

Let us look at this process of molecular dismantling and reconstruction more closely. The term biologists use to describe the pathways that molecules follow inside the cell is *metabolism* ("turning over"), and Figure 2-5 represents a gross oversimplification. In a cell as

---

3. We are talking here about replacement of the average molecule in a cell. Some molecules are replaced more quickly than this, while others, such as DNA, are not replaced at all as long as the cell retains its individual existence. However, even in this case the cell may perform minor repairs to damaged sections of the DNA molecule.

complex as a mouse cell, thousands of separate chemical reactions are taking place at the same time, representing hundreds of separate metabolic pathways. Thus the mouse cell has many more options than are shown in Figure 2-5. The glucose used as a raw material in mouse starch synthesis need not necessarily come from plant starch. Inside the cell, glucose can be synthesized from breakdown products of lipids, proteins, and other food sources. Likewise, the energy necessary to resynthesize monomeric glucose into mouse starch may come from a wide variety of other sources.

Complex as metabolic pathways are, there are three rigid criteria all such pathways must satisfy. First, *the end products of the pathway must contain less energy than the starting materials*. This energy loss constitutes the driving force for the overall pathway. Second, *all metabolic change occurs via many small chemical steps* instead of a few major changes. To use a sculptural analogy, new molecules are chiseled out of old ones instead of being die-cast out of virgin materials. To give an example, glucose (a molecule containing six carbon atoms) must be cleaved into two three-carbon fragments before its internal store of chemical energy can be tapped to any appreciable degree for cellular needs (Appendix A). This act of splitting requires 11 separate chemical steps, instead of the single step one might reasonably expect. The advantage to the cell is that numerous small steps provide greater opportunities for control and regulation than would a single large step. Finally, *each small chemical step is mediated through the agency of a specific enzyme*.

The word *enzyme* means "in yeast," because at one time enzymes were thought to be restricted to yeast cells, where they were intensively studied by French scientists interested in making better wines. Enzymes today are known to occur in every living cell, and life as we know it would be impossible without them. All enzymes are proteins and, in terms of cellular performance, they influence metabolism in three important ways. They speed up chemical reactions, they make chemical reactions more specific, and they represent the control valves used to regulate metabolic pathways.

Enzymes as Catalysts

Enzymes can be used to speed up chemical reactions, but with one important qualification: the reaction can be accelerated only if it would have taken place anyway, without the enzyme. To use a musical analogy, an enzyme is to a chemical reaction what a good piano teacher is to a pupil—the teacher can greatly accelerate the progress of a talented pupil, but can never make a Glenn Gould out of an Aaron Slick. Likewise, the presence of enzymes in a metabolic pathway has no effect on the direction the pathway will take. Whether or not a series of reactions *can* take place depends not on the presence or absence of enzymes but on energy balance; if energy is released in the overall process, the sequence will take place. A simple example illustrates this fact. Paper (which is almost pure cellulose) burns quickly, releasing energy in the form of heat and light and producing carbon dioxide and water as end products. Chemically, the reaction is written as follows:

$$\text{cellulose} + O_2 \longrightarrow CO_2 + H_2O + \text{energy (heat + light)}$$

The starting materials (cellulose plus oxygen) are almost totally converted into end products because of the large amounts of energy lost by the products. The energy released represents chemical energy locked up inside the starting materials. The products therefore represent a more stable arrangement of atoms than do the starting materials.

Cellulose can also be "burned" via the agency of enzymes. All we have to do is feed a sheet of paper to a termite. Inside the gut and deep tissues of the termite, cellulose will again be converted to carbon dioxide, water, and energy:

$$\text{cellulose} + O_2 \xrightarrow{\text{termite enzymes}} CO_2 + H_2O + \text{energy}$$

What is significant is that the total amount of energy released in the termite body is exactly the same as would have been released by burning, though of course the termite carries out the process in a more leisurely fashion and therefore neither glows nor overheats.[4] But there is no enzyme or combination of enzymes that can take carbon dioxide and water and convert them spontaneously to cellulose and oxygen, because energy for such a process would have to be absorbed instead of being released.

Enzymes are therefore used to speed up chemical reactions that would have taken place anyway, given sufficient time. But a reaction

---

4. It may seem paradoxical that the nonenzymic reaction (burning) occurs at a faster rate than the enzymic reaction. To make a valid comparison of rates, the two reactions must be observed at the same temperature. At temperatures observed inside the body of the termite, the nonenzymic reaction is immeasurably slow.

that is possible is not necessarily a reaction that is probable. It is possible for a tree to fall (and impossible for a fallen tree to right itself again). But while all trees eventually fall, one may walk for many years through a forest before actually catching one in the act. Likewise, all metabolic pathways in the cell run in only one direction—the direction representing the release of energy. But in the cell these reactions occur at rates that are thousands (sometimes millions) of times faster than would be observed spontaneously. This high speed is one of the distinguishing features of the living process. The intensity of life is a consequence of enzymes.

Enzymes—Specificity

Another property of enzymes is *specificity*. Imagine a chemist given the task of preparing some glucose-6-phosphate.[5] One method he could use would be to heat a solution of glucose and pyrophosphate, with the reaction going as shown in Figure 2-6. A phosphate group from pyrophosphate would displace an —OH (hydroxyl) group in the number-6 position of glucose, forming glucose-6-phosphate. Unfortunately, the pyrophosphate is not choosy about the —OH group it displaces with a phosphate. The chemist will discover in his flask roughly comparable amounts of glucose-1-phosphate, glucose-2-phosphate, and so on (Fig. 2-6). To use a nonchemical analogy, the problem is comparable to that of throwing a hat onto a six-pegged hatrack. If the pegs are all of the same size, then which peg the hat lands on will be dictated largely by chance.

Chemists have long been used to such a situation; they avoid the problem by taking the mixture of reaction products and separating out the one product that is of interest. Purification is generally the hardest task in the preparation of a synthetic compound.

However, in a living cell such a situation would be intolerable. Imagine a metabolic sequence as follows:

$$A \rightarrow B \rightarrow C \rightarrow D \rightarrow E \rightarrow F$$

Let us consider the first step, $A \rightarrow B$. If only half of A is converted to

---

5. As its name implies, glucose-6-phosphate is in fact a sugar molecule carrying an attached phosphate group. Glucose-6-phosphate is a very common and very important metabolic intermediate found in all cells of the human body.

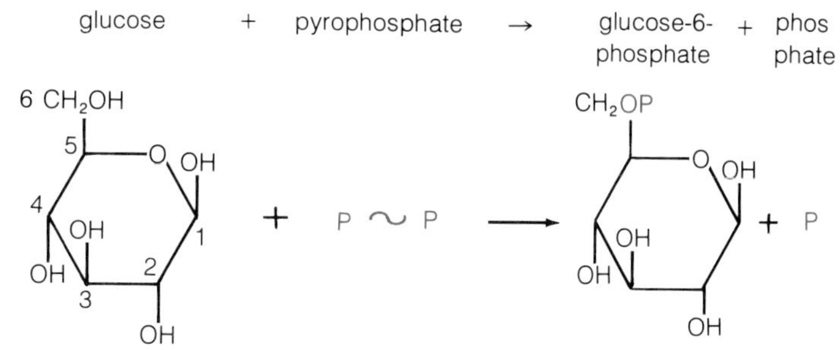

ENZYMES—SPECIFICITY / 43

glucose + pyrophosphate → glucose-6-phosphate + phosphate

Figure 2-6 CHEMICAL SYNTHESIS OF GLUCOSE-6-PHOSPHATE. Glucose is a sugar and pyrophosphate a molecule containing two phosphate groups bound together by a high energy chemical bond. Only a few of the possible side products are shown.

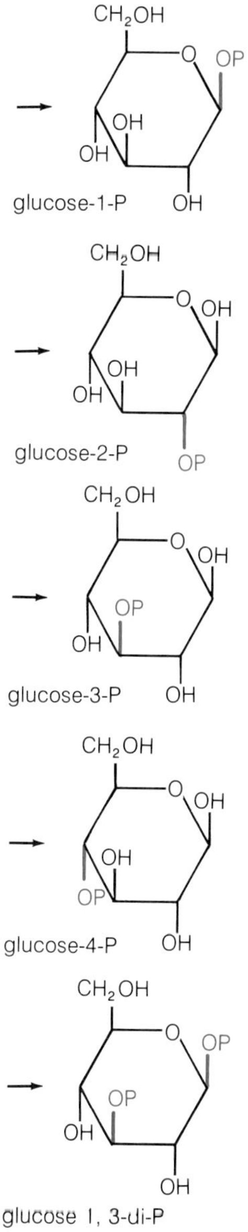

the desired product B, with the rest winding up as miscellaneous by-products, and if this 50 percent efficiency is also true for B → C, C → D, D → E, and E → F, then the overall efficiency for the entire sequence will be $\frac{1}{2} \times \frac{1}{2} \times \frac{1}{2} \times \frac{1}{2} \times \frac{1}{2}$ or $\frac{1}{32}$ (approximately 3 percent). For the cell, such a situation would be as chaotic as for a bank customer who deposited $100 and discovered that only $3.16 wound up in his account, the rest being squandered by tellers and bank officers. The reaction we mentioned before, the formation of glucose-6-phosphate (glucose-6-P) is a very important reaction in the cell and is carried out by the enzyme hexokinase as follows:

$$\text{glucose} + \text{R—P} \sim \text{P} \xrightarrow{\text{hexokinase}} \text{glucose-6-P} + \text{R—P} + \text{energy}$$

R—P~P is a modified form of pyrophosphate called ATP. The specificity of this reaction as catalyzed by hexokinase is 100 percent. Absolutely *no* glucose is converted into glucose-1-P, glucose-2-P, and so on. Presumably the first enzymes that appeared in evolutionary history had lower degrees of specificity and were eliminated in the evolutionary struggle that followed. Without the 100 percent specificity of enzymes, the long metabolic pathways observed in cells would not be possible. Furthermore, it is precisely these same enzymes that dictate the detailed nature of the metabolic pathways. Whether the path from compound A to N proceeds through the intermediates A → B → C → N or the intermediates A → L → M → N depends not on energy differentials but on the nature of the enzymes present.

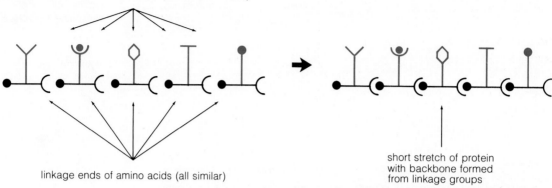

Figure 2-7 POLYMERIZATION OF AMINO ACIDS INTO PROTEIN. Note that only the linkage ends of the amino acids participate to form the backbone of the protein. Linkage ends are similar in essentially all natural amino acids, while the function ends come in 20 different varieties.

Enzymes as Proteins

How do enzymes achieve these feats of speed and specificity? To answer briefly, they owe it all to the fact that they are proteins. Proteins, as you recall, are one of the major classes of macromolecules and are found only in living systems. Proteins, like polysaccharides, are made up of simpler molecules, and for proteins these subunits are called amino acids. All are relatively small molecules which come chemically equipped for two purposes. One end of the amino acid serves to link it to its neighboring amino acids in the protein chain. This is called the linkage end, and this end is identical in every amino acid. The other end of the molecule is variable and is called the functional end (Fig. 2-7). Functional ends come in 20 different varieties (hence there are 20 different amino acids), and make up the working surface of the protein molecule. Between them, the 20 different functional ends represent the entire spectrum of chemical possibilities. Some are acidic, others basic. Some are hydrophilic, others hydrophobic. Some are aromatic, others aliphatic. Some are oxidizable, others are not. Some can ionize, others cannot, and so on.[6] An infinite number of chemically varied protein molecules can be assembled from the 20 different amino acids, just as an infinite number of paintings can be created from 20 different pigments.

---

6. These are all expressions of molecular structure or molecular behavior. Aromatic molecules typically contain ring structures, derived from benzene, while aliphatic molecules contain branched or straight chains. Oxidizable molecules can lose electron pairs, and ionizable molecules can lose charged fragments such as H^+.

ENZYMES AS PROTEINS / 45

Figure 2-8 TERTIARY STRUCTURE OF A PROTEIN. Only the backbone is shown for simplicity. Regions with a regular repetitive structure, such as helical twisting or complementary pleating, are common. Such regions are referred to as the secondary structure of the protein. The protein shown is the enzyme lysozyme. This enzyme breaks down bacterial cell walls and therefore has an antibiotic-like activity.

Protein molecules do not have a fixed length. The simplest proteins contain about 60 amino acids, while others may contain over 1,000. But even in a protein containing thousands of amino acids, the position in the chain of each amino acid is rigidly specified. In making copies of proteins, essentially every amino acid must be in its place. Hundreds of examples can be quoted in which a single amino acid substitution (out of hundreds or thousands) leads to partial or complete loss of function in the protein.

The exact sequence of the amino acids in a protein is defined as the *primary structure*. In their working state, proteins do not exist as long spaghettilike strands, but are coiled up in intricate twirls and loops. This is referred to as the *tertiary structure* of a protein (Fig. 2-8).

Again, the tertiary structure of a protein is no accident. No matter how seemingly bizarre the twists and turns, every molecule of a particular protein twists and turns in exactly the same way as every other molecule of that particular protein. If it does not, it will not be able to perform its specific function. The tertiary structure is maintained by chemical interactions between adjacent segments of the protein strand, and it is largely the functional groups of the protein that enter into these chemical interactions. There is one important difference between the chemical bonds forming the primary structure and the chemical bonds forming the tertiary structure. Bonds forming the primary structure are strong (covalent) bonds. Bonds forming the tertiary structure are weak bonds—each being only 2 to 5 percent as strong as a covalent bond. (See Appendix B for a fuller discussion of weak bonds.) Since an individual protein molecule may have hundreds of such weak bonds, the tertiary structure can maintain its integrity under normal conditions. But even a mild stress imposed on the protein destroys its tertiary structure. Increased temperature causes molecules to twist and vibrate violently. Moderate temperatures cannot rip covalent bonds apart, but they can destroy the weak bonds.

When the albumen (white) of an egg is heated in a pan, it turns white and becomes solid. Heat has destroyed the tertiary structure of the egg proteins and the egg dies, even though the primary structure is still intact. If a little demon could twist the now-chaotic protein strands back into their correct conformations, the proteins would once more become functional and presumably life would return. An example of milder protein damage is observed in heat stroke or during a high fever. Normal body temperature is 98°F. If it rises to 106°F, permanent brain damage can result because proteins of the brain cells are "cooked," leading to the death of cells. (Brain cells cannot be replaced.) If the body temperature rises to 110°F, death comes in hours. The difference between 98°F (normal) and 108°F (lethal) is not very large and even at 98°F, temperature vibration is constantly destroying large amounts of cellular protein. The reason why cells do not die at this temperature is because they can replace damaged proteins as soon as the damage occurs. At 108°F, the rate of destruction exceeds the rate of replacement. Proteins undergo similar stress with relatively small changes in acid-base balance or in salt concentration.

The Advantages of Instability

The question is, why did living organisms entrust such a vital function to such a fragile class of molecules? It would not be very difficult to stabilize the tertiary structure of a protein—the job could be done by inserting several covalent linkages between adjacent strands to restrict movement of the strands. In fact, many proteins have already "thought" of this solution and carry interstrand covalent bonds. But

THE ADVANTAGES OF INSTABILITY / 47

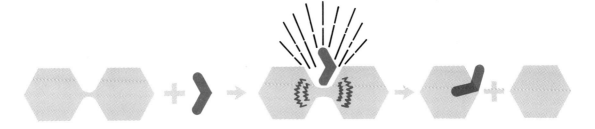

schematic reaction:
A + B → AB* → C + D

Figure 2-9 A HYPOTHETICAL UNCATALYZED CHEMICAL REACTION. If the desired product is to be formed, the reactants must collide with each other at a proper angle and with sufficient energy.

even where such strong connections are found, there are not enough present in the molecule to give it more than partial protection.

But there is a positive aspect to a loose, unstable method of construction. A molecule held together by weak forces can change its shape easily. Such internal movement is necessary to enzymes for two reasons. First, internal movement is often involved in catalysis (the speeding up of chemical reactions) and, second, internal movement can be used to control the rate of enzyme activity by a process called *allosterism* (literally, "alternate shape"). Incidentally, the type of internal movement observed in enzymes is of a precise and predetermined sort. Enzymes do not move in the chaotic way that might characterize a vanilla pudding or a bagful of jellybeans.

How does an enzyme speed up a chemical reaction? The details are complex and chemically sophisticated, but there are some universal things that enzymes do. Let us suppose we want to carry out the reaction shown in Figure 2-9.

The only way to achieve a reaction without the enzyme would be to apply heat to a mixture of A and B. This would cause A and B to fly about wildly, leading to many high-velocity collisions between the two. Some collisions would lead to a chemical reaction and, depending on the point of impact, several different products would form. Only a collision "amidships" would yield the desired reaction. Because this is an improbable event, the rate of reaction would probably be slow even with an increase in temperature.

With an enzyme present, no significant temperature rise is necessary. The enzyme involved will have a specific surface configuration that recognizes and binds reactants A and B. This is called the *active site* of the enzyme. See Figure 2-10. The reactants bind to the active

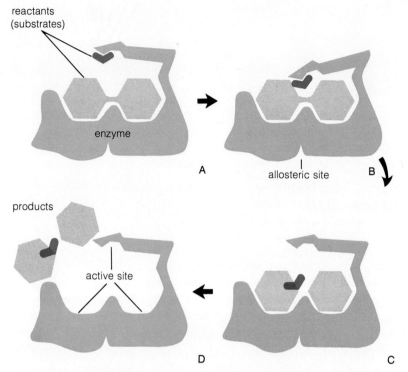

Figure 2-10 A HYPOTHETICAL ENZYME–CATALYZED REACTION: SEQUENCE OF EVENTS. (A) Binding of reactants to enzyme. (B) Internal movement of enzyme to bring reactants together. (C) Formation of products; return of enzyme to original conformation. (D) Products leave enzyme surface; enzyme ready to bind more reactants.

site, the enzyme moves to bring them together at the correct angle, the reaction occurs, and the products leave the enzyme surface, freeing the active site for further reactant molecules.

Enzymes as Control Valves

A number of enzymes, in addition to an active site, have a second site called the *allosteric site*. When this site binds the allosteric inhibitor (generally the end product of a long metabolic pathway), the tertiary structure of the enzyme is sufficiently distorted so that the active site can no longer bind the reactant. In this way, the cell exercises a fine control over its internal materials. Materials not needed at the time are not manufactured. Usually only one enzyme in a metabolic pathway is susceptible to allosteric control. This enzyme is typically the first enzyme after a branch point in a pathway (Fig. 2-12).

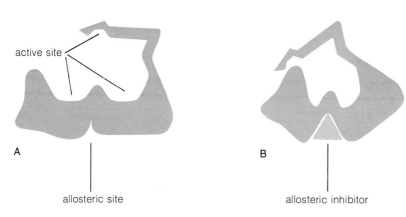

Figure 2-11 ALLOSTERIC INHIBITION OF AN ENZYME. (A) Normal conformation of enzyme; allosteric inhibitor not bound. (B) Active site of enzyme distorted by binding of allosteric inhibitor.

Birth Defects

It is estimated that the human body is specified by perhaps 1,000,000 different genes, most of which can manufacture a specific protein. The vast majority of these proteins function as enzymes, and the absence or malfunction of any constitutes a birth defect. A look at a single such birth defect illustrates the far-reaching consequences of enzyme malfunction to the body.

The amino acid phenylalanine serves as a starting point for the synthesis of the black pigment melanin (Fig. 2-13). Babies suffering from the birth defect *phenylketonuria* (PKU) lack the enzyme needed for the first step of the pathway leading to melanin. As a result, such babies grow up to be fair, blond, blue-eyed, and mentally retarded (if left without treatment). The fairness and blondness are due to the lack of melanin. Brain damage results from high levels of phenylpyruvate, a side product formed from phenylalanine. As phenylalanine begins to pile up because its outlet via the melanin pathway is blocked, it spills over into an alternate pathway involving the enzyme

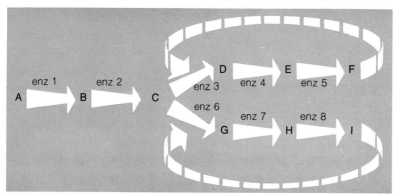

Figure 2-12 REGULATION OF METABOLIC PATHWAYS VIA ALLOSTERIC INHIBITION.

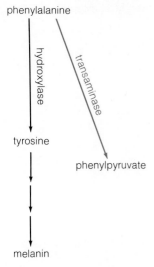

Figure 2-13
ENZYME DEFECT IN PKU. The PKU pathway is shown in color. It becomes activated if hydroxylose is nonfunctional.

transaminase, an enzyme that is normally inactive. The transaminase converts phenylalanine into phenylpyruvate which is then excreted in the urine. Fortunately, serious mental retardation can be avoided by restricting the diet so as to eliminate phenylalanine. The root causes of most other birth defects are not understood in this kind of detail. Considering the many opportunities for error, it is amazing that humans function as well as they generally do.

We started this chapter by asking what processes distinguish a living from a nonliving system. A living system, though highly organized, does not disintegrate into chaos as required by the second law of thermodynamics, and it accomplishes this through a continual input of energy and raw materials. The processes whereby food inputs are transformed into living substances, and living substances once more degraded into waste materials, is called metabolism, and metabolism is guided and controlled by enzymes.

Appendix A
Energy Metabolism

Respiratory Energy Metabolism: Glycolysis, the Krebs Cycle, and Electron Transport

The energy supply of an organized society is vital to its survival. Nowadays it is all too easy to imagine the chaos that would follow if all supplies of coal, gas, oil, uranium, and electric power were shut off. The most highly organized areas, the cities, would disappear first as mass transport and communications fell apart and food deliveries and industrial production halted. The reverberations would spread even to farmers whose tractors and harvesters would lie useless, to commercial fishermen whose diesel engines would stop, and in fact would affect everyone who could not grow or hunt his own food.

A very similar situation exists in the cell. There are a number of powerful poisons (cyanide and arsenic, to name two) that destroy a cell because they block its energy production. The elucidation of this process of energy production was the first aim of the early biochemists, beginning with Louis Pasteur and including others, such as Büchner, Harden and Young, Warburg, Embden, Meyerhof, Krebs, Cori, and

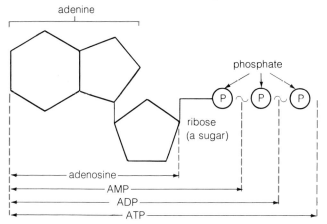

Figure 2-14 THE STRUCTURE OF ATP AND ITS DERIVATIVES. The phosphate groups are linked to each other by means of high-energy chemical bonds. All of the derivatives of adenosine play important roles in cell metabolism.

Lipmann. These great scientists realized that all further understanding of metabolic processes would depend on an elucidation of energy processes, and they were spectacularly successful in their work. Minor details must still be clarified, but the guiding principles are now well established.

First, in contrast with the variety of energy sources used in human society, the cell uses only one form of energy to perform all its tasks, ranging from chemical synthesis to physical movement, generation of electric impulses, and pumping against concentration gradients. The form of energy used is chemical energy, and the specific chemical employed is a molecule called *ATP*, for *adenosine triphosphate*. ATP is used by every known form of life on earth: bacteria, trees, insects, sponges, fish, birds, and of course humans. The structure of ATP is shown in Figure 2-14.

ATP can transfer one or both of its terminal phosphate units to other molecules, a reaction accompanied by the release of a considerable amount of energy. As a consequence, the two terminal phosphate groups are said to be linked by means of high-energy bonds (indicated as squiggles).

Coupled Reactions

To appreciate how this allows ATP to accomplish things inside the cell, let us consider a simple task: the linking of molecules A and B into the compound AB:

$$A + B \rightarrow AB \quad (\text{not spontaneous})$$

Let us assume that the process requires a modest amount of energy. Therefore the reaction cannot occur spontaneously, since all spontaneous reactions occur with a release of energy. In order to "drive" the reaction, ATP is brought in. First, ATP donates a terminal phosphate group (P) to A:

$$A + ATP \xrightarrow{enzyme} A{-}P + ADP \qquad \text{(spontaneous)}$$

The reaction releases energy, hence it occurs spontaneously. Second, A—P reacts with B. Again, energy is released, hence the reaction occurs spontaneously:

$$A{-}P + B \xrightarrow{enzyme} AB + P \qquad \text{(spontaneous)}$$

We have now achieved our goal—the linking of A and B by means of two spontaneous (energy-releasing) reactions. The energy cost of the original reaction has been paid by ATP, which has consequently been degraded to the less energy-rich state of adenosine diphosphate (ADP) + phosphate. Because the degradation of ATP to ADP + P and the synthesis of AB from A + B occur by means of common intermediates, we say that the two processes are chemically *coupled*. If two processes are coupled, the energy surplus of one process may be used to pay the energy deficit of the other process. A simple physical analogy is a siphon, in which a column of water rises in the ascending arm because it is physically pulled by the water column in the descending arm. (Fig. 2-15) As long as distance Y exceeds distance X, water will rise in the ascending column, against the force of gravity. But as soon as we uncouple the two processes (for instance, by introducing an air hole at the top of the column), no water rises in column X, no matter how long Y is or how much water flows in it. Likewise, if ATP breaks down to ADP + P by itself, without touching another molecule, it cannot do any useful work no matter how much energy is released in the process.

The next question is, how does the cell manufacture its ATP? The principle involved is the same as before, the use of coupled reactions. In this case, the reaction that absorbs energy is the phosphorylation of ADP to form ATP:

$$ADP + P \xrightarrow{energy} ATP$$

The overall reaction that supplies energy is the oxidation of glucose to form carbon dioxide and water:

$$glucose + 6O_2 \longrightarrow 6CO_2 + 6H_2O + energy$$

Glucose, you recall, is the simple sugar that polymerizes to form starch. It represents the central energy source in our diet. However, the

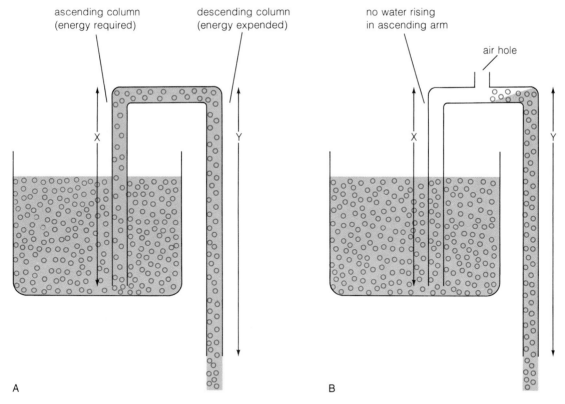

Figure 2-15 A SIPHON AS AN ANALOG OF COUPLED AND UNCOUPLED CHEMICAL REACTIONS. (A) Fluid rises in arm X against the pull of gravity, but only if arm Y is longer than arm X. (B) The energy drop in arm Y cannot be used to pull fluid up in arm X due to air hole.

amount of energy released in the oxidation of a molecule of glucose is about 100 times greater than is needed to form one molecule of ATP from ADP. Therefore the oxidation process is carried out gradually, in several dozen coupled steps which release energy in smaller packets and allow not 1, but 38 molecules of ATP to be formed for every molecule of glucose oxidized. These numerous small steps are of course specified, catalyzed, and controlled by the agency of individual enzymes.

Biological Oxidation

Since it is the stepwise oxidation of glucose that provides the energy for coupled formation of ATP, we should ask what is meant by oxidation. Oxidation is a familiar process in everyday life; the rusting of steel, the burning of wood, and the explosion of gasoline vapor are all examples of oxidations. It happens that in each of these cases the oxidizing agent is atmospheric oxygen. The same is true in the cellular oxidation of glucose, but only in an overall sense. Most of the intermediate oxidations occurring inside the cell do not involve oxygen at all, but involve instead the transfer of electrons from one molecule to another. This, then, is the fundamental definition of *oxidation:* the loss of electrons by a molecule or ion. These lost electrons cannot just be thrown away, but must be given suitable accommodations in a new molecule or ion. Therefore each oxidation is accompanied by a complementary process called *reduction* —the gain of electrons by a different molecule or ion. When glucose reacts with oxygen, the carbon atoms of glucose lose a total of 24 electrons and become oxidized to carbon dioxide:

$$\overset{24\ e-}{\overline{glucose + 6O_2}} \longrightarrow 6CO_2 + 6H_2O$$

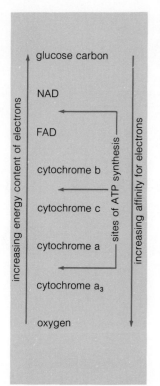

Figure 2-16
THE ELECTRON TRANSPORT CHAIN.

Consequently, oxygen atoms gain a total of 24 electrons and become reduced to water. As mentioned before, this oxidation process releases a great deal of energy and therefore occurs spontaneously. Electrons flow spontaneously from glucose carbon to oxygen, and not the other way around, because oxygen has a stronger affinity for electrons than does carbon. Another way to express this is to say that electrons associated with oxygen contain less energy than electrons associated with glucose carbon. The cell contains a number of molecules whose electron affinity is intermediate between that of glucose carbon (lowest affinity) and that of oxygen (highest affinity). Figure 2-16 shows how these substances are ranked.

Electrons are passed down this chain (called the *electron transport chain*) in pairs, in a stepwise fashion, and, as they travel down the chain, energy is released. Part of this energy is used in coupled reactions to drive ATP synthesis. Each pair of electrons that travels down the electron transport chain drives the synthesis of three molecules of ATP. Since a total of 12 pairs make the trip for each glucose molecule, 36 molecules of ATP should be generated in this manner. The sites of the coupled reactions are indicated by arrows. In fact, two of the 12 electron pairs enter the electron transport chain at the level of FAD instead of NAD. Because they enter the chain at a lower energy level, they generate only 2 ATP's per pair and the overall ATP yield derived from the electron transport chain is 34 instead of 36.

The Dismantling of Glucose

Finally, let us briefly follow the fate of the carbon atoms of glucose. Glucose is dismantled into carbon dioxide in two stages. Stage 1 is called *glycolysis* ("to dissolve sugar"). Glycolysis is a

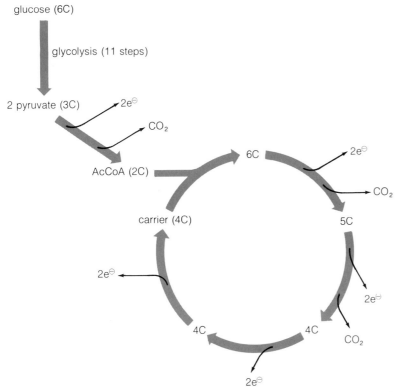

Figure 2-17 GLYCOLYSIS AND THE KREBS CYCLE (SIMPLIFIED). The glycolytic pathway converts a 6-carbon molecule (glucose) into two 3-carbon molecules (pyruvate) by means of 11 chemical steps. The Krebs cycle then systematically dismantles pyruvate into three 1-carbon fragments (CO_2) and at the same time oxidizes it, causing release of electrons. The 2-carbon fragment (AcCoA) can also be produced by the breakdown of fats and proteins.

process involving 11 separate reactions, and during this stage the six-carbon molecule glucose is split into two three-carbon pyruvate fragments. The two pyruvate fragments contain less energy than the glucose, and part of this energy is trapped in the form of two ATPs. At the same time, two pairs of electrons are removed from glucose and passed on to two NAD molecules which feed them into the electron transport chain.

Stage 2 is called the *Krebs cycle*. This stage takes place inside the soluble portion (matrix) of the mitochondrion. During this stage, the three-carbon fragments of pyruvate are completely degraded to carbon dioxide. Each pyruvate also gives up five pairs of electrons, which enter the electron transport chain (located in the inner membrane of the mitochondrion). The dismantling of pyru-

TABLE 2
Energy yield in respiratory breakdown of glucose

	ATP'S PRODUCED DIRECTLY	ATP'S FROM ELECTRONIC TRANSPORT CHAIN	
		ELECTRON PAIRS REMOVED	EQUIVALENT ATP'S FROM ELECTRON TRANSPORT CHAIN
Glycolysis	2	2	6[b]
Krebs cycle	2	2 × 5 = 10	28[a]
Total	4		34
Grand total	38 ATP's (4 + 34)		

[a] Two pairs of electrons enter the electron transport chain at a lower energy level and yield 2 ATP's per pair instead of 3.

[b] Under some conditions the two electron pairs removed during glycolysis (as reduced NAD) may yield only 2 ATPs per electron pair, reducing the ATP equivalents from 6 to 4.

vate involves a kind of cyclic piggyback ride on the back of a carrier molecule as shown in Figure 2-17. Table 2 summarizes this. It can be seen that the great bulk of the ATP is extracted via the electron transport chain of the mitochondria. Interestingly, there are organisms that are able to live in the absence of oxygen by using glycolysis alone. Yeast used in alcoholic fermentation is one example. It is thought that this represents the most primitive form of energy metabolism that evolved on earth. Because the primitive earth contained no atmospheric oxygen, direct ATP from glycolysis may have represented the only option open to these early organisms. The much greater energy riches of the electron transport chain had to await the appearance of free oxygen in the atmosphere. It was put there by another living process, photosynthesis.

Photosynthesis

A friend of mine carries a bumper sticker on his car which says, "Have you thanked a green plant today?" He is a botanist, but the sentiment is valid for everyone. Green plants are engines which produce chemical energy that suffices not only for their own needs (for night as well as day) but leaves enough of an excess to provide for all animal needs as well. Not only are these engines pollution-free, but they give off the molecular oxygen on which all animal life depends. Would it not be marvelous to have an engine like this to power the family car? Unfortunately, under the best of conditions, a

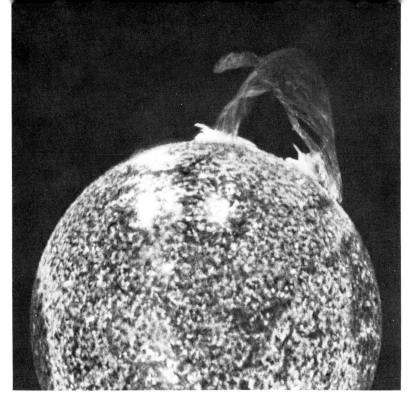

Figure 2-18 A SOLAR FLARE. This flare is made up of incandescent gases thrown 260 million miles (32,000 earth diameters) into space by thermonuclear processes inside the sun. A minute fraction of this energy is trapped on Earth by the process of photosynthesis.

small car running on photosynthesis could travel at most only $\frac{1}{3}$ of a mile per day. Plants do things in a beautiful way, but they are not in a hurry to go anywhere. In green plants, photosynthesis occurs in the green-colored organelles called *chloroplasts* (Fig. 2–19).

TABLE 3
Comparison of glucose oxidation and photosynthesis

	GLUCOSE OXIDATION	PHOTOSYNTHESIS
Overall reaction	Glu + $6O_2 \rightarrow$ $6CO_2 + 6H_2O$	$6H_2O + 6CO_2 \rightarrow$ glu + $6O_2$
Electron flow	Glu-C $\rightarrow$ O_2	$H_2O \rightarrow CO_2$
Organelle	Mitochondrion	Chloroplast
Energy relationships	Energy released	Energy absorbed (from sunlight)

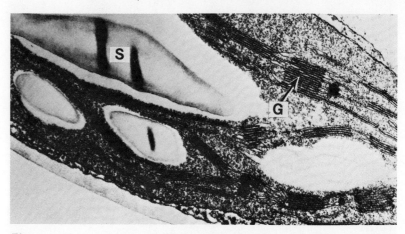

Figure 2-19 CHLOROPLAST. View inside a single chloroplast, showing photosynthetic membranes stacked into grana (G). Also visible is a starch grain (S).

Photosynthesis is fundamentally a mirror image of the glucose oxidation carried out in mitochondria (see Table 3). Whereas carbon is oxidized in the mitochondrion, it is reduced in the chloroplast. The source of the electrons needed to reduce carbon dioxide to the glucose level is the oxygen of water. The obvious problem is that the electrons sitting on the oxygen of water are at a very low-energy level. To transfer them to glucose carbon involves a steep rise in energy content, hence the reaction cannot occur spontaneously. A molecule called *chlorophyll* helps solve this problem. The chlorophylls are large, rigidly flat, extremely complicated molecules, and have magnesium atoms in their centers. The behavior of chlorophyll comes closest to resembling the behavior of a photoelectric cell (for example, the kind used in the exposure meter of a camera). In a photoelectric cell, packets of light (called quanta) strike a suitable metallic surface, causing electrons to be ejected from the metallic surface. The electrons are carried off by a wire and can do electrical work, such as moving a needle attached to a galvanometer, before returning to the emitting surface (Fig. 2-20).

In the most primitive kind of photosynthesis, such as that carried out in photosynthetic bacteria, a very similar picture is observed. A quantum of light strikes the chlorophyll molecule, causing it to emit an electron. The electron, instead of traveling through a wire as in a photoelectric cell, travels through a series of molecular carriers before returning to the chlorophyll molecule at a lowered energy level. As the electron passes through the carrier molecules, it performs work via coupled reactions which drive ATP synthesis. This type of photosynthesis is called *cyclic photophosphorylation*.

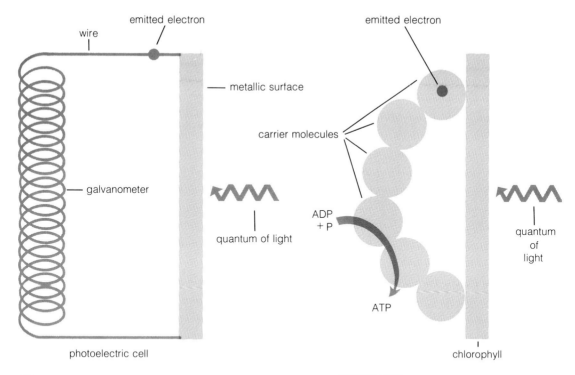

Figure 2-20 COMPARISON OF PHOTOELECTRIC CELL AND PRIMITIVE PHOTOSYNTHESIS (CYCLIC PHOTOPHOSPHORYLATION). In both cases the electron makes a complete loop to return to its original starting material.

In green plants a somewhat more complex sequence of events is observed. A variety of chlorophylls and accessory pigments are involved, and these are organized into two distinct kinds of *pigment systems*. Each pigment system consists of approximately 300 light-absorbing molecules (mostly made up of two chemical forms of chlorophyll) plus 1 chlorophyll molecule which functions photoelectrically, emitting electrons in response to light. These are called the *reaction centers* (Fig. 2-21). The light-absorbing molecules of the pigment systems act as light traps, absorbing photons of light and conducting them to the single chlorophyll molecule located in the reaction center. Here one photon of light causes the photoelectric emission of one electron from the chlorophyll ring structure.

The two kinds of pigment systems differ in two respects. First, the reaction centers of the two systems respond to light of slightly different wavelength (system I responds to light of 700 mμ, system II to light of 680 mμ). Second, the electrons emitted by the two systems have different fates.

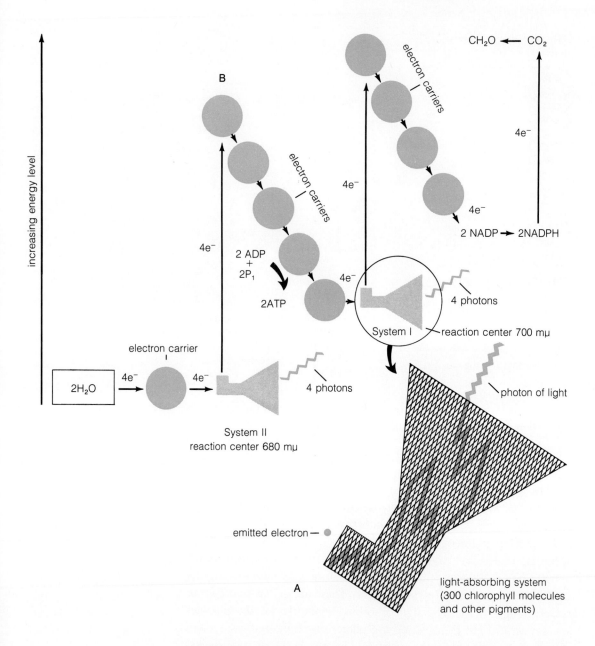

Figure 2-21 PHOTOSYNTHESIS IN GREEN PLANTS. (A) Path of a photon of light inside a pigment system. The pigment system acts as a light funnel, trapping photons and conducting them to the single chlorophyll molecule in the reaction center. The reaction emits one electron for one photon absorbed. (B) Overall path of photons and electrons in photosynthesis of green plants. The electrons emitted by system I are ultimately incorporated into NADP and then CO_2. Electrons emitted by system II ultimately replace electrons emitted by system I.

The electron emitted by system I is emitted at a high-energy level and passed on to carrier molecules which deposit them on a molecule called NADP (a phosphorylated version of the NAD used in glucose oxidation). NADP thus becomes reduced. It is this NADP which, with the help of ATP, reduces carbon dioxide to the glucose level of oxidation. However, the chlorophyll of the system I (700 mμ) reaction center is now missing an electron, and this electron deficit must be made good. The necessary electron is supplied by system II chlorophyll (680 mμ) reaction center (Fig. 2-21). As the system II reaction center absorbs a photon of light, it emits an electron which first passes through a series of carrier molecules and drives ATP synthesis via coupled reactions and then is finally transferred to the system I reaction center.

But we are now left with an "electron hole" in system II. This electron hole has such a strong affinity for electrons that it is able to take an electron away from reduced oxygen (water). Water is thus oxidized to molecular oxygen, and system II is restored to electric neutrality. Note that the system II electron hole acts as an enormously powerful oxidizing agent; it has a stronger affinity for electrons than does molecular oxygen, normally the most powerful oxidant in nature. But this electron hole has the most fleeting of lifetimes and can be induced in chlorophyll only by the action of light.

In summary, two photons of light are necessary to move a single electron all the way from water to system II to system I to carbon dioxide. Since the reduction of one molecule of carbon dioxide to the carbohydrate level requires four electrons, the overall reaction may be summarized as follows[7]:

$$2H_2O + CO_2 \xrightarrow[4e-]{8 \text{ photons}} O_2 + CH_2O$$

---

[7]. For the sake of clarity, the reaction as written here is unbalanced. One of two oxygen atoms associated with carbon dioxide is liberated as water. The oxygen atoms associated with carbon dioxide are neither oxidized nor reduced.

Appendix B
Weak Chemical Bonds

Sugar is a solid at room temperature, water is a liquid, and air is a gas. What accounts for these differences in physical state? The answer is weak interactions between molecules. Air molecules show almost no interaction, and each molecule moves independently. Water molecules interact more strongly, hence the molecules are packed in the close yet disordered manner typical of liquids. Sugar molecules interact most strongly and are stacked in the close regular way associated with solids. Yet these close-packed, regularly disposed sugar molecules each retain their molecular identity. The proof comes when we add water without changing the temperature; the solid crystal of sugar dissolves, and each sugar molecule wriggles off into a crowd of water molecules. In terms of physical behavior, it now most closely resembles a gas. And the same kinds of weak chemical bonds that operated between sugar molecule and sugar molecule in the crystalline state are now operating between sugar molecule and water molecule in the dissolved state. These weak bonds are of crucial importance in the functioning of the living cell. They control such diverse phenomena as water solubility, membrane assembly, enzyme catalysis, cellular self-recognition mechanisms, and the replication of genetic material.

All weak bonds depend in some way on electrostatic attraction between molecules. In living systems, the important kinds of weak bonds include ionic bonds, hydrogen bonds, and Van der Waals bonds. *Ionic bonds* are established between full electric charges of opposite sign. For example, the functional group of the amino acid lysine normally carries a positive charge, and the functional group of the amino acid glutamic acid normally carries a negative charge. If these two amino acids are incorporated into a single strand of protein, they can twist the protein strand into a loop (Fig. 2-22).

A *hydrogen bond* operates with partial electric charge differences. Many biological molecules contain groupings of the type R—O—H and $\genfrac{}{}{0pt}{}{R}{R}$⟩N—H, where the solid lines represent covalent bonds and R represents the remaining portion of the molecule. A covalent bond between two atoms is established by the sharing of a

WEAK CHEMICAL BONDS / 63

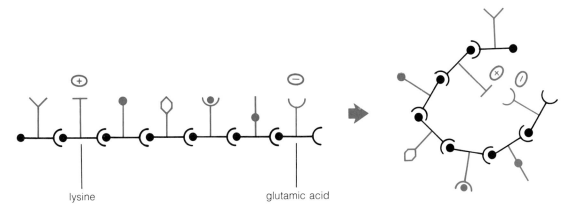

Figure 2-22 IONIC BONDS. A short stretch of protein is twisted into a loop due to mutual attraction between oppositely-charged functional groups of two amino acids.

pair of orbital electrons.[8] Thus the orbital electrons linking oxygen and hydrogen may be indicated as follows:

$$-\text{O:H}$$

The orbital electrons (represented by dots) lie closer to the oxygen than they do to the hydrogen, because the oxygen nucleus pulls on them with a positive charge of 8, while the hydrogen nucleus can only pull with a positive charge of 1. The result is that oxygen receives more than its "fair share" of electrons and becomes partially negative, and hydrogen receives less and becomes partially positive:

$$\overset{\delta-}{-\text{O}}\overset{\delta+}{-\text{H}}$$

Here the symbol δ indicates a partial electric charge, as opposed to the full charges found, for example, on the functional groups of the amino acid glutamic acid or lysine. The same reasoning applies to linkage between nitrogen and hydrogen:

$$\overset{\delta-}{\underset{}{\rangle\text{N}}}\overset{\delta+}{-\text{H}}$$

Oxygen and nitrogen generally carry partial negative charges in biological linkages, and can therefore interact with the electron-poor hydrogens of —OH and =N—H groups. Because an electropositive hydrogen is always involved in this type of linkage, it is called a *hydrogen bond*. Figure 2-23 shows three hydrogen bonds holding

---

8. The orbital electrons of an atom are typically its outermost electrons.

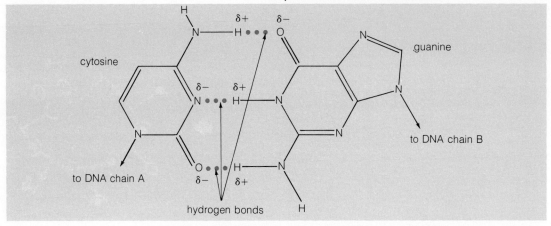

Figure 2-23 HYDROGEN BONDING BETWEEN TWO COMPLEMENTARY BASES IN DNA. Note that in each case an electropositive hydrogen is weakly attracted to an electronegative oxygen or nitrogen.

together the complementary bases guanine and cytosine in DNA. This type of specific hydrogen bonding is responsible for keeping the two chains of DNA together. The resulting specificity of base pairing leads to the mechanism responsible for DNA self-replication, as we shall discover in Chapter 3.

A *Van der Waals bond* arises as a statistical improbability. Consider two neon atoms (Fig. 2-24). Each carries 10 positive charges on its nucleus and is surrounded by 10 negatively charged electrons, making the atom as a whole electrically neutral. Furthermore, the whirling electrons are as likely to be found on one side of the atom as on the other side. On an average, there should be no areas more electron-dense than any others. But this is only a long-term probability. If we look at the atoms instantaneously, we might, for example, discover the arrangement shown in situation B. Here the electrons show a temporary bunching on the left side of the atom, making this side electronegative and the right side correspondingly electropositive. Since opposite charges attract, our temporarily polarized atom will induce a partial negative charge on the side facing its right-hand neighbor, and therefore an instantaneous attractive force will be set up between the two atoms. Such charge inequalities fluctuate from instant to instant, but the induced charges in neighboring atoms keep pace with the fluctuations. The larger the total number of electrons in each atom or molecule, the stronger the attractive forces that are set up. However, the attractive force drops off very rapidly with increasing distance between the two groups (it is inversely proportional to the sixth power of the distance). Therefore, good Van der

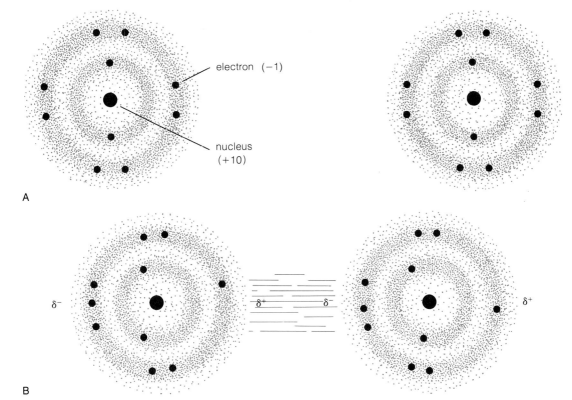

Figure 2-24 VAN DER WAALS BONDING BETWEEN TWO NEON ATOMS. (A) When the 10 negative electrons of neon are symmetrically placed around the positive nucleus, each atom is electrically neutral and symmetrical. No attraction results between adjacent neon atoms. (B) A momentary electron asymmetry around the nucleus produces a momentary negative and positive pole in each atom. This results in an electrostatic attraction between the two atoms, even though each atom remains electrically neutral as a whole.

Waals bonding requires a close surface "fit" between two molecules. The close fit between an enzyme and its substrate is a necessity if the enzyme is to bind the substrate with effective Van der Waals bonding.

All weak bonds represent bonding energies much lower than those typical of covalent bonds. Their energy content ranges from 2 to about 10 percent of the energy content of a covalent bond. At room temperature, molecules have enough kinetic energy to break such weak bonds constantly, while covalent bonds are almost never broken as a result of simple kinetic pulling and thumping. Large biological molecules may contain hundreds or thousands of such internal weak bonds per molecule, hence the breaking of one or two bonds at a time does not destroy the molecule. The broken bonds reestablish themselves spontaneously at the proper locations. But the ease of formation and destruction of the weak bonds makes possible the uniquely biological life-styles of the large molecules—the unzippering of DNA during replication, the dissociation of reaction products from enzymes, and the pinching-off and resealing of membranes. All these events reflect the advantages of using bonds that can be broken without using a chemical crowbar.

Bibliography

GENERAL READING

Isaac Asimov, *The Intelligent Man's Guide to the Biological Sciences,* Washington Square Press, 1964. Asimov is the unsurpassed master of popular writing on scientific topics, and this book is only one of the over 200 titles he has published. He combines a lucid, unjargonistic writing style with a sense of astonishment at the spectacle of life. The following brief excerpt is part of his discussion of the molecular structure of polysaccharides.

> Glucose and galactose combine to form lactose, which occurs in nature only in milk.
>
> There is no reason why such condensations cannot continue indefinitely, and in starch and cellulose they do. Each consists of long chains of glucose units, condensed in a particular pattern.
>
> The details of the pattern are important, because although both compounds are built up of the same unit, they are profoundly different. Starch in one form or another forms the major portion of humanity's diet, while cellulose is completely inedible. The difference in the pattern of condensation, as painstakingly worked out by chemists, is analogous to the following: Suppose a glucose molecule is viewed as either right-side-up (when it may be symbolized as "u") or upside down (symbolized as "n"). The starch molecule can then be viewed as consisting of a string of glucose molecules after this fashion ".... uuuuuuuuu ....," while cellulose consists of ".... ununun unun. . . ." The body's digestive juices possess the ability to hydrolyze the "uu" linkage of starch, breaking it up into glucose, which we can then absorb and from which we can obtain energy. Those same juices are helpless to touch the "un" linkage of cellulose, and any cellulose we ingest travels through the alimentary canal and out.

Albert L. Lehninger, *Bioenergetics,* 2nd ed., W. A. Benjamin, 1971. An intelligent, incisive account of energy processes within the cell. Lehninger starts with a discussion of the very subtle concept of energy itself. Not easy, but worth it.

TEXTBOOKS

James D. Watson, *The Molecular Biology of the Gene,* 3rd ed., W. A. Benjamin, 1975. This has been one of the most spectacularly successful books on cell genetics and metabolism ever published. Written by a Nobel Prize winner who also happens to be a very talented writer.

Albert L. Lehninger, *Biochemistry,* 2nd ed., Worth Publishers, 1975. Perhaps the best all-around textbook of biochemistry on the market today. This is an enormous and changing field, not easy to cover or to understand. Lehninger's text has the necessary scope, authority, and clarity to make an approach possible.

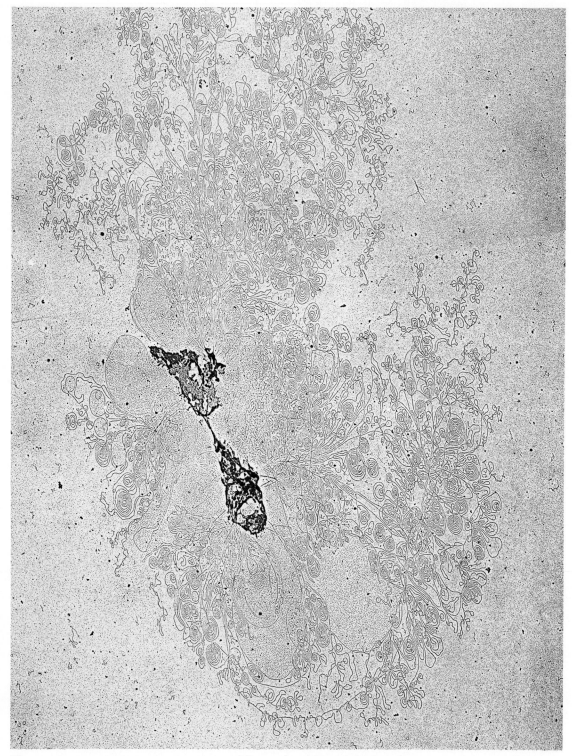
Figure 3-1 OSMOTICALLY SHOCKED E. COLI. The tangle of bacterial DNA forms a single continuous circle.

chapter 3

At the time of this writing, there is living somewhere in a hospital in New York State a very special baby. He is now a full year old and is able to stand up, smile, and recognize his parents and those who care for him, but has never touched a human being directly or been touched or hugged in return. He was delivered by caesarean section, picked up by a surgeon wearing sterile rubber gloves, placed in a sterile enclosure of transparent plastic, and has remained there ever since. All his food enters through

DNA and Protein Synthesis

germproof portholes, all his air is filtered, and all of his human contact takes place through an intervening sheet of plastic. He cannot come out of his prison and live because his body contains no functional lymphocytes—the white blood cells that defend the body against foreign organisms such as bacteria. Our primary antibacterial defense involves the synthesis of circulating plasma proteins called *antibodies*, which recognize and immobilize foreign organisms. Antibodies are synthesized by lymphocytes. This baby is suffering from combined immune deficiency, a birth defect like PKU, which reflects a fundamental error in the architecture of the body. In each case, the birth defect expresses itself as the lack of a functional protein—the enzyme hydroxylase in PKU, or plasma antibody proteins in combined immune deficiency. In each case, the consequences are devastating. The living organism is so complex that we often become aware of its component parts only when they break down, and the number of potential trouble spots runs into the hundreds of thousands.

Let us now ask the next question, *why* is hydroxylase nonfunctional in a baby with PKU? *Why* are antibody proteins missing in a baby with combined immune deficiency? The problem extends back to the very first cell that started the baby's development—the fertilized egg. Biologically speaking, life does not start with the fertilized egg. The mother's egg and the father's sperm were both living cells, the offspring of other living cells. Life arises from life and is continuous along an immense chain back to the very first life. However, with the fertilized egg arises a unique set of blueprints. These blueprints dictate the development of every new individual, who will be different from all other individuals who have ever lived before. To be sure, he or she will share innumerable similarities with others of the species: number of fingers, number of teeth, number of vertebrae in the backbone, and so on. But when each of the perhaps 1 million different character traits is examined, it is certain that differences will be found. (The only exceptions are identical twins.) In the case of the baby mentioned above, the blueprint was flawed.

DNA and Information

Now, as we have seen, the language of the cell is the language of chemistry. Therefore the blueprints are written out in a chemical language, the language of *deoxyribonucleic acid* (DNA). The DNA molecule is the longest naturally occurring molecule and, before its secret was discovered, apparently one of the most monotonous. It consists of endless stretches of repeating units called *nucleotides*. There are only four different kinds of nucleotides in DNA, constituting the "letters" of its "alphabet." These letters are combined

FROM DNA TO CELL / 71

into "words," all of which are only three letters long. The words are then combined into "sentences" or "statements" averaging 500 words in length. Biologists call such statements *genes*. Human cells carry between 1 and 2 million such genes per cell.[1] For comparison, a complete set of the *Encyclopaedia Britannica* contains about 25 million words. Therefore each human cell carries within it the information content of 20 to 40 such sets of the encyclopedia.

Besides the staggering mass of information our DNA carries, perhaps the second most astonishing thing is its miniaturization. Spread out as a layer one molecule thick, the DNA contained in a single cell occupies only about one two-hundredth of the area of the head of a pin. Even in this era of miniaturization, the most advanced computer memory core seems gross in comparison, in spite of the fact that advanced computer memory storage is approaching molecular dimensions (Fig. 3-2).

Figure 3-2
ADVANCED, MINIATURIZED COMPUTER MEMORY STORAGE. The vertical, light-colored lines are aluminum interconnectors about 2.3 microns wide. Each short cross line on the aluminum contacts one memory cell, of which 8 are visible in the photo. The total field shown is big enough to contain 4 average-sized human body cells, whose DNA would carry $\frac{1}{2}$–1 billion codons ("words") per cell.

From DNA to Cell

Amazing though it is, we must remember one thing about our DNA: it is not alive. It contains all the information for building a living cell, but by itself it is no different from a complicated chemical sitting in a bottle on the shelf. To go from a chemical, DNA, to a living camel or elephant or opera singer requires two kinds of processes. First, DNA information must be *translated* into terms of protein structure: enzymes, membrane proteins, antibodies, and so on. These proteins, by themselves and by their action on other kinds of molecules, translate information about life into existing life. In other words, they create a cell.

The second process is something biologists call *embryological development*. We all start our individual lives as fertilized eggs. The fertilized egg is a single cell and, even though alive, is capable of none of the things we hold to be human—it cannot reason, it cannot see or hear or touch, it cannot feel pain or pleasure, and it cannot even move of its own volition. At this time nothing distinguishes it from the fertilized egg of a mouse or an antelope or a whale, other than future potential. To go from a fertilized egg to a living organism requires, first, a tremendous *multiplication* of cells (the human body contains an estimated 5 trillion cells, or 5×10^{12} in mathematical shorthand). Each of these cells is specified by its own package of DNA, and this means that a DNA molecule must be capable of many, many extremely accurate acts of self-replication.

---

1. The actual number of genes in a human cell is unknown. The figure of 1 to 2 million is obtained by dividing the weight of total cellular DNA by the average weight of a gene.

To go from a fertilized egg to a living organism further requires a process of cell *differentiation*. Among the 5 trillion cells that make up our bodies, there are many different kinds, such as muscle cells, thyroid cells, and retinal cells. Once more, it is DNA control that underlies the process of differentiation. Without giving away the whole story just now, we will say that a thyroid cell, for example, contains not just the DNA information needed to make a thyroid cell, but also DNA information for nerve, muscle, bone, retina, and every other type of cell in the body. But all this other information is not being *genetically expressed*. We see, then, that the informational (or genetic) substance DNA must be capable of at least three separate kinds of activity:

1. It must be able to replicate itself (make exact copies).
2. It must direct protein synthesis.
3. It must be capable of selective genetic expression.

The Replication of DNA

We now discuss each of these aspects in greater detail, beginning with the question of DNA replication, both historically and logically the first to be approached. The understanding of DNA replication flows directly from the understanding of DNA structure, and here we mean its three-dimensional structure. By now every schoolboy knows that this structure was worked out in 1953 by Watson and Crick. James Watson and Francis Crick built on the data of Maurice Wilkins and Rosalind Franklin, and some critical data of Erwin Chargaff. This is what they knew when they started their work on the three-dimensional structure of DNA:

1. They knew that DNA was a polymer of four different kinds of nucleotides, each nucleotide in turn consisting of a

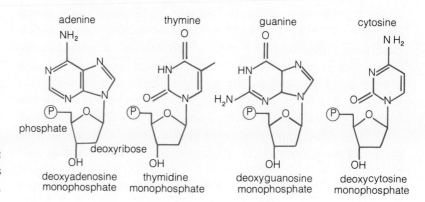

Figure 3-3
THE FOUR NUCLEOTIDES OF DNA.

phosphate group, the sugar deoxyribose, and a base (either adenine, guanine, cytosine, or thymine). See Figure 3-3.

2. They knew that the process of polymerization involved only the sugar and phosphate groups, with the bases dangling free and with no apparent restriction on base sequence. See Figure 3-4.

3. The x-ray diffraction photographs of Wilkins and Franklin showed an unusual "X" arrangement of reflection spots, which Crick interpreted as showing helical structure in DNA.

4. The data of Chargaff on the base composition of different DNAs showed that the fraction of adenine was always very close to the fraction of thymine, and that the fraction of guanine was always close to that of cytosine. See Table 1.

TABLE 1
Base ratios from DNAs of various species

	SPECIES Base proportion, %			
	A	T	G	C
Alcaligenes faecalis (bacterium)	16.5	16.8	33.9	32.8
Salmonella paratyphi (bacterium)	24.8	25.3	24.9	25.0
Wheat germ	27.3	27.1	22.7	22.8[a]
Human liver	30.3	30.3	19.5	19.9
Saccharomyces cerevisiae (yeast)	31.7	32.6	18.3	17.4

[a] Cytosine plus methylcytosine.

In an attempt to reconcile all these data, Watson and Crick published a memorable paper in *Nature*. Less than a page long, it proposed the now-famous double-helical structure for DNA. The DNA molecule contains two strands. Each strand consists of alternating sugar-phosphate units as described before. These form the "backbone" of the strand. Dangling free from each sugar is a nucleotide base. The strands are held together by weak forces (hydrogen bonds) between opposing bases. These hydrogen bonds form only between adenine and thymine on the one hand, and guanine and cytosine on the other (see Fig. 2-23 on page 64). Thus, if the base sequence along one strand reads AGCTG, the sequence along the

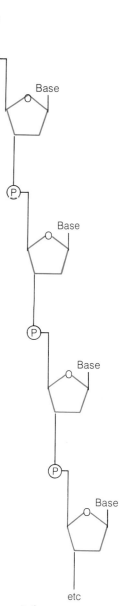

Figure 3-4
POLYMERIZATION OF NUCLEOTIDES MAKING UP DNA.

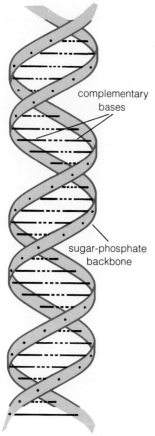

Figure 3-5
DOUBLE HELICAL STRUCTURE OF DNA, AS PROPOSED BY WATSON AND CRICK.

other strand must read TCGAC, in order to maintain the A-T and G-C pairing. Furthermore, the bases are quite hydrophobic, and this kind of close stacking with their flat faces pressed against each other like poker chips minimizes their exposure to the water medium outside and further stabilizes the structure (Fig. 3-6).

Watson started his collaboration with Crick at Cambridge, England in 1951, and they published the *Nature* paper in 1953. Though he originally came from the Midwest, he had by now acquired some British characteristics, and modestly concluded the paper by saying, "It has not escaped our notice that the specific pairing we have postulated immediately suggests a possible copying mechanism for the genetic material."

The mode of DNA replication, on which Watson and Crick elaborated in later publications, involves separation of the two strands, followed by the laying-down of new complementary strands along each open arm. See Figure 3-7.

The replication process requires activated[2] free nucleotides as precursors for the new strands, and it also requires specific enzymes.

The importance of a scientific idea is probably inversely proportional to its complexity. Both the structure and the mode of replication proposed for DNA by Watson and Crick were so economical and so beautiful that they were immediately accepted by the scientific world at large. The double-helical structure explained many previously puzzling facts about DNA (such as the base ratios). It also opened a broad channel for further research into the fundamental processes of genetics. Instead of drying up with the passage of years, this channel has become progressively broader and has absorbed the creative energies of a generation of biologists. The results of these studies constitute the cathedrals of the twentieth century.

---

2. Nucleotides are activated by means of ATP. This provides the necessary energy input needed for the process of polymerization.

Figure 3-6
STABILIZATION OF DNA HELIX BY HYDROPHOBIC BONDS. Flat sides of bases are pressed to similar faces above and below, excluding water. The hydrophobic (Van der Waals) bonding is shown in light color, hydrogen bonds in dark color. Only the latter are responsible for specific base pairing.

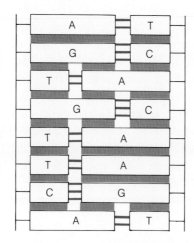

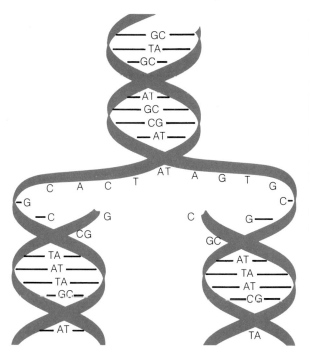

Figure 3-7
MODEL OF DNA
REPLICATION.
(A) Old strands, with complementary bases held together by hydrogen bonding. (B) Region of strand opening. (C) Free activated nucleotides lining up against complementary bases in opened old strand. (D) Polymerization of free nucleotides to form sugar-phosphate backbone.

Proof for the Watson–Crick Mechanism

The immediate result of the Watson–Crick paper was that cell biologists were stimulated to find convincing proof for the new model of DNA. Watson and Crick had built their model on purely logical (and esthetic) considerations, working with molecular models constrained by x ray diffraction data. They offered no proof that DNA actually replicated in the manner proposed. Proofs were quickly forthcoming from many different directions. Perhaps the loveliest experiment of this type was the Meselson-Stahl density gradient experiment. The thrust of this experiment was to distinguish between three possible modes of DNA replication, only one of which was consistent with the Watson–Crick proposal. The three modes are shown in Table 2.

But how is one to distinguish an original DNA molecule from one that has been newly synthesized? Chemically they are identical, and so the problem could not have been solved by any known chemical method. But by the late 1940s and early 1950s methods had become available for the separation and purification of a wide variety of naturally occurring *isotopes*.[3] One such isotope was nitrogen-15, or

---

3. The word "isotope" means "same place" and refers to the fact that two isotopes occupy the same place in the periodic table. Thus they constitute two forms of a given element that do not differ in chemical properties but do differ in atomic weight.

TABLE 2
Possible modes of DNA replication

METHODS OF REPLICATION	PARENT DNA	NEW DNA	
Conservative	‖	‖ ‖	Only one offspring DNA constructed from new materials
Semiconservative	‖	‖ ‖	Each offspring DNA contains 50% old material and 50% new construction. Consistent with Watson–Crick proposal
Nonconservative	‖	‖ ‖	Each offspring DNA constructed of new materials; parental DNA discarded

"heavy nitrogen." It has chemical properties essentially identical to those of the normal "light nitrogen" (nitrogen-14), but is about 7 percent heavier by weight. Therefore bacteria growing in a medium containing heavy nitrogen synthesize a DNA that is more dense than normal DNA. (A DNA molecule is rich in nitrogen content.) Meselson and Stahl allowed bacteria to grow in heavy nitrogen and then transferred them to a medium containing light nitrogen and allowed the DNA to replicate. All newly synthesized DNA would have to be made from light precursors and would therefore be less dense than the parental DNA (Fig. 3-8).

To separate DNAs of different densities, Meselson and Stahl prepared a density gradient. When cesium chloride (a salt) is centrifuged at very high speed for long periods of time, a more dense salt solution is created at the bottom of the tube than at the top. DNA introduced into such a centrifuge tube migrates down the tube until it encounters a region with its own density, as shown in Figure 3-9. The observed results were clearly in agreement with the Watson–Crick hypothesis.[4]

---

4. In all fairness, the mechanism by which the complementary strands are laid down is more complex than had been assumed in the beginning. For a discussion, see J. D. Watson, *The Molecular Biology of the Gene,* 3d ed., W. A. Benjamin, New York, 1975.

PROOF FOR THE WATSON-CRICK MECHANISM / 77

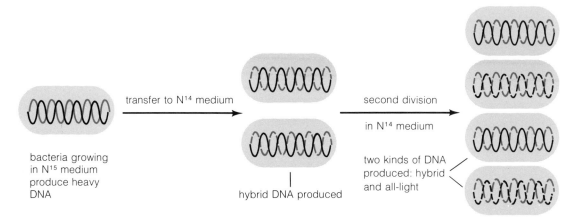

Figure 3-8 ISOTOPE DISTRIBUTION IN BACTERIAL DNA EXPECTED FROM WATSON–CRICK MODEL OF DNA REPLICATION.

Figure 3-9 THE MESELSON-STAHL EXPERIMENT. A cesium chloride density gradient is set up by centrifugation. When DNA is introduced into the tubes and centrifuged, it sediments toward the bottom of the tube until it reaches its own density. The results obtained are shown below.

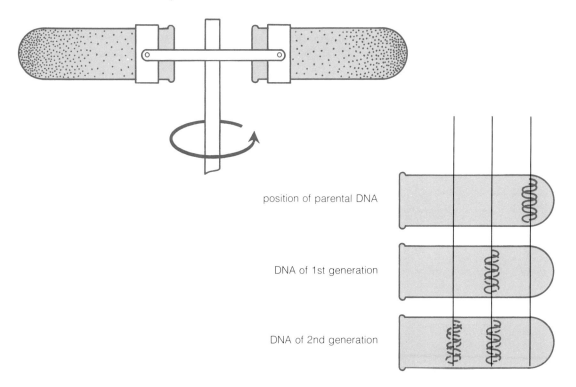

The DNA synthesized in the first generation was a hybrid molecule containing 50 percent old material and 50 percent newly synthesized material. Watson and Crick had predicted that first-generation DNA would contain one old strand and one new strand. DNA synthesized in the second generation was of two density classes. Half was a hybrid, as in the first generation, and the other half was made entirely of new materials. This result had again been predicted by the Watson–Crick hypothesis (Fig. 3-9).

Before leaving the topic of DNA replication, we should point out one of the most impressive aspects of this process, its precision. There is a game that children play called Rumor. They form a circle, and the first one whispers a sentence into the ear of the second one. Number 2 then repeats what he heard to number 3, and so on around the circle. When the whispered sentence finally comes back to its originator, it is invariably garbled and usually hilarious. Now, if DNA molecules were in the habit of playing Rumor, it would be no fun, since the last statement would always be identical to the first. DNA replication is so accurate that, on the average, a single mistake is made only in one case out of 10,000 base pair replications. That's a high standard even for adults.

DNA and Protein Synthesis

Protein synthesis is the second major sphere of activity for DNA and, in fact, its "reason for being." As we saw earlier (p. 40-49), it is the cellular proteins that define the cell. Unfortunately, proteins are unable to replicate themselves. That is to say, in a string of amino acids, there is no chemical reason why, say, a valine side group should pair with a free valine molecule in solution, or a serine with a serine, and so on. If they did, cellular existence might have been a lot simpler. This is where DNA steps in. Safe inside its nuclear shelter, away from the dangerous enzymes of the cytoplasm, it spins out orders for the day-to-day activities of the cell. At the same time, it is a responsive kind of ruler, keeping in touch with cytoplasmic events via reverse messengers and changing its own orders as called for by circumstances.

While the ultimate orders for protein building emanate from DNA, the immediate act of protein assembly is entrusted to another class of nucleic acid called *ribonucleic acid* (RNA). The differences between DNA and RNA are so small that they are obvious only to enzymes, which must be able to make this crucial distinction. Like DNA, RNA is a polymer consisting of four kinds of nucleotides.[5] Three of these are the same ones found in DNA: adenine, guanine, and cytosine. The fourth is called uracil and is a close chemical

---

5. Certain classes of RNA contain small amounts of bases other than the four major ones.

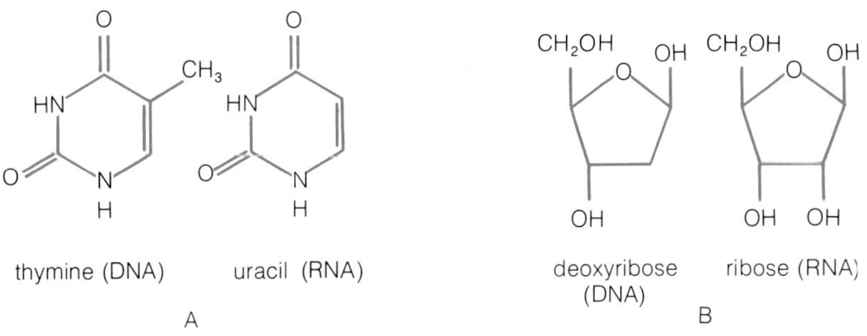

Figure 3-10
CHEMICAL DIFFERENCES BETWEEN DNA AND RNA. (A) Base differences: thymine (DNA) versus uracil (RNA). (B) Sugar differences: deoxyribose (DNA) versus ribose (RNA).

cousin of thymine. Like thymine, uracil pairs specifically with adenine. The sugar present in RNA is ribose, again a close chemical cousin of the deoxyribose of DNA (Fig. 3-10).

Finally, while RNA molecules are perfectly capable of base pairing, they do not in general exist as simple double helices. They are single-stranded molecules which are either extended or more commonly folded into complex tertiary structures analogous to those of proteins. These tertiary structures contain stretches of double helix formed by the interaction of neighboring nucleotide bases. Three kinds of RNA molecules are known: messenger RNA, transfer RNA, and ribosomal RNA. Table 3 summarizes the differences between DNA and RNA.

TABLE 3
Summary of the differences between DNA and RNA

	DNA	RNA
Bases	Adenine	Adenine
	Guanine	Guanine
	Cytosine	Cytosine
	Thymine	Uracil
Sugar	Deoxyribose	Ribose
Structure	Double helix	Single chain — may be extended or folded into partially helical regions
Made by	DNA	DNA
Location in cell	99.99% in the nucleus	Nucleus and cytoplasm
Function	Makes DNA and RNA	Makes protein

Participation of RNA in Protein Synthesis: Messenger RNA

The train of events whereby the base sequence of DNA is translated into the amino acid sequence of a protein begins with another protein. An enzyme called RNA polymerase partially separates the two strands

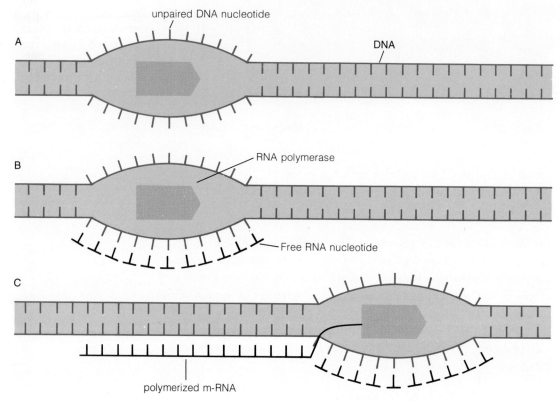

Figure 3-11 SYNTHESIS OF RNA BY RNA POLYMERASE. (A) Enzyme attaches to promoter region of DNA, causing separation of the two DNA strands. (B) Activated RNA nucleotides specifically base pair to exposed DNA bases. The RNA nucleotides are then polymerized (chemically linked) by the polymerase. (C) RNA polymerase has moved further down the DNA helix, trailing a growing RNA chain.

of DNA so as to expose the pairing edges of the bases. The same enzyme then makes a complementary RNA copy of one of the two strands (Fig. 3-11). For example, a DNA sequence that reads TTAGCTA produces the complementary RNA sequence AAUCGCU. This requires activated RNA nucleotides as precursors. These nucleotides are synthesized on a continuous basis by cellular machinery, using simpler molecules as raw materials. The freshly made RNA is called *messenger RNA* (mRNA), because it carries a genetic message from the nucleus, where the original message is stored, to the cytoplasm, where it is translated into protein. The use of mRNA (as opposed to the direct translation of DNA into protein) offers several advantages. First, it represents a method of amplification. A given stretch of DNA can make many copies of mRNA, each of which can then proceed to make protein. Second, by separating the area where DNA is stored (the nucleus) from the area where protein is synthesized (the cytoplasm), the cell lowers the amount of accidental damage its DNA can suffer. Third, the use of mRNA gives the cell another area of control—a kind of fine-tuning knob for the regulation of protein synthesis.

The Genetic Code

The alphabet of nucleic acids contains 4 letters (the bases), while the alphabet of proteins contains 20 (the amino acids). Hence there cannot be a letter-for-letter translation between the two. Twenty different symbols composed of only four kinds of nucleotide bases can be created by reading the bases in clusters of three (the triplet code). The individual base clusters are called *codons*. This mechanism allows 64 different symbols, more than the 20 required.[6] Some of the excess is used as punctuation. (There are three separate "stop" signals plus a single "start" signal.)

By now, the meaning of all 64 triplets has been deduced, and a sample selection is shown in Table 4.

TABLE 4
A sampling of the genetic code

CODON	AMINO ACID
UUU	Phenylalanine
UUC	Phenylalanine
UUA	Leucine
UUG	Leucine
UAA	Termination
UGA	Termination
UGU	Cysteine
CUU	Leucine
CUC	Leucine
CCG	Proline
CAU	Histidine
AUG	Methionine (and chain initiation)
GGG	Glycine
GCU	Alanine
GCC	Alanine

6. To see how 64 different symbols are generated by arranging 4 different bases into groups of 3, imagine the 4 bases as cards of 4 different suits. If only one card is dealt, there are 4 possible hands. If two cards are dealt, there are $4 \times 4 = 16$ possible hands since, for each possible card received in the first deal, there are 4 further possibilities in the second deal. If three cards are dealt, the possible hands are $4 \times 4 \times 4 = 64$.

It is obvious, first, that the same amino acid may be coded for by several codons. Thus UUU and UUC both code for phenylalanine. Further, the *order* of nucleotides within a codon is important; UUC stands for phenylalanine, while CUU gives leucine. This is quite analogous to the rules of English, where the words "top" and "pot" have different meanings. However, there is one amazing difference between the genetic code and human language. The word "top," for instance, has a meaning in Latvian and in Turkish. Its Latvian translation is "becomes," and the Turkish is "cannonball." Both are rather far removed from the English meaning. But the codon CUU means leucine to a bacterium, an artichoke, a camel, or a sea robin. The genetic code is universal to all living forms—surely one of the most impressive demonstrations of the unity of all life on earth.

RNA and Protein Synthesis: Transfer RNA

Now, to go from CUU to leucine is not something that can be done directly. That is to say, there is nothing about the chemical properties of the codon and the amino acid that make a direct union possible. Chemically speaking, leucine has no more interest in the CUU codon than it has in the UUU codon. Therefore the cell uses an *adaptor* to link CUU with leucine. One end of the adaptor interacts specifically with CUU, while the other end binds leucine. This adaptor is a small molecule of RNA called *transfer RNA* (tRNA) (at least small compared to mRNA). There are approximately 40 different kinds of adaptors per cell. The three-dimensional structure of the tRNA for phenylalanine in yeast has been recently elucidated. It is a curious L-shaped structure formed from a single strand of RNA 76 nucleotides long (Fig. 3-12). The strand bends back on itself, allowing 42 bases to pair with each other in the familiar double-helical pattern. The bases that cannot form complementary pairs are forced into a series of loops containing exposed bases. One of these loops contains the sequence AAGm.[7] This base triplet pairs specifically with the codon UUC and is therefore called an *anticodon*. The anticodon loop attaches the tRNA to the mRNA. The other end of the tRNA carries the unpaired base sequence CCA. It is to this end that the amino acid is attached. The other loops on the molecule may represent attachment sites for the enzyme needed to attach the amino acid to its specific tRNA.

---

7. Gm stands for methylguanine, a modified form of guanine having the same pairing properties as the parent molecule.

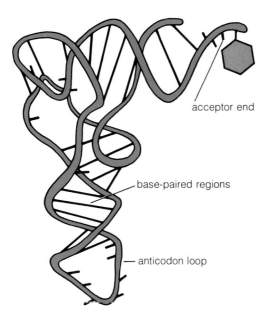

Figure 3-12
THREE-DIMENSIONAL STRUCTURE OF PHENYL-ALANINE TRANSFER RNA FROM YEAST.

RNA and Protein Synthesis: Ribosomal RNA

Before we can transform mRNA plus tRNA, which is carrying an amino acid, into protein, we need one more crucial component, the *ribosome* (see Chapter 1, page 18). The ribosome does for protein synthesis what a tape deck does for tape-recorded music; while the information for the music resides in the magnetic tape, its translation into sound requires the machinery of the tape deck. Likewise, the ribosome contains all the machinery needed to read off mRNA codons in an orderly sequence and to string together the incoming amino acids into a sequence making up a protein. The ribosome contains roughly 60 percent RNA and 40 percent protein. It is manufactured inside the nucleus in a specific region called the *nucleolus* but does its work in the cytoplasm. In bacteria, in which ribosomes have been best studied, there are perhaps 20,000 of these small bodies per cell, and they account for approximately one-fourth of the total cell mass. The ribosome easily dissociates into two fragments, a light fragment and a heavy one. These two subunits have separate functions in the intact ribosome. The light subunit binds mRNA, while the heavy one binds incoming tRNAs (each with its cargo of one amino acid). The heavy subunit also carries the growing peptide chain.[8] Figure 3-13 represents the series of events occurring along a growing polypeptide chain.

---

8. A peptide is a short or incomplete protein.

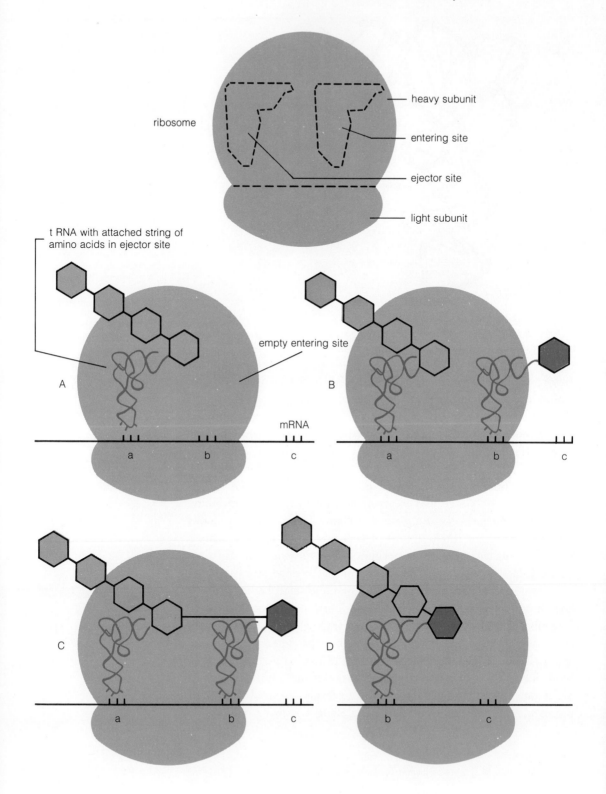

Figure 3-13 THE MECHANISM OF PROTEIN SYNTHESIS. (A) Ribosome with mRNA bound to light subunit. The tRNA carrying a growing peptide is bound to the ejector site of the heavy subunit. Binding of tRNA to the ribosome requires energy plus a protein (transfer factor 2). (B) Attachment of second tRNA to entering site, where it specifically base-pairs to codon b of mRNA. This requires energy plus transfer factor 1. (C) Peptide bond formation between incoming amino acid and growing peptide. As a result, the tRNA in the ejection site breaks contact with its peptide and is ejected. The process of peptide bond formation requires the enzyme peptidyl transferase, an integral protein of the heavy ribosomal subunit. (D) Movement of growing peptide and attached tRNA from entering site to ejection site. This requires energy plus transfer factor 2. There is simultaneous movement of the mRNA relative to the ribosome, so as to place codon c at the entering site. The ribosome complex is now ready to accept its next amino-acid-carrying tRNA.

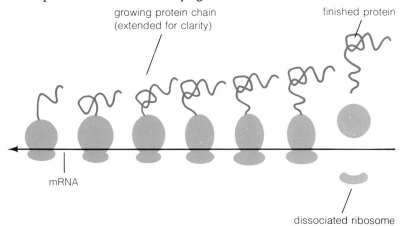

Figure 3-14 POLYRIBOSOMES IN PROTEIN SYNTHESIS. Note that the growing proteins spontaneously assume their final and specific 3-dimensional folding.

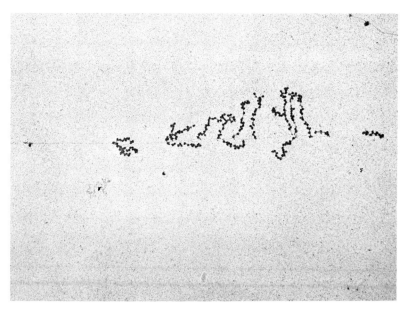

Figure 3-15 ELECTRON MICROSCOPE PHOTO OF POLYRIBOSOMES. As soon as mRNA is transcribed from DNA, it is translated into protein by means of polysomes. In this bacterial system the mRNA is rapidly degraded as shown by decreasing size of the polysome units toward the right of the figure. Transcription of DNA runs from left to right. Polysomes at extreme right are shorter because of degradation of mRNA.

The process of protein synthesis can be seen to be a very busy one: tRNAs loading and unloading, transfer factors shuttling in and out, peptide bonds snapping into place, and mRNA clicking along, codon by codon. The emerging protein molecule grows longer, amino acid by amino acid. At the same time, it folds up into its characteristic tertiary structure. Finally, a chain termination codon enters the ribosome and, with the help of a special releasing factor, the finished protein dissociates from its tRNA carrier and floats off to assume its new duties in the cell. For a typical bacterial protein with a chain of 300 amino acid residues, the entire process, from start to finish, takes about 10 seconds. The cell improves on this record by letting several ribosomes "read" a stretch of mRNA simultaneously. This means that the same stretch of mRNA can produce a finished protein molecule in less than 2 seconds (Figs. 3-14 and 3-15).

Control of DNA Function

Now, the cell is very much a planned economy. Just as a human industrialized state decides how much of its resources it will invest in the manufacture of consumer goods, tanks, highways, and medical services, so a cell decides how much of its energy resources will be spent on the manufacture of specific kinds of proteins. The number of different kinds of proteins an organism *can* make is always much higher than the number it does make. For instance, it is estimated that the human body can produce in the neighborhood of 1 million different kinds of antibodies. The baby we mentioned at the beginning of this chapter is unable to synthesize any of these proteins, and thus his life is in jeopardy. But his survival would be equally threatened if he suddenly began to produce *all* 1 million potential antibodies, if for no other reason than the drain on the body's energy stores. Such a situation would be analogous to a particularly vicious terminal-stage cancer. Likewise, it can be calculated that a typical bacterium, *Escherichia coli,* can make 2,000 to 4,000 different types of proteins (most of which have not been discovered yet). But during normal growth only 600 to 800 different proteins are present in the *E. coli* cell. The rest simply are not made. They represent the store of adaptability *E. coli* has for adjusting to changing environmental circumstances.

How does the cell control its protein synthesis? We should offer two words of caution. Most of our knowledge of protein synthesis comes from a study of bacterial cells, which are much easier subjects for study than the vastly more complex mammalian cells. It would be extraordinary if no new principles were discovered in the mammalian system. Second, even in the bacterial system, most of the attention has been focused on mRNA synthesis, where obvious control is exercised. Much still remains mysterious about ribosome function, and again it would be surprising if additional control points were not discovered here. In spite of these possibilities, the basic relatedness of all living forms argues against a major revolution in our understanding of this area.

Much of the work on control of bacterial protein synthesis was carried out in the Paris laboratories of Jacques Monod and François Jacob. Their experiments are among the most elegant in the literature of molecular biology and offer clear proof that the aesthetic sense functions as strongly in science as it does in the arts. Jacob and Monod investigated the problem of galactosidase synthesis in *E. coli*.

The Control of Lactose Metabolism in E. coli

The colon bacterium *E. coli* can, under suitable conditions, produce an enzyme called β-galactosidase. This enzyme is most important to the bacterium when part of its diet includes the compound sugar lactose. (For example, cow's milk contains 5 percent lactose, and human milk 7 percent. Lactose does not have a sweet taste, so we are generally not aware of its presence.) Lactose by itself cannot be utilized by the metabolic pathways of *E. coli* and must be broken down by β-galactosidase into its constituent parts before the cell can use it (Fig. 3-16).

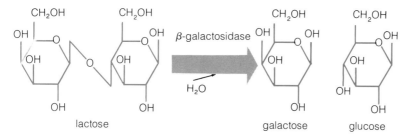

Figure 3-16 THE HYDROLYSIS OF LACTOSE BY β-GALACTOSIDASE.

Escherichia coli acts very rationally. When no lactose is present in the environment, it contains less than three molecules of β-galactosidase per cell. Should lactose appear in the environment, the number of β-galactosidase molecules per cell jumps to about 3,000. The enzyme is said to be *induced* in the presence of lactose. It was this phenomenon that Jacob and Monod set out to investigate. They showed that the induction of a protein results from an increase in synthesis of the necessary mRNA.

Once the mRNA is made, it is translated forthwith into protein. What is the nature of the on-off switch controlling the transcription of DNA into mRNA? In fact, two separate switches appear to exist. To use an electrical analogy from the home, these may be compared to a normal on-off switch and to a dimmer switch. The on-off switch responds in an all-or-none way, whereas the dimmer switch allows for a graduated response. The function of the on-off switch is carried out by a specific protein called a *repressor*. The repressor physically binds to the DNA and prevents it from manufacturing the necessary mRNA. The function of the dimmer switch is carried out by a specific region of DNA called the *promoter*. The promoter is the DNA region to which the enzyme RNA polymerase must bind in order to initiate mRNA synthesis. Because the promoter can bind RNA polymerase with varying degrees of affinity, it can control the rate of mRNA synthesis.

Let us first look at the nature of the interaction between the repressor and DNA. The repressor protein binds to a specific region of DNA called the *operator*. When the repressor is bound to the operator, no synthesis of β-galactosidase takes place, because no mRNA is being synthesized. The repressor that blocks β-galactosidase synthesis also blocks the synthesis of two other proteins necessary for the utilization of lactose. One is a permease (Chapter 1) which sits inside the cell membrane and makes it possible for the bulky lactose molecule to enter the cell. The other is an enzyme called acetylase whose role in lactose utilization is still obscure. The genes for β-galactosidase, the permease, and the acetylase sit next to each other on the bacterial chromosome. They are called the structural genes. The operator region of the DNA is much shorter in length and sits immediately in front of the three structural genes (Fig. 3-17).

With the operator blocked by the repressor, the RNA polymerase cannot reach the structural genes and no mRNA synthesis can begin. This is the normal state of affairs when lactose is absent from the environment. The three structural genes are said to be *repressed*. The cell does not manufacture metabolically expensive proteins for which it has no need. However, as soon as lactose appears in the environment, it throws the control switch into the "on" position. It does this by entering the cell,[9] combining with the repressor, and inactivating it by an allosteric mechanism (Chapter 2) so the repressor can no

---

9. Even in the absence of lactose, the cell carries a few copies of the permease which allow a small number of lactose molecules to enter should any be encountered.

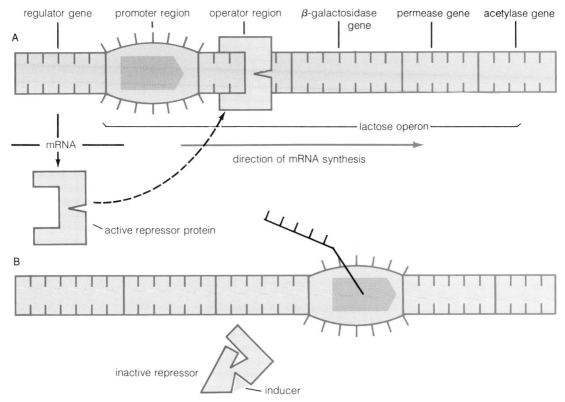

Figure 3-17 MODEL OF THE LACTOSE OPERON. The numbers indicate the base pairs in each gene. (A) Regulator gene codes for repressor protein, which interacts with operator to block passage of RNA polymerase. (B) Repressor protein allosterically inactivated by interaction with inducer (lactose), allowing polymerase to transcribe structural genes.

longer bind to the operator (Fig. 3-18). The result is the *induction* of β-galactosidase, permease, and acetylase synthesis.

Let us next look at the "dimmer" switch mechanism in the control of lactose metabolism in *E. coli*. If the cells are grown on glucose as an energy source, no synthesis of the β-galactosidase, permease, or acetylase takes place. If lactose is now added to the glucose medium, the three proteins are induced, but at a lower rate than if lactose alone were present as an energy source. The higher the ratio of glucose to lactose in the environment, the lower the degree of induction of the three enzymes. This variable response is mediated by the promoter region of DNA. The promoter region is a short stretch of DNA just in front of the operator region (Fig. 3-17). The base composition of the promoter region shows great enrichment with respect to A-T base pairs and diminution with respect to G-C base pairs. The hydrogen

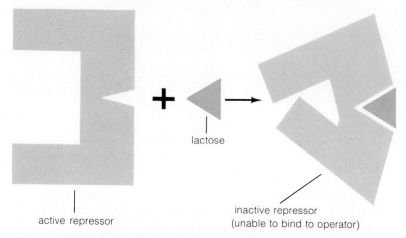

Figure 3-18 THE PROCESS OF INDUCTION: ALLOSTERIC INACTIVATION OF THE LACTOSE REPRESSOR BY COMBINATION WITH LACTOSE. The process is similar to the allosteric inactivation of enzymes (p. 48).

bond forces holding A-T base pairs together are weaker than the forces holding G-C base pairs together. Hence stretches of DNA with an abundance of A-T sequences can be pried open more easily. The promoter region is the site of the attachment of RNA polymerase, which involves strand separation of the DNA (Fig. 3-19).

Every structural gene (or cluster of genes) carries its own promoter. Thus the regulator gene, which synthesizes mRNA for the repressor protein, carries its own promoter, and so do the three structural genes involved in lactose metabolism. The complex of structural genes, operator region, and promoter region constitutes an *operon*—a group of genes that can be controlled as a unit.

We mentioned before that, in the presence of glucose, the lactose operon is only weakly induced. This is the result of a decreased affinity of its promoter for RNA polymerase. In order for the polymerase to bind efficiently, the promoter must interact with a facilitating protein called the CAP protein (Fig. 3-19). In a complex way, the amount of CAP protein available responds to other environmental variables such as the presence of glucose, which is a preferred food source for *E. coli*. This interaction involves a small cyclic nucleotide, cyclic AMP (Chapter 5).

Other promoters, such as the promoter for the regulator gene, lack this ability to respond in a graded way. These promoters are "preset" for a certain level of affinity. If their structural gene codes for a protein are needed in large amount by the cell, the promoter binds polymerase with high efficiency. Genes whose gene product is needed in small amounts have less efficient promoters. The regulator gene is an example of the latter class.

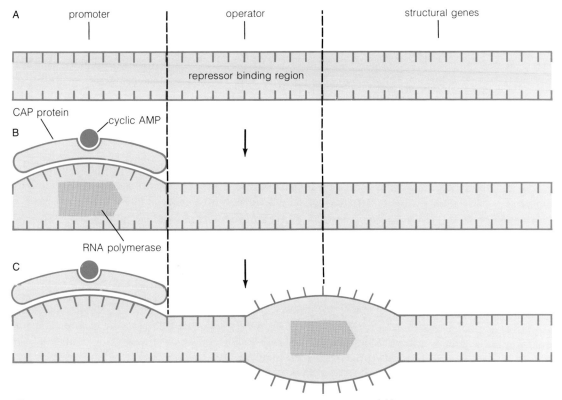

Figure 3-19 THE PROMOTER AND THE ROLE OF THE CAP PROTEIN. (A) Inactive gene. Two strands of DNA are held together by hydrogen bonding between complementary bases. (B) Interaction of CAP protein (cyclic AMP binding protein) with promoter region causes weakening of hydrogen bonds between base pairs in the promoter region. RNA polymerase can insert itself between the opened strands to start the transcription process. (C) RNA polymerase begins transcription of the structural gene. The transcription process can proceed only if the operator is not blocked by the repressor.

To summarize, the lactose operon represents a group of structural genes whose expression can be controlled in a rational and subtle way by the cell. The induction-repression response corresponds to an on-off switch that responds to the presence or absence of lactose in the environment. This response is mediated by the binding of the protein repressor to the operator region. The other response is a graded one, responding to the presence or absence of alternate food sources in the environment. This response is mediatd by the binding of RNA polymerase to the promoter region.

Other control systems based on the operon principle are known in bacteria. In all of them, a cluster of genes coding for a single metabolic pathway can be turned on or off as cellular needs dictate. In all cases, control is exercised at the level of RNA transcription. When we add to these the control mechanisms exercised at the level of enzyme function (such as allosteric control), and the suspected mechanisms operating at the level of protein synthesis, we begin to appreciate the subtleness of response available to the living cell in meeting environmental challenges. Our genetic complement is far from a passive, static cargo that can change only through creeping evolution. It responds actively to the changing circumstances enveloping its host.

These same control mechanisms are the ones responsible for the process of cell differentiation—the gradual unmasking of specialized functions in cells of a developing embryo. The early stages of an embryo cannot see because there are no retinal cells, cannot think or feel because there are no nerve cells, and cannot move actively because there are no muscle cells. The specific regions of DNA required for these functions are present, but they have not sprung into activity. Only with gradual development do the specific DNA fractions begin to make mRNA, and mRNA begins to make the specific proteins. Once more we see an example of the opportunism of life. Genetic control mechanisms evolved originally to allow simple cells like *E. coli* to adapt quickly to changing external circumstances. These same control mechanisms (with much elaboration and addition) were then transformed by advanced organisms into the agents of embryological development.

Bibliography

GENERAL
READING

James D. Watson, *The Double Helix,* Atheneum Publishers, 1968. This slim volume is an account of the work that led to the elucidation of DNA structure. It is written by one of the two principals involved. Besides being high science, it makes an extremely important psychological point: science is an activity that involves human beings. In the excerpt here Watson superbly conveys the almost druglike excitement of scientific discovery at its best. The Cavendish Laboratory, where Watson and Crick were working, is at Cambridge University.

Custom then locked the doors of the Cavendish at 10:00 P.M. Though the porter had a flat next to the gate, no one disturbed him after the closing hour. Rutherford had believed in discouraging students from night work, since the summer evenings were more suitable for tennis. Even fifteen years after his death there was only one key available for late workers. This was now pre-empted by Hugh Huxley, who argued that muscle fibers were living and hence not subject to rules for physicists. When necessary, he lent me the key or walked down the stair to unlock the heavy doors that led out onto Free School Lane.

Hugh was not in the lab when late on a midsummer June night I went back to shut down the X-ray tube and to develop the photograph of a new TMV sample. It was tilted at about 25 degrees, so that if I were lucky I'd find the helical reflections. The moment I held the still-wet negative against the light box, I knew we had it. The tell-tale helical markings were unmistakable. Now there should be no problem in persuading Luria and Delbrück that my staying in Cambridge made sense. Despite the midnight hour, I had no desire to go back to my room on Tennis Court Road, and happily I walked along the backs for over an hour.

TEXTBOOKS

James D. Watson, *The Molecular Biology of the Gene,* 3rd ed., W. A. Benjamin, 1975. This engaging and lucid advanced treatment does not require a Ph.D. to understand it.

Part II

The Organism

Figure 4-1 RED BLOOD CELL ENMESHED IN A FIBRIN CLOT.

chapter 4

As I write this, my daughter has just turned 4 years old, and her knowledge of physiology does not extend to such areas as liver function (or, indeed, the mere fact of its existence) or the mode of operation of the lungs. But she does sometimes point to the left side of her breast and say with certainty, "This is where my heart is!" She is probably typical of all of us in becoming aware of the heart as the first internal body organ. It is also probably

The Circulatory System

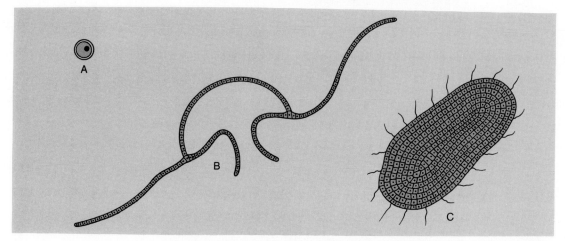

Figure 4-2 EVOLUTION OF MULTICELLULARITY. (A) A single cell in equilibrium with external ocean (free exchange of O_2, CO_2, food, etc.). (B) Jelly fish as a simple multicellular aggregate—all cells have direct frontage on the ocean. (C) More complex multicelluar aggregate—cells in interior unable to gain easy access to O_2, food, and other necessities.

fair to say that the heart retains this preeminent position throughout life. A baby first hears its mother's heart as it lies against her breast. A youth of 19 or 20 knows the heart from its wild pounding when his beloved suddenly appears. And when we are old, we listen to the heart anxiously for the slight tremors and hesitations that forecast the final stillness.

Such appreciation of the role of the heart is universal and intuitive. Yet the full role of the circulatory system, which the heart is part of, is almost always greatly underestimated. The fact is that this role is as subtle and far-reaching as it is essential to our survival, and perhaps the place to start our discussion is with those organisms that survive without a circulatory system. Such organisms are not easy to find—it is probably safe to say that every living thing we are likely to see in day-to-day existence has a circulatory system; this includes all insects, earthworms, fish, and even trees, grasses, flowers, and ferns. We are most likely to discover an acirculatory organism by bringing out a microscope and looking at a drop of rich pond water or a drop of water from the upper layers of the ocean. This organism will be most typically represented by a single cell. Such single cells may offer little as visual spectacles but, in terms of biological time, they are the gnarled old oaks of the evolutionary forest. Organisms that look very much like modern-day bacteria are thought to have inhabited the oceans 3 billion years ago. Such cells have nothing resembling a heart or arteries or capillaries. But, as Isaac Asimov has pointed out, they do not need them, because they have a circulatory system trillions or quadrillions of times larger than ours. In effect, the whole ocean serves as their blood, because the whole ocean can serve to exchange their respiratory gases, absorb their waste products, carry their food supplies, and protect them from drastic changes in the environment.

All this began to change when single cells started to aggregate into differentiated multicellular organisms. The early stage of the aggregation process, as represented by the jellyfish, brought little external change to individual cells, since every member cell of the aggregation still had "frontage" on the ocean, either directly or by means of diffusion through a thin intervening layer of cells (Fig. 4-2). Representatives of these simple forms are still with us but, in the animal world at least, they tend to be small and play a relatively unimportant role in the affairs of the world. Their simple body plans imply a relative lack of specialized function among component cells, and specialization has been the thrust of multicellular evolution.

Thus, first the oceans, and secondarily the dry continents, were inherited by organisms so complex that their innermost cells no longer had access to the life-giving waters outside. The way out of this impasse was ingenious; if the inner cells could not have access to the ocean outside, then the solution was to pinch off a little puddle of the ocean and carry it around inside, as the jellyfish had already begun to do in a primitive way. But by adding an internal pump, and an efficient diffusion surface for the exchange of small molecules, even the most deeply buried cell could be placed in effective contact with the great outside. Other refinements were added: internal sensors responsive to changes in osmotic pressure and levels of body wastes, along with compensating machinery. Added also was an absorptive surface which functioned in the uptake of food molecules and, finally, there were added organs for internal communication and control (Fig. 4-3).

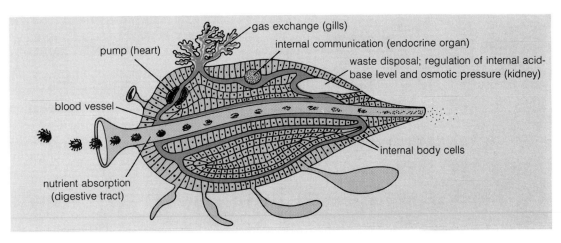

Figure 4-3 THE CIRCULATORY SYSTEM AND SATELLITE ORGANS.

In effect, such a multicellular organism became a well-run city, with a population far larger than any human city, but with perfect municipal services. The city is composed of many distinct types of individuals (cells) carrying out specific tasks. In spite of their specialization, their needs differ little from those of their remote and undifferentiated ancestors; they need food, oxygen, and a clean, safe environment that does not fluctuate very much. No human city can provide these needs for all its citizens, but the living organism can and does, via the circulatory system and its satellite organs. As a consequence, each constituent cell works at its optimum, contributing its specific skills toward the welfare of all the other cells to which it is bound. To appreciate some of the intricacy of these processes, let us look first at the component parts of the circulatory system—the blood, the heart, and the blood vessels—and then at some of the satellite organs.

The Blood

On superficial observation, it seems that we are filled to the brim with blood, since a slight nick anywhere on the body produces a temporary flow. In fact, blood is not such a common substance at all—we contain only an average of 5 quarts each, and our readiness to bleed is simply a reflection of the widespread distribution of this fluid. When blood is first drawn from the body, it is a thick red fluid. If we treat it to prevent clotting, and then let it sit quietly, blood separates into two portions: a clear upper portion called the *plasma,* and an opaque red lower portion, containing the red blood cells and other cells.

Plasma

The Role of Water. Far from serving only as a neutral suspension medium for the red blood cells, the plasma determines most of the basic properties of blood. About 90 percent of plasma is made up of the miracle compound, *water,* and it is this compound more than any other that shapes the functional behavior of the blood.

Water is such a strange molecule that, if it did not exist, it would have to be invented. In fact, after over a century of effort leading to more than a million synthetic compounds, organic chemists have come up with only one or two molecules (such as dimethyl sulfoxide) that approach the virtuoso performance of water in the natural world.

The first, unduplicable property of water is its immense ability to act as a solvent (Ch. 1, p. 8). Given time, water dissolves mountains made of the hardest rock. Of more significance to living organisms, it functions as a good solvent for many cellular components such as salts, sugars, proteins, acids and bases, nucleic acids, and many polysaccharides. This list includes almost all the molecules used by cells as raw materials or as external agents of communication (hormones). It is the job of the circulatory system to carry such raw materials from their point of entry in the digestive system to the site of utilization by individual body cells. Likewise, the circulatory system becomes a route of communication and control between individual cells of the body, mediated by trace amounts of specific hormones which float dissolved in the water of plasma and affect individual target cells. If these were the only functions performed by the circulatory system, we would already consider it a very vital system. (Although of course the beating of the heart would not be such a life and death question as it is now.)

Water has another outstanding property; it can hold extraordinarily large amounts of heat (above five times more than most other liquids). To use the language of physics, it has a *high heat capacity*. Everyone knows of the moderating influence oceans exert on the climate of coastal regions. The blood coursing through our bodies has a similar moderating effect on the temperature of internal organs. Some organs, such as the liver, normally work harder than others, such as the intestines. Other body tissues, such as the muscles of the thigh, become very active during a quick run. Yet such differences in activity produce very little difference in internal temperature, because the blood absorbs extra heat and carries it away to the surface of the body, where it is dissipated directly by radiation or indirectly through the vaporization of sweat. (This makes use of another special property of water, its *high heat of vaporization*.) The result is an internal temperature control system which allows internal body cells to function in an optimal and unchanging environment even though the work load and the external conditions may be changing from minute to minute.

Let us now consider the roles of the other plasma constituents. On a weight basis, the second most important plasma constituents are the proteins, as seen in Table 1.

TABLE 1
Plasma constituents

Water	90%
Proteins	7%
Albumin (4.5%)	
Globulin (2.5%)	
Fibrinogen (0.3%)	
Salts	1%
Miscellaneous	2%

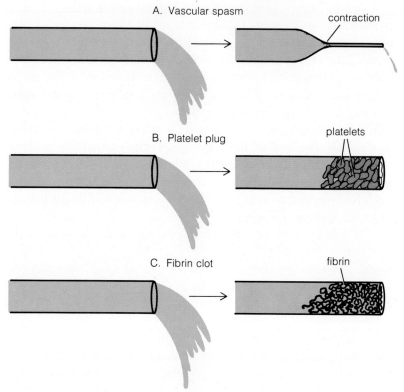

Figure 4-4 TERMINATION OF BLEEDING.

The Role of Proteins. Albumins are small proteins and therefore contribute importantly to maintaining the internal osmotic pressure of the blood.[1] Along with *globulins,* they also function as piggyback carriers for lipids and their derivatives. Lipids are molecules that dissolve poorly in water and travel in the blood in bound form, temporarily attached to protein fractions. One fraction of globulins, the *gamma globulins,* makes up the body's circulating antibody fraction. Antibodies are molecules that specifically react with foreign objects in the body and form a part of the immune system (see page 115).

Fibrinogen is a protein designed for emergency use only. During normal body function, it has little to do but circulate. However, should a break occur in a blood vessel, it assumes a critical role. The body has evolved several alternate methods of stopping bleeding, and in all of these fibrinogen plays a primary role (Fig. 4-4).

---

1. The osmotic pressure of a solution is proportional to the total number of osmotically active particles per unit volume. A given weight of a small protein contains more particles (molecules) than an equivalent weight of a larger protein. For a discussion of osmotic pressure, see Chapter 1.

Perhaps the body's most direct response to a broken blood vessel is a *vascular spasm*. The broken vessel constricts and shuts off blood flow. This is made possible by circular smooth muscles embedded in the walls of all larger blood vessels. The most important stimulus to vascular spasm is pain—the stronger the pain, the stronger the spasm. For instance, the loss of a leg by crushing can result in almost no blood loss, even though the major artery of the leg has been cut. However, loss of the same leg to high-speed shrapnel might produce a cleaner cut with less pain but much more bleeding.

Fortunately, few individuals living in North America today need ever cope with a severed leg. But even the most sedentary existence leads to the rupture of literally dozens of small blood vessels every day. Activities such as scratching, brushing of teeth, bumping into furniture, and picking up a heavy object, lead to rupture of the smallest blood vessels, the *capillaries,* in the affected region. At the site of such damage, the inner lining of the capillary loses its normal properties and attracts a layer of *platelets,* cell-like fragments present in the blood in large numbers (about 300,000 per cubic millimeter). As the platelets lay down a cobblestonelike surface, they acquire new properties and begin to attract other layers of platelets. Thus a plug made of platelets quickly forms and blocks the ruptured vessel.

While the vascular spasm and the platelet plug represent rapid responses to injury, both are always backed up by the slower but longer-lasting mechanism of blood clotting. A blood clot is made up of very long, tangled strands of the protein *fibrin.* Fibrin is derived from fibrinogen by a mechanism about to be described. The fibrin clot begins to form in 15 to 20 seconds in severe cases, and in 1 to 2 minutes with a minor cut. Within 3 to 6 minutes the entire blood vessel is filled with a clot. Later such a clot is either dissolved by body enzymes (as in the slow disappearance of a "black eye"), or is invaded by fibroblast cells and converted into scar tissue. The blood vessel is then abandoned, and new bypass routes are created in the region. A fibrin clot also forms behind the region where a vascular spasm occurs and behind a platelet plug (Fig. 4-1).

Formation of a fibrin clot involves one of the most complex sequences of chemical steps known in organismic biology. Over 30 different substances are now thought to participate in blood clot formation, and more are certain to be discovered with the passage of time. The immediate trigger for clot formation involves the enzymic transformation of fibrinogen into fibrin. The enzyme involved (*thrombin*) is a protein-splitting enzyme which cleaves two small fragments from fibrinogen to change it into a longer, more asymmetric molecule, fibrin (see Fig. 4-5). Fibrin monomers polymerize in the presence of calcium to form the long, tangled threads of the final

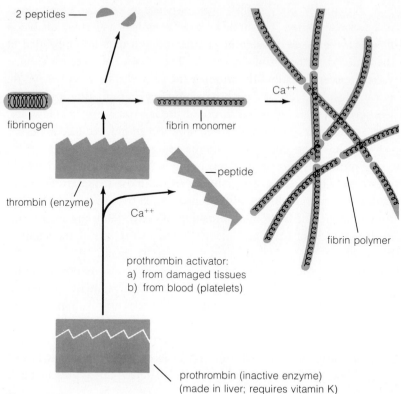

Figure 4-5 BLOOD CLOTTING.

clot. Of course, this means that thrombin must not be present in freely circulating blood. It is synthesized as needed by an activator enzyme which cleaves a peptide from an inactive precursor, *prothrombin*. Formation of the activator enzyme involves a similar cascade of precursors and active products and represents the rate-limiting step in clot formation. The ultimate initiating factor is thought to be released by damaged tissues or, less frequently, via the breakdown of platelets in the blood.

A timely blood clot can save a life, but a clot in the wrong place can lose it. A clot that forms inside the body can break loose and travel to the brain or the heart, where blockage of the narrow blood vessels can kill a patch of the brain (a *stroke*), or the heart (a *heart attack*). Similar damage can occur to the lungs, kidneys, retina, bone, and other parts of the body, but here the threat to life is less immediate. Of the many factors that contribute to such internal clot formation, one deserves mention. This is *atherosclerosis*—the roughening of the smooth inner lining of the arteries due to calcified deposits of cholesterol. Such roughening can damage the platelets and start clot formation.

The Role of Salts. Blood has a salty taste. This is a reflection of its high content of inorganic salts such as sodium phosphates and carbonates. Salts contribute to two essential functions: they have major responsibility for maintaining the osmotic pressure of the blood, and they help regulate the acid-base balance of the blood through their buffer action. For example, violent muscular exercise leads to a release of lactic acid into the blood. This acid dissociates (ionizes) to produce hydrogen ions and lactate ions:

$$H\, Lac \rightleftharpoons H^+ + Lac^-$$

The resultant hydrogen ions, if allowed to accumulate, would acidify the blood to the point where intracellular enzymes would cease to function. Coma and death would result. The buildup of hydrogen ions is resisted by the buffer action of salts such as sodium carbonate:

$$H^+ + Na_2CO_3 \rightleftharpoons Na^+ + NaHCO_3$$

As a result, large amounts of acid or base may be released into the bloodstream with only a minimal change in hydrogen ion concentration. Once these natural stabilizing mechanisms are overwhelmed, as in unchecked diabetes, death follows.

The Red Blood Cells

Let us now consider the cellular fraction of whole blood. There is a story about a saintly man who was set upon by a gang and left bleeding and broken on the ground. As he lay on the ground and watched his blood ebbing away, his last thoughts were for his red blood cells. "Farewell, dear friends," he called out. "How sad it is that you must die too!" Needless to say, such saintliness is rare, but perhaps no more rare than an appreciation of the blood as a living tissue made up of individual living cells. Long before people were receiving heart transplants, they received blood transplants, and literally millions of lives have been saved as a result.

Numerically, by far the largest component of the cellular fraction is the *red blood cells,* or *RBCs.* Some biologists refuse to give full cellular status to the RBC because in its mature, functional state it lacks a nucleus, mitochondria, and other typical cell organelles. However, the RBC maintains itself in an organized state and to accomplish this requires food in the form of glucose. And when it is first formed in the bone marrow, the RBC contains a nucleus and mitochondria like other cells of the body. It sheds these during the process of maturation, because they would interfere with its physiological function in the body.

The life and structure of the RBC are formed around a single task: the transport of hemoglobin. Less than 5 percent of its internal dry weight consists of other constituents, such as those required for its own metabolism. Since hemoglobin is a protein with very high solubility in plasma, why carry it around in a cellular package? In fact, some animals such as the earthworm carry their hemoglobin in a direct solution in the blood. However, the blood of vertebrates contains such high concentrations of hemoglobin that, if it were dissolved directly in blood, the blood would become a viscous syrup instead of a free-flowing liquid. Furthermore, hemoglobin in free solution would exert considerable osmotic pressure, sufficient to draw water out of body tissues through which it was circulating.

The mammalian RBC is smaller than most body cells—about 7 microns in diameter instead of 20 microns of a typical body cell. It also has a very atypical shape; it resembles a disk with a thin center and thicker edge (Fig. 4-1). This shape is made possible because of the absence of a cell nucleus. The small size and unusual shape of the mammalian RBC give it a relatively large surface area. As a matter of fact, the total surface area of our RBCs is about 2,000 times larger than the surface area of the body itself.

This large surface area of the RBC allows for a rapid flow of oxygen into and out of the cell. It is ironic that none of the oxygen that enters the RBC is destined for use by the RBC itself. All of it is earmarked for use by other body cells—those with mitochondria and an oxygen-dependent energy-generating system.

The Role of Hemoglobin. Once inside the RBC, oxygen is bound by *hemoglobin* which fills the interior. A fixed amount of oxygen can be carried in solution by the plasma; the use of hemoglobin increases 70-fold the amount of oxygen carried. In practical terms, this means that we can go without breathing for about 5 minutes before all circulating oxygen is exhausted and the brain cells begin to die. Without hemoglobin in our blood, this safety margin would be reduced to about 5 seconds.

Hemoglobin does more than pick up oxygen readily—it also knows when to let it go. There is a subtle accommodation to the slight partial pressure differences of oxygen in the lungs versus oxygen in other body tissues. Such subtlety of response is possible only with subtlety of structure, and hemoglobin is one of the most sophisticated molecules known. In the living world, the most complicated molecules are almost always proteins, and hemoglobin is not an exception. The working hemoglobin molecule contains four subunits. Each subunit consists of a protein portion (globin) and a nonprotein, iron-containing portion (heme) (see Fig. 4-6). It is to the heme that an oxygen

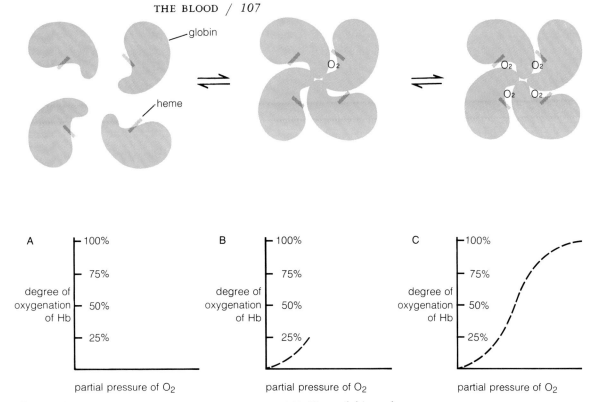

Figure 4-6 HEMOGLOBIN AND OXYGEN TRANSPORT. (A) Hemoglobin molecule with no oxygen bound to heme. The four subunits are widely spaced, and hemoglobin is a poor oxygen acceptor—i.e., it takes a relatively strong "push" to attach an oxygen molecule to one of the heme groups. (B) Hemoglobin molecule with one oxygen molecule attached to a heme group. A chemical (allosteric) transition has occurred as a result. The four subunits are more tightly bound to each other, and the Hb molecule shows greater affinity for oxygen molecules. A slight increase in oxygen pressure now results in greater uptake of O_2 by Hb. (C) Fully oxygenated hemoglobin molecule. The four subunits are still tightly bound and all heme groups are occupied, hence further increase in oxygen pressure results in no net increase of oxygenation. However, a slight drop in oxygen pressure will allow Hb to release much of its bound O_2.

molecule is loosely bound. As the hemoglobin molecule picks up oxygen in the lungs and loses it in the deep body tissues, the molecule appears to be "breathing," expanding as oxygen is discharged and contracting as it is picked up. This is an example of the kind of allosteric transition dsecribed in Chapter 2, p. 48.

Physical changes are accompanied by chemical changes with far-reaching consequences both for oxygen and carbon dioxide transport.[2] The net result is that in the lungs hemoglobin acts to release carbon dioxide from the blood and pick up oxygen, while in other body tissues it acts to release oxygen and pick up carbon dioxide.

The hemoglobin molecule, complex as it is, can be damaged in the course of its activity, and some of the common physiological "accidents" (for example, a change in the oxidation state of its constituent iron) can be repaired by machinery found in the RBCs. However, modern living conditions can expose the hemoglobin molecule to threats for which it has had no evolutionary preparation. One of the most deadly of these is carbon monoxide (CO). This colorless, odorless gas is generated at the rate of 1 pound per hour by a single car inching along in city traffic. If breathed in pure form, this pound of carbon monoxide would be sufficient to kill around 350 people. Carbon monoxide can work in such minute quantities because it binds to the heme of hemoglobin about 200 times more strongly than oxygen does. The hemoglobin, loaded with the wrong molecule, becomes useless for oxygen or carbon dioxide transport. Another important source of carbon monoxide is cigarette smoke. It has long been known that cigarette smokers are more susceptible to heart attacks and other heart disease than are nonsmokers. The cause is now thought to be the carbon monoxide content of cigarette smoke, not its nicotine content. The heart tissue is literally starved of oxygen, and the result is degeneration of the heart muscle, development of atherosclerosis, and an increased chance of heart attack.

The White Blood Cells

Living among the RBCs are the far less numerous *white blood cells* (*WBCs*). In the blood, RBCs outnumber WBCs by about 700 to 1. However, the importance of the WBCs is far greater than such numbers show. They are a mobile component of the body's defense system, and many of them lead active, even violent lives.

Consider a microbiologist who wants to raise a crop of bacteria. In order to provide the typical bacterium with ideal living conditions, the microbiologist takes a fluid medium and warms it to about 37°C. He adjusts the pH to 7 and keeps it there with buffers. He adds minerals such as phosphate and ammonium ion, a food source such as glucose, and aerates the medium vigorously to provide the bacteria with oxygen. If these conditions do not sound familiar, they should,

---

2. Hemoglobin does not bind the bulk of transported carbon dioxide directly as it does oxygen, but makes it possible for carbon dioxide to be carried in the plasma in bicarbonate (HCO_3-) form.

because these are the exact conditions of the human bloodstream. Once a single bacterium enters such an environment, it multiplies without limit until the host organism is dead. It would, that is, if it were not for the WBCs which are also in the blood. The WBCs make the blood a decidedly nonideal environment for bacteria or any other foreign body.

All WBCs have certain properties in common. All are formed in the bone marrow, from which they emerge to take up final residence in various parts of the body. This means that agents that destroy the bone marrow can also destroy the body's natural defense system. One of the most notorious of such agents is strontium-90, a long-lived radioactive isotope formed as a result of nuclear explosions. Strontium has the chemical properties of calcium, hence becomes incorporated into bone tissue. Large amounts can knock out WBC production completely, while smaller amounts can induce leukemia, a disease characterized by the cancerous growth of WBCs.

All WBCs have the unusual ability of *diapedesis* ("walking through"). That is, their movements are not restricted by the walls

Figure 4-7 DIAPEDESIS.

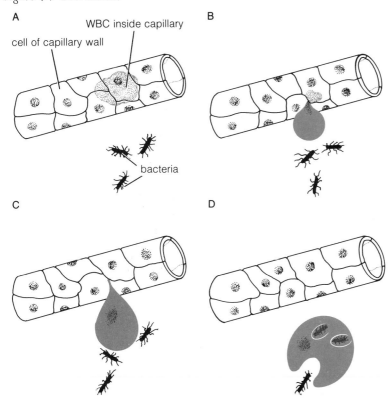

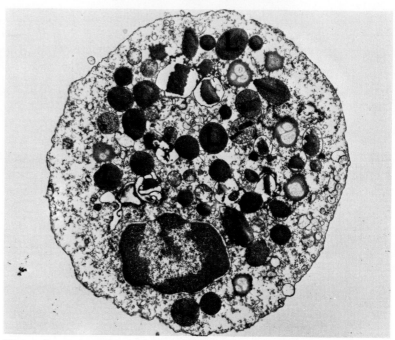

Figure 4-8 ELECTRON MICROSCOPE PHOTO OF A GRANULOCYTE. The dark bodies (L) are lysosomes used in phagocytosis.

of the capillaries, as are those of RBCs. WBCs can squeeze through cracks in the capillary walls much smaller than the WBCs. They do this by an oozing kind of motion called *ameboid motion*. WBCs are therefore able to move by voluntary movement, instead of being swept passively along as are the RBCs (Fig. 4-7).

The two most common kinds of WBCs are *granulocytes* ("cells with granules") and *lymphocytes* ("cells in the lymphatics") (Figs. 4-8, 4-9).

Granulocytes are extremely active cells, moving about the body in search of foreign bodies. A chemical sense guides them to their prey and, once contact is made, a granulocyte engulfs and literally eats its victim. The process was described in the discussion of lysosome function and phagocytosis (Chapter 1).

Lymphocytes, though simpler than granulocytes in terms of structure, are much more complex functionally. The large central nucleus is suggestive of cells found in an embryo and, in fact, lymphocytes are *pluripotent*—they can differentiate into several different cell types just as other embryonic cells can. For instance, lymphocytes can differentiate into *macrophages* ("large eaters"). These are large cells which engage in phagocytosis. They are more voracious feeders than granulocytes are and become important in the later stages of an infection, when a stronger body response is called for. Lymphocytes

THE BLOOD / 111

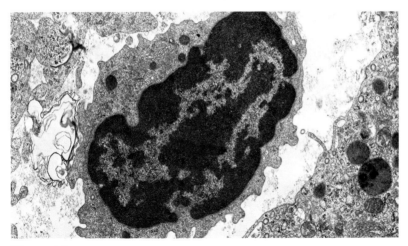

Figure 4-9 ELECTRON MICROSCOPE PHOTO OF LYMPHOCYTE.

also differentiate into *fibroblasts* ("fiber formers"). Anyone who has ever had a deep cut anywhere on the body knows about the action of fibroblasts. As the wound heals, the damaged area is not replaced by the original skin cells but by fibroblasts which form scar tissue that binds the cut edges together.

Immune Cells. Finally, lymphocytes can differentiate into *immune cells.* All immune cells are formed in the bone marrow, but thereafter they may follow two separate paths (Fig. 4-10). The smaller fraction is released directly into the circulatory system, from which the cells migrate to such lymphatic tissue as the lymph nodes, spleen, tonsils, and gut. Such cells are called *B cells,* and they are capable of forming the soluble *antibodies* that circulate in the plasma as a result of challenge by a foreign *antigen.* Antibodies are proteins of the gamma-globulin fraction and look superficially like the letter "Y". The stem of the Y is identical among all antibodies. The arms vary from one antibody to another, though one arm is always identical to the other. Each B cell manufactures only one kind of antibody. Literally millions of variants are possible among these variable arms, and it is their job to combine specifically with a challenging antigen. The antigen may be a protein, polysaccharide, or other complex molecule usually of foreign origin.

In a rather circular definition, an antigen is defined as any large molecule that elicits antibody production by an immune cell. The sequence of events leading to the production of circulating antibodies is illustrated in Figure 4-11. Such circulating antibodies may be extremely long-lived; people who are now over 80 years old and who were exposed to the influenza epidemic of 1890 still show the specific antibodies in their blood.

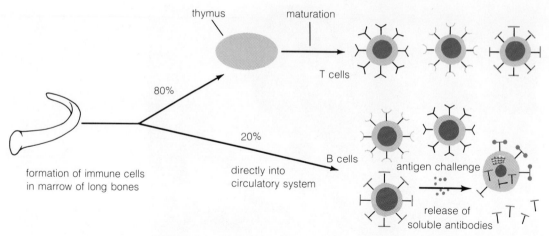

Figure 4-10 ORIGIN OF IMMUNE CELLS. The ultimate origin of both T and B cells is the bone marrow. T cells are matured in the thymus. B cells in chickens are matured in the Bursa of Fabricius. In humans no analog of the Bursa has been found.

However, most immune cells do not enter the circulatory system directly. These cells are routed first through the *thymus* gland, where they undergo a subtle transformation (hence the name *T cells*). The thymus then "seeds" the T cells throughout the body's lymph nodes and other lymphatic tissues. T cells do not produce soluble antibodies, but they can destroy foreign cells nonetheless. Their most important function is to differentiate between "self" and "nonself." The implications of such differentiation are enormous, and T cells may be involved in such diverse phenomena as the rejection of transplanted organs and the development of leprosy, rheumatoid arthritis, and cancer. Cancer cells are "nonself"; they carry surface antigens different from those carried by other cells of the body. Such cells may arise by somatic mutations (mutations in cells other than sperm or eggs), by viral entry, and by other ways. The T cells normally patrol the body and immediately destroy any cancer cells they encounter. It is significant that children with immune-deficiency diseases contract cancer at a rate 100 to 1000 times higher than normal children of the same age. The rising incidence of cancer with age may simply represent the gradual loss of function by the body's T-cell system.

The Heart and Blood Vessels

Having considered the cellular and noncellular components of blood, let us now take a look at the external portion of the circulatory system—the heart and blood vessels. Hearts do not show much variation

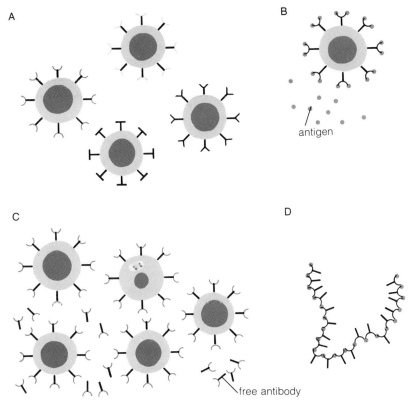

Figure 4-11 THE ANTIBODY RESPONSE. (A) Many B cells, each with unique surface antibody. (B) Challenge by foreign antigen. The antigen reacts with only one kind of B cell. (C) Challenged cell begins to divide and forms a clone of similar cells all of the original type. These cells release soluble antibodies into the plasma. (D) Soluble antibodies react with antigen to form a high-molecular weight, insoluble complex that inactivates the antigen and makes it more susceptible to granulocyte attack.

in the animal world—they are always hollow pumps with one-way valves. However, blood conduits do vary. There are two fundamentally different ways of moving blood through the body. The blood may be forced directly through the spaces separating body cells (an *open system*), or it may be confined to a *closed system* of blood vessels (Fig. 4-12). An open system, such as that of insects, crustaceans, and snails, has the advantage of direct contact between blood and body cells. However, such open systems must carry any respiratory pigments (such as hemoglobin) in direct solution in plasma, and they offer such great resistance to blood flow that the circulation must of

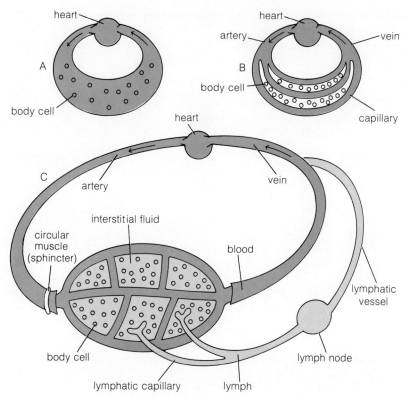

Figure 4-12. OPEN AND CLOSED CIRCULATORY SYSTEMS. (A) Open system: blood flows directly among body cells. (B) Closed system: blood restricted to closed blood vessels. (C) Mixed system of vertebrates: restricted circulation of blood proper, partial open circulation of lymph. Flow of blood into the capillary bed may be shut off by contraction of the circular smooth muscle at the arteriolar end. The relative dimensions of the lymphatic vessel and lymph node are exaggerated—they are much larger than the capillary bed.

necessity be sluggish. A closed system allows much higher velocities of blood flow, but pays a penalty in the form of an added diffusional barrier between the blood and the body cells being supplied. The vertebrate system is largely a closed system. Blood is restricted to fixed blood vessels. However, one fraction of the circulating fluid, the lymph, flows through an open system (p. 117).

The Heart

This hard-working muscle is never at rest for long. In a life-span of 70 years, the human heart beats an average of 2½ billion times. The heart of an individual 19 years old has already contracted

600,000,000 times and has only 1.9 billion beats left. This represents an immense amount of work. If the heart could be coupled to a mechanical energy transducer, it would drive a 2,000-pound car about 2½ miles in city traffic for every 24-hour period. Such major energy expenditure calls for special structural provisions on the part of heart muscle (Fig. 5-26). In comparison with smooth or skeletal muscle, heart muscle is supplied more abundantly with mitochondria, the transducers of chemical energy in the cell.

Another unusual structural feature of heart muscle is the meshwork arrangement of the individual muscle cells. This means that, once a contraction starts in any part of the heart, a wave of electrical depolarization will spread out in all directions and cause every cell in the heart to contract in a coordinated manner. Only such a smooth, coordinated response can lead to useful work being done by the heart. Heart muscle is also unusual in that it contracts spontaneously in a rhythmic way—it needs no nervous or hormonal stimulus as do smooth or skeletal muscle. Even a tiny piece of the heart containing just one or two cells will continue to contract rhythmically if kept alive in a culture medium.

Blood Vessels

Arteries. Blood vessels carrying blood away from the heart are called *arteries;* those carrying blood toward the heart are called *veins.* Arterial blood is always under much higher pressure than venous blood since it receives the full force of the heart pumping. Therefore arteries tend to have much thicker walls but otherwise their structure is basically the same as venous structure (Fig. 4-13). In both cases there is a smooth inner lining called the *endothelium* ("inner skin"), formed of smooth, close-fitting, pancakelike cells. These cells rest on

Figure *4-13* ARTERIAL AND VENOUS STRUCTURE.

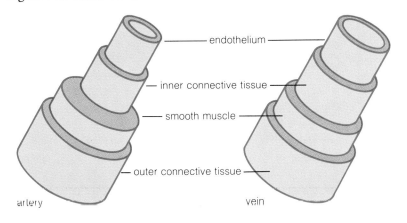

a thin support of elastic connective tissue. The middle layer consists of smooth muscle and is much thicker in arteries than in veins. The outer layer again consists of connective tissue made up mostly of loose collagenous fibers. The thick, flexible walls of the arteries serve to contain the blood that is forced in under high pressure. However, the arterial walls do not always hold, especially if the blood pressure is abnormally high. A sudden surge in pressure can balloon out and burst an arterial wall. The result is internal bleeding, or hemorrhage. A hemorrhage occurring in the brain produces effects similar to those of a stroke and constitutes the single most important medical threat to individuals with high blood pressure.

The thick, elastic arterial walls also dampen out the pulsating nature of blood flow. With each contraction of the heart, the diameter of the arteries swells, and this sudden swelling can be picked up as the pulse in the wrist, upper arm, ankle, neck, and other places. When such an artery is cut, the blood flows out in spurts instead of in an even flow. As arteries travel further from the heart, they break up into increasingly finer branches called *arterioles,* and here the variations in blood pressure become smaller and smaller, until blood flow becomes smooth as the arterioles disappear into a bed of *capillaries.* If the arteries lose their elasticity, as happens in atherosclerosis, there follows a rise in blood pressure, with possible effects such as a stroke or heart attack.

Capillaries. The capillaries are the working segment of the circulatory system, where the exchange of food, respiratory gases, and wastes takes place between the blood and the body cells. To make such an exchange possible, the thickness of the capillary wall is reduced to the theoretical minimum—the walls are one cell layer thick, and they represent an extension of the endothelial layer of the arteries. Because the capillaries are so narrow, the red and white cells of the blood must squeeze through in single file, and this slow progress gives the RBCs ample time to exchange their load of respiratory gases. As the separate bloodstreams coalesce again into small and then larger veins, leading back to the heart, the flow of blood speeds up again.

The capillary system of the human body is vast. If all the capillaries were laid end to end, they would circle the earth 2½ times. Their internal surface area would occupy a full city block. Their internal volume is so large that, if all the capillary beds were to open at the same time and fill with blood, a kind of internal bleeding to death would result. The blood pressure would fall, shock would set in, and death would follow. As a consequence, most capillary beds remain closed at any one time. The opening and closing of the capillary bed is controlled by a ring of smooth muscle at the arteriolar end called a *sphincter* muscle. Arteriolar sphincters are controlled externally as well as internally (nervous and hormonal control). The opening and closing of such sphincters (*vasodilation* and *vasoconstriction*) allow the body to shift the available blood flow to those organs that need it most.

The capillary beds of the body do not maintain fixed positions. Old capillary beds may be resorbed as a result of blood clots or injury, and new ones may be created in a continuing adaptation to current needs.

Lymphatics. The capillary walls are not absolutely watertight, and their permeability may rise sharply under the influence of oxygen lack, injury, or hormones such as histamine. The lump raised by a mosquito bite, or the skin rash resulting from an allergic reaction are both examples of capillary leakage. However, even under normal conditions some fluid is forced out of the capillary walls as a result of blood pressure. This fluid is called the *interstitial fluid*, and it is in direct contact with the body cells (Fig. 4-12). The interstitial fluid closely resembles blood plasma, though the concentration of dissolved proteins is lower because of the retarding effect of the capillary walls. It contains no RBCs. Because of continued inflow from the capillaries, the interstitial fluid is under a slight pressure. It is drained by very thin-walled lymph capillaries which coalesce and return the fluid, now called *lymph*, to large veins near the shoulder. The lymphatic vessels must pass through filtering devices called *lymph nodes*. Here the lymph is forced in close contact with masses of WBCs[3] which examine it for foreign bodies and mount an attack if any are present. Swollen lymph nodes are thus a sign of infection upstream in the portion of the body that lies above them. If the lymph nodes are for any reason blocked, the interstitial fluid in the drainage area builds up pressure, connective tissue proliferates, and the area of drainage may grow to monstrous proportions. A particularly tragic example of this results from the tropical disease called elephantiasis. Here the lymph nodes are blocked by a microscopic parasite, the filaria worm.

Veins. Veins are defined as the blood vessels that carry blood away from a capillary bed and toward the heart. Most veins carry deoxygenated blood, but the veins returning from the lungs transport blood in the oxygenated state. Blood pressure reaches a minimum in the veins. In fact, the blood entering the heart is normally sucked in, not pushed in. One may estimate the pressure of venous blood in the veins of the hand as shown in Figure 4-14.

---

3. WBCs migrate to the lymph nodes via the familiar process of diapedesis. A very large fraction of the body's WBCs reside not in the blood but in the lymphatic system and interstitial spaces.

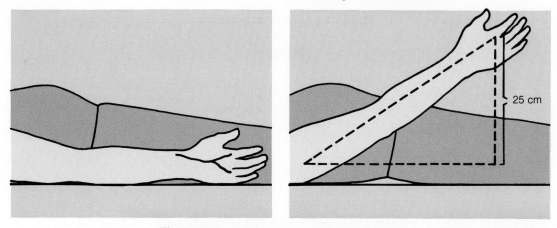

Figure 4-14 ESTIMATION OF VENOUS PRESSURE IN VEINS OF THE HAND. The venous pressure may be estimated by measuring the distance above the heart that the hand must be raised to collapse the superficial veins of the hand.

Because of the low blood pressure in the veins, venous return is assisted by internal one-way valves, especially in the veins of the arms and legs (Fig. 4-15). The valves work in conjunction with muscle action. As body muscles squeeze the veins during normal exercise, blood is forced toward the heart and cannot move backward because of the one-way valves. Such muscle action is especially important in the legs. This is why standing at attention for a long period of time can result in fainting, as venous return from the legs slows down and blood pressure to the head falls.

Because of the low internal pressure they must contain, veins have much thinner walls than arteries, and under normal conditions the walls are adequate for their task. However, there is one condition that normally increases venous pressure in the legs and can produce damage to venous walls. This condition is pregnancy. As the growing fetus presses against the large vein running up to the heart (the vena cava), venous pressure in the legs rises. One of the first symptoms is swelling of the ankles. This results from *edema*—the increased outflow of fluid across capillary walls and into the interstitial space, as a result of venous back-pressure. If the condition persists, the walls of the veins themselves may balloon out, resulting in an unsightly condition called *varicose veins*.

Regulatory Mechanisms in the Circulatory System

No discussion of the circulatory system would be complete without a discussion of regulatory mechanisms. The heart and the blood vessels do not work blindly—they adjust blood supply to the second-to-second needs of the body. A trained athlete can increase blood flow

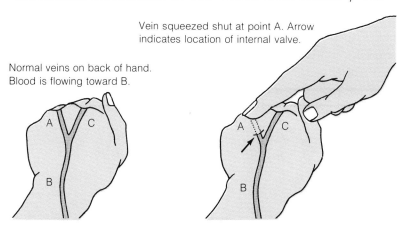

Figure 4-15 DEMONSTRATION OF ONE-WAY VALVE IN SUPERFICIAL VEIN OF THE BACK OF THE HAND. Apply finger at point B and "milk" vein up towards A. Blood is now flowing from C to B but can't back up from B to A because of one-way valve at arrow.

to muscles by a factor of 20 during a peak effort. The heart can reach a pulse rate of 300 per minute (compared to a normal of 72) during sexual orgasm. On the other hand, concentrated thought seems to result in no change in blood flow to the brain—it takes as much energy to daydream as it does to solve a problem in differential equations. Incidentally, such responsiveness of the heart is one of the chief obstacles to the manufacture of an artificial heart and, until this problem is solved, the transplanted heart, with all its problems, must remain the final option for the desperately sick heart patient.

There are two different ways in which the heart and the blood vessels can respond to changing conditions in the body. An *intrinsic* response is one involving a direct response of the heart or blood vessel to changes in its environment. An *extrinsic* response is one that is mediated by a nervous or hormonal stimulus.

Intrinsic Mechanisms

Regulation of the Heart. The most important intrinsic response of the heart is known as Starling's law of the heart. Essentially, this principle states that the heart pumps out as much blood as it receives from the large veins. Under normal conditions, a single contraction of the heart ejects about 70 milliliters of blood. During exercise, muscular contractions of the body force more venous blood into the heart. This causes the heart to increase in size as it becomes engorged with blood. The stretching of heart muscle (just like the stretching of skeletal muscle) causes the muscle to contract more vigorously. The result is an increase in the amount of blood ejected per contraction (stroke volume). The stroke volume may rise to 140 milliliters during exercise.

Regulation of the Blood Vessels. The intrinsic response of the blood vessels is called *vasomotion*. It occurs largely at the level of the arterioles and involves shunting the available blood supply from one body region to another. The normal blood budget of the body is given in Table 2.

TABLE 2
Normal distribution of blood in the body

BODY ORGAN OR TISSUE	PERCENT OF TOTAL BLOOD FLOW	
	AT REST	DURING EXERCISE
Brain	14	6
Heart	3	3
Kidney	22	4
Liver	10	8
Intestinal tract	13	14
Skeletal muscle	15	53
Skin	4	9
Miscellaneous (fat, bone, gonads, etc.)	19	3

As mentioned before, during exercise a trained athlete can increase the amount of blood flowing to muscle by a factor of 20. Part of this results from increased stroke volume and part from increase in heart rate, but a significant factor is the opening of capillary beds in muscle. Though other tissues also receive more blood in absolute terms, their increases are smaller and they wind up with a small percentage of the total. This opening or closing of capillary beds involves opening or closing of the sphincter muscles in the arterioles. The intrinsic signal for opening of these muscles is a local drop in the oxygen saturation of blood. As a tissue uses up available oxygen because of increased activity, it stimulates the local arterioles to open and flood the tissue with fresh blood.

Extrinsic Mechanisms

The same arterioles and the heart can also respond extrinsically to nervous or hormonal stimuli. The nerves that run to the heart are of two kinds: the parasympathetic nerves, which slow it down, and the sympathetic nerves, which accelerate it (see Ch. 5, p. 148). Of these, the parasympathetic system is the more important. Parasympathetic impulses, arriving through the vagus nerve, exert a braking effect on the heart even under normal conditions. The heart rate is therefore more likely to be accelerated by a decrease in parasympathetic impulses than by an increase in sympathetic ones, just as a car can be accelerated by taking one's foot off the brake. What triggers the nerve impulses? Stretch receptors in the large arteries (the aortic arch and carotid bodies) are very sensitive to slight changes in arterial pressure. Increased pressure elicits vagal activation, while decreased pressure leads to lessening of vagus activity and a speed-up of the heart.

To return to the extrinsic regulation of local blood flow, the arterioles of the body show a duality of innervation similar to that of the heart. Parasympathetic and sympathetic nerves produce opposite effects. Vasodilation (opening of sphincters) is also caused by the hormones histamine and bradykinin, as well as by local elevation of the carbon dioxide level or an increase in acidity.

What is the end result of all these regulatory processes? Claude Bernard, the great French physiologist of the nineteenth century invented the term *homeostasis* ("staying the same") to describe the process. The circulatory system serves to shield the individual cells of the body from all the extremes of environment and activity encountered by the organism as a whole. Whether the total individual is lounging by a pool or straining every muscle in hand-to-hand combat, whether he is sweating in blazing heat or shivering in subzero cold, his trillions of internal body cells remain bathed in a serene, unchanging, and close-to-perfect fluid environment.

Appendix A
The Digestive System

Surely one of the more dismal discoveries of physiology is that a fast-order roadside hamburger and a superb 2-inch steak barbecued over mesquite coals are indistinguishable as far as your body cells are concerned. This is because your body cells cannot eat steak or potatoes, or even milk or orange juice. All they can eat is small molecules, and it is the job of the digestive system to convert the food available on the table into food that is usable by the body cells: glucose, amino acids, fatty acids, vitamins, and other small molecules and ions. As a matter of fact, people can be kept alive for months or years by intra-

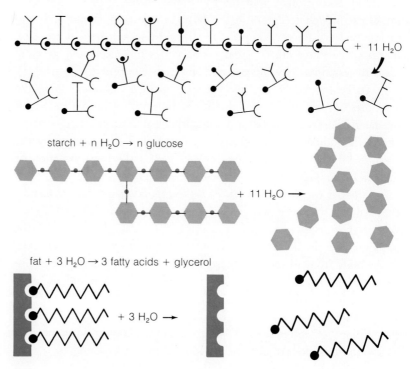

Figure 4-16 DIGESTION AT THE CHEMICAL LEVEL. The digestive process requires the chemical hydrolysis of polymers. The nature of the hydrolysis process is shown for a protein, a polysaccharide (starch), and a fat.

venous feeding. A dilute soup of the molecules mentioned is dripped directly into the bloodstream by means of a catheter. The transformation of "table food" into "cellular food" is partly a physical process and partly a chemical one. The physical processes may begin in the kitchen, as meat is cooked, potatoes are mashed, and peas are boiled. They continue in the mouth and in the stomach. Chemical processes carry the general term of *hydrolysis* ("dissolve by means of water"). The large bulk of our food is chemical polymers (see Ch. 2, p. 37), and they are depolymerized by the addition of the elements of water to each subunit. All such hydrolyses would occur impossibly slowly under natural conditions and are therefore speeded up by the release of specific enzymes. (See Ch. 2, p. 40.) These enzymes are released in an assembly-line fashion. Thus proteins are first broken down enzymically into shorter fragments called peptides, by the use of the enzyme pepsin. Later, the enzymes trypsin and chymotrypsin reduce the resultant peptides into smaller peptides. Finally, the enzymes aminopeptidase and carboxypeptidase (and various dipeptidases) hy-

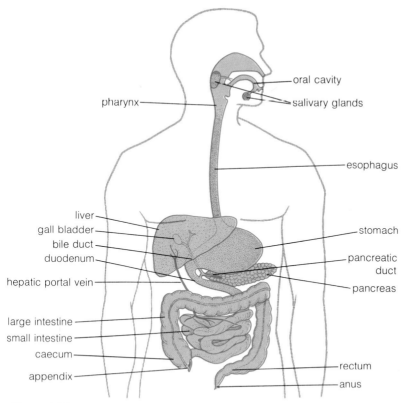

Figure 4-17 THE DIGESTIVE SYSTEM.

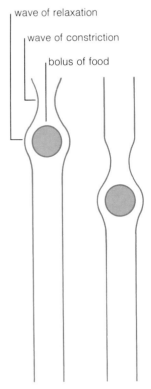

Figure 4-18
PERISTALSIS. In the esophagus food is moved by a coordinated response of the circular and longitudinal smooth muscle layers. Contraction of the former narrows the diameter of the esophagus; contraction of the latter widens the diameter. A similar mechanism operates throughout the digestive tract.

drolyze the remaining small peptides into free amino acids which then pass into the circulatory system through the walls of the small intestine. The following is a brief summary of the roles of the different digestive organs in the digestive process (Fig. 4-17).

The mouth contains teeth for the physical reduction of food, salivary glands which release the enzyme amylase (which breaks down starch into two-sugar fragments called maltose) and the tongue which is used in the swallowing of food.

The *esophagus* conducts chewed and partially digested food to the stomach by a process called *peristalsis* (Fig. 4-18).

The *stomach* acts as an organ of temporary storage, allowing us to get by with only a few meals a day. It releases the enzyme pepsin (which converts protein into peptides). The stomach also secretes large amounts of hydrochloric acid, which is essential for pepsin activity but lethal to most bacteria. The stomach thus acts as a sterilizing organ which kills the bulk of ingested bacteria.

The *duodenum* is the first part of the small intestine and the most important region in terms of digestive processes. It receives secretions from the liver and gall bladder, the pancreas, and its own microscopic digestive glands.

The *pancreas* secretes a string of digestive enzymes plus sodium bicarbonate (used to neutralize the hydrochloric acid entering from the stomach). The digestive enzymes include trypsin and chymotrypsin (protein → peptides), pancreatic amylase (starch → maltose), lipase (fats → fatty acids plus glycerol), plus carboxypeptidase, amino peptidase, and dipeptidase (peptides → amino acids).

Duodenal enzymes complete the hydrolysis of small carbohydrate fragments. Thus maltase converts maltose to two glucoses, sucrase converts table sugar to glucose plus fructose, and lactase converts milk sugar to galactose plus glucose.

The *gall bladder* acts to store and concentrate the bile secreted by the liver. Most important of these bile constituents are the bile salts, which act as detergents to keep lipids in suspension long enough for the water-soluble enzyme lipase to hydrolyze them.

The *small intestine,* along most of its length, acts as an organ of absorption of digestive products. To aid in the process, it has an enormous surface area which results from infoldings of the intestinal wall called villi (Fig. 4-19) and microscopic infoldings of the cell membrane called microvilli (Fig. 1-15).

Figure 4-19
VILLI OF THE SMALL INTESTINE.

After absorption across the walls of the small intestine, the sugars and amino acids resulting from the digestive process are not dumped directly into the general circulation, but pass first to the *liver* by means of the hepatic portal vein. Here the products are temporarily stored, to be released into the general circulation at a steady rate that matches body demands. The hormones insulin and glucagon play a major role in this process.

The *large intestine* acts to reabsorb most of the 8 quarts of water secreted daily into the digestive tract by the various digestive organs. If the contents of the large intestine move through too quickly (diarrhea), the body may be threatened by water loss. About 50 percent of the contents of the large intestine is made up of bacteria. In herbivores, such as horses and rabbits, the intestinal bacteria contribute significant amounts of nutrients by hydrolyzing cellulose (the animals themselves lack the necessary enzyme). These animals have a large *caecum,* or blind end, in the large intestine to slow down transit time and thus increase the efficiency of cellulose hydrolysis. The human large intestine is too short to make cellulose a practical source of food. However, the resident bacteria are not entirely useless—they contribute a major share of the body's vitamin K. This vitamin is essential for the process of blood clotting. The remaining undigested foods (mostly cellulose) and bacteria are eliminated through the rectum by the process of defecation.

Appendix B
The Kidney

Few body systems are more important than the kidney. In fact, the kidney is so essential that nature has seen fit to provide us with two copies of this organ. Even when one of the two is totally removed, the remaining kidney has sufficient reserve capacity to allow the individual a normal, active life. (I have lost many a game of tennis to a friend with only one kidney.) The kidney assists in many ways in the maintenance of internal homeostasis. It removes accumulations of toxic waste products, helps maintain a proper acid-base balance, and contributes to regulation of the blood pressure. Malfunction of the kidneys can therefore reflect on all these body functions.

The kidney works like a housekeeper who cleans house by throwing out everything except the people and then going out and carrying back all the valuables. This work is carried on in each kidney by about 1 million *nephrons*. Each nephron functions as an independent unit and contains a *glomerulus* ("little ball") with a surrounding capsule, a *kidney tubule* (made up of a proximal tubule, a loop of Henle, or distal tubule, and a collecting duct). This connects to the bladder, and is surrounded by a capillary bed (Fig. 4-20). The glomerulus is a ball-shaped tuft of capillaries closely enveloped by a capsule. Blood enters the capillaries under high pressure from the renal arteries, and the pressure forces some of the fluid through the capillary walls. Since the cellular elements and large proteins are too large to cross the capillary walls, the filtered fluid resembles protein-free plasma. This filtrate crosses the inner wall of the surrounding capsule and enters the first part of the tubule (the *proximal tubule*). Now begins the job of transforming the glomerular filtrate into urine. Glucose and salts, valuable commodities to the body, are pumped out of the tubule by active transport and into the capillary network that surrounds the tubule. Thus they are returned to the circulatory system. There is no machinery for active salvage of waste products such as urea, hence they are targeted for inclusion in the urine. But perhaps the most important component to be salvaged is water. The average

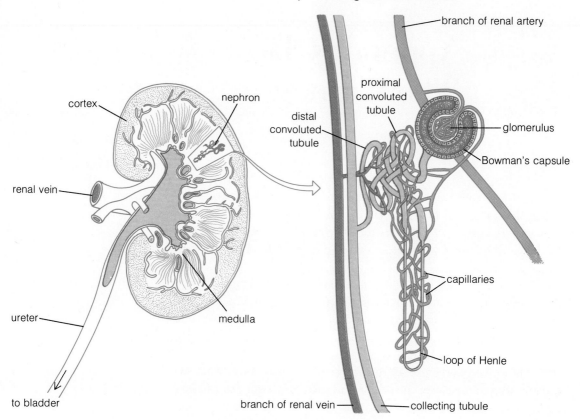

Figure 4-20 THE KIDNEY (LEFT) AND THE NEPHRON (RIGHT).

individual forms 45 gallons of filtrate per day. If all this water were lost through the bladder, one would have to drink approximately a bathtubful of fluid a day to make up the loss. Needless to say, the kidney works hard to resorb all the water it can. As a result, less than 1 percent of the filtrate volume winds up in the form of urine. The method by which water is resorbed is not a simple matter of active transport, as in the case of glucose or salt resorption. If nothing else, the ATP needed for the active transport of 45 gallons of water a day would be beyond the reach of the body's energy metabolism. Consequently, a slightly less efficient but far less "expensive" mechanism of water resorption has been evolved, one based on an osmotic principle.

In the kidney, there is a gradient in salt concentration, hence osmotic pressure. The greatest concentration of salt (NaCl) is found in the inner part of the kidney (the *medulla*), and the concentration tapers off toward the outside (the *cortex*). Now, the individual nephrons of the kidney are positioned in such a way that each kidney tubule originates in the cortex, sends a loop (the loop of Henle) into the medulla, rises back to the cortex, and then, merging with other tubules, descends once more into the medulla (Fig. 4-20).

THE KIDNEY / 127

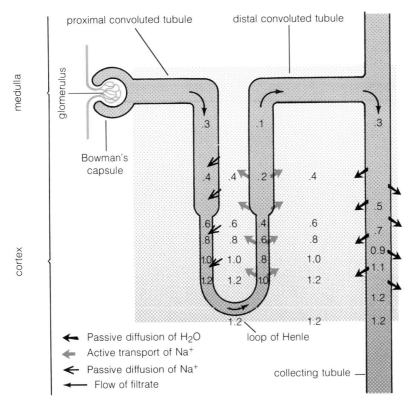

Figure 4-21 THE COUNTERCURRENT MECHANISM IN THE KIDNEY. Figures indicate solute concentration in osmols per liter.

The key component in establishing the osmotic gradient is the ascending arm of the loop of Henle. It is here that sodium ions are actively pumped out of the tubule and into the external tissue fluid (chloride ions follow by electrostatic attraction). No water molecules leave the ascending arm, since the walls are impermeable to water. After the filtrate passes through the distal convoluted tubule and begins to move down the collecting tubule, water flows out passively into the external tissue fluid because of the osmotic pressure differential. (See Ch. 1, p. 13.)

The walls of the collecting tubule are permeable to water. In fact, urine has a higher osmotic pressure than blood plasma. The success of the kidney in resorbing water depends on its ability to establish a high osmotic pressure in the external tissue fluid. The high pressure actually established (1.2 osmolar[4] versus 0.3 osmolar in

---

4. The osmolarity of a solution is proportional to its osmotic pressure. It is a measure of the concentration of osmotically active particles. A nonionic molecule such as glucose produces one osmotically active particle per molecule. An ionic compound such as sodium chloride produces two osmotically active particles (NaCl → $Na^+ + Cl^-$).

plasma) could not be established by an uncomplicated sodium pumping mechanism. The sodium pump of the ascending arm of the loop of Henle is only able to establish a concentration gradient of about 0.2 osmolar between the inside of the tubule and the outside. A simple pump would therefore be able to take an initial filtrate of 0.3 osmolar concentration and boost it to 0.5 osmolar outside the tubule. This would constitute a rather ineffective water retrieval mechanism. The much higher salt concentration actually achieved depends on the countercurrent mechanism illustrated in Figure 4-21. The sodium ions actively pumped out by the ascending arm are able to reenter the parallel and closely juxtaposed descending arm by a process of passive diffusion. As a result, at the bottom of the loop of Henle, the sodium ions being pumped out are 1.2 osmolar. The sodium pump now has a much higher initial ion concentration to work with. The same pumping effort therefore results in a much higher external salt concentration, hence more effective water retrieval from the collecting tubule.

Two other points should be mentioned. At the beginning of this discussion we mentioned that the kidney, besides disposing of waste products, also helps regulate blood acid-base balance and blood pressure. The acid-base balance is regulated by an ionic pump similar to the sodium pump described. This pump is located in the walls of the proximal and distal tubules, and it can take sodium ions destined for excretion in the urine and exchange them for hydrogen or ammonium ions. If hydrogen or ammonium ions are lost, the urine becomes acidic, hence the blood remaining behind becomes more basic. If sodium or potassium ions are lost in the urine, the urine becomes basic and the blood more acidic. Under normal conditions, the urine is very mildly acidic.

Maintenance of blood pressure by the kidney involves an element of "self-interest" on the part of the kidney. Without a high filtration pressure in the glomerulus, the kidney ceases to function. Therefore falling blood pressure stimulates the kidney to release the protein hormone *renin*. Renin acts as an enzyme in converting the protein *angiotensinogen* into the shorter peptide *angiotensin*. Angiotensin causes strong vasoconstriction of the arterioles, resulting in increased blood pressure because of increased resistance to blood flow.

The Artificial Kidney

Unlike the heart, an artificial replacement organ for the kidney may be just around the corner. Before the mid-1950s, all individuals with failure of both kidneys died. At this time, the first artificial kidney dialysis units came into use. The treatment required taking

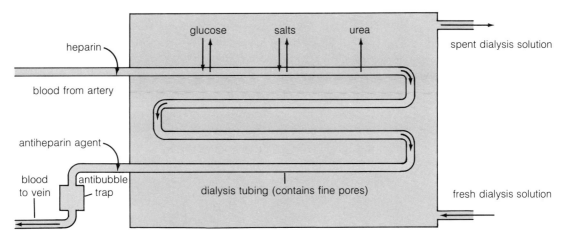

Figure 4-22 THE ARTIFICIAL KIDNEY. The dialysis solution is kept at body temperature and contains glucose and salts in the same concentration as found in blood plasma, hence there is no net change in these components in dialyzed blood. But since the dialysis solution contains no urea, uric acid, creatinine, or other wastes, these are rapidly lost from the blood. Heparin is added to prevent clot formation. Antiheparin agent is added to prevent internal bleeding.

blood from an artery and passing it through plastic or cellophane dialysis tubing immersed in a solution of salts and glucose identical to that of normal filtrate fluid. The dialysis tubing functioned as the walls of glomerular capillaries by allowing free movement of small molecules (including urea), but not proteins or red blood cells. The blood was returned to the vein, containing all the original proteins, blood cells, and a proper mixture of salts and glucose, but no urea or other waste products (Fig. 4-22).

Patients who had been condemned to die suddenly found themselves restored to life. Since very few machines were available in the first years, doctors had to make heartbreaking decisions about which patients would be given access out of the many who needed it. Today dialysis centers have enough machines for all who need them, but the cost of a machine remains sufficiently high that each one must be shared by many patients.

Another development in kidney therapy has been the use of transplanted kidneys to replace kidneys destroyed by disease or injury. These were first attempted in the mid-1960s. For those lucky enough to have an identical twin, transplants of kidneys present no problems, but for others there is a problem of tissue rejection. Transplants must be matched for antigenic groups and may come from close relatives or from cadavers. If the kidney is not rejected within the first 2 years, there is a good chance that it will last for many more years—patients now alive have carried transplanted kidneys for as long as 10 years. The two problems at this time are tissue rejection and lack of sufficient cadaveric kidneys. A wider public appreciation of the importance of donating kidneys and other organs to organ banks after death would solve the second problem.

A solution for the near future (5 to 10 years) is the expected development of a wearable artificial kidney. The kidney dialysis machine is a hospital-based facility about the size of a small refrigerator. It requires medical supervision, causes great discomfort in usage, and cannot be used for more than 12 hours every 3 to 4 days. A portable artificial kidney will be based on space technology; astronauts in long space journeys now detoxify their urine by the use of sorbent materials and can thus use the water over again. Otherwise, each astronaut would produce in the neighborhood of 240 pounds of unusable urine during 80 days of space travel.

Bibliography

GENERAL READING

Isaac Asimov, *The Human Body,* New American Library, 1964. A small classic by one of the masters of science writing. The following brief excerpt deals with the human intestines.

> In man, what is left of the caecum (a vestige perhaps of herbivorous ancestors) is of no particular use and can actually be a source of trouble. To the bottom of the caecum is attached a small appendage, or appendix (the two are essentially the same word), which is a further remnant of a once sizable and usable caecum. This appendage is about 2 to 4 inches long and is shaped very much like a worm. In fact it is called the *vermiform appendix* (vur'mi-form; "worm-shaped" L). An insignificant foreign body, an orange-pit, perhaps, that has survived digestion, can find its way by ill chance into this narrow blind alley set into a wider blind alley. This may set up first an irritation and then a dangerous inflammation (*appendicitis*). It is only in the last century that removing the appendix in such cases (*appendectomy*) has become a simple operation without much danger of peritonitis.

Homer W. Smith, *From Fish to Philosopher,* Doubleday, 1961. A delightful treatment of the evolutionary role of the kidney in regulation of the internal environment of vertebrates.

F. Macfarlane Burnet, *Self and Not-Self: Cellular Immunology,* Cambridge University Press, 1969. Immunology as discussed by a Nobel Prize winner and pioneer in the field.

TEXTBOOKS

Russell Myles DeCoursey, *The Human Organism,* 4th ed., McGraw-Hill, 1974. Generally clear and readable.

Arthur C. Guyton, *Function of the Human Body,* 4th ed., W. B. Saunders, 1974. This is the shorter version of a comprehensive text.

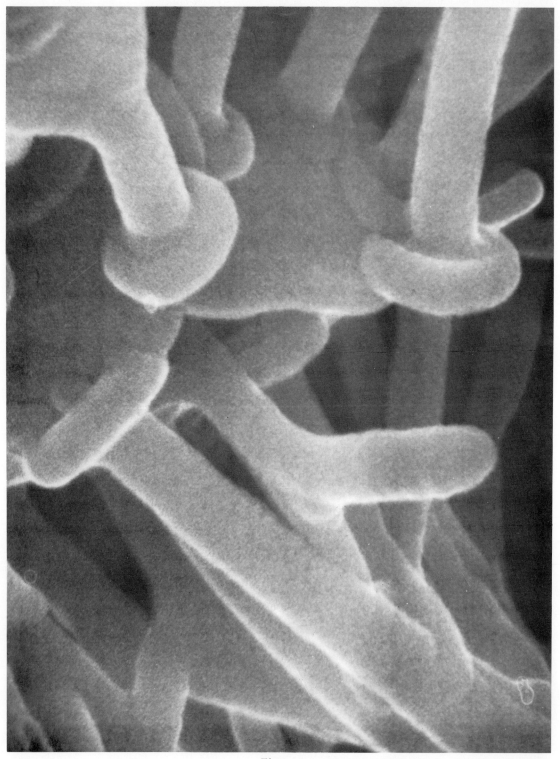
Figure 5-1 SYNAPTIC KNOBS IN APLYSIA, A SEA SNAIL.

chapter 5

The Nervous System

T he human brain has a strange relationship to the body in which it is housed. The brain is the core of the self, yet all of this "I" develops only in response to the experience of the eyes, the ears, the fingers, the skin, and so on. A severed, unimprinted brain would be as silent as an empty city. And conversely, the life of the body itself revolves indirectly around the subtle inputs entering the brain. To read a book, to meet a person, to buy a car, to run a mile—all of these represent, ultimately, attempted improvements in the contents of the brain.

This intimate coupling between the body and the brain reflects the functions of the nervous system as a whole. These functions include:

1. The collection of information about the external as well as the internal environment. This information is filtered through the organs of the sensory system, such as the eyes, the ears, the pain receptors, and the osmotic pressure receptors. It is conveyed to the brain by the *afferent* or ingoing nerve fibers.

2. The processing and integration of received information. This function is carried out by the central nervous system (the brain plus the spinal cord). The nature of the information processing involves both built-in rules (reflexes) and acquired information (memory and learning).

3. Acting on the information received. This function is mediated by the *efferent* or outgoing nerve fibers which control specific target organs such as glands, skeletal muscles, and the smooth muscles of arterioles and internal organs.

These general functions of the nervous system are the same in an insect, a sparrow, or a human being. But the big difference between the human nervous system and all other nervous systems lies in the complexity of the integrating portion, the brain.

In this chapter we first look at the properties of neurons and their mode of interaction with each other. There are about 10,000 million neurons in the human nervous system, and it is probably safe to say that each is connected more or less directly to every other neuron. Then we look at the larger structural relationships of the human nervous system and the specialized functions of the peripheral and central nervous systems. Table 1 lists the various parts of the vertebrate nervous system.

TABLE 1
Classification of the vertebrate nervous system

I. PERIPHERAL NERVOUS SYSTEM
   A. Voluntary
   B. Autonomic
      1. Sympathetic
      2. Parasympathetic

II. CENTRAL NERVOUS SYSTEM
   A. Brain
      1. Forebrain
         Thalamus
         Hypothalamus
         Cerebrum
      2. Midbrain
      3. Hindbrain
         Medulla & Pons
         Cerebellum
   B. Spinal cord

The Neuron

It is strange to think that the complex, mysterious organ that is the brain has been assembled out of a single basic kind of cell, the *neuron* (Fig. 5-2). Clearly, the brain is more than the sum of its parts, yet it is still valuable to examine the structure and function of these basic subunits of the nervous system. Neurons originate from skin cells on the back of the developing embryo. These cells fold inward and form a hollow tube. (See Chapter 6.) In this phase, the cells look much like any other generalized body cell and divide rapidly. Next comes a phase of cell differentiation. The embryonic cells stop dividing and begin to reach out long probes which will ramify through every part of the body (Fig. 5-3). Each neuron is genetically programmed to find a specific target cell or organ, such as a retinal cell, a muscle fiber or, most commonly, another neuron. A mature neuron typically consists of an *axon* (which may or may not contain an insulating sheath of myelin), a *cell body,* and several *dendrites* (Fig. 5-2). The axon is usually longer than the dendrites (it may reach 3 feet in length); it conducts nerve impulses away from the cell body, while the dendrites conduct nerve impulses toward the cell. In terms of function, neurons do two things: they conduct nerve impulses, and they can secrete specific chemicals at their axon terminals.

The Nerve Impulse

The nerve impulse is an electrical phenomenon, but the electric current involved is different from the one that lights up a light bulb. The electric current in a light bulb rushes through at close to 186,000 miles per second and consists of electrons. The electric current in a neuron lumbers along at no faster than 180 miles per hour and is made up of ions, not electrons. (As you may recall, ions are electrically charged atoms or molecules. For instance, table salt in water dissociates into sodium ions and chloride ions: $NaCl \rightarrow Na^+ + Cl^-$.)

We can use ions to generate electric potentials. Let us consider a very simple experiment (Fig. 5-4). Here we use a semipermeable membrane such as cellophane to separate two solutions of sodium and protein ions. The more concentrated solution on the right will lose some of its sodium ions to the weaker solution on the left. The negatively charged proteins, however, will not be able to follow, because they are too large to squeeze through the pores. The result is an electric potential which is set up across the membrane as shown. This

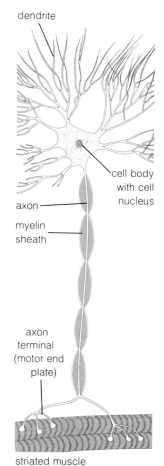

Figure 5-2 AN ADULT MOTOR NEURON. The target organs of this neuron are a pair of striated muscle fibers. Not all neurons possess a myelin sheath.

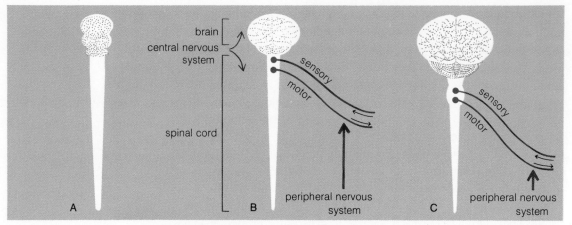

Figure 5-3 DEVELOPMENT OF THE HUMAN BRAIN AND NERVOUS SYSTEM. (A) Embryo: individual cells relatively unspecialized and still capable of cell division. Brain and spinal cord form as a hollow tube. (B) Newborn: individual cells (neurons) now fully differentiated and no longer capable of cell division. (C) Child (two years): arborization occurs in the brain as individual neurons establish more complex interconnections. There is no increase in total number of cells, but there is an increase in white matter.

electric potential can be measured by means of a suitable meter, and the greater the concentration difference, the greater the electric potential measured. Essentially all living cells show a difference in electric potential across their cell membranes, and differential ion permeabilities of this sort play a major role in establishing membrane potentials. However, of all body tissues, only nerve and muscle tissues can alter their membrane potentials in an explosive way, and this ability becomes the basis of both the nerve impulse and the stimulus for muscle contraction.

Let us look at a segment of a neuron at rest (Fig. 5-5). Although all positively charged ions must be balanced by negatively charged ions, we have ignored the latter, since they play only a passive role in the generation of the nerve impulse. Two ions are involved in the generation of the nerve impulse—sodium (Na^+) and potassium (K^+). These ions produce opposing electrical effects. As shown, both have large concentration gradients between the inside of the neuron and the outside. Na^+ is present in 14-fold excess outside, while K^+ is present in 28-fold excess inside the neuron. These large concentration gradients are maintained by a metabolic pump which selectively pumps sodium ions out and potassium ions in. If the energy-producing machinery of a neuron is poisoned, the pumps stop working, the two ionic concentrations approach equivalence inside and out, and all nervous activity disappears.

Figure 5-4 ELECTRIC POTENTIAL SET UP ACROSS A SEMIPERMEABLE MEMBRANE. Both solutions are electrically neutral at the start, with numbers of positive charges exactly counterbalancing numbers of negative charges. The situation changes as the semipermeable membrane allows small, positive sodium ions to escape from the right-hand compartment. The bulky, negatively-charged proteinate ions are left behind.

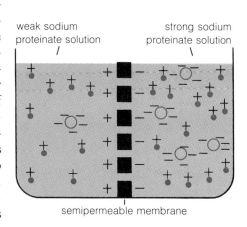

If the two cations (positive ions) are balanced across opposing concentration gradients, why doesn't the neuronal membrane show a net electric charge of zero? Instead, the outside face of the membrane is electropositive with respect to the inside. We can explain this anomaly by considering only the behavior of the potassium ions. The high excess of internal K^+ is built up as a result of metabolic pumping. Next, the K^+ begins to diffuse back outward across the cell membrane. As the K^+ ions diffuse outward, they are trailed by negatively charged chloride (Cl^-) ions. It is this trailing tail of Cl^- that renders the inside of the cell membrane electronegative with respect to the outside.

But why doesn't the inward diffusion of Na^+ ions counteract the outward flow of K^+ and cancel the electric charge generated? Because the membrane in the resting state is relatively impermeable to Na^+ ions. Both the K^+ ions diffusing outward and the Na^+ ions making their way inward must pass through individual "gates" in the cell membrane. During the resting state, the ionic gates for Na^+ are al-

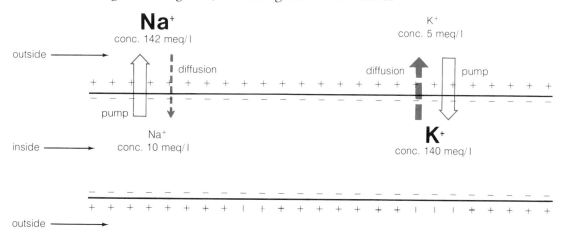

Figure 5-5 SEGMENT OF A NEURON AT REST.

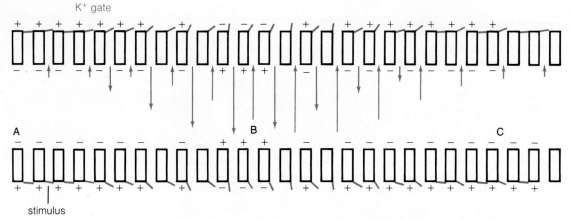

Figure 5-6 FLUX OF IONS DURING A NERVE IMPULSE. (A) Region of rest. Na+ gates almost closed, K+ gates partly open. (B) Region of nerve impulse. During first half of impulse, Na+ gates open maximally, allowing Na+ to rush in and reverse charge on membrane. During second half of impulse, K+ gates open maximally, K+ rushes out and restores original polarization after a brief overshoot (hyperpolarization). Entire process lasts 1.5 milliseconds. (C) Region of recovery. Na+ and K+ gates as in (A).

most entirely closed, while the gates for K+ are opened more widely by a factor of about 100. Thus it is the outward diffusion of the less-restricted K+ ion that largely dictates the electric charge on the resting nerve cell membrane. Incidentally, the overall voltage difference between the inside and outside of the nerve cell is about 0.08 volt. This voltage is about one-twentieth of that generated by a single flashlight battery. And the total electric power generated by the human brain is about 10 watts.

During the nerve impulse, all of these relationships undergo violent but very short-lived changes (Fig. 5-6). The stimulus for initiation of the nerve impulse may be electrical or chemical. We will discuss chemical stimuli later. An electrical stimulus requires the application of a negative charge to the cell surface. This results in membrane *depolarization,* or loss of charge difference between the inside and the outside of the cell. This electrical change leads to a physical change in the membrane; it causes the sodium gates to open rapidly. In less than half of a millisecond, the rate of sodium entry rises by a factor of several thousand over the resting rate. This results in a temporary reversal of charge, with the inside of the neuron now being positive and the outside negative. As quickly as they opened, the sodium gates begin to close, and the potassium gates now open to their maximum extent, allowing K+ to spill outward and driving the cell membrane back toward its original polarization. In fact, there is

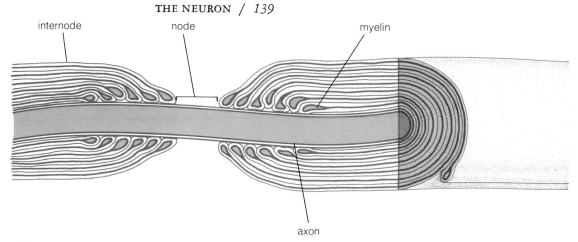

Figure 5-7 THE MYELIN SHEATH. The myelinated lamina (internodal region) are formed by the spiral wrapping of a Schwann cell around the axon. Nodal regions are free of Schwann cells. Nerve impulses "jump" from one node to the next.

a little overshoot, resulting in a temporary (0.5-millisecond) hyperpolarization until the K^+ gates return to their resting state. During the resting state, the sodium pump rids the neuron of all excess sodium ions, and the potassium pump replenishes it with any lost potassium ions.[1]

Two aspects of this process are worth mentioning. First, a nerve impulse is an all-or-none phenomenon. An electrical (or chemical) stimulus is either strong enough to initiate the nerve impulse, or is not. Any increase in the strength of the stimulus above the necessary threshold produces no change in the nature of the response. But if the nature of the nerve impulse cannot be modified, how does the nervous system convey information about the strength of an external stimulus such as light, sound, or temperature? The answer is frequency coding. A nerve encodes information about the strength of a stimulus by changing its rate of firing. For instance a strong stimulus produces rapid, repetitive firing of the nerve, while a weak stimulus elicits a much slower train of nerve impulses. To use the analogy of John Eccles, the nervous system works in a Morse code that is all dots!

Another characteristic of the nerve impulse worth mentioning is the fact that it is self-propagating. Whether the impulse travels 3 feet or 3 millimeters to the end of a neuron, it arrives with none of its force diminished. This is because the inrush of sodium causes electrical depolarization of the adjacent region of the cell membrane

---

1. The amount of sodium that enters (or the amount of potassium that leaves) during a single action potential is negligible. A neuron with its sodium-potassium pump shut off can respond tens of thousands of times before rising sodium levels (and falling potassium levels) destroy membrane polarization.

causing further sodium inrush, which in turn causes further depolarization, and so on. Without such a self-propagating mechanism, the low-voltage signals would quickly fade and die out.

Elegant though the ion flux mechanism is, it has one serious drawback: it is a very slow method of sending a message. This is a disadvantage that is built into the basic design of the nerve fiber. A small fiber may have a conduction speed of only 0.5 mile per hour—a rate slower than a leisurely walk. For example, after you have stubbed your toe, you become aware immediately of the fact that there has been a mechanical collision. But it takes about a second for the sense of pain to register, because the pain conductor nerves have the lowest conduction speed in the body. Invertebrates (such as the squid) have coped with this problem by building larger neurons, since larger neurons have faster conduction rates. But vertebrates hit on the brilliant invention of the *myelin sheath* (Fig. 5-7). The myelin sheath is a spiral wrapping of a fatty material, myelin, around the neuron. The wrapping is discontinuous, resulting in a series of naked nodes and wrapped internodes. The discontinuous nature of the sheath arises from the fact that it is formed from enveloping cells called *Schwann cells* which twist around the neuron many times. Ultimately all the cytoplasm of the Schwann cell is squeezed out, leaving behind only the fatty insulation layer.

In a myelinated fiber, the wave of electrical depolarization does not creep along the surface of the neuron, but can jump from node to node. The result is greatly increased speed of transmission (about 50 times faster). This use of myelin has undoubtedly been a major factor in the success of vertebrates on earth.

The Synapse

What happens to the nerve impulse when it reaches the end of the nerve fiber? In the vast majority of cases, the adjacent cell is another neuron, and not a muscle or endocrine gland. Electron microscope photographs show that when two neurons meet, they do not typically come in direct contact. Instead, there is a close approach with a distinct physical separation in between (Fig. 5-8). Such a junction is called a *synapse* ("to clasp closely"). Transmission of nerve impulses across the synaptic space (also called the *synaptic cleft*) is not electrical but *chemical,* and this has very important consequences for the nervous system. The ability to secrete chemicals is one of the fundamental attributes of the neuron. While many other cells have this ability, the neuron secretes its chemical in precise dosages in response to a signal, and uses it to activate an adjacent neuron or target cell.

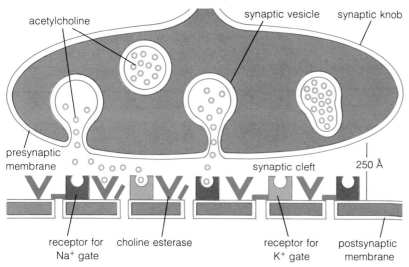

Figure 5-8 SYNAPTIC TRANSMISSION BY ACETYLCHOLINE. Arrival of a nerve impulse at the presynaptic membrane causes discharge of acetylcholine into the synaptic cleft. The acetylcholine can either interact with Na^+ and K^+ gate receptors or with the choline esterase molecule, which destroys acetylcholine. The former interaction initiates a nerve impulse in the postsynaptic membrane. The latter interaction terminates it. A scanning electron microscope view of synaptic knobs is shown in Figure 5-1.

An electron microscope photograph of the synapse (Fig. 5-1) shows little knobs at the end of each axon branch as it impinges on a dendrite or a cell body. These *synaptic knobs* are filled with tiny round *synaptic vesicles* which carry the transmitter substance. A variety of different transmitters is known in the human nervous system; all are small molecules and contain nitrogen. Some neurotransmitters (acetylcholine, norepinephrine, glutamic acid) work by depolarizing the postsynaptic nerve membrane. Others produce the opposite effect —they hyperpolarize the postsynaptic membrane. This group includes glycine and γ-aminobutyric acid (GABA).

The action of neurotransmitters has been extremely difficult to study. They are released in minute amounts. For instance, a single synaptic vesicle contains only about 2,000 molecules of acetylcholine. If that sounds like a lot, then compare it to the number of water molecules in a single drop of water: 3,300,000,000,000,000,000,000, or 3.3×10^{21} in exponential notation. If that's not enough to give the neurochemist a headache, there is the further difficulty that neurotransmitters have very short lifetimes—they are typically destroyed in less than 1/1000 second after release. No wonder that our list of neurotransmitters is still very incomplete and very tentative.

How can such minute and evanescent packages of materials accomplish their task? Neurotransmitters work as gate openers in the postsynaptic membrane, allowing selected ions to enter or leave and change the polarization of the membrane. In the case of acetylcholine (Fig. 5-8) the arrival of a nerve impulse at the synaptic knob causes a number of vesicles to migrate to the surface of the cell membrane and spill out their acetylcholine into the synaptic cleft. The acetylcholine can then combine with two kinds of receptors: gate openers or acetylcholine destroyers. The gate-opening receptors open ionic gates for both Na^+ and K^+. The inrush of Na^+ initiates a new nerve impulse in the postsynaptic membrane. The other kind of receptor is an enzyme called acetylcholine esterase (or cholinesterase) which destroys the neurotransmitter by hydrolysis. But before all the acetylcholine can be destroyed, it has already accomplished its task of starting a new nerve impulse.

It is significant that some of the most powerful poisons known work by interfering with this system. The deadly nerve gases synthesized during World War II but never used react chemically with cholinesterase and inactivate it. As a result, acetylcholine is not destroyed and keeps the ionic gates open. Instead of a brief nerve impulse, the body experiences a continuous seizure. Death occurs by drowning in one's own saliva, or by the aspiration of vomited material. Organophosphorus insecticides such as malathion and parathion are chemical derivatives of the nerve gases, and every year farm workers and sprayer pilots experience fatalities as a result of accidental overexposure. The amount of insecticide needed for a lethal dose is surprisingly low, especially for parathion. One foolish chemist died after self-injection of 0.004 ounces during a study of parathion toxicity. He had prepared an atropine-filled syringe as an antidote, but died before he could reach for it.[2]

If too much acetylcholine is deadly, too little can produce the same result. *Botulinus toxin,* a protein produced by an anaerobic bacterium, prevents the release of acetylcholine from the presynaptic terminal. The result is paralysis of all skeletal muscles and death as breathing stops.

The Inhibitory Synapse. The nervous system contains many neurons whose function it is to inhibit other neurons. These *inhibitory neurons* (Fig. 5-9) form inhibitory synapses which release either glycine or GABA in discrete packets as with acetylcholine. (Each neuron releases only one kind of neurotransmitter.) Inhibitory neuro-

---

2. Atropine binds to acetylcholine gate-opening receptors without initiating a nerve impulse itself.

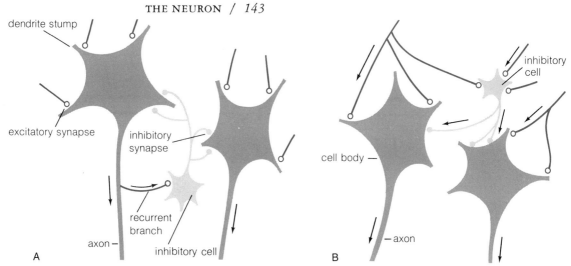

Figure 5-9 TWO TYPES OF INHIBITORY PATHWAYS. (A) Feedback inhibition. (B) Feed-forward inhibition. Inhibitory cells and synaptic knobs are shown in black.

transmitters produce *hyperpolarization* of the postsynaptic membrane, and they do this by opening the ionic gates for K^+ but not for Na^+.

The important point to keep in mind is that a single neuron receives synapses from thousands or tens of thousands of other neurons. The known record is held by the Purkinje cells of the brain. Each Purkinje cell receives in the neighborhood of 80,000 synapses from other cells in the brain. The synapses impinging on a single neuron represent a combination of excitatory and inhibitory inputs. There is thus a battle between the two kinds of input, which decides whether the neuron will be activated, and how intensely.

Now let us see how the inhibitory neurons are tied in with the excitatory ones. In Figure 5-9 the neuron on the left gives rise to a recurrent (or returning) branch from its axon output. The recurrent branch stimulates an inhibitory cell, which then inhibits the neuron by means of inhibitory synapses on its cell body. Such *feedback inhibition* is found, for instance, in the motor neurons that control the movement of skeletal muscle. The *feed-forward inhibition* shown in Figure 5-9 is likewise very widely distributed among the body's neuronal circuitry. Now, obviously, the more intensely the left-hand neuron in Figure 5-9 fires, the more strongly it inhibits itself. Isn't this an example of wasteful design? Wouldn't a brain incorporating this principle be irrational? To see the logic of the design, we have to remember that there are thousands of other neurons running in parallel with the one we are discussing innervating related parts of the body. Our diagram cannot show them. Each neuron sends out a recurrent fiber to a specific inhibitory cell, and each inhibitory cell inhibits many neurons in close proximity. Thus when one group of neurons fires strongly, it generates a field of strong feedback inhibition on all other neurons in the neighborhood, no matter what they are doing. Thus only the strongest impulses survive this barrage of inhibition. The result is that neuronal performance is made much more specific and selective and does not reflect the general background "noise" of the brain. Thus, for instance, the sense of touch can distinguish two sharp points lying very close to each other, even though the fingertip is almost equally indented between them. There are at least two synaptic centers (called *nuclei*) on the path from the fingertip to the cerebral cortex, and each contributes a further sharpening or "sculpturing" of the signal by means of inhibitory synapses.

To summarize, the synapses of the nervous system act as agents of integration and discrimination, through a summation of excitatory and inhibitory impulses. Synapses, moreover, act as one-way gates in the nervous system. A nerve impulse can travel in either direction down a neuron, but synaptic transmission is always one-way.

We take one-way nerve transmission for granted. For instance, a burning sensation in a finger results in quick withdrawal of the hand. If nerve impulses could run either way, then withdrawal of the hand might produce a burning sensation in the finger!

Further, the synapse may act as an amplifier. For instance, the synapse between a nerve and a muscle is about 100 times more effective in depolarizing the muscle membrane and thus starting a contraction than is the nerve by itself.

Finally, in a drug-conscious age it is perhaps worth pointing out that essentially all the hallucinogenic drugs, such as LSD and mescaline, act at the level of the synapse, either by mimicking natural neurotransmitters or by interfering with their function. And there are some neurologists who feel that schizophrenia, a disease responsible for filling two-thirds of all beds in mental hospitals, may ultimately be explained as a biochemical defect of neurotransmitters.

Fast Neuron Transport

The secretion of neurotransmitters at synaptic endings is perhaps the most obvious illustration of the secretory nature of the neuron. But there is a second, more subtle kind of nervous secretion. It has been known for a long time that, if the nervous supply to a muscle is cut, the muscle atrophies even though there is no interference with its blood supply. The shriveled legs of the polio victim reflect the destruction of their motor neurons. Interestingly, when a neuron has most of the synapses impinging on it destroyed, it atrophies and dies just as the muscle cells do. While some of these changes reflect a lack of normal activation by nerve impulses, the more severe effects must be explained on a different basis.

This explanation comes from the remarkable experiments of Sidney Ochs of Indiana University Medical School. Ochs was studying fast transport of molecules in the sciatic nerve of a cat. The sciatic is a long nerve running from the spinal cord down the hind leg. It contains many neurons which conduct impulses both ways. Those neurons conducting impulses from the foot up to the spinal cord are very long, originate in the foot, have their cell bodies just outside the spinal cord, and terminate inside the spinal cord. By using a radioactive tracer injected into the cell bodies, Ochs found that there is active protein synthesis in the cell bodies. The newly synthesized protein was being transported in both directions down the neuron, away from the cell body. The rate of transport was about 400 millimeters per day, many hundreds of times faster than possible by simple diffusion. Energy was required for the transport. If a local region of the neuron was deprived of energy by treatment with cyanide or anoxia, the protein could not cross the region and piled up behind the block as behind a dam.

What happens to the protein when it reaches the neuron terminus? There is evidence that it crosses the synaptic cleft and enters the muscle or neuron that is innervated, where it exerts a trophic or "feeding" function. Neurons are incredibly active protein synthesizers—a single neuron synthesizes 30 percent of its own weight of protein per day. Thus we arrive at the concept of a nervous system whose components are interlocked not only by means of nerve impulses, but also by a vast process of chemical communication which involves proteins and perhaps other macromolecules as well. Our present understanding of this trophic interaction is still embryonic, but its significance cannot be overestimated.

The Peripheral Nervous System

So far we have examined the general properties of neurons and their mode of interaction with each other. We now take a look at the larger structural relationships of the human nervous system and the specialized functions of the various parts. As is obvious from Figure 5-3, there is a large and natural division of the nervous system into the central nervous system (CNS, brain plus spinal cord) and the peripheral nervous system. As is also obvious from Figure 5-3, the CNS is the first to originate in embryonic life, in the form of a hollow tube formed by skin folds along the back. The peripheral nervous system is a later outgrowth of the CNS and never loses its connection with the CNS.

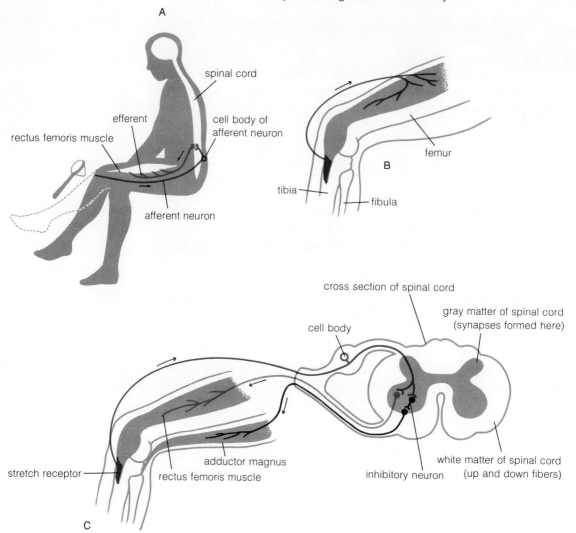

Figure 5-10 THE KNEE JERK REFLEX. (A) Excitatory elements of knee jerk reflex. (B) A hypothetical, "simplified" reflex. (C) More complete representation of the knee jerk reflex.

Traditionally, neuroanatomists have recognized two divisions of the peripheral nervous system: the *voluntary* nervous system and the *autonomic* nervous system. Voluntary nerves are usually described as peripheral nerves that are under conscious control, while autonomic nerves are described as nerves beyond the reach of such conscious control. Voluntary nerves exist in two distinct functional forms. *Sensory* or *afferent* fibers relay information to the CNS about the external environment: heat, pain, pressure, and so on. The CNS correlates this information and then sends impulses down the *motor* or *efferent* fibers of the voluntary nerves which terminate on the skeletal muscles of the body.

Voluntary Nerves and the Nerve Reflex

The complex of afferent neuron, CNS correlation, and efferent neuron often functions in a predetermined, predictable manner. Such a complex is called a *nerve reflex*. The archetypal example of a reflex is the knee jerk. The basic elements of this reflex are illustrated in Figure 5-10A. If the leg is allowed to dangle freely and a sudden stretch is given to the tendon just under the kneecap, the leg jerks forward. The tendon contains a receptor sensitive to stretch, which activates the afferent neuron. The afferent neuron conducts an impulse to the spinal cord, where the impulse crosses a single synapse to activate the efferent neuron which in turn stimulates the upper thigh muscle (the *rectus femoris*), causing the leg to jerk forward.

Now, if you were an engineer told to design a leg with a knee jerk, you might well object that this is a roundabout and inefficient way of causing the knee to jerk. Why not build the much more economical system shown in Figure 5-10B, in which the stretch receptor activates a neuron with a direct connection to the upper thigh muscle? The answer of course is that the knee jerk was not built into the leg to be elicited by tapping with a rubber hammer. The knee jerk per se is a trivial response of a reflex primarily designed for the act of crouching. When you crouch, the tendon with its stretch receptor is stretched, and reflex contraction of the rectus femoris keeps you from collapsing downward into a squatting position. However, notice that the rectus femoris has an antagonistic muscle running along the inside and back of the thigh (the *adductor magnus*). This is the muscle that would be used to flex the knee and, while the rectus femoris is putting forth a major effort, any contraction by its antagonist would be inappropriate. Hence part of the knee jerk reflex involves the inhibition of the adductor magnus as shown in Figure 5-10C. At the same time that the afferent fiber makes an excitatory synapse with the efferent to the rectus femoris, it also makes a synapse with an inhibitory neuron in the spinal cord which in turn synapses with the efferent to the adductor magnus and inhibits it. In fact, there are additional synapses not shown in Figure 5-10C that convey impulses up to the brain and out to other parts of the body. This sophisticated circuitry makes it essential that all afferent impulses go first to the spinal cord, where the available switchboard facilities can be used to notify all relevant tissues of the development.

Autonomic Nervous System

When we look at the autonomic nervous system, we find a similar system of afferents and efferents organized into reflex arcs. However, both the inputs and outputs are concerned with events *inside* the body, such as the action of the heart, the intestines, and the lungs. Both anatomically and functionally, the autonomic nervous system responds to two body states: war and peace. The *sympathetic* nerves of the autonomic nervous system enter the spinal cord in the thoracic (chest) and lumbar (abdominal) areas. This system dominates during emergency situations: a state trooper's flashing signal appearing in your rearview mirror, awakening to see a dark shadow standing over your bed, or facing a sudden charge by a German shepherd. There is a common and unmistakable response: your heart pounds, you take deep breaths, and your stomach seems tied into a knot. There are other changes in your body: the level of blood sugar rises, the bronchioles (small air ducts) of the lungs open wide, and the arterioles of the skeletal muscles dilate. Your body is preparing you for physical combat or for immediate flight. In a civilized society, the sympathetic response is not always adaptive. A pounding heart does not improve a public speech, a job interview, or an exam.

The *parasympathetic* nerves enter the CNS in two widely separated areas: the cranium (skull) and the sacral area (pelvic part of the spinal cord). The parasympathetic nerves antagonize the effects of the sympathetic nerves. The heart is slowed down, blood pressure falls, and the stomach and intestines are activated. This system is designed for the conservation of energy. Perhaps the best time to observe it in action is after a large Thanksgiving meal (assuming you like the people at the table). A languid, full, sleepy feeling steals over the body and the nicest thing to do is remain seated in your chair and listen to someone else telling a story. It is perhaps significant that in so many cultures the feast is the central act of a peace treaty.

Biofeedback. The word "autonomic" means self-governing, and it was at one time thought that the autonomic nervous system worked independently of the central nervous system, although it was anatomically obvious that the two were interlinked. Nevertheless, it was conventional wisdom that body functions such as heart rate, control of blood pressure, and rate of intestinal contractions were beyond conscious control. Then, during the late 1960s, Neal Miller at Rockefeller University astonished the scientific world with reports that he had taught rats to change all these functions on demand. Finally, in a tour de force, he trained his rats to blush at will—in one ear but not in the other!

Miller's reasoning was simple: the rat can be rewarded for responding in the desired way only if the experimenter knows from minute to minute how its internal organs are responding to the rat's commands. Consequently, Miller used instrumentation that monitored the performance of internal organs and displayed this information in the form of flashing lights or buzzing sounds which could be coupled to suitable rewards for the rat. This technique is called *biofeedback*, and has now been widely extended to people. (Of course, humans generally do not need a further reward.) Various research groups have succeeded in such medically important tasks as reduction of blood pressure, decrease of pulmonary air resistance (important in asthma), and reduction of the pain of migraine headaches. The last-mentioned was an unexpected side effect following training in raising the temperature of the hands. Temperature sensors are attached to the hands and forehead. (The hands are generally about 10°F cooler.) The subject is told to repeat autosuggestive phrases such as, "I feel very quiet. . . . My arms and hands are heavy and warm. . . . My whole body is relaxed and my hands are relaxed and warm. . . ." These phrases are designed to help the patient concentrate on the task at hand. Success is signaled by a quiet buzzer tone. It was discovered that as the temperature of the hands was raised, the headaches disappeared or became less acute in 75 percent of the patients. After several months of practice, the patients could achieve the same result without the aid of the biofeedback device.

Many other internal functions show a similar susceptibility to control, from brain waves to intestinal contractions. One of the most astonishing has involved training male volunteers to raise their scrotal temperatures by several degrees. This immediately suggests a novel method of contraception, since a prolonged testicular temperature rise of a few degrees is sufficient to inactivate or kill sperm-producing cells. In the future, hot pants may not be an exclusively feminine fashion!

The Central Nervous System—Vegetative Functions

It is a truism that the human brain is far and away the most complex organ in the human body. Perhaps less widely appreciated is the enormous biological effort that goes into the making of the human brain. My daughter is 4½ years old, and today she proudly received her first diploma, the one from nursery school. She does not realize yet how many more graduations there are in store for her. Though she is already in many ways a complex human being, she could not survive for a week if she were abandoned. Yet a deer, a robin, or a silversides of the same age would be in its prime. The reason a human being is so helpless at 4 is because it is so omnipotent at 24. The human brain in its fully developed form simply cannot be built in much less than 20 years, and of course in functional terms it never stops developing.

Brain size increases rapidly after birth; its growth slows down after age 10, and maximum brain size is reached at about age 20.

Newborn brain	350 grams
1 year	700 grams
3 years	1,000 grams
10 years	1,200 grams
20 years	1,350 grams

Thereafter there is a gradual decrease in the size of the brain, both in terms of total weight and number of neurons present.

Brain weight is proportional to total body weight, and therefore tends to be higher in men than in women. This is not reflected in intelligence. Within large limits, high intelligence may be associated with great variation in brain size. Thus the Russian writer Ivan Turgenev had a brain capacity of just over 2,000 grams, while the French writer Anatole France, equally well known, had a brain capacity just short of 1,200 grams. These represent the extremes. It seems that, once brain size falls below 1,000 grams, mental retardation follows, though again size by itself is not a reliable indicator of brain function. Many mental retardates have brains of normal or above-average size.

The basic plan of the central nervous system is already well-established in a 5-week-old embryo (Fig. 5-11). The brain is seen as a tubular, flexed extension of the spinal cord and is divided into three sections, the forebrain, midbrain, and hindbrain. While the brain of a vertebrate such as a reptile clearly retains these divisions in its fully developed form, the adult human brain obscures these fundamental divisions because of a monstrous development of the forebrain. The cerebral hemispheres, arising as lateral outgrowths of the forebrain, push out sideways and to the back to such an extent that the

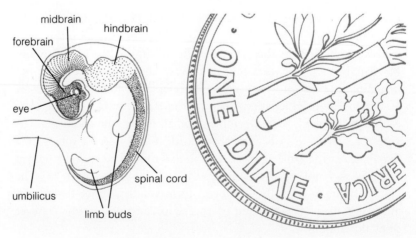

Figure 5-11 THE BRAIN AND SPINAL CORD IN A 5-WEEK HUMAN EMBRYO. A dime is drawn for comparison.

THE CENTRAL NERVOUS SYSTEM—VEGETATIVE FUNCTIONS / 151

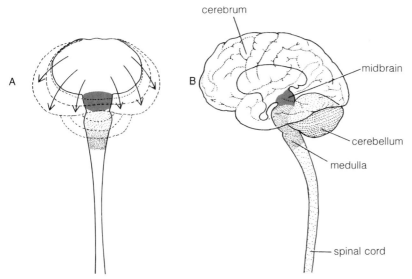

Figure 5-12 DEVELOPMENT OF THE HUMAN BRAIN. (A) Top view of developing brain. (B) Adult brain in hemispherical section.

rest of the brain is hidden underneath (Fig. 5-12). The adult brain still retains its tubular organization—it contains inner chambers called *ventricles* which are filled with fluid, the *cerebrospinal fluid*. The human brain is 85 percent water (it contains more water than the same volume of blood does) and is so jellylike that it collapses of its own weight when standing free. One of the functions of the cerebrospinal fluid is to act as an internal shock absorber.

The Hindbrain

In general, the hindbrain has retained more of its ancient functions than have other parts of the brain. The hindbrain consists of the *medulla* ("inner part"), basically a thickened extension of the spinal cord, and the *cerebellum* ("little brain"), an outgrowth on the back. Although the hindbrain of mammals still does approximately the same things it did in reptiles and fish, it has come under greater forebrain control.

The Medulla and Pons. The medulla carries a lot of through traffic of neurons coming up from the spinal cord and going to the upper centers. Along with adjacent portions of the brainstem, it also sends out 12 pairs of *cranial nerves*. Four of these nerves are part of the parasympathetic system. The cranial nerves are concerned with basic body functions such as transmission of sensory information for sight, smell, and hearing. They are involved in the act of swallowing,

152 / *The Organism: Nervous System*

in tongue movement and speech, in regulation of heart rate and vasomotion, in moving the eyeball in its socket, and so on. Besides sending out such cranial nerves, the medulla also contains control centers which regulate breathing, heart rate, and vasoconstriction. Because of the enormous importance of these body functions, major injury to this part of the brain leads inevitably to death.

The Cerebellum. Pushing outward from the brainstem is the cerebellum. It has been called the silent brain, because electrical stimulation of this area produces neither sensation nor movement. However, any damage to this region immediately shows up as a disturbance of balance and muscular control. The cerebellum is the organ that refines and coordinates all muscular movement. Watch your hand as you sign your name. This apparently simple act requires the orchestration of dozens of muscles of hand and arm. With split-second timing they accelerate, decelerate, cooperate, and antagonize, and do it all in a way that is characteristic of you and of no one else. Immediate orders for muscular movement emanate from the motor cortex of the cerebrum (which we will discuss shortly), but these orders undergo continuous review and, if necessary, correction by the cerebellum. The internal circuitry of the cerebellum is now relatively well understood, and neurophysiologists feel that the cerebellum will be the first section of the brain to be understood in its entirety.

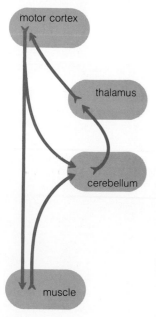

Figure 5-13
ROLE OF CEREBELLUM IN CONTROL OF MUSCLE MOVEMENT.

A simplified picture of the cerebellar role is given in Figure 5-13. A motor neuron in the motor cortex fires, sending an impulse to the appropriate muscle. However, the neuron also sends a collateral impulse to the cerebellum. When the muscle responds to the signal, its muscle spindle reports back to the cerebellum about its state of contraction. (The muscle spindle is an internal receptor analogous to the stretch receptor illustrated in Figure 5-10.) Furthermore, these signals go to exactly the same region of the cerebellum that is being stimulated by the motor cortex. The cerebellum can thus compare the "intention" and the "performance," calculate the error, and send back correcting impulses to the motor cortex, via the thalamus. In a cat, the whole circuit takes about 1/100 second, but in a human it is slower, about 1/50 second. Such high-speed cerebellar function is characteristic of birds and mammals, but not of reptiles and amphibians, and has unquestionably contributed to the success of the former on earth.

The cerebellum thus modifies an evolving train of muscle movement. This is why we can bring a coffee cup rapidly to our lips without crashing it into our front teeth, or shake a hand in greeting without dislocating a thumb or finger. The principle involved is one of continuous feedback—it is the same principle that is used by a target-finding missile. The cerebellum never initiates a muscular movement, just as a missile, hopefully, does not decide by itself when to take off.

THE CENTRAL NERVOUS SYSTEM—VEGETATIVE FUNCTIONS / *153*

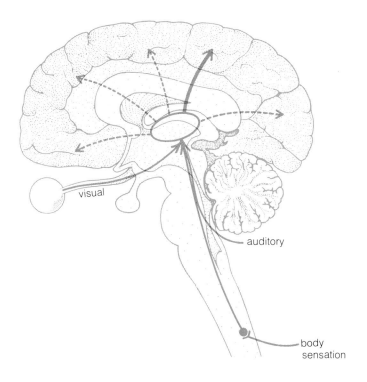

Figure 5-14 THALAMUS.

The Midbrain

The midbrain has fared poorly in the evolutionary sweepstakes. It represents the highest level of consciousness in fish, in which the cerebrum is concerned mostly with olfaction and the midbrain with visual stimuli. In mammals, the midbrain largely contains ascending and descending tracts carrying neuronal traffic to centers above and below. In mammals, it is still involved with the sense of sight in that it gives rise to cranial nerves controlling the movements of the eyeball in its socket. It also contains relay stations for visual impulses. However, the actual interpretation of visual stimuli has been moved to the cerebral cortex in the forebrain.

The Forebrain

The forebrain is by far the largest and most complex portion of the human brain. The full understanding of this shadowy structure will require generations of work by neurophysiologists and may perhaps never be achieved. We will look in very general terms at the functions of three areas of the forebrain: the thalamus and reticular activating system, the hypothalamus, and the cerebral cortex.

Thalamus. The *thalamus* ("enclosed chamber") is an area of gray matter located deep in the center of the forebrain (Fig. 5-14). The gray matter of the CNS is made up of neuronal cell bodies and is rich in synapses, while the white matter consists of myelinated axon fibers running from one place to another. The *reticular activating system* or RAS (reticular means "netlike") is a diffuse system of fibers and cell bodies which runs through the brainstem from the medulla up to the lower thalamus (Fig. 5-15). The thalamus and the RAS serve, to some extent, similar functions.

The thalamus acts as a sensory screen which is essential for the act of concentration. Think of yourself at this moment. You are sitting down some place, reading this book. Perhaps the radio or the television set is on, and you can hear people talking. There is furniture in your room which you can see. You are aware of a tremendous number of tactile sensations: the feel of the chair beneath you, the feel of the clothes on your body, the feel of the book in your hand. Perhaps there is an odor of cooking in the room. And yet, if this book has been successful, you have not been conscious of all these external stimuli, since your mind has been concentrating on an object that is neither visible nor audible. This is an example of concentration, or discrimination, and it is an ability that is literally essential to survival. A driver who window-shops while negotiating a busy street is a menace to both himself and to others.

How does the thalamus accomplish its task of screening out irrelevant information? All sensory information, such as touch, heat and cold, pain, sight and sound, reaches our consciousness via nerve terminations in the cerebral cortex. But before a sensory nerve such as a touch receptor can reach the cortex, it makes a synapse in the thalamus (Fig. 5-14). As pointed out earlier, such synaptic pathways are very susceptible to either inhibition or augmentation. The degree of suppression can be sometimes extreme. For instance, a soldier in the midst of battle may be unaware of a shrapnel wound which under calmer conditions would cause agonizing pain. Traditional medicine has long exploited this property of the thalamus in using counter-irritants to relieve pain. It is possible that acupuncture may be explained in similar terms; the needles may cause stimulation of pain pathways from the skin that depress other pain pathways via inhibitory action at synaptic relays in the thalamus.

The RAS of the brainstem is concerned with the related functions of sleep and wakefulness. When you are asleep, normally adequate stimuli such as quiet conversation, music, or subdued light do not register consciously. This is because of synaptic inhibition in the RAS centers. However, strong stimuli such as the sound of an alarm clock or a vigorous shake of the shoulder can overcome the

THE CENTRAL NERVOUS SYSTEM—VEGETATIVE FUNCTIONS / 155

RAS inhibition and reach your consciousness, leading to awakening. In the conscious or awake state, the RAS allows passage of essentially all external stimuli, because the RAS itself is activated. Activation of the RAS may come either from below, by strong sensory stimuli, or from above, from the cortex itself.[3] An example of the latter is the individual who tosses and turns in bed and cannot fall asleep because of vivid thoughts roiling his mind. The opposite problem may plague the turnpike driver; the mind is lulled by a hypnotic movement of the road, and there isn't sufficient physical sensation to counteract the mental becalming. Under such conditions it is good to remember that muscular movement is one of the most powerful activators of the RAS. Stopping the car and taking a short walk, or even the physical act of eating a fruit or a sandwich, generally provide sufficient activation to restore the wakeful state. The same measures should also be effective during a boring lecture!

Extensive physical damage to the RAS (as occurs in sleeping sickness), results in an incurable coma. With less severe damage, a

3. The cortex referred to is the cerebral cortex, discussed on page 160ff.

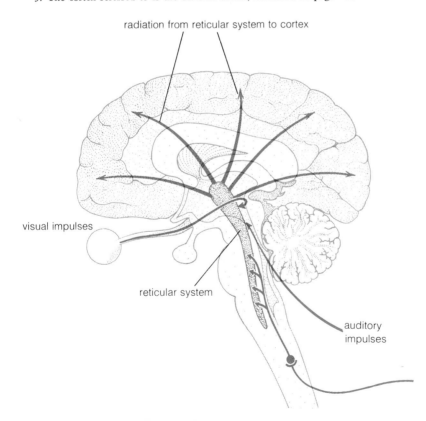

Figure 5-15 THE RETICULAR ACTIVATING SYSTEM.

strong stimulus still results in arousal. Barbiturates cause sleep by depressing the RAS. Thus, in summary, both the thalamus and the RAS act as gates for sensory input into the cortex. The thalamus is a selective gate which allows for discrimination and concentration, while the RAS is an all-or-none gate which allows for either sleep or wakefulness. Why is sleep necessary? It is said that Thomas Alva Edison made do with only 4 hours a day, and that this perhaps gave him the extra time needed to invent the light bulb. The fact is that no one knows why we need sleep, beyond the fact that it is the brain and not the rest of the body that suffers during sleep deprivation. Prolonged sleep deprivation produces a psychotic state.

Hypothalamus. The *hypothalamus* ("below the thalamus") has been given an unfortunate name. Though physically it lies just below the thalamus, functionally the two have very little relation to each other (Fig. 5-16). While the thalamus functions as a kind of receptionist for the higher centers, the hypothalamus runs a domain of its own. It is responsible for three related areas of activity: (1) it controls an assortment of vegetative functions of the body, (2) it is the source of a number of basic drives and emotions, and (3) it is an important organ of endocrine secretion.

Hypothalamic control of vegetative functions is exercised through the autonomic nervous system and is shared with major centers in the medulla (such as the breathing center, the heart rate center, and the vasoconstriction center). Some of the vegetative functions controlled by the hypothalamus include temperature regulation, the light response of the pupil, and contraction of the urinary bladder. All these functions are mediated by the autonomic nervous system (for example, vasodilation of skin arterioles as a response to heat).

The drives and emotions mediated by the hypothalamus include the basics: fright, anger, pain, pleasure, the sex drive, thirst, and hunger and satiation. Missing are such allegedly human instincts as altruism, curiosity, aesthetic appreciation, and romantic love. Evidence for such localization of drives and emotions is based on electrical stimulation of selected areas of the hypothalamus. For example, electrical stimulation of the lateral hypothalamus produces a full-blown rage in an animal. Thus, a dog will attack and bite the experimenter and anything else within reach. A cat will savagely attack a rat that it would have ignored before the delivery of the stimulus. It can be shown that this is not just a defensive reaction to an unpleasant or painful stimulus.

Such aggressive behavior can be inhibited by electrical stimulation of an adjacent region of the forebrain, the *caudate nucleus*. One experimenter implanted a radio-controlled electrode in the caudate nucleus of a bull and allowed the bull to charge. When the bull was a

THE CENTRAL NERVOUS SYSTEM—VEGETATIVE FUNCTIONS / 157

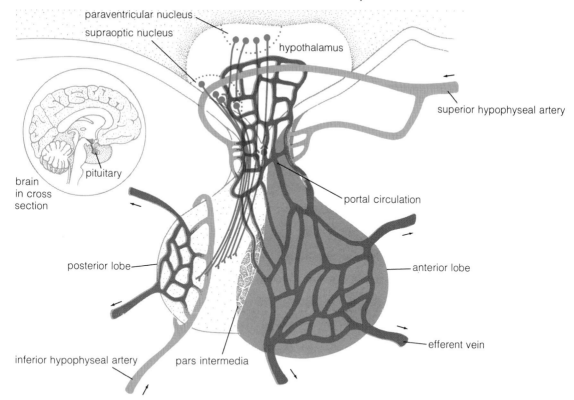

Figure 5-16 HYPOTHALAMIC-PITUITARY CONNECTIONS. Neuronal connections: the posterior lobe receives neurons from the hypothalamus. The neurons transport the hormones oxytocin and vasopressin via axonal fast transport. The hormones are then stored in the posterior lobe prior to release. Circulatory connections: note that posterior lobe has a separate blood supply. All blood reaching the hypothalamus exits via veins from the pituitary. The portal circulation carries hypothalamic hormones.

few feet away, the scientist delivered a radio signal to the brain of the bull, causing the animal to screech to a stop.[4] Similar experiments have been performed with free-ranging baboons. Remote-control interference in the brains of such animals has resulted in permanent changes in their dominance relationships within the troop.

One of the most interesting hypothalamic areas was discovered by James Olds in 1953. This is the so-called *self-stimulation area*. Technically, this area is found in the lateral hypothalamus, only millimeters away from the area that induces rage behavior. When rats were allowed to press a lever that delivered a mild stimulus to the self-stimu-

---

4. Caudate stimulation does not function as a specific inhibitor of aggressive behavior. When this area of the brain is stimulated, the animal stops whatever it has been doing and holds its position, like an arrested movie. The function of the caudate nucleus in humans is obscure. Nevertheless, it illustrates the strong veto power that higher centers of the brain have over basic emotions such as rage.

lation area, they behaved as though such stimulation was the ultimate reward. Rats would press the lever 5000 times an hour (more than once a second) for many hours at a time. They would ignore food even when desperately hungry. While it is hard to delve into the feelings of a rat, its behavior suggests that the ultimate feeling of pleasure is expressed through the region of the hypothalamus.

The hypothalamus contains a separate hunger and satiation center. If the satiation center of a rat is destroyed, the animal starts to eat and never becomes satiated. As this behavior continues, the animal may reach monstrous proportions, so that its legs barely touch the ground. Conversely, destruction of the hunger center produces an animal that loses all interest in food, even though it may waste away to a fur-covered skeleton.

Cells in the hypothalamic thirst center respond to increased osmotic pressure of local circulating blood and give rise to the feeling of thirst. As water is drunk, it enters the circulatory system, lowers the osmotic pressure, and thus abolishes the feeling of thirst. The picture of the human being under hypothalamic control is not an attractive one, but it must be remembered that all these drives are essential for physical survival. And we have a cortex which exercises a powerful restraint on the hypothalamus. In a nuclear age, this restraint must be strengthened.

Finally, the hypothalamus is more than a neurophysiological phenomenon—it is also an organ of hormonal secretion.[5] This should not come as a surprise, if we remember that all neurons normally secrete synaptic transmitters as well as higher-molecular-weight proteins which act as trophic substances. The hormones of the hypothalamus are polypeptides—short chains of amino acids. Once more, their concern is with the regulation of basic body processes: sex, growth, metabolic level, water retention, birth, milk production, and resistance to stress. It has been known for a long time that emotional states can have an effect on body functions. For example, an emotionally deprived infant grows poorly and may be stunted. All sorts of emotional factors influence menstruation, fertility, and the sex drive. Likewise, environmental influences may affect body function, for example, prolonged exposure to cold increases the body's basal metabolic rate. But it has been only very recently that the central role of the hypothalamus in all such phenomena has been understood. Various difficulties have stood in the way of experimental work. One was the peculiar nature of the hypothalamic blood supply (which will be described shortly). Another has been the extremely minute amounts of hormone involved. Thyrotropin releasing factor (TRF) one of

---

5. Hormones are molecules that act as internal messengers, affecting the physiological response of appropriate target organs.

the hypothalamic hormones, has recently been purified and synthesized. In order to isolate and purify this substance, no less than 5 million sheep hypothalamuses were required. The yield was only 1 milligram of TRF, approximately the weight of one leg of a fly!

The Pituitary. A central feature of the hormonal role of the hypothalamus is the fact that it is mediated by means of a second brain appendage, the *pituitary* (Fig. 5-16). When the pituitary was discovered by Vesalius in the 1500s, he thought it had something to do with the phlegm or mucus of the nose and throat. The word "pituitary" means "phlegm." Of course, the pituitary has nothing to do with phlegm at all. Biologists, who are very sensitive to such inaccurate terminology, have substituted the more accurate if noncommittal term *hypophysis,* which means "to grow under." However, we will stick with the earlier term just because it is more deeply entrenched in common usage.

The pituitary is actually a compound structure. The anterior part (the adenohypophysis) is glandular in nature and is embryologically derived from the roof of the mouth. The posterior part (the neurohypophysis) is neuronal in nature and develops as an outpocketing of the brain.

As can be seen from Figure 5-16, the pituitary is anatomically pressed under the hypothalamus, and for once anatomical juxtaposition reflects a functional linkage as well. But the two parts of the pituitary are linked to the hypothalamus by very different mechanisms. The anterior lobe is linked by means of the circulatory system. The arterial blood reaching the lower regions of the hypothalamus does not rejoin the circulatory system directly by means of an efferent vein. Instead, it flows to the pituitary by means of small *portal veins* which break up into capillaries on reaching the anterior lobe. Thus they act as gates between two capillary systems. Anatomically speaking, this is a very "privileged" circulatory connection between two organs, and the only other place such a linkage occurs in the human body is between the small intestine and the liver (see Chapter 4, page 128). The portal veins carry hypothalamic hormones to the anterior lobe. While all hormones of the body spend at least part of their lifetime in the blood, it is only the hypothalamic hormones that are carried in the blood directly to their target organs. One result of such direct transmission is a great increase in efficiency. All the hypothalamic hormones reach their target cells directly and in full force, instead of being diluted in the general circulation.

None of the hypothalamic blood reaches the posterior lobe of the pituitary. Instead, the posterior lobe is connected to the hypothalamus by means of the nervous system (Fig. 5-16). Two hormones, oxytocin and vasopressin, are made in the cell bodies of the neurons located in the hypothalamus and are then transported by fast axonal transport to the posterior lobe.

The hormonal role of the hypothalamus is one of the most actively researched areas of physiology today, and therefore the list given in Table 2 (page 180) may be expected to change, perhaps even before this book is printed. For a closer look at the hypothalamic hormones, see Appendix C, pages 178–186.

The Central Nervous System—The Cerebral Cortex

The cerebral cortex, more than any other attribute, distinguishes human life from all other life on earth. It is by far the largest structure in our brain and, though there is universal appreciation of its importance, it remains the most mysterious part of our nervous system.

The evolutionary trend among the vertebrates has been to assign an ever more central role to the cerebral cortex. A decerebrate frog can hop around, feed itself, and behave in a way that is hard to tell from normal. A decerebrate cat can stand up, walk, and chew food placed in its mouth, but behaves in a robotlike, unresponsive way. A decerebrate human is blind, deaf, and unable to move a muscle in its body spontaneously. Such an individual is like a newborn infant, without the infant's responsiveness. With proper care, it can live for 30 years or longer, at a massive cost to its family or to society, and without ever achieving more than a dim approximation of consciousness.

Structurally, the cerebrum contains white matter in the interior regions and gray matter (cell bodies) on its surface, or cortex. In order to pack in more cortical material within a fixed space, the surface of the human cerebrum shows deep folds and fissures (Fig. 5-17). The larger fissures act as anatomical markers, dividing the cortex into several lobes. While many of the most important cortical functions cannot be assigned to a fixed anatomical area, some functions can be so located. Thus the back of the occipital lobe gives rise to the sense of sight, a region on the temporal lobe is involved with hearing, and an interior area is involved with the sense of smell. Another area receives sensory information from the skin. Next to it lies an area that controls voluntary muscle movements.

If one stimulates the sight area in the occipital lobe with a fine electrode, the individual sees flashes of light or simple colors but no complete visual images. Yet destruction of this area leads to complete blindness, as though the eyes had been plucked out. The visual cortex receives information from nerves originating in the retina, and it is here that electrical information from the eyes is interpreted as the

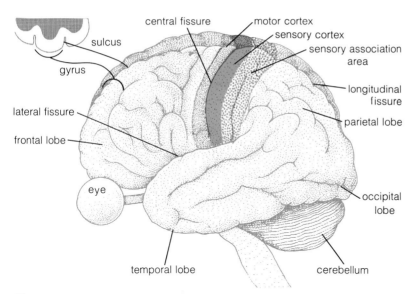

Figure 5-17 ANATOMY OF THE CEREBRAL CORTEX.

sense of sight. A similar function is served by the hearing and smell areas. It is worthwhile remembering that there is no physical correspondence between, say, our sense of sight and the physical object impinging on the retina, the photon. Thus, if we could anatomically connect the nerves from the ear with the visual cortex and the nerves from the eye with the auditory cortex we would "see" sounds and "hear" light!

Motor and Sensory Cortex

The fissure between the frontal and parietal lobes (called the central sulcus) separates the *motor cortex,* to the front, from the *sensory cortex,* to the rear (Fig. 5-17). The motor cortex gives rise to the neurons (called *motor neurons*) that run to the skeletal muscles and execute voluntary movements. This does not mean that, say, the decision to eat an apple originates in the motor cortex. Nor does it deny that other parts of the brain (such as the cerebellum) contribute importantly to voluntary movements. However, all the necessary impulses sent to muscles of the hand, jaw, and tongue must leave the cortex from this area. If a tiny area of the motor cortex is stimulated electrically, a specific muscle or group of muscles will contract.

Interestingly, the area of the motor cortex that controls the index finger is right next to the area that controls the middle finger, which in turn lies next to the area that controls the ring finger, and so on. As a result, one can map the motor cortex in the form of a "little man," or homunculus, with the size of the body parts proportional to

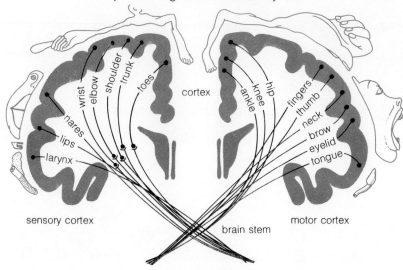

Figure 5-18 THE MOTOR AND SENSORY HOMUNCULI. Motor and senory nerves to different areas of the body follow separate pathways through brain stem and spinal cord. Adjacent areas on motor and sensory cortex correlate with adjacent regions of the body.

the motor area controlling them (Fig. 5-18). Such a homunculus has grossly swollen hands, lips, tongue, and facial areas. This reflects the enormous human importance of the hands as organs of manipulation and the lips and tongue as organs of speech. Electrical stimulation of the hand and finger area results in the twitch of a single muscle or even a bundle of fibers within the muscle. Stimulation of the areas controlling the upper arm or trunk causes gross movements involving up to 30 to 50 muscles acting as a group.

In human evolution, the muscles of the hands and fingers have been designed for mechanical tasks requiring the greatest skill and coordination. The muscles of the face, lips, and tongue play a similar important role in the tasks of communication. But the human cortex is remarkably plastic and, if necessary, it can learn to adapt other portions of the body musculature to accomplish a given task. An interesting example can be seen in the art of Matisse (Fig. 5-19). The etching on the left is about 5 inches high and was done in the 1930s. It shows the intelligence, vibrancy, and nervous energy characteristic of Matisse. The picture on the right was painted in 1952, when Matisse was an old man but still breaking new ground artistically. The picture is about 5 feet tall and was painted with large brushes moved by muscles of the arm. The oranges were cut out of paper and glued on. It is remarkable that the Matisse style and intelligence are still unmistakably present, in spite of the fact that entirely different muscles and cortical areas have been called on to express them. This

THE CENTRAL NERVOUS SYSTEM—THE CEREBRAL CORTEX / 163

Figure 5-19 Left: Matisse etching, 5″, 1930s. Right: Matisse drawing, 5′, 1952.

illustrates the subservient nature of the motor cortex to other centers of the brain.

The sensory cortex, lying just behind the central sulcus, provides us with important information on which part of the body is being stimulated. As with the motor cortex, the sensory cortex can be mapped in the form of a homunculus (Fig. 5-18). The large lips and hands and fingers of the sensory homunculus indicate that these regions of the body are the most sensitive to external stimuli (especially for the sensations of light touch, pressure, and position). This localization of sensation is made possible by the fact that sensory neurons from adjacent parts of the body are kept separate as they go up the spinal cord, synapse in the thalamus, and enter the sensory cortex. If the sensory cortex is totally destroyed, one can still distinguish the type of sensation being received, for example, pain, heat, cold, touch. However, there is little discrimination with regard to intensity and point of origin of the stimulus. This residual diffuse sensation is mediated by the thalamus.

Behind the sensory cortex lies the *sensory association area* (Fig. 5-17). Here some of the more complex qualities of sensation are appreciated. Impulses from the thalamus, from the brainstem, and from the sensory cortex might say something like, "The fingers and hand are being stimulated." The sensory association area may go on to say, "The arm and fingers are extended away from the body. The fingers are touching two round objects. The objects are sitting on top of each other." Finally, with the help of other information and other association areas of the brain, one might arrive at the concept, "The hand is holding two tennis balls." The nature of such an interpretative process remains one of the most inscrutable aspects of the brain. It involves a synthesis of present sensory information, past memories and experiences, and rational faculties of the brain in a mix that the most sophisticated electronic computer can only faintly approximate. The various association areas responsible for this occupy a major portion of the cerebral cortex.

Memory and Learning

Only slightly less mysterious are the related functions of *memory* and *learning*. The ability to remember underlies a great deal of higher mental function, and there is no doubt that the accuracy, speed, and range of our memory influences our entire personality structure through our perception of and response to any given situation. Both experience and psychological experiments tell us that there are two types of memory: short-term memory and long-term memory. Short-term memory may involve looking up a phone number and remembering it long enough to dial it, or walking down a crowded street and briefly remembering all the faces, store windows, parked cars, pigeons, and street noises that make up a kind of stream of consciousness. The great bulk of such impressions seems to vanish from the mind without a trace, but sometimes an impression registers more intensely and can be recalled months and years afterward. Thus the great psychiatrist Carl Jung describes his earliest memory:

> I am lying in a pram, in the shadow of a tree. It is a fine, warm summer day, the sky blue, and golden sunlight darting through green leaves. The hood of the pram has been left up. I have just awakened to the glorious beauty of the day, and have a sense of indescribable well-being. I see the sun glittering through the leaves and blossoms of the bushes. Everything is wholly wonderful, colorful and splendid.[6]

---

6. C. G. Jung, *Memories, Dreams, Reflections*. Pantheon, New York, 1961, p. 6.

At the time he wrote of this memory, Jung was an old man. He had had the initial experience when he was 2 or 3 years old. There is nothing extraordinary about the scene that Jung remembers—what seems more striking is the intensity of his reaction to it. It thus appears that, once a memory trace has been "fixed" in the brain, it can persist there for a lifetime.

But why should the brain make this distinction between short-term memory and long-term memory? Wouldn't it be good to retain everything we experience? We would have to read a textbook or hear a poem only once to retain a perfect impression. Perhaps such a brain would not be as desirable as it sounds, and our ability to forget may act as a protective mechanism. The Russian neurophysiologist A. R. Luria describes the case of a man who had extraordinary powers of memory; there was nothing he could not remember, and nothing he could ever forget. In 2 or 3 minutes, he could learn a table of 50 numbers or a list of 70 words, which he could repeat or equally easily present in the reverse order. If given a word from the middle of the list, he could recite the others in a backward or forward direction from this point. Without difficulty he memorized the nonsensical formula

$$N \times \sqrt{d^2 \times \frac{85}{vx}} \times \sqrt[3]{\frac{276^2 \cdot 86x}{n^2 v \cdot \pi 264}} \times n^2 b = sv \frac{1624}{32^2} \cdot r^2 s$$

Fifteen years later, without warning or intervening exposure, he was asked to reproduce it and did so faultlessly. Jerome Bruner, in a foreword to Luria's book,[7] suggests that this man behaved as though his short-term memory process was metabolically defective and all impressions were fed directly to his long-term memory. Did this man achieve fame in some intellectual field? He failed as a musician and as a journalist and wound up as a performing mnemonist. He had great difficulties in ordinary life. He lived in a world of intense visual imagery which overwhelmed him with an endless stream of perceptions. He was unable to think in abstract or general terms. He had great difficulty in planning, because he could not withdraw sufficiently from ongoing reality. He had a faulty sense of time, since events of many years ago seemed to him as vivid as those of yesterday. Thus it appears that our ability to forget may be as important to intellectual function as our ability to remember.

What is the physiological basis of memory? The evidence available on which to build a generalized theory is skimpy at best, but

---

7. A. R. Luria, *The Mind of a Mnemonist*. Basic Books, 1968.

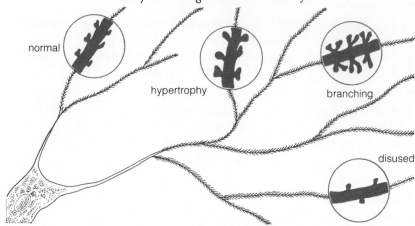

Figure 5-20 PLASTICITY OF DENDRITIC SPINE SYNAPSES. On the left is shown a dendrite of cortical neuron with normal dendrite spines. These are thought to function as synaptic junctions. On the right are shown postulated changes in dendritic spine synapses.

suggestive leads have been found. One is that short-term memory may be based on establishment of reverberating (circulating) circuits in the cortical neuron. This is consistent both with anatomical observations, which show that such circuits exist, and with electrophysiological observations which show that when a localized area of the cortex is stimulated electrically increased electrical activity continues in the area for some time afterward.

The basis of long-term memory appears to be somewhat different. S. H. Barondes, then of Albert Einstein School of Medicine, has shown that if a learned behavior is to be transferred to long-term memory, protein synthesis must take place. Barondes used mice, which were taught a simple light-associated maneuver. After the mice had learned the maneuver, Barondes administered acetoxycycloheximide, a chemical that interferes with protein synthesis. If the chemical was administered before or immediately after learning, the maneuver was forgotten when the mice were tested 7 days later. However, if the inhibitor was administered more than 30 minutes after learning, the memory had been already consolidated and could not be erased.

How is protein synthesis related to the memory trace? The most economical hypothesis is that the newly synthesized protein is used in the building of new synapses or in the modification of existing ones (Fig. 5-20). The most likely places for such synapses are the dendritic spines of cortical neurons. While no microscopic evidence exists that new synapses are formed with the usage of a particular neuron, there is evidence that synapses are lost with disuse. Thus, when mice are raised in darkness, neurons in the visual cortex show regression of dendritic synapses.

The human brain is a spectacular creation. There are altogether too many topics we have not been able to touch on in this brief chapter—language, the nature of consciousness, the differences between the right and left brains, disease states such as epilepsy and schizophrenia, comparisons with other high intelligences such as that of the dolphin. To quote John Eccles, "It is often said that all the good scientific questions have been answered. . . . For the brain at least this is not true. I think that all the best scientific questions are still waiting."

Appendix A
The Sense Organs

The Eye

We are visual creatures who receive more environmental information through the eyes than through any other sense organ. A single human eye contains an estimated 100 million light-receptor cells (called *rods* and *cones*) in its retina (Figs. 5-21 and 5-22). In addition, the retina

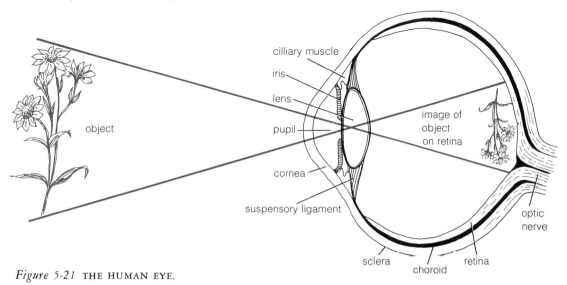

Figure 5-21 THE HUMAN EYE.

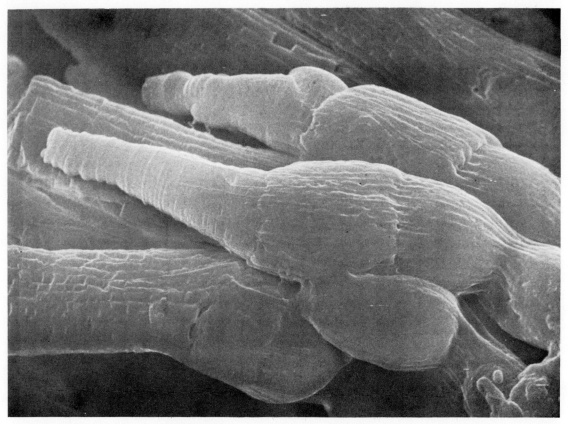

Figure 5-22 RODS AND CONES OF THE RETINA. Two cones with tapering cell bodies are visible in the front. The rods are in the background.

contains a built-in *signal-processing system* made up of neurons which are both excitatory and inhibitory in nature. Activity in one rod or cone influences responses in adjacent pathways, so that there is enhanced information about edges, shadows, movements, and the forms of objects. Of course, further information processing occurs in the brain. Finally, the eye resembles a camera in that it contains a light-focusing device, the *lens,* and a light-metering device, the *iris.*

Retinal cones are far less numerous than rods, but they respond to light of various wavelengths and are therefore responsible for color vision. The rods respond only to light or dark. They are phenomenally sensitive; a single photon of light triggers a response in a rod. Consequently, rods are used in night vision.

The chemistry of the visual process has been worked out by George Wald of Harvard University and his co-workers. A photon of light interacts with a conjugated protein called *rhodopsin* and causes it to undergo a chemical change. Rhodopsin is present in the membrane of the receptor cell and, after undergoing the chemical change, it opens an ionic gate in the membrane, which triggers a nerve impulse. The nerve impulse must then cross at least two synapses before reaching the optic nerve and passing on to the brain. During this time it undergoes synaptic processing.

The *lens* of the eye is quite different from any lens found on a camera. The camera lens focuses by moving backward and forward. The lens of the eye focuses by changing its own shape through lateral muscular action (Fig. 5-21). The thicker the lens, the closer the object being brought into focus.

The *iris* of the eye acts like the diaphragm of a camera in controlling the amount of incoming light. The small opening in the middle is called the *pupil* and, as the pupil moves from pinpoint size to its maximum opening, the amount of light entering is increased by a factor of 40. The opening or contraction of the iris occurs in a fraction of a second and is controlled by the autonomic nerves. In addition to the iris, the eye has a slower mechanism for light and dark adaptation which can adapt it for much greater variations in light and dark. This mechanism involves the retina itself and requires several minutes. It is the mechanism you use when entering a dark movie theater. After a few minutes, your retina has increased the amount of rhodopsin in each visual rod, making your eyes more light-sensitive and allowing you to see an empty seat.

The Ear

In animals such as dolphins and bats, the ear is the dominant sensory organ, supplying the kind of dense information flow that the eyes do in humans. However, this does not necessarily mean that their hearing is more acute than ours. Sound consists of alternate waves of compression and decompression traveling through the air. The human ear can respond to compression-decompression rates as slow as 20 cycles per second (very low pitch) to as high as 20,000 cycles per second (very high pitch). The human ear shows its greatest sensitivity in the range of 1,700 cycles per second. In this range, a sound may be audible if it displaces the eardrum only 1×10^{-9} centimeter. That is much less than the width of a molecule. On theoretical grounds, it seems unlikely that any other animal's ear could be more sensitive than this because if it were, the molecular vibrations of the hearing apparatus itself would constitute a stimulus, and the animal would hear a constant hum. Thus again, as in the case of the eye, the human ear approaches the theoretical limits of sensitivity.

The human ear consists of three parts: the outer and middle ears are filled with air, while the inner ear is filled with fluid (Fig. 5-23). The middle ear is not a closed chamber but connects via the *eustachian*

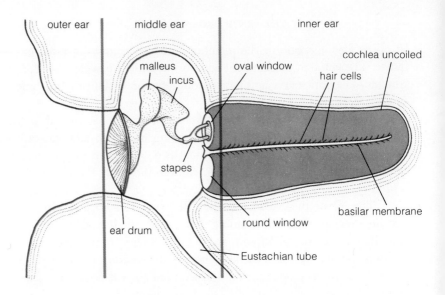

Figure 5-23 SCHEMATIC VIEW OF THE HUMAN EAR. The cochlea has been drawn straight instead of coiled for clarity. The three semicircular canals of the inner ear are not shown.

tube to the upper throat. This tube opens every time one swallows, and thus keeps the pressure in the inner ear equal to atmospheric pressure. Otherwise the pressure differential across the eardrum could stretch it or even cause a rupture.

The *eardrum* is connected by means of three tiny bones (malleus or hammer, incus or anvil, and stapes or stirrup) to the *cochlea*. The cochlea constitutes the inner ear and generates the nerve impulses that travel to the brain. As a sound wave arrives at the eardrum, the area of compression pushes it inward, and the subsequent area of decompression pulls it out again. This sets up a vibration in the eardrum, which is efficiently transferred by the ear bones to a sort of inner eardrum on the cochlea, called the *oval window*. Vibrations of the oval window set up vibrations in the fluid filling the cochlea. The cochlea is divided longitudinally by a long *basilar membrane* which does not quite reach the end of the cochlea. Since fluids are incompressible, the pressure waves in the cochlear fluid must be relieved by pushing out on a second membrane called the *round window*. If the vibrations have a long wavelength (corresponding to a low frequency), they travel easily through the fluid, reach the end of the cochlea, and cross through the basilar membrane at its furthest end. The basilar membrane contains *hair cells* which translate mechanical vibrations into nerve impulses. Impulses from the far end of the basilar membrane are interpreted in the brain as low pitched tones. Higher-pitched sounds have a shorter wavelength. When they strike the oval window, the vibrations induced in the cochlear fluid travel only a short distance before crossing the basilar membrane and traveling back to the round window. As the hair cells in this region are stimulated, they again give rise to nerve impulses which are interpreted as high-pitched sounds in the brain. The intensity of sound is measured by the amplitude of displacement of the hair cells.

One of the proofs for the above theory is the fact that high-intensity sounds can destroy hair cells in the basilar membrane. If the sound is low-pitched, destruction of hair cells occurs at the far end. The animal becomes deaf to low-pitched sound. A high-pitched sound likewise destroys the proximal region of the basilar membrane, and the animal becomes deaf to high-pitched sounds. There is evidence that prolonged exposure to such noise sources as jet planes, heavy trucks, chain saws, rock bands, and pneumatic hammers can lead to partial hearing loss. Few urban dwellers can escape such hazards.

The inner ear also contains a series of *circular canals*. These have nothing to do with hearing but function in balance and in conveying information about body position.

Taste and Smell

These are called the chemical senses, because their receptors are excited by chemical stimuli. Taste receptors are excited by water-soluble chemicals in food, while smell receptors are excited by volatile chemicals in the air. In practice, the two show a certain degree of overlap; for example, in drinking a cup of fine coffee we are aware of taste as well as aroma.

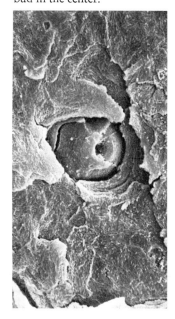

Figure 5-24
A surface view of the tongue of a rat showing a papilla with a single taste bud in the center.

The taste receptors are arranged in *taste buds* on the tongue. Each taste bud contains numerous receptor cells which project very thin hairlike receptors called microvilli through a central pore into the mouth. The olfactory receptors are very similar in structure and function but are scattered singly across the moist inner surfaces of the nose.

There are four basic taste sensations: sweet, salty, sour, and bitter. Different taste buds show relative though not absolute differences in sensitivity to these sensations. The overall taste of a food reflects the ratio of four basic tastes. The four primary taste sensations provide chemical information about the food. Sweetness receptors are stimulated by sugars, and essentially all natural foods that taste sweet are safe and nutritious. Salt receptors respond to ionic compounds such as table salt. Sourness is a measure of acid content. The bitter taste buds have a protective function, since most of the poisonous components of wild herbs have a bitter taste.

The study of smell is in a less advanced state than that of any other sensory modality, but it has been proposed that smells, like tastes, are also composed of a few fundamental types. One proposed list includes the following fundamental smells:

		EXAMPLES
1.	camphorlike	mothballs
2.	musklike	sex organs
3.	flowerlike	lilac
4.	peppermintlike	peppermint
5.	etherlike	gasoline
6.	pungent	garlic
7.	putrid	rotten meat

In humans, the sense of smell shows a high degree of accommodation; when you enter a room filled with cooking odors or a room with tobacco smoke, you are at first highly conscious of the smell but quickly lose the ability to detect it. This accommodation occurs at the level of the brain neural pathways, not in the olfactory receptors themselves. In humans, odors also show strong masking effects. This is the idea behind the use of perfume.

Other Body Senses

There is a popular misconception that humans possess only five senses (sight, hearing, taste, smell, touch), though some psychic individuals presumably possess a "sixth" sense. While physiologists disagree about the true number of human senses, there is universal agreement that it is much higher than five. Some of the more important body senses include:

1. The proprioceptive sense ("beginning in the self"). This is the sense that allows you to know where every part of your body is located at any given moment, even though your eyes may be closed. It is mediated by stretch receptors located in the tendons and muscles of your body (Fig. 5-25).

2. Pain. The receptors for this sense are free nerve endings (Fig. 5-25). Pain receptors respond to substances released from cells when the cells are damaged. There is surprisingly close agreement among different individuals as to what constitutes a

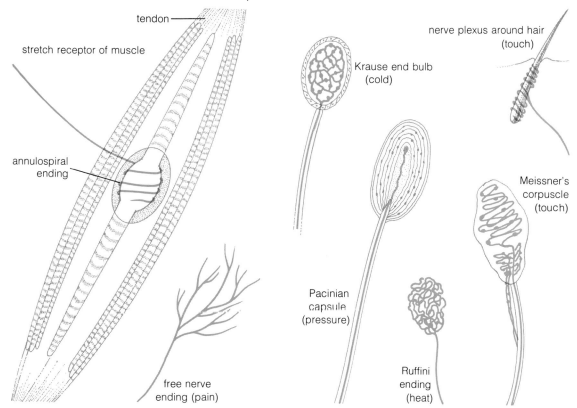

Figure 5-25 THE SKIN AND MUSCLE SENSES.

painful stimulus. For instance, when heat was used to cause tissue damage, almost all persons tested felt pain when tissue temperatures reached 44° to 46°C (111° to 115°F). This is a very narrow range. Where people differ enormously is in their reactions to a stimulus perceived as painful.

3. The skin senses. Each of the following has been shown to be transmitted by means of anatomically different receptors (Fig. 5-25): heat, cold, pressure, and touch.

Thus the human body has evolved to respond to a tremendous range of environmental information. Unfortunately, we have changed the environment by introducing entirely new stresses for which our evolutionary history has provided no receptors. We can receive a lethal dose of radiation and not feel it. We can breathe lethal concentrations of carbon monoxide and not be aware of it. The list of silent killers is a long one, and it is not diminishing.

Appendix B
Muscle

Some animals, like the electric eel, have done "stunning" things with muscle, converting it into an electric shock generator, but human muscle does essentially one thing—it contracts.[8] While, evolutionarily speaking, our muscle may be rather noncreative tissue, such evolutionary stability may be a sign of perfection. Muscle makes up 40 percent of our bodies by weight, and lies at the very basis of our animal existence. It is the one basic tissue that plants do not possess, and it has made all the difference.

While all muscle in our body is contractile, three distinct types may be found that show some specialization of function (Fig. 5-26). *Smooth muscle* is found in internal organs such as the intestines and the uterus. It has a simple cellular structure and is capable of slow but prolonged contraction without fatigue. *Striated muscle* arises by a fusion process, as many precursor cells merge into a single long *muscle fiber* containing many nuclei; such muscle is capable of extremely rapid contraction, but it fatigues quickly. It is used to move all bones, thus providing locomotion. *Cardiac muscle* is found only in the heart and has properties that are in some ways intermediate between smooth and striated muscle. It does not show fatigue, yet contracts rapidly. Though still composed of identifiable single cells, each cell is so tightly linked to its neighbors that the entire mass contracts as a unit. Cardiac muscle contracts spontaneously, without a nervous or hormonal stimulus.

A basic feature of muscle contraction is that it can exert force in only one direction. Therefore, in order to achieve movement of the skeleton, muscles are always arranged in antagonistic pairs (Fig. 5-10). How does the contraction process take place? Our basic ideas come from the work of H. E. Huxley of Cambridge University. It appears in all cases that the contraction process results from two kinds of proteins sliding past each other (Fig. 5-28). Each muscle fiber

---

8. A few, such as the Purkinje fibers of the heart, have become specialized for the conduction of action potentials.

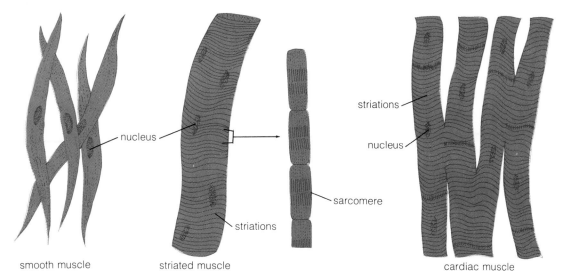

Figure 5-26 THREE TYPES OF MUSCLE TISSUE.

consists of innumerable contractile units called *sarcomeres* ("flesh units") laid side by side. The central portion of the sarcomere is called the A band and consists of filaments of the thick protein *myosin*. The peripheral areas of the sarcomere are called the I bands and consist of filaments of the thinner protein *actin*. Actin and myosin interdigitate. When both energy (for example ATP) and a starting signal (for example, calcium ions) are present, the two kinds of protein begin to slide past each other. As a result, the A band retains its original dimensions, but the I bands shrink, producing overall contraction (Fig. 5-28).

What causes the sliding motion of actin and myosin? Under very high electron microscope magnification, myosin fibrils are seen to have small knobs projecting out from their axis. It is proposed that these knobs or bridges interact with specific attachment points on the actin fibers, moving them along by a ratchet mechanism. During the process, ATP is broken down and its energy is used to fuel physical movement.

The contractile process is initiated by a nerve impulse. Since the motor neuron branches at its axonic tip, a single nerve may make contact with as many as 100 muscle fibers. The point of contact is called a *neuromuscular junction*. Like the nerve-nerve junction, this junction is a synaptic junction. Muscle cells, like nerve cells, have electrically charged cell membranes which can be depolarized. The incoming axon terminal releases acetylcholine which depolarizes the muscle cell membrane and starts an action potential. The key result of the action potential is that Ca^{2+} ions are allowed to cross the cell membrane and enter the muscle fiber. It is the Ca^{2+} ions that constitute the ultimate trigger for the contractile process—they make it

Figure 5-27
PHOTO OF GUINEA PIG STRIATED MUSCLE.

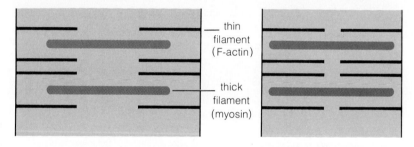

Figure 5-28
ROLE OF ACTIN AND MYOSIN IN MUSCLE CONTRACTION.

possible for myosin to use ATP to power the sliding process. Now, the muscle fiber is relatively enormous in diameter (up to 100 microns). If it contained a simple cell membrane, the Ca^{2+} ions could not diffuse into the center of the fiber very quickly to cause contraction of internal sarcomeres. Consequently, the membrane of the muscle fiber has an elaborate system of invaginating tubules called the T system (Fig. 5-29). Tubules of the T system reach deep into the center of the muscle fiber, coming into close proximity to cisternae (blind sacs) of the sarcoplasmic reticulum. These cisternae are filled with Ca^{2+} ions. When the cell membrane is depolarized, the depolarization travels along the T system deep into the fiber and triggers the release of Ca^{2+} from the cisternae. The Ca^{2+} ions are then further distributed internally, allowing all sarcomeres to contract at almost the same time. After the action potential is over, Ca^{2+} ions are rapidly pumped back into the cisternae, the sarcomeres relax, and the muscle is prepared for a new contraction.

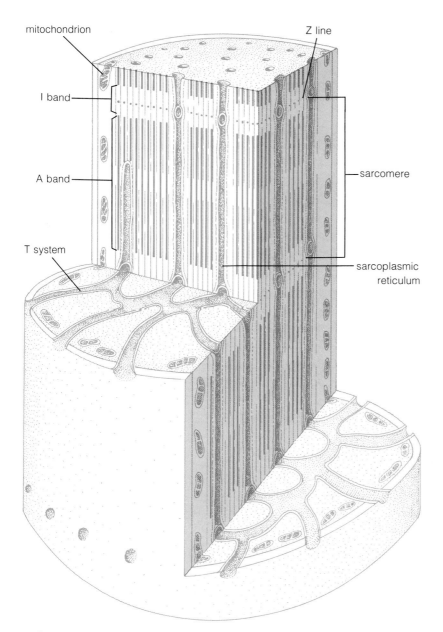

Figure 5-29 THE T-SYSTEM AND THE SARCOPLASMIC RETICULUM. Tubules of the T-system run transversely and open to the outside of the muscle fibril. The sarcoplasmic reticulum consists of tubules running lengthwise within the muscle fibril. It represents a modified endoplasmic reticulum. The T-system and the sarcoplasmic reticulum come in very close contact with each other but are not continuous. Both function in the distribution of calcium ions needed during muscle contraction.

Appendix C
The Hormonal System

Hormones are chemical messengers, "agents of excitation" as indicated by the Greek derivation of the term. The hormonal system shares with the nervous system the task of integration and coordination of body function. Both systems function in an "intelligent" way—that is, both incorporate the principle of feedback regulation in their operation. A message is dispatched from a control center and elicits a change in the activity of the target system. Information about the activity of the target system is fed back to the control center, causing it to increase or decrease its stimulus intensity.

The two systems differ largely in their speed of response. The nervous system controls rapidly changing activities such as striated muscle movements. Hormones in general regulate metabolic activities of a more leisurely sort: growth, digestive processes, mineral balance, metabolic level, and so on. A plant, such as an oak tree, lacks muscle tissue and also lacks a nervous system. But after one has begun to appreciate how truly sophisticated and extensive an oak tree's hormonal system is, one might well be reminded of the animist belief that the oak tree has a soul. We will look very briefly at plant hormones in Chapter 9. Here we restrict ourselves to the major hormones of the human body.

The concept of the hormone has been traditionally restricted to chemical messengers carried in the bloodstream, into which they have been secreted by the *endocrine,* or ductless glands (as opposed to *exocrine* glands, such as the salivary glands, whose secretory product does not enter the bloodstream). The major endocrine glands are shown in Figure 5-30. However, there are two other classes of compounds that meet the definition of a hormone. (1) The neurohumors are compounds such as acetylcholine, norepinephrine, serotonin, and

THE HORMONAL SYSTEM / 179

GABA. They are released from the axon terminal of cell A and stimulate the postsynaptic membrane of cell B. (2) The "second messengers" such as cyclic AMP, carry messages not from one cell to another but from one cellular region to another, all within the same cell. Table 2 lists some of the major hormones of the human body.

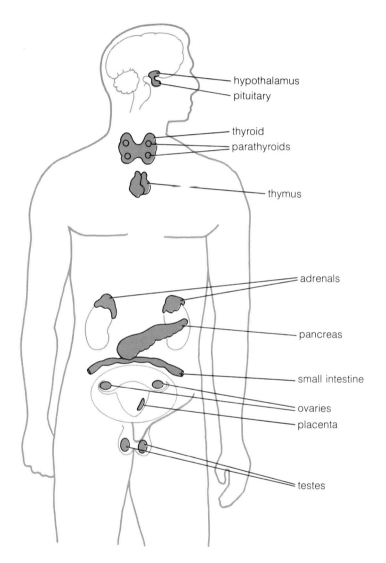

Figure 5-30 MAJOR ENDOCRINE GLANDS.

TABLE 2 *Hormones of the endocrine glands*

HORMONE	ORIGIN	FUNCTION
CRF (corticotropin releasing factor)	Hypothalamus	Stimulates corticotropin release from anterior pituitary
FSH-RF (FSH-releasing factor)	Hypothalamus	Stimulates FSH release from anterior pituitary
GRF (growth hormone releasing factor)	Hypothalamus	Stimulates GH release from anterior pituitary
LRF (luteinizing hormone releasing factor)	Hypothalamus	Stimulates LH release from anterior pituitary
PRF (prolactin releasing factor)	Hypothalamus	Stimulates prolactin release from anterior pituitary
TRF (thyrotropin releasing factor)	Hypothalamus	Stimulates thyrotropin release from anterior pituitary
GRIF or somatostatin	Hypothalamus	Antagonizes GRF
PRIF	Hypothalamus	Antagonizes PRF
Oxytocin	Hypothalamus	Uterine contractions, milk let-down
ADH antidiuretic hormone (or vasopressin)	Hypothalamus	Water conservation through reduced urine output
ACTH or corticotropin (adrenocorticotropic hormone)	Anterior pituitary	Stimulates synthesis of steroid hormones by adrenal cortex
FSH (follicle stimulating hormone)	Anterior pituitary	Sperm production in males, growth of Graafian follicle in females
GH (growth hormone)	Anterior pituitary	Stimulates growth, elevates blood sugar, stimulates protein synthesis
LH (luteinizing hormone or ICSH)	Anterior pituitary	In males, stimulates testosterone production; in females, induces ovulation, converts follicle to corpus luteum, stimulates secretion of progesterone by corpus luteum
Prolactin	Anterior pituitary	Milk production
TSH thyrotropin (thyroid stimulating hormone)	Anterior pituitary	Stimulates thyroxin production by thyroid gland
MSH (melanocyte-stimulating hormone)	Pars intermedia of anterior pituitary	Skin pigmentation
Thyroxin (T_3 and T_4) or thyroid hormone	Thyroid	Increases metabolic rate; required for normal growth
Thyrocalcitonin (calcitonin)	Thyroid	Lowers blood Ca^{2+} level
PTH parathormone (parathyroid hormone)	Parathyroids	Raises blood Ca^{2+} level, lowers blood PO_4 level

HORMONE	ORIGIN	FUNCTION
Pancreozymin	Small intestine	Stimulates pancreas to produce digestive enzymes
Secretin	Small intestine	Stimulates pancreas to produce $NaHCO_3$
Aldosterone	Adrenal cortex	Necessary for Na^+ retention by kidney, increases body fluid
Cortisone	Adrenal cortex	Elevates blood sugar, counters inflammation
Epinephrine	Adrenal medulla	Alarm reaction: increased heart rate, blood sugar, BMR
Norepinephrine	Adrenal medulla, nerve endings	Vasoconstriction
Glucagon	Alpha cells of pancreas	Raises blood sugar, decreases liver glycogen
Insulin	Beta cells of pancreas	Necessary for entry of glucose into body cells
Renin	Kidney	Elevation of blood pressure via angiotensin production
Estrogen	Graafian follicle of ovary, placenta	Development of female secondary sexual characteristics
Progesterone	Corpus luteum of ovary, placenta	Thickens uterine lining, development of breast alveoli, prevents ovulation, relaxes uterus
Testosterone	Testes	Development of male secondary sexual characteristics
Chorionic gonadotropin	Placenta (chorionic membrane)	Maintains corpus luteum during pregnancy
Prostaglandins	Seminal vesicle	Contraction of smooth muscle
The neurohumors		
Acetylcholine	Presynaptic nerve terminals	Membrane depolarization for postsynaptic nerve, skeletal and smooth muscle; membrane hyperpolarization for heart muscle
Norepinephrine	Presynaptic nerve terminals	Membrane depolarization for heart muscle; membrane hyperpolarization for intestinal smooth muscle
GABA	Presynaptic nerve terminals	Membrane hyperpolarization for postsynaptic nerve
The second messengers		
Cyclic AMP	Inner edge of cell membrane	Activates the phosphorylating enzyme protein kinase; effect depends on nature of target cell; for adrenal cortex → cortisone synthesis, corpus luteum → progesterone synthesis
Cyclic GMP	Inner edge of cell membrane	Counteracts effects of cyclic AMP in competent cells

Instead of examining all of the above,[7] we shall take time here to look only at three hypothalamic hormones. The three hormones involved (GRF, TRF, and oxytocin) illustrate the central role of the hypothalamus in the control of vegetative functions of the body. We will also see that this hypothalamic control is exercised through the agency of other hormones and endocrine glands.

If the list of abbreviations for hypothalamic hormones looks like a United States government phone directory, let us make some generalizations. (1) All hypothalamic hormones are small peptides. (2) The hormones which control the anterior lobe of the pituitary are called *releasing factors* or release-inhibiting factors, and they either stimulate or inhibit the production of separate and distinct hormones by cells of the anterior lobe. (3) Hormones synthesized by the anterior lobe are large proteins. While some, like GH, prolactin, and MSH, produce direct effects on target cells, most act by stimulating still other hormone-producing organs. (4) Hormones associated with the posterior lobe are released from the posterior lobe by a nerve stimulus, and then produce direct physiological effects on target organs. We will now look at three hypothalamic hormones in greater detail to illustrate these points.

GH is a protein with a molecular weight of 21,500. It is made up of 188 amino acids joined in a chain with two cross-links; in 1970 this protein was totally synthesized in a laboratory. It is the largest protein ever synthesized.

As implied by its name, GH stimulates growth, probably through increased protein synthesis. The synthetic product has the same activity in the body as native GH. Though it is obviously not yet an item on the drugstore shelf, synthetic GH means that a body height of 4½ or 5 feet need no longer be accepted as an act of fate. Conversely, gigantism can now be forestalled by treatment with the sex hormone DES, as described in Chapter 6.

The levels of GH, rather unexpectedly, fluctuate greatly in the body. They are highest during sleep and during protein starvation. The normal plasma GH level is about 2 milligrams per liter during the day. This rises to 10 to 15 milligrams per liter during sleep, and even higher during protein starvation. (A milligram is one thousandth of a gram.) Thus it appears that, in terms of growth, 1 hour of sleep equals about 6 hours of wakefulness. The higher levels of GH present during protein starvation do not necessarily lead to increased growth, since protein is needed as a raw material.

GH secretion is controlled by two hypothalamic hormones: GRF and GRIF. But exactly what stimuli the hypothalamus responds to remains a mystery. As mentioned before, emotional deprivation in-

---

7. The sex hormones will be further considered in Chapter 6.

hibits GH production. Protein deficiency increases it. However, normal growth requires other hormones as well, particularly thyroid hormone and insulin.

Having considered the relationship between the hypothalamus and body growth, let us now discuss the relationship between the hypothalamus and *basal metabolism*. (Basal metabolism is the rate at which the resting body consumes food and produces energy. It represents the energy expenditure required to stay alive.) The responsible hypothalamic hormone is TRF, which, like GRF, is transported to the anterior pituitary by the portal circulation. TRF is one of the smallest hypothalamic peptides known—it consists of only three amino acids linked together. TRF represents the beginning of a cascade of stimulus and response which reaches from the hypothalamus through the pituitary and the thyroid until ultimately every cell of the body is brought into response (Fig. 5-31).

In response to TRF stimulation, the anterior pituitary produces the hormone *thyrotropin* ("turning on the thyroid"). Thyrotropin is one of a class of hormones from the anterior pituitary that stimulates other endocrine glands to produce hormones. Thyrotropin, as well as the trophic hormones FSH and LH, are large proteins which contain carbohydrate side groups.

Thyrotropin leaves the pituitary in the general circulation and is carried to the *thyroid* ("shieldlike," named in reference to the thyroid cartilage which looks like a shield and lies above the thyroid gland). The thyroid is a butterfly-shaped gland which lies just below the Adam's apple and wraps around the windpipe.

Since thyrotropin is a large protein, it cannot cross the cell membrane of the thyroid cells (Fig. 5-32). Instead, it complexes with a specific receptor on the surface of the cell membrane. The complexed receptor then increases the activity of an enzyme, *adenyl cyclase*. Though the exact mechanism of this activation is unknown, a transducer protein is required. Figure 5-32 proposes a simple mechanical analogy. The transducer pushes the adenyl cyclase into contact with ATP when the receptor protein is complexed with thyrotropin. After activation, the adenyl cyclase converts cytoplasmic ATP (the same molecule that mediates most cellular energy transformations) into a cyclic derivative, cyclic AMP. Cyclic AMP then turns on the intracellular machinery that results in the synthesis of a new hormone, *thyroxin*.

The importance of the mechanism illustrated is enormous. Cyclic AMP functions not only in the thyrotropin-thyroxin interaction, but has been shown to be essential for dozens of other specific cellular functions, ranging from glycogen breakdown to muscle contraction to DNA and RNA synthesis. Cyclic AMP has become known as a

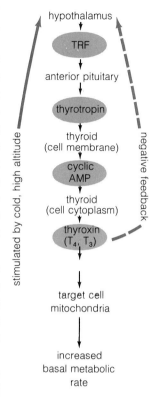

Figure 5-31
HYPOTHALAMIC CONTROL OF BASAL METABOLISM.

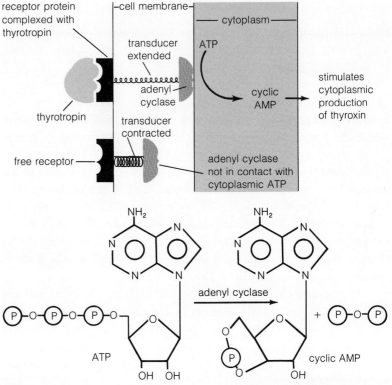

Figure 5-32 THE ROLE OF CYCLIC AMP IN THYROXIN PRODUCTION.

"*second messenger.*" In this instance, thyrotropin is the first messenger that can do no more than knock on the door and say, "I am here!" However, by the activities of adenyl cyclase, a single molecule of thyrotropin can generate hundreds of molecules or cyclic AMP per second, thus producing a tremendous biochemical amplification of the original signal. Of course, there also exist cytoplastic mechanisms for terminating the signal by destruction of the cyclic AMP.

The ultimate response to the signal depends on the intrinsic genetic properties of the cell in question; for example, a thyroid cell produces thyroxin, a muscle cell contracts, and a liver cell breaks down glycogen. The basic work on the cyclic AMP system was done by Earl Sutherland of Case Western Reserve University and Vanderbilt University, and in 1971 he was awarded the Nobel Prize in recognition.

To finish our story, the thyroxin produced by cells of the thyroid now leaves the cells and enters the general circulation as one more hormone. Chemically, thyroxin is unique in that it is an amino acid containing iodine. For this reason minute amounts of iodine are needed in our diet. Since local abundance may fall below the biological minimum, small amounts of iodine are added to salt to ensure that everyone receives an adequate supply.

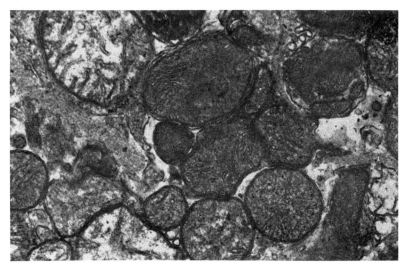

Figure 5-33 INTERACTION BETWEEN THYROXIN AND MITOCHONDRIA. Two types of mitochondria are seen. The ones with darker-colored cristae are in a more active metabolic state. Exposure of a cell to thyroxin results in the production of more mitochondria of this type.

The thyroxin now enters all metabolically active cells and speeds up their rate of energy production. Though the evidence is still incomplete, thyroxin seems to act on the mitochondria, causing them to speed up oxygen consumption and increase ATP production (see Chapter 1 and Fig. 5-33).

We have now seen a kind of hormonal "house that Jack built": This was the hypothalmus that produced the TRF that caused the anterior pituitary to produce the thyrotropin that turned on the thyroid cell membrane that caused the transducer to activate the adenyl cyclase that converted the ATP to cyclic AMP that carried the message to cytoplasmic machinery to produce the thyroxin that caused the target cell mitochondria to speed up their rate of energy production! Why is it all so complicated? Why doesn't the hypothalamus make thyroxin directly, as long as it is producing an amino acid derivative anyway? We see here one of the differences between a living system and an engineering system—the living system is a "wise" system in that it responds to many kinds of variables and then integrates this information to produce a response that best fits the situation. Each step in such a complex sequence represents a potential locus for regulation and integration. Unfortunately, the specific regulatory mechanisms affecting the basal metabolic rate (BMR) are only dimly understood. The BMR rises after prolonged exposure to cold or high altitudes. Rising levels of thyroxin inhibit thyroid function, probably by an effect on the hypothalamus (Fig. 5-31). High thyroxin levels lead to CNS stimulation—everyone knows the thin, jumpy, nervous person who can eat more than anyone else at the table and yet gains no weight because of a very high BMR. Presumably the "nervousness" is a manifestation of the negative feedback effect on the CNS and the hypothalamus. However, this is not the only site of control. A class of hormones known as *prostaglandins* affect a wide range of body functions by interfering with adenylate cyclase, the enzyme necessary for cyclic AMP formation. Presumably, as research proceeds, still further control sites will be discovered in the regulation of basal metabolism of the body.

Now let us consider one more hypothalamic hormone, *oxytocin* ("rapid birth"). Chemically oxytocin is similar to GRF and TRF in being a small peptide. It differs in two respects. It is transported to the posterior pituitary via axonal transport, and it exerts a direct effect on its target organs, the smooth muscle of the breasts and the uterus. No intermediate hormone is involved. This means that CNS control of oxytocin production is very pronounced. After being transported to the posterior lobe, the oxytocin is thought to be bound to some substance which prevents diffusion out of the lobe. When nerve fibers from the hypothalamus fire, a transmitter substance releases the oxytocin from its bondage and allows it to diffuse throughout the body. In the breasts, oxytocin causes contraction of smooth muscle around the alveoli, causing milk to be let down. To a nursing mother with full breasts, a baby's cry may be sufficient stimulus to release milk from the nipples (Fig. 6-1). There are many ways in which these fibers may be activated, for example, the act of suckling itself. This happens even in the woman who is not nursing. Thus, during sexual intercourse, stimulation of the breasts and other erogenous areas produces the same result. Of course, in this case the breasts usually contain no milk that can be ejected.

Oxytocin also causes contraction of the smooth muscles of the uterus, and during orgasm the high levels of blood oxytocin are thought to cause the rhythmic contractions of the uterus that help in the transport of sperm up to the oviducts. This effect of oxytocin on the uterus plays a vital role during the birth process. Though oxytocin levels in the blood are fairly high as pregnancy draws to a close, the effects of oxytocin are countered by even higher levels of blood progesterone (Chapter 6). As the progesterone level falls at the end of pregnancy, oxytocin starts to predominate and causes initial weak contractions of the uterus. As the baby's head is forced toward the cervix, the cervical ligaments are stretched and a strong nerve impulse is sent to the hypothalamus, causing even greater release of oxytocin, hence stronger contractions until birth results.

Incidentally, oxytocin is produced in males as well as females, but since males lack uteri and alveolized breasts, their hormone has no known function. Though we have not discussed all the hypothalamic hormones, the three we have looked at illustrate the range and importance of this organ, both as a hormonal and neuronal control center.

Bibliography

John C. Eccles, *The Understanding of the Brain,* McGraw-Hill, 1973. A Nobel-Prize-winning neurologist and a humanist looks at the brain. This little booklet is difficult at times, but rewarding.

Lewis Thomas, *The Lives of a Cell,* Viking Press, 1974. This small collection of essays touches on all areas of biology. The following excerpt comes from the discussion of the autonomic nervous system.

> Working a typewriter by touch, like riding a bicycle or strolling on a path, is best done by not giving it a glancing thought. Once you do, your fingers fumble and hit the wrong keys. To do things involving practiced skills, you need to turn loose the systems of muscles and nerves responsible for each maneuver, place them on their own, and stay out of it. There is no real loss of authority in this, since you get to decide whether to do the thing or not, and you can intervene and embellish the technique any time you like; if you want to ride a bicycle backward, or walk with an eccentric loping gait giving a little skip every fourth step, whistling at the same time, you can do that. But if you concentrate your attention on the details, keeping in touch with each muscle, thrusting yourself into a free fall with each step and catching yourself at the last moment by sticking out the other foot in time to break the fall, you will end up immobilized, vibrating with fatigue.
>
> It is a blessing to have options for choice and change in the learning of such unconsciously coordinated acts. If we were born with all these knacks inbuilt, automated like ants, we would surely miss the variety. It would be a less interesting world if we all walked and skipped alike, and never fell from bicycles. If we were all genetically programmed to play the piano deftly from birth, we might never learn to understand music.

Richard F. Thompson, *Introduction to Physiological Psychology,* Harper and Row, 1975. A very lucid, research-oriented text. There are frequent quotes from the original literature.

GENERAL READING

TEXTBOOKS

Figure 6-1 NURSING MOTHERS. Top: Peru; Bottom left: United States; Bottom right: Guinea.

chapter 6

The nineteenth-century French novelist Honoré de Balzac, though a great lover of women, was an even greater lover of literature. During his sexual activities he worked hard to arrive just at the border of bliss, without reaching an actual ejaculation. The reason, Balzac explained, was that every man was given only a finite amount of energy, and any energy spent on an orgasm was energy taken away from one of his books. Skilled though Balzac was, he was not always able to keep his resolution, and then he would stalk into the cafe, pound the table angrily, and shout, "I've just lost another book!"

Reproduction

Balzac was doing consciously something that a great deal of Western culture and society encourages us to do unconsciously—he was trying to sublimate his sexual energies by diverting them into other channels. Possibly a high-water mark in this regard was reached in Victorian England, where the young men and women were so savagely repressed that they were practically driven to go out and establish an empire.

The sex drive is unquestionably one of the most powerful drives in the human species. But *why* is so much biological energy invested in sex? The significance of sex is not to be found in the simple process of reproduction, in the sense of replacing an aged body with a new one. Such simple reproduction does not require the sexual process at all, and thousands, perhaps hundreds of thousands, of species in all five kingdoms can routinely reproduce asexually. Sex becomes significant in an evolutionary perspective because the sexual process more than anything else is responsible for generating the new variants on which evolutionary selection can act. These new variants are generated by the process of *meiosis,* or reductional cell division (Fig. 6-2), which we describe more fully in Chapter 7.

In this chapter we will look at sex and the reproductive process first on a cellular level, and then in the human reproductive organs. We will examine the production of sperm and eggs, and the process of fertilization, development, and birth. And finally, because sex is important to humans not just as a means of reproduction, we will look at some means of birth control.

Sex on the Cellular Level

Meiosis has been called the "chromosome chopper," since it generates new assortments of genes on the chromosomes. Thus the chromosomes you inherited from your parents are not the same ones you will pass on to your children. And from the point of view of the species, some chromosomal arrangements exist that endow their bearers with a better chance of survival and reproduction in their specific environment. This is therefore the evolutionary value of sex.

But if meiosis reduces the number of chromosomes per cell, there must exist a second mechanism that restores the original number. This is the process of *fertilization.* It follows that in cellular terms, the sexual process consists of a cycle of meiosis followed by fertilization. Meiosis takes place within the *gonads* (testes or ovaries); it establishes new gene assortments but reduces the chromosome number to half. Fertilization among the vertebrates requires two separate individuals and restores the original chromosome number. Two naked

SEX ON THE CELLULAR LEVEL / *191*

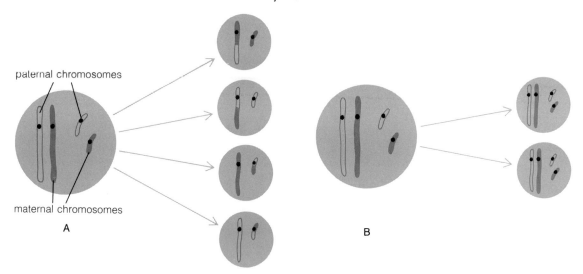

Figure 6-2 COMPARISON OF MEIOTIC AND MITOTIC CELL DIVISION. (A) Meiosis (reductional division). Note that the products of a meiotic division contain only half as many chromosomes as the parent cell, and their genetic content represents a scrambling of maternal and paternal inheritance. This type of cell division is used in sexual reproduction. (B) Mitosis (equational division). The products of mitotic cell division are genetically identical to the parental cell. This type of cell division is used in vegetative (asexual) reproduction.

cells or *gametes* ("spouses") are brought together and fused into a single cell, the *zygote* ("yoked together").

Now, this is not always a simple task. We saw in Chapter 4 that the cells of our body live in an environment that is tranquil and protective compared to the environment outside. Therefore the trick in fertilization is to arrange things so that the gametes, even of a land organism, experience a minimum of environmental shock during the time they are separated from the parental body. Since the gametes are single cells and usually miniscule in size, it is also desirable to provide them with some sort of homing device so that they can find each other. Such hormonal homing mechanisms are known in ferns, mosses, algae, and some primitive animals such as *Hydra*. There is no conclusive proof that they are used in human reproduction.

Meiosis is a universal process which occurs in essentially the same way in all organisms that reproduce sexually. But the manner in which different organisms have solved the twin problems of environmental shock and homing mechanisms during fertilization can differ enormously from one organism to another. It seems that evolution

Figure 6-3 AN EXAMPLE OF EXTERNAL FERTILIZATION: SIAMESE FIGHTING FISH SPAWNING. Prior to mating, the male (light color) builds a nest of saliva-covered bubbles. Eggs and sperm are released directly into the water. After the eggs have been fertilized in the water, the male picks them up in his mouth and spits them into the nest.

Figure 6-4 AN EXAMPLE OF INTERNAL FERTILIZATION: WALKING STICK INSECTS COPULATING. The smaller male aligns his genitalia with those of female by means of claspers. The process of sperm transfer requires several hours for completion.

SEX ON THE CELLULAR LEVEL / 193

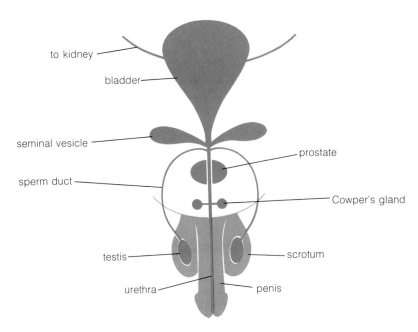

Figure 6-5 MALE REPRODUCTIVE ORGANS IN FRONT VIEW (SCHEMATIC).

has poured all its creativity into this aspect of the sexual cycle. The results often profoundly affect the overall life-style of the organism in question.

Perhaps the most severe environmental shock a gamete can experience is lack of water. Since the internal environment of every living organism is a watery one, a naked gamete plunged into the dry world of our land environment would immediately perish. Some of the most creative solutions in this area have been made by land plants (see Chapter 9). Among the vertebrates, two basic approaches are used. The older and simpler method is called *external fertilization* and involves the simultaneoues discharge of sperm and eggs into a body of water, where the sperm then fertilizes the eggs (Fig. 6-3). This is the method followed by most of the aquatic vertebrates—the fish and the amphibians. Even a garden toad (an amphibian), who spends the bulk of its life on land, must find a pond or other body of water for the act of fertilization.

All higher vertebrates and anthropods (even such unlikely ones as snakes, whales, and water bugs) practice *internal fertilization* (Fig. 6-4). The male introduces a copulatory organ into the body of the female and deposits the sperm deep inside, where the gametes find themselves in an environment little different from that of other body cells. We now look at the two aspects of the sexual cycle: the production of gametes in the gonads and the process of fertilization in the human body.

Production of Gametes and Sex Hormones

Testes

The *testes* of the human male develop deep inside the body, near the kidneys (Fig. 6-5). At birth, they descend through the *inguinal canal* into the *scrotum*, where they remain throughout adult life. The slightly lower temperature of the scrotum is essential for normal sperm production. The testes of a very young boy are already competent to produce sperm, but they lack the hormonal signal to do so. The signal to begin sperm production comes from the hypothalamus, and it first appears during puberty (ages 12 to 14). At this time, the hypothalamus begins to secrete releasing factors which stimulate the anterior pituitary to release the hormones FSH and LH (follicle-stimulating hormone and luteinizing hormone) (Fig. 6-6). FSH is then carried in the blood to the testes, where it stimulates testicular cells to begin sperm production. Sperm production is a typical meiotic process which results in the formation of four *spermatids* from a single germinal cell. Spermatids have half the number of chromosomes of the germinal cell, and they mark the completion of the meiotic process (Fig. 6-7). But in order to form viable sperm, the spermatids must first undergo a process of maturation which greatly reduces the amount of cytoplasm present and adds a long tail and a midpiece, resulting in a mature sperm. No other changes in nuclear content occur during the maturation process.

During the maturation process the sperm is redesigned for the long journey to meet and fertilize the egg. Head cytoplasm is reduced to a minimum, and the head is equipped with an *acrosome* ("apical body"), an organelle loaded with hydrolytic enzymes to allow the sperm to chemically penetrate a variety of tissues. (In fact, under the unnatural conditions of the test tube, sperm can penetrate a tremendous variety of cells other than egg cells.) The *midpiece* is the motor of the sperm loaded with mitochondria for ATP production. The *tail* of course is the organ of propulsion, pushing the sperm through the fluid film of the uterus and oviduct.

The sites of sperm production within the testes are the *seminiferous tubules* (Fig. 6-8). Germinal cells are positioned around the outside of each tubule, cells of later meiotic stages toward the middle, and mature sperm fill the centers of the tubules. The tubules coalesce into an *epididymis* (see page 207) and then a *sperm duct* or vas deferens. The seminiferous tubules have the curious property of intolerance to normal body temperature. They need the slightly cooler temperatures of the scrotum to function in sperm production. In many mammals with a definite breeding season, the testes descend into the

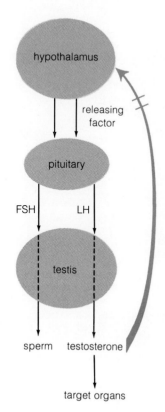

Figure 6-6
HYPOTHALAMIC CONTROL
OF SPERM PRODUCTION
AND TESTOSTERONE
SYNTHESIS.

PRODUCTION OF GAMETES AND SEX HORMONES / 195

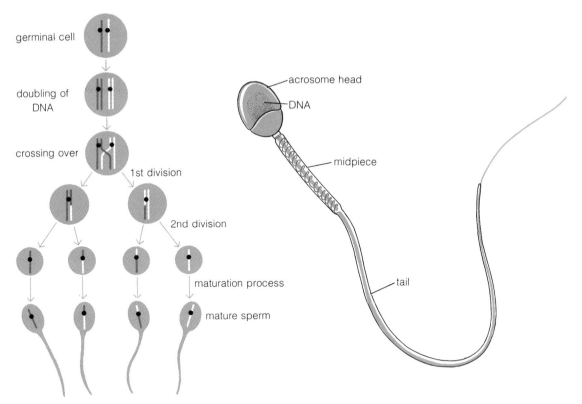

Figure 6-7 MEIOSIS AND SPERM PRODUCTION. The conversion of a spermatid into a sperm is a maturational process involving no change in nuclear material.

Figure 6-8 SEMINIFEROUS TUBULES AND INTERSTITIAL CELLS OF THE TESTIS.

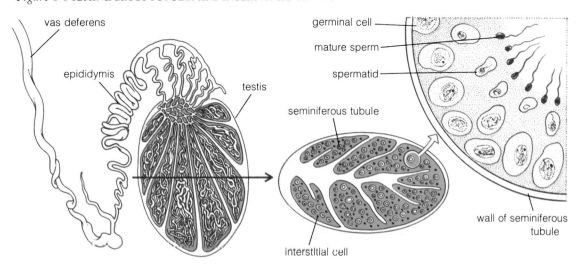

scrotum only during the breeding season and then return to the inside of the body for the rest of the year. Humans are sexually active continuously, and the testes stay put in the scrotum once they have descended. Undescended testes show atrophy of the seminiferous tubules; in this case, the male is sterile. The seminiferous tubules can also be destroyed by other agents, such as mumps virus, typhus fever, and other disease processes.

The numbers of sperm produced by a single male are phenomenal: each milliliter of semen contains about 120 million sperm, and there are perhaps 4 milliliters of semen in an ejaculate. Thus, at least in theory, a single male could fertilize every fertile woman on earth in one or two tries. Impressive as this performance may be, it is a poor showing in comparison with the male pig. A boar's orgasm lasts for a half hour, during which time he ejaculates about 500 milliliters of semen!

Such high numbers are not really an example of waste. It is true that ultimately a single sperm fertilizes the egg. But if that single sperm were placed alone in the vagina, it would have no chance at all of reaching the egg in the oviduct and fertilizing it. When the sperm count in a male drops to one-fourth of its normal value (30 million per milliliter instead of 120 million per milliliter), the male is effectively sterile. It takes the cooperative efforts of many sperm to reach the egg and fertilize it, and we will see why when we take a closer look at the process of fertilization.

Once sperm production is initiated during puberty, it continues throughout life, although the rate of production slows down after age 60. The factors that control the rate of sperm production are poorly understood. FSH is required for the process, but there is no known feedback mechanism that limits overproduction or speeds up underproduction.[1] In fact, the total amounts produced seem to vary with sexual activity. Very high sexual activity reduces the number of sperm in an ejaculate.

Testosterone Production. The testes have a second function besides producing sperm: They produce the male sex hormone *testosterone* from puberty onwards. The site of production is the *interstitial cells* of the testes (Fig. 6-8). These cells are not heat-sensitive, and they are not affected by a *vasectomy* (cutting of the vas deferens for purposes of contraception). In either case, normal testosterone production continues, and the hormone is carried to all parts of the body via the circulatory system. When testosterone reaches target organs scattered throughout the body, it induces *secondary sexual characteristics.*

---

1. FSH levels are under negative feedback control of circulating testosterone. But the rates of sperm production and testosterone production need not be related.

PRODUCTION OF GAMETES AND SEX HORMONES / 197

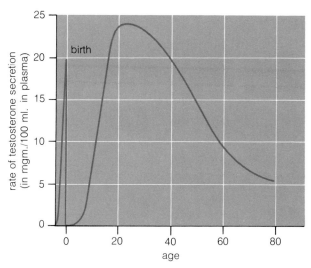

Figure 6-9 TESTOSTERONE PRODUCTION IN MALES AS A FUNCTION OF AGE.

In the male, testosterone production shows two great surges. One occurs before birth, during fetal life, and the other occurs during puberty (Fig. 6-9). Both surges produce major restructuring of the body. The first surge begins as soon as the indifferent gonads are transformed into testes at about 8 weeks of gestation. The testes immediately begin to produce testosterone under the influence of the placental hormone *chorionic gonadotropin* (the chorion is part of the placenta, and gonadotropin means "turner-on of the gonads"). The fetal testosterone then causes development of the external male genitalia, the penis and scrotum. If the testes fail to mature at this point or are surgically removed, the external genitalia develop into a clitoris and labia even though the embryo is a genetic male. Later, at about the time of birth, testosterone stimulates the testes to descend into the scrotum, and failure to descend usually means that insufficient testosterone is being produced.[2]

After birth and throughout boyhood, testosterone production is very low. This is because chorionic gonadotropin, being a placental hormone, no longer stimulates the testes. During these years there is little difference in body form between a boy and a girl, aside from the differences in external genitalia. Then, at puberty, a new controlling hormone appears on the scene. This is *luteinizing hormone,* or LH (named for its function in the female body). LH is a hormone of the posterior pituitary whose production is stimulated by a factor

---

2. When Hitler's body was found and autopsied by the Russians at the end of World War II, the only remarkable finding was that one testis had failed to descend. It thus appears that the man who was obsessed with sexual purity was himself suffering from disturbed reproductive function.

198 / *The Organism: Reproduction*

from the hypothalamus. In males, LH is sometimes called ICSH (interstitial cell stimulating hormone) although the LH of males and females is identical in chemical structure.

As LH incites the testes to testosterone production, massive amounts of testosterone flood the body and induce a series of secondary sexual characteristics:

1. There is increased *hair growth*. Men and women have the same numbers of hair follicles, but the hair of males grows thicker because of testosterone stimulation.[3] In fact, it has been demonstrated that a man's beard grows more rapidly during periods of sexual arousal, when body testosterone levels are elevated.

2. The *voice* quality changes because of an increase in the size of the larynx (Adam's apple).

3. There is a *thickening of the skin,* as well as *stimulation of the sebaceous glands*. The latter effect manifests itself as the terrible acne that affects so many boys in their teens. Fortunately, with the passage of time the sebaceous glands adjust to the new and increased levels of testosterone and the acne disappears spontaneously.

4. Testosterone produces an *increase in muscle strength*. This is why testosterone is administered as a form of treatment for people with gross muscular weakness.

5. There is also *increased calcium retention* in the bones. Again, testosterone is used medically in the treatment of osteoporosis—a disease of old age resulting from calcium loss and characterized by brittle, honeycombed bones. Increased calcium retention is also the factor that ultimately brings an end to bone growth. The long bones of the body are formed initially as cartilage which is then progressively calcified and transformed into bone (Fig. 6-10). The growing regions of the bone are regions of cartilage located near the tips. These growing regions are continually invaded from behind by bone. Thus there is a "race" between the growing cartilage and the growing bone and, once the bone has overtaken the cartilage, the race is over and further elongation of the bone is impossible. This accounts for the spurt in growth during puberty, followed by the achievement of fixed body size.

How does testosterone exert its varied effects on the body? Its basic mode of action is to increase protein synthesis in target organs, such as muscle and bone. Testosterone appears to do this by interacting with nuclear DNA.

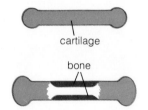
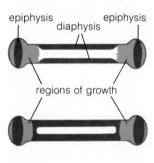
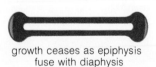

Figure 6-10
PATTERNS OF GROWTH IN BONE. The cartilaginous regions of bones are the only areas capable of elongation. Once the growing cartilage has become calcified, further elongation of the bone is impossible.

---

3. Paradoxically, the hormone that gives is the same hormone that takes away, since *baldness* is also a testosterone-dependent phenomenon in the genetically predisposed male (that is why so few women become bald).

PRODUCTION OF GAMETES AND SEX HORMONES / 199

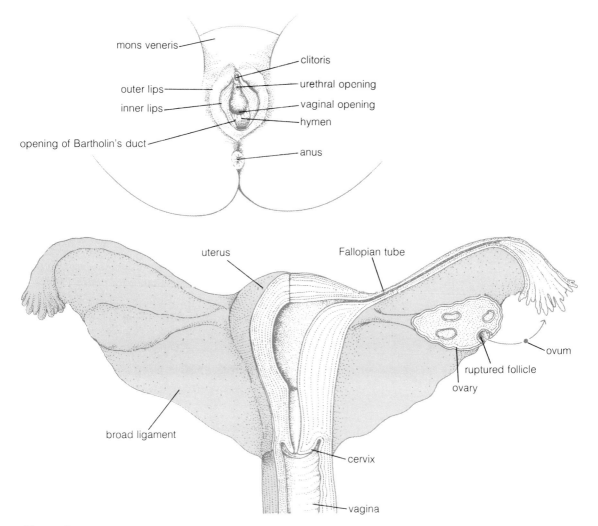

Figure 6-11 EXTERNAL AND INTERNAL FEMALE GENITALIA, FRONTAL VIEW.

At the behavioral level, testosterone is also responsible for the *sex drive,* as well as for *aggressiveness.* There is reason to think that testosterone stimulates the sex drive in females as well as in males. The hormone is found in much smaller amounts in women than in men, but the small amount present plays a key role. In the female, it is the adrenal glands that synthesize testosterone and related steroids.

Body levels of testosterone are controlled by a feedback loop involving the hypothalamus (Fig. 6-6). As hormone levels rise, the hypothalamus produces less LH releasing factor, leading to less LH and testosterone. All sorts of nervous factors may raise or lower body testosterone levels via influences on the hypothalamus. The result is that absolute levels of testosterone tend to vary wildly among males, yet normal sexual function is possible within very wide limits.

The Ovaries

Let us now look at the function of the gonads in women. Though the anatomical relationships are simpler than in males, the hormonal interactions are more complex. The *ovaries* are bean-shaped structures, slightly smaller than the testes, and lie deep inside the abdominal cavity (Fig. 6-11). As in the case of testes, the ovaries produce both gametes (ova) and hormones (estrogen and progesterone).

Egg Production and Ovulation. Nature has been economical and uses the same hormone to stimulate gamete production in females as in males: FSH. Again, the supply of FSH is miniscule during childhood, and large-scale production begins at puberty. But beyond this, there are striking dissimilarities between the male and female pattern.

First, whereas males produce astronomical numbers of gametes throughout their reproductive life-span, the female system contains a fixed number of potential gametes. At birth, a female carries about 400,000 potential ova. However, no more than about 500 of these will ever develop into eggs capable of fertilization during her reproductive lifespan, which extends from the early teens to the late forties or early fifties.

Second, the meiotic process leading to egg cell production shows gross differences in timing and cell division when compared with the male system (Fig. 6-12). Meiosis is arrested at birth prior to the first

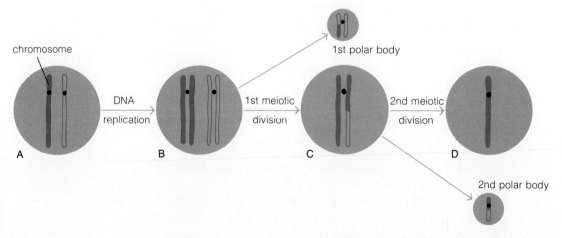

Figure 6-12 MEIOSIS AND OVUM FORMATION. (A) Precursor germinal cell. (B) Cell with nuclear material replicated. Meiosis reaches this stage at birth. (C) Cell at end of first meiotic division, with first polar body. This cell is produced at the time of ovulation. (D) Mature ovum with second polar body. The ovum is produced by the second meiotic division, which occurs at the time of fertilization. The polar bodies do not play a further reproductive role.

PRODUCTION OF GAMETES AND SEX HORMONES / 201

meiotic division. The rest of the meiotic cycle is completed for one potential germ cell at a time, at a rate of approximately one cell per month. The two meiotic divisions occur with unequal division of cytoplasm, resulting in one ovum and two *polar bodies* (so called because they remain stuck to the surface of the ovum and establish the "poles" of the first cell division after fertilization). It is important that the ovum receive as much cytoplasm as possible because, for the first few days after fertilization, the developing embryo has to draw most of its nourishment from within its own tissues. The eggs of birds, reptiles, amphibians, and fish are among the largest cells known in the biological world. The eggs of mammals are much smaller—smaller than the size of the dot on this letter "i." This is true whether the mammal is a whale, a human, or a field mouse. Nevertheless, a cell in this size range is still gigantic compared with other body cells.

Ovulation means the discharge of the ovum from the ovary. In humans, the act of ovulation is typically completely unknown to the woman. This is unfortunate since, if the time were known, contraception would be a much simpler and more effective practice than it is now. But things could be worse—in animals such as the rat and the rabbit, the act of copulation itself acts as a trigger for ovulation, via a nervous signal to the hypothalamus. Thus every sex act leads to impregnation.

In the human ovary, the potential egg begins its development inside a sheath of cells called the *Graafian follicle* ("little container of Graaf" after the Dutch physiologist who first described the structure) (Fig. 6-13). Each of the 400,000 or so potential ova carries

Figure 6-13
THE MENSTRUAL CYCLE. The menstrual cycle is controlled by hormones of the hypothalamus and anterior pituitary. The sex hormones are produced by two ovarian structures: the Graafian follicle and the corpus luteum.

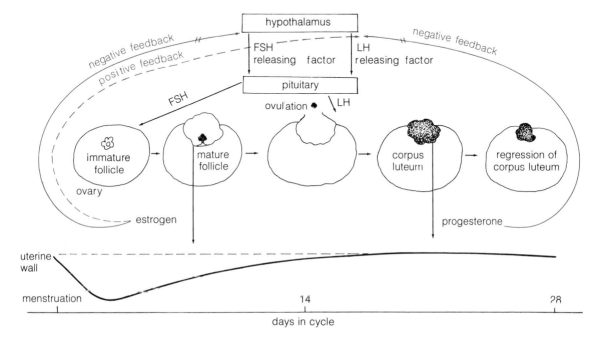

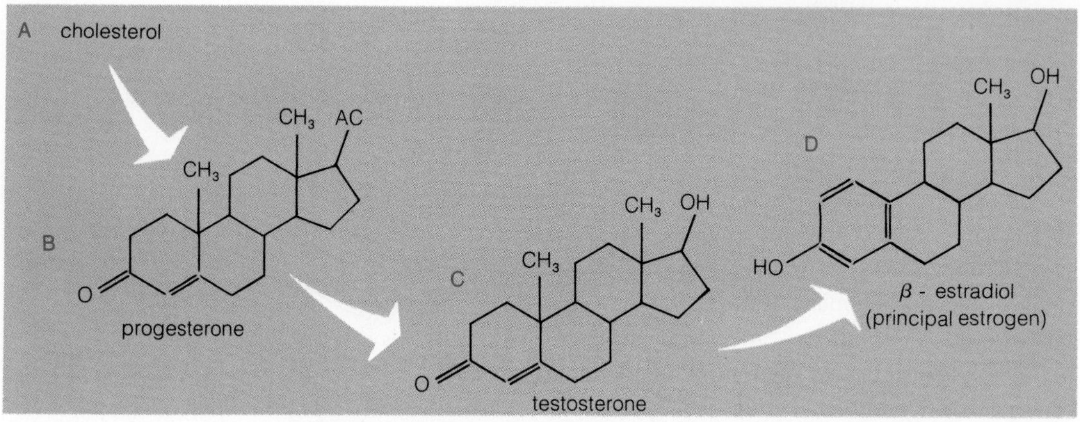

Figure 6-14 CHEMICAL SIMILARITIES BETWEEN MALE AND FEMALE SEX HORMONES. All sex hormones are steroids, derived from the basic steroid cholesterol.

this sheath of cells. However, at the beginning of each menstrual cycle, perhaps 30 or 40 of these tiny follicles begin to enlarge and develop in response to the hormone FSH, which pours out of the anterior pituitary. This is the same hormone used by males to produce sperm. Under stimulation from FSH, the follicle becomes so swollen in size that it can be seen with the naked eye as a blister beneath the surface of the ovary. At the same time, the Graafian follicle becomes an organ of internal secretion, producing a class of female sex hormones called *estrogens* ("generation of frenzy"). The concentration of estrogens begins to rise in the circulatory system and, once the level is sufficiently high, negative feedback is exerted on the hypothalamus, which controls FSH levels via FSH releasing factor. At the same time estrogen exerts positive feedback on the LH releasing factor machinery of the hypothalamus. As a result the hypothalamus begins to produce high levels of LH releasing factor. The result is a surge of LH, which acts as the stimulus for ovulation. In general, only the most highly developed follicle ruptures. All the others regress and are resorbed by ovary tissue, since FSH is no longer present for their development. The scar of the ruptured follicle is converted into another hormone—producing structure, the *corpus luteum* ("yellow body," because it has a yellowish color in cows, where it was first studied).

After the ovum is released by the ovary, it finds itself cast adrift in the body cavity and must find its way into the opening of an *oviduct* ("egg duct") if it is to start a successful pregnancy. The ovum has no means of self-propulsion like the sperm does. It appears to be swept into the oviduct passively, by means of currents set up by cilia lining the oviduct. The mechanism shows a remarkable efficiency. There is a medical example of a woman who had only one ovary and only one open oviduct, on the opposite side of the body. Therefore the ovum had to make its way to the other side of her body before entering the open oviduct. Yet this woman produced four children.

Ovaries and Estrogen Production. We have already seen that the Graafian follicle of the ovary produces estrogen under FSH stimulation. The corpus luteum is also an endocrine organ—it produces the hormone *progesterone* ("steroid in anticipation of pregnancy"). The estrogens (chiefly β-estradiol) and progesterone bear a remarkable chemical similarity to their male counterpart, testosterone. In fact, testosterone and the two female hormones can be metabolically interconverted in the gonads and the adrenal glands (Fig. 6-14). Given their minute chemical differences, it is surprising that the hormones exert such drastically different effects on the body.

Estrogens are responsible for the establishment of secondary sex characteristics in women:

1. There is an increase in the size of the internal genitalia (*oviducts, uterus,* and *vagina*).

2. Development of the external genitalia occurs: deposition of fat in the *mons pubis* and outer lips (*labia majora*) and enlargement of the inner lips (*labia minora*) (Fig. 6-11).

3. There is a change in the *vaginal epithelium* from a thin layer to a much thicker layer of stratified squamous cells. The latter is considerably more resistant to trauma and infection and makes sexual intercourse possible.

4. The breasts develop fatty deposits and a network of internal ducts (Fig. 6-15). This represents the first stage in preparing the breasts for milk production.

5. There is a broadening of the pelvis which, during the birth process, makes the baby's passage easier.

6. Bone growth is stimulated, but there is an even greater stimulation of the process of calcification. The result is a rapid fusion of the epiphyses with the diaphyses, and earlier cessation of growth (Fig. 6-10). This is why women tend to be shorter than men.[4]

7. Perhaps the most pronounced effect of estrogen is an increase in fat deposition beneath the skin, especially in the area

---

4. In Australia, the synthetic estrogen diethylstilbestrol (DES) is being used to check anticipated excessive body growth in girls of pubertal or prepubertal age. DES mimicks natural estrogens in stimulating closure of the epiphyses. It is administered only in cases in which normal growth processes are expected to lead to gigantism.

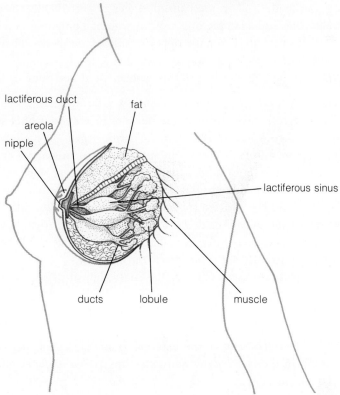

Figure 6-15 EXTERNAL AND INTERNAL VIEWS OF THE MATURE BREAST. The network of ducts and fat develops in response to estrogen stimulation, while lobules develop in response to progesterone. Artificially administered estrogen and progesterone will elicit the same changes in the male breast.

of the buttocks and thighs. This is why superficial muscles and blood vessels are more easily seen in men than in women, and this is also why women have a lower specific gravity than men— they tend to float better.

The mode of action of estrogen is very similar to that of testosterone: an increase in protein synthesis in the target areas. However, the target areas are different: uterus, breasts, skeleton, and certain fatty tissues.

Ovaries and Progesterone Production. In the nonpregnant woman, progesterone most importantly affects the breasts and the uterus. The breasts develop lobules which are necessary for milk production. The lining of the uterus becomes thickened and filled with blood vessels, secretory glands, and stored foods (Fig. 6-16). A rich table is laid in expectation of the arrival of the developing

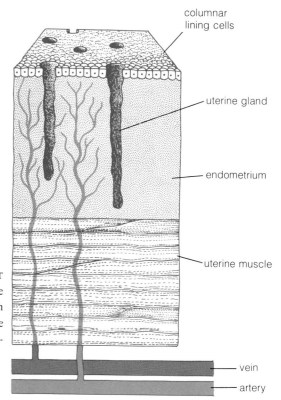

Figure 6-16 THICKENED UTERINE WALL AS A RESULT OF ESTROGEN AND PROGESTERONE STIMULATION. The microscopic uterine glands secrete uterine milk, which serves as a food source for the early embryo. The endometrium is largely sloughed off during menstruation and rebuilt during the next cycle.

embryo, since the embryo will implant itself in the wall of the uterus and start drawing nourishment from it.

The important difference between the secretion of male and female sex hormones is that in the male, hormone levels are innately stable. In the female, there is an ebb and flow of hormones, with estrogen first rising and then declining, and then progesterone rising and again declining. This inner rhythm of the woman's body is called the *menstrual cycle,* and it persists throughout her child-bearing years. The explanation for the tidal nature of the hormone levels is the fact that both estrogen and progesterone exert a negative feedback on the hypothalamus. As soon as each hormone reaches an adequate circulatory level, it begins to cut back on its own synthesis. The feedback of progesterone is weaker than that of estrogen, and it must be accompanied by estrogen to succeed in hypothalamic inhibition. Once progesterone has succeeded in reducing LH output, the corpus luteum loses its signal for producing more progesterone. As a result, the uterine wall cannot maintain itself in its thickened, at-ready state and sloughs off the excess by means of menstrual flow ("monthly flow"). See Figure 6-13.

Female Sex Hormones and Aging. The uterine environment is a maelstrom of estrogen and progesterone. The fetus is exposed to levels of these hormones far higher than those it will experience in

206 / *The Organism: Reproduction*

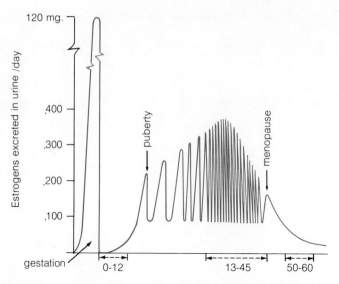

Figure 6-17 BODY ESTROGEN LEVELS AS A FUNCTION OF AGE.

adult life. In fact, a woman will experience such high levels again only if she becomes pregnant. However, the lowest body levels she will experience extend from the time of birth up to puberty (12 to 15 years) (Fig. 6-17). During childhood, estrogen and progesterone exert a violent negative feedback on the hypothalamic-pituitary axis, resulting in very low body levels of gonadotropins (FSH and LH). At the time of puberty, the feedback effect is weakened; both gonadotropin and sex hormone levels shoot up, and the cyclic changes are initiated. The menstrual cycles are irregular at first but usually settle down to a regular pattern during the twenties and thirties. At about age 45 to 50, the menstrual cycles become irregular again. This is due to a "burning out" of the ovaries. By this time, the ovaries have lost most of their primordial follicles as a result of resorption of the ova. Only a small number of primordial follicles remain to be stimulated by gonadotropins, and the production of estrogens and progesterone decreases as the number of remaining follicles approaches zero. This process is called *menopause* and may last from several months to several years.

Menopause may be a very difficult experience for a woman. The body, which for decades has been accustomed to a regular ebb and flow of sex hormones, is suddenly thrown back into a state it has not known since childhood. A variety of physiological and psychological disturbances may result: "hot flashes" (characterized by extreme flushing of the skin), disturbances in breathing, irritability, fatigue, and occasionally various psychotic states. Some of the more severe symptoms may be eased by daily administration of small amounts of estrogen.

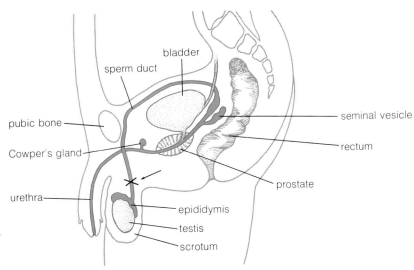

Figure 6-18 MALE SEX ORGANS, SIDE VIEW. The arrow indicates region of sperm duct cut during a vasectomy.

Fertilization

So far we have looked at the processes leading to the production of sperm and eggs. The reproductive organs have the further task of bringing these gametes into contact with each other, by the process of *copulation*. Most of the sexual anatomy is in fact concerned with this function and not gamete production per se.

Role of Male Sex Organs

The path of sperm through the male body is a very circuitous one (Fig. 6-18). After production in the seminiferous tubules of the testes, the sperm enter the *epididymis* ("upon the testes"), a long (about 20 feet), very thin (0.04 millimeter diameter), coiled tube. When the sperm enters the epididymis, it is unable to move. At this point it acquires the capacity for movement and moves on to the *sperm duct* to be stored until ejaculation. During sexual excitement, the penis undergoes an *erection*. This results from the inflow of blood into the *erectile tissues*—spongy, hollow bodies that run the length of the penis. In the normal, unexcited state, blood flows slowly through these tissues. During erection, the incoming arteries become dilated, while the outgoing veins become constricted. Internal blood pressure builds up, causing the penis to increase many times in size.

The male orgasm, or climax, involves a series of peristaltic contractions which begin in the epididymis and sweep down the sperm duct. At the same time, the accessory sexual organs (the prostate and the seminal vesicles) add their secretions to the sperm to create *semen*.

Technically speaking, the process up to this point is called *emission*. It may occasionally take place during sleep (the wet dream). But during the normal sex act there, occurs in addition a rhythmic contraction of the muscles that enclose the base of the erectile tissue. Semen is forced outward through the penile portion of the *urethra,* a process called *ejaculation*.

Role of Male Accessory Sex Organs. The prostate and the seminal vesicles contribute neither to the sex drive nor to the production of gametes. However, their secretions are essential for successful fertilization.

The *seminal vesicles* ("semen containers") were named before their true function was understood. They add a variety of components to the semen, the two most important being *fructose* and a variety of *prostaglandins*. Fructose is a simple sugar, and it is the chief food source for the sperm on its long journey to the ovum. This food source is important because the sperm contains no stored food of its own and must depend on the outside for its energy. The prostaglandins are hormones which also have been misnamed. When first discovered, they were thought to be produced by the prostate gland. It is thought that they contribute successful fertilization by inducing contractions in the uterus.

The *prostate* ("to stand before") is a muscular gland which encircles the sperm duct at the point where it joins the urethra descending from the bladder. It adds to semen a milky fluid containing alkaline buffers and protein-splitting enzymes (proteases). Semen is mildly alkaline (pH 7.5). The optimal acid-base range for sperm motility is a mildly acid environment (pH 6 to 6.5). However, the vaginal secretions of the female are more strongly acidic (pH 3.5 to 4.0). These acid secretions would rapidly kill sperm.[5] In fact, sperm can survive in the female reproductive tract for 1 to 3 days, largely because of the neutralizing effect of the alkaline buffers of the prostate.

The prostate is often a source of grief in later years. It may become enlarged and constrict the urethra, causing difficulties in urination. It may also become cancerous—approximately 1 man out of 40 dies of cancer of the prostate.

---

5. The acid secretions of the vagina have two functions. First, they reduce fungal and bacterial infections in this area. Second, the vinegary odors of these secretions have played an evolutionary role as female sex attractants. They still play this role among primates such as the rhesus monkey. In humans, cultural factors have overridden this natural function, as witness the vaginal sprays and deodorants now on the market.

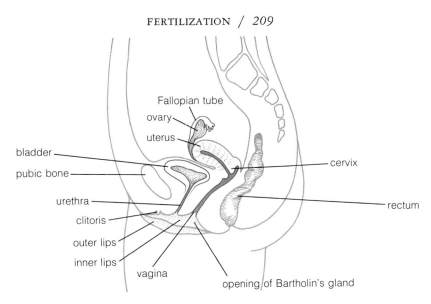

Figure 6-19 FEMALE SEX ORGANS IN SIDE VIEW. Note homologies to the male system (Fig. 6-18). The homologous structures include penis and clitoris, scrotal sac and outer lips, Cowper's glands and Bartholin's glands, testes and ovaries, and a portion of the prostate and uterus.

Cowper's glands are small glands at the base of the penis which secrete a lubricating mucus. This mucus is largely secreted during the excitement phase prior to ejaculation and, by oozing out and covering the head of the penis, it contributes in a small way to necessary lubrication during the sexual act. Most of the lubrication is contributed by the female.

The entire process of erection and climax is under control of the autonomic nervous system. Any degeneration of the peripheral nerves, as occurs for example in diabetes, can lead to impotence. (About 50 percent of male diabetics become impotent.) However, the great majority of difficulties of sexual function are not physiological but psychological in origin. The central nervous system carries a strong veto power over the sexual process.

Female Genitalia

Just as the male and female sex hormones differ only in fine chemical points, so the male and female genitalia have more in common than superficial differences suggest (Fig. 6-19).

The *clitoris* ("to hide") constitutes the chief erectile tissue in the female. It becomes turgid during sexual excitement in the same manner as the comparable male tissue. However, the clitoris has only a sensory function. It contains large numbers of sensory nerve endings which respond to tactile stimulation. The clitoris appears to be the only organ whose sole purpose is pleasure. There is also additional erectile tissue encircling the vagina. Tumescence of this tissue increases contact between the penis and the vaginal walls during intercourse.

The walls of the vagina contain countless microscopic glands which secrete a lubricating mucus during sexual excitement. The same function is served by the twin *Bartholin's glands,* located beneath the inner lips of the vagina, which secrete their lubricant immediately inside the entrance to the vagina. These glands are similar in structure and function to the Cowper's glands of the male. The female thus produces the great bulk of the lubricant necessary during sexual intercourse. If the walls of the vagina are dry, intercourse produces an irritating or painful sensation instead of the massaging-type stimulation necessary for sexual pleasure.

Orgasm in the female, as in the male, consists of a series of rhythmic contractions of the reproductive tract. The contractions begin in the vagina and sweep up the uterus and oviducts. As in the male, emotional aspects play a major role. But unlike the case in the male, the female orgasm is not essential for fertilization. A woman can become pregnant even by artificial insemination, in the total absence of any stimulation. Nevertheless, orgasm in the woman is thought to serve a biological function. It has been shown in the case of the cow that the rushing contractions of orgasm are of great help in sperm transport. With orgasm, sperm make the long trip up the uterus and into the oviducts in only 5 minutes. In the absence of orgasm, sperm cannot make the trip on their own in much less than an hour. It is probable that a similar effect takes place in humans. Women who undergo natural fertilization are somewhat more fertile than those who undergo artificial insemination.

The Path of Sperm

As a result of ejaculation, sperm are deposited close to the *cervix* ("the neck"), the opening of the uterus into the vagina. In order to fertilize the ovum, the sperm must make their way through the cervix, up the uterus (the Latin word for the womb), and into the upper third of the oviduct, where the ovum is slowly making its own way downward (Fig. 6-19). However, fertilization is possible even if the sperm is deposited around the opening of the vagina, if the woman has normal vaginal secretions at the time. During their trip, the sperm must cross two barriers. One is a mucus plug blocking the cervix. At the time of ovulation, this mucus plug assumes a clearer, watery consistency and therefore forms less of a barrier to sperm. Only a minority of the sperm make their way through the cervix and into the uterus. Of those that get this far, no more than half will choose the correct oviduct, since only one contains the ovum. And when the few survivors finally reach the ovum, they find their goal surrounded by a layer of smaller cells derived from the Graafian follicle, the *corona radiata* ("radiant crown") (Fig. 6-20). Though this layer is partially disrupted by the time the ovum reaches the oviduct, it is thought that the sperm cooperatively disrupt the remaining cells through the use of hydrolytic enzymes present in their acrosomes. Finally a single sperm penetrates the ovum, and immediately a fertilization membrane lifts up from the surface of the ovum and excludes all further sperm. The nucleus of the penetrating sperm migrates inward and fuses with the nucleus of the ovum, and a zygote (fertilized egg) has been created.

The original sperm count, seemingly so high, is required to ensure that fertilization takes place at the end of this long and obstacle-filled path to the ovum.

Figure 6-20
A SECTION THROUGH AN OVARY SHOWING A DEVELOPING GRAAFIAN FOLLICLE AND THE CORONA RADIATA ENCLOSING THE OVUM.

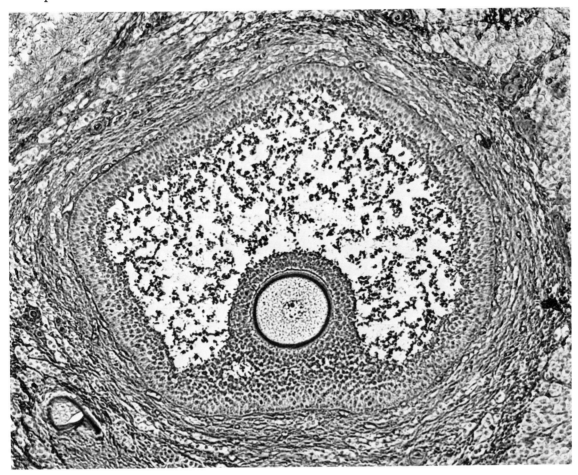

Courtesy of Carolina Biological Supply Company

Pregnancy

Pregnancy begins in the oviduct. While it takes less than an hour for the sperm to travel up the oviduct, it takes about 3 days for the fertilized egg to travel down to the uterus. The egg is moved along by means of a fluid current set up by cilia lining the walls of the oviduct. However, obstructions in the oviduct retard the movement of the fertilized egg. This provides sufficient time for it to undergo several stages of cell division before reaching the uterus. At the same time, the oviducts form secretions that nourish the developing embryo and stimulate cell division. The first few cell divisions of the human egg have been studied outside the human body, and it appears that secretions from the oviducts are essential for these cell divisions.

Role of Progesterone

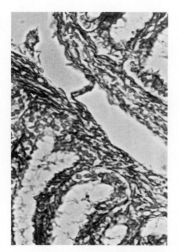

Figure 6-21
CROSS SECTION OF THE WALL OF A MOUSE OVIDUCT. Note the long ciliated projections of the oviduct wall.

After a 3-day trip down the oviduct, the dividing mass of cells arrives in the uterus. It stays there for 9 months, gradually taking on human form and function. The key hormone during this period is progesterone. It is progesterone that maintains the inner walls of the uterus in a thickened and nutrient-rich state. It is also progesterone that keeps the muscular tissue of the uterus in a relaxed state. Should these muscles contract, the embryo would be expelled and a spontaneous abortion would result. And finally, it is progesterone that suppresses ovulation during pregnancy, so that no second pregnancy is started while the first one is still in progress. It does this by feedback inhibition of LH production, which is required for ovulation.

But as we pointed out before, this very feedback inhibition of LH production also leads to degeneration of the corpus luteum and to a decline in progesterone levels. The resultant menstruation would immediately terminate the pregnancy by loss of the developing embryo. This dilemma is resolved by the introduction of a new hormone made by the embryo but affecting the body of the mother. This hormone is chorionic gonadotropin. mentioned on page 197 (Fig. 6-22). Chorionic gonadotropin behaves in all respects like LH, with one important difference: it suffers no feedback inhibition from high progesterone levels. It therefore maintains the corpus luteum in an active state, causing it to produce the required progesterone. Chorionic gonadotropin also forms the basis of the common urine test for pregnancy. Two weeks after a missed menstrual period, the hormone is produced in such large amounts that some spills over into the urine, where it is easily detectable by immunological techniques.

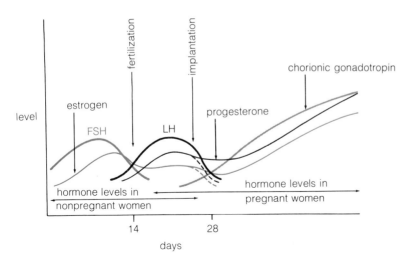

Figure 6-22 HORMONE LEVELS IN THE BODY OF THE PREGNANT AND NONPREGNANT WOMAN. Pregnancy begins within one day of ovulation (about day 14 of the normal menstrual cycle).

The corpus luteum remains as the major source of progesterone only for the first 3 months of pregnancy. After that the demand becomes so great that the corpus luteum cannot satisfy it, and a new source is established: the *placenta* (from the Greek word for flat cake, which it resembles). The placenta is the organ through which the fetus receives nourishment and oxygen from the mother's circulatory system, and through which it excretes wastes. If the transition from corpus luteum to placental progesterone is not a smooth one, a spontaneous abortion may occur, hence this is a critical period in pregnancy.

Embryonic Development

The transition from a single cell (the fertilized egg) to a multicellular organism (such as a human being) is one of the most astonishing and still least understood spectacles in biology. It appears to be a difficult process in itself, since an estimated one-third of all human pregnancies wind up as spontaneous abortions, mostly within the first few days. The causes may involve faulty genetic instructions, an inhospitable environment in the mother's body, or some combination of the two.

The embryo is concerned with two sorts of development. First, it must develop the permanent structures that will allow it to survive and function as an adult. Evolutionarily speaking, these are very old

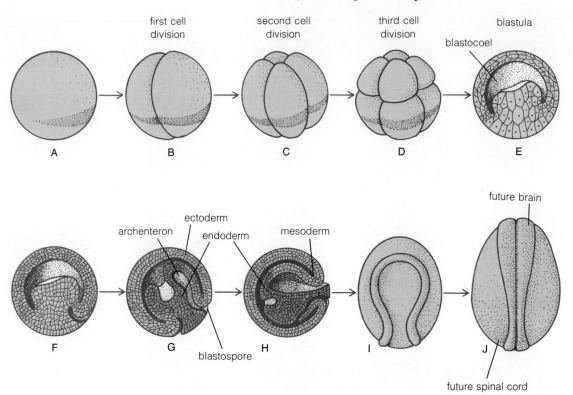

Figure 6-23 CLEAVAGE AND EMBRYOLOGICAL DEVELOPMENT IN THE FROG. A–D and I–J are external views. E–H are cross-sectional views showing internal developments.

processes whose basic outlines were established more than 500 million years ago. We share these processes with all other species of chordates and of echinoderms (such as the starfish). Second, the embryo must develop temporary structures which function only during embryonic life and allow it to interact with its mother's body. These structures form as a series of specialized membranes around the embryo and can be evolutionarily traced back to the reptiles, who invented them about 300 million years ago. In humans, they have been transformed into such structures as the placenta, the umbilical cord, and the amniotic sac.

Development of Permanent Structures. We already touched on this subject in Chapter 3, when we discussed the nature of differentiation and control of DNA expression. Briefly, the problem is that all cells can arise only from other cells, by replication of the total

DNA content. Yet ways must also be found of making cells that differ from each other. This is done via unequal division of the cytoplasm.[6] Cytoplasmic factors then selectively activate or inactivate genetic information present in the nucleus. The cytoplasmic factors may also affect neighboring cells. In the case of the frog egg (Fig. 6-23), there is a strong cytoplasmic gradient extending from the top (animal pole) to the bottom (vegetal pole) of the egg. The first two cell divisions are meridional and partition the cytoplasm equally. If the four resultant cells are separated, they will develop into identical quadruplets. It is thought that a similar mechanism is responsible for the birth of identical twins in humans.[7] However, the third cell division in the frog embryo is equatorial and results in very unequal partitioning of the cytoplasm. Separation of the cells at this time no longer results in the birth of identical offspring. Instead, the top and bottom cells have already diverged in their embryological potential. The top cells will form the future skin, muscle, and nervous system, while the bottom cells will form the digestive tract and derived organs.

The development of an embryo is quite different from the factory assembly of an automobile. The components going into the car are already in their final form when they are bolted or welded on to the unfinished car. But the unfinished embryo never consists, say, of connective tissue linking together a little hand and a spleen, with lungs scheduled to be added next. During embryonic development, a complete animal is potentially present at all times. Embryological development perhaps resembles most of all a miniature restatement of the evolutionary process. This observation was not lost on nineteenth-century biologists, who expressed it with the aphorism, "ontogeny recapitulates phylogeny"; that is, individual development restates evolutionary development. While this statement requires a great deal of qualification and exception, it contains a strong kernel of truth.

As we will see subsequently, the chordate line (of which humans form a part) has its closest living relatives among the echinoderms (such as the starfish and the sea urchin). Both chordates and echinoderms are animals with a basically tubular, three-ply body form.

---

6. The cytoplasmic inequality need not be quantitative in nature. Even when two daughter cells have very similar volumes of cytoplasm, they may contain different amounts of mitochondria, yolk granules, ribosomes, and so on.

7. In humans, the formation of identical twins may also occur by separation at later stages of development. Such later separation processes may result in the sharing of a single placenta, chorion, or even amnion.

Both evolved from an animal with a two-ply body construction, as typified by the coelenterates (for example, *Hydra,* jellyfish). These in turn are thought to have evolved from a spherical form with a single tissue layer. No animals of the latter type are alive today, nor have fossil remnants been discovered.

We find the same intermediate stages in the embryological development of echinoderms and chordates. Embryologists out of necessity have concentrated their studies on the animals with the largest and most accessible eggs. Since mammalian eggs are tiny and develop deep in the inner recesses of the oviduct and uterus, mammalian embryology has remained a poorly studied area. Most studies on embryological development have featured species such as the sea urchin, the frog, and the chicken. In each case, the egg is large enough to be easily manipulated and develops outside the mother's body.

Comparison of the sea urchin and the frog shows that each egg develops into a hollow ball of cells called a *blastula* ("little sprout"). There has been no increase in overall size, but a single large cell has been transformed into a number of much smaller, metabolically more active cells. The greater the cytoplasmic polarity of the egg, the more asymmetric the resultant blastula. The extreme is reached in the egg and blastula of birds. Here the blastula is a tiny mass of cells sitting on top of a large mass of a cellular yolk, which will serve as the food source for the growing embryo.

The blastula contains a single tissue layer (though not always a single cell layer). The formation of successive tissue layers involves active migration of cells. The process is different in each case, but the result is a structure with two tissue layers called a *gastrula* ("stomach") (Fig. 6-23). The inner layer is called the *endoderm* ("inner skin"), and the outer layer the *ectoderm* ("outer skin"). The third layer, the *mesoderm* ("middle skin") forms by further migration of cells that initially come from the region where the endoderm and ectoderm meet. In the frog, the region is called the dorsal lip of the blastopore. Each tissue layer contributes specific tissues to the adult body. The endoderm supplies the digestive tract, lungs, liver, and pancreas. The ectoderm supplies the skin and nervous system. The mesoderm forms the bulk of the body: bone, muscle, blood, heart and blood vessels, kidney, and gonads.

Embryological development beyond the gastrula stage becomes increasingly complex. However, one of the major events in chordate development involves formation of the nervous system by means of a furrowing of the ectoderm on the dorsal (back) side. The edges of

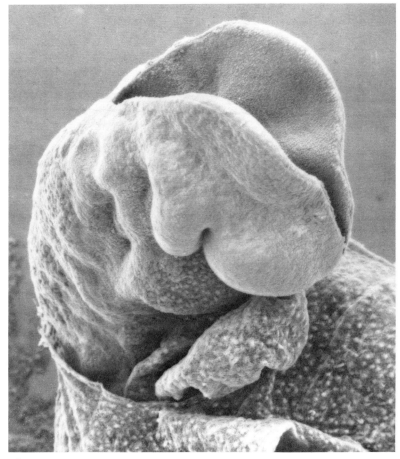

Figure 6-24 SCANNING ELECTRON MICROSCOPE VIEW OF NEURAL FOLDS IN A MOUSE EMBRYO. The area on top is the future brain.

the furrow become elevated, meet to form a tube, and sink again below the ectoderm (Fig. 6-24). At the same time, the mesoderm immediately underlying the neural tube forms a stiff rod, the *notochord* ("chord in the back"), which becomes surrounded by the future vertebral column (Fig. 6-23).

The later stages of embryological development can perhaps be fully grasped only by employing a four-dimensional plot—the three spatial dimensions plus time. And yet the actors on the stage, the cells of the frog or bird or leopard or human, seldom seem to be confused. Monsters arise in peoples' imaginations, seldom in natural development. One of the reasons is that the individual cells of the embryo are not independent but interact to form mutually compatible structures. For example, why do we form only two eyes—why not have some extras in the back of the head, or even in the shoulders or chest?

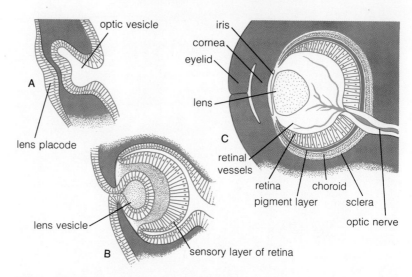

Figure 6-25 FORMATION OF THE LENS AND RETINA.

The formation of the lens of the eye begins as a process of *induction* by the optic vesicle, an outgrowth of the forebrain (Fig. 6-25). The optic cup comes in close contact with the overlying ectoderm. It releases soluble factors which are absorbed by the ectodermal cells and cause them to begin synthesis of massive amounts of crystallins, the transparent proteins of the lens. If a piece of cellophane is placed between the optic cup and the ectoderm, a lens does not form in the ectoderm. Conversely, if the optic cup is transplanted to some other region, say, below the ectoderm of the belly, a lens may form in the alien ectoderm. Further, the relationship is reciprocal; the developing lens in turn releases inducers which cause the optic vesicle to develop into the retina. However, the experimental eye in the belly will be blind, since it will lack the necessary nerve connections to the brain. Induction is a very general, always present process in development. It involves multiple contacts through time by a variety of tissues with changing signals and reciprocation of influences.

The Fetal Membranes. The human embryo must develop a set of specialized organs that will allow it to interact with the body of the mother, whose blood provides food and oxygen and a means of removing metabolic wastes. The fetal membranes of the human (and of all mammals and birds) represent a remarkable "steal" from the reptiles (Fig. 6-26). In fact, the reptile egg may well be the most important evolutionary invention of this class of vertebrates. For the reptiles, the amniote egg represented an escape from the need for external fertilization and aqueous development. The egg contains a

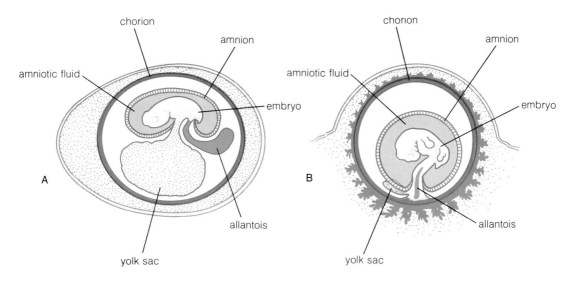

Figure 6-26 EMBRYOLOGICAL MEMBRANES OF THE REPTILE AND HUMAN. The reptile egg is almost identical to the bird's egg. (A) Reptile egg. (B) Human fetal membranes, first month of pregnancy.

food source (the *yolk sac*), an internal aqueous environment (the *amniotic fluid*), a receptacle for waste disposal (the *allantois*), and an oxygen–carbon dioxide exchange system (diffusion through the shell, the *chorion,* and the allantois).

All the same membranes are present in the human embryo—the only structure not present is the shell. But the relative proportions of the membranes are grossly distorted, and they have assumed new functions. It is important to remember that fetal membranes are living tissues which are permeated with blood vessels and work actively to maintain a successful pregnancy.

The chorion ("skin") is the outermost membrane. It begins to play a vital role during the very first days of pregnancy. After fertilization, the early embryo spends 3 days traveling down the oviduct to the uterus. On arrival in the uterus, it spends an additional 5 days clinging passively to the uterine wall and being nourished by "milk" formed by innumerable small glands in the uterus. Toward the end of this period, the chorion[8] has developed sufficiently and begins literally to eat its way into the uterine wall. The process is one of phagocytosis, and the digested part of the uterine wall serves as food

---

8. At this time the chorion is called the *trophoblast,* and the feeding process described is called trophoblastic feeding.

for the embryo. This type of feeding is important during the first 10 weeks of pregnancy. The process is called *implantation.* Shortly after implantation, the chorion also begins to form chorionic gonadotropin, as discussed before. This ensures maintenance of the thickened uterine lining. With time, the chorion develops into the major portion of the placenta. The fetal placenta interdigitates with the uterine wall, and now the developing embryo can feed itself not by digesting maternal tissues but by absorption of nutrients from the maternal blood. This is a much more efficient method of feeding. At the end of the third month the embryo weighs about 14 grams (½ ounce), but at the end of the fourth month, with placental feeding, its weight shoots up to 110 grams (4 ounces). In the placenta, the maternal and fetal circulations remain separate but are separated by only a thin layer of tissue. This thin separating membrane allows the exchange of foods, respiratory gases, and waste products which are carried back and forth through the umbilical cord. The placenta, which becomes the dominant food source for the embryo by the third month of pregnancy, also manufactures large amounts of estrogen and progesterone, as discusssed before.

The allantois and the yolk sac are vestigial structures in the human embryo.[9] The allantoic blood vessels permeate the chorion, and this membrane thus forms part of the umbilical cord and placenta. The yolk sac serves a nutritive function during the first few days of pregnancy. Its further functions, if any, are obscure.

The amnion is a very strange membrane. Perhaps it explains the universal human attraction to the ocean. We spend the first 9 months of our lives swimming in a fluid. We do not drown, of course, because the placenta, not the lungs, is the source of oxygen. The fluid we swim in is the amniotic fluid, enclosed by the amnion. It is the original water bed; it allows the fetus to move freely, cushions it against shocks, and may ease the trauma of delivery by helping to dilate the cervix.

The amniotic fluid, besides salts and body metabolites, also contains small numbers of free-floating cells which have been shed by the skin and amniotic membrane of the embryo. By the third to fourth month of pregnancy, sufficient amniotic fluid has formed that some may be withdrawn by a needle inserted through the abdominal wall (Fig. 6-27). The technique is called *amniocentesis* ("puncture of the amnion") and has been applied to tens of thousands of women. This procedure is important because the small sample of fetal cells and

---

9. The allantois is well-developed in birds and reptiles and in some mammals such as the pig. In humans, the allantois is present only as an umbilical stalk. However, the allantoic blood vessels provide most of the blood supply to the fetal part of the placenta.

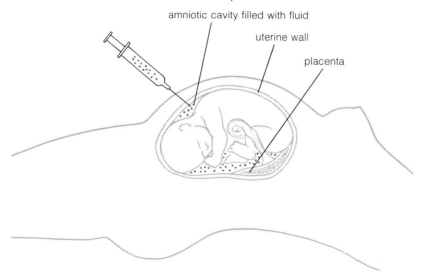

Figure 6-27 AMNIOCENTESIS. The amniotic fluid withdrawn by a syringe is centrifuged to separate the cells. The cells are then cultured to provide sufficient material for chromosomal and biochemical (enzymatic) analyses which will confirm or disprove a suspected birth defect.

amniotic fluid obtained can tell an obstetrician a great deal about the developing embryo, such as its sex (by chromosome staining), any Rh reaction (a pathology in which the mother develops antibodies to her own fetus) or, most important, the presence or absence of a number of birth defects. For example, a mother who has given birth to a baby with Down's syndrome (mongolism) has a greatly increased chance of giving birth to a second Downie. Amniocentesis can tell her whether her second child will be normal (by a simple chromosome count). Tay-Sachs disease is another birth defect, even more tragic than Down's syndrome. The baby is born normal but, as the result of an enzyme defect, progressive neurological degeneration sets in, leading in a few years to blindness and invariably to death. As in the case of Down's syndrome, parents who are known carriers of the trait may elect to abort an affected fetus and try again to have a normal child.

Amniocentesis at present can be used to detect only a small number of potential birth defects. For example, diabetes cannot be detected because it develops later in life and at any rate is not expressed in skin cells. Neither can one detect potential mental retardation, sickle-cell anemia, galactosemia, or a host of other genetic defects, because they are not expressed in skin cells. But the list of genetic diseases that can be detected is constantly growing, as human choice replaces the hand of fate.

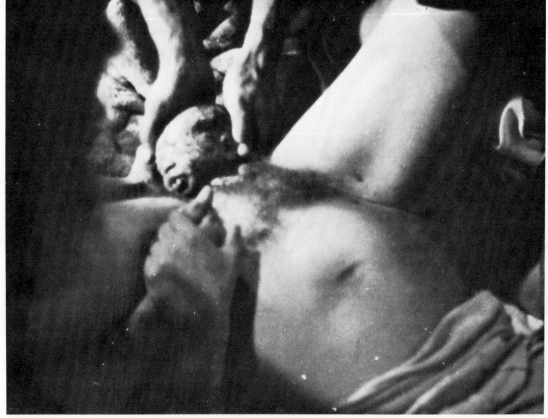

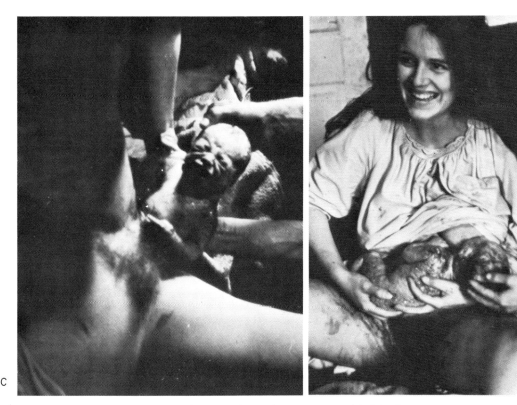

Figure 6-28 HUMAN BIRTH. This birth is taking place at home, with friends holding the mother's hand and the father delivering the baby. (A) Labor. (B) Delivery of the head. (C) Delivery of shoulders and chest. (D) Minutes later, the baby seeks its mother's breast. The umbilical cord is still intact.

Birth

In the great majority of Western societies, the twin rites of birth and death have been removed from the common range of human experience. The justification used is medical necessity, and there is no question that Western medicine has made births safer and premature deaths more rare. But as commonly practiced today, medical authority has also removed a great dimension from life experience. In the case of birth, even the mother is often so drugged that there is only a dim recollection of the process afterward. More recent years have seen the beginning of a reversal in such attitudes. The beauty of the natural birth process is shown in Figure 6-28.

In terms of the fetus, birth marks a changeover from an aquatic to a terrestrial existence, accomplished within a span of less than 5 minutes. During these crucial first minutes, the following events must take place:

1. The circulatory system must abandon its circuit through the umbilical cord and placenta. The completion of this step serves as a trigger for the two following steps.

2. The lungs must draw in the first breaths of air. In the fetal state, the lungs are collapsed and largely bypassed by the circulatory system.

3. The circulatory system must abandon the lung-bypass channels it used in fetal life and send blood to the lungs. Failure to abandon these bypass channels produces a "blue baby" with an overworked heart and inadequate oxygen delivery to the body tissues.

In terms of the mother, birth involves a change in the quiescent state of the uterus. Toward the end of a typical 9-month pregnancy, the mother's circulating progesterone levels begin to drop. This is the hormone that has kept the uterine walls relaxed. As the uterine muscles become more irritable, the process of *labor* begins. The central role of oxytocin in this process was described in Chapter 5. Uterine contractions, initially gentle and widely spaced, press the head of the fetus against the cervix. As the cervix dilates (widens), a nerve reflex elicits oxytocin release from the posterior pituitary. A positive feedback is set up as the released oxytocin stimulates uterine muscles to press harder, causing increased cervical dilation and oxytocin release.[10] From the first gentle contractions to the final delivery 8 hours or longer may go by. The longest and most difficult portion of labor involves forcing the baby's head through the cervix and vagina.

Delivery is difficult both for the baby (whose head may be temporarily "sculptured" by the process of passage) and for the mother (whose vagina may be torn). Only 100 years ago, before the introduction of sterile technique in hospitals, 1 woman in 10 lost her life during a hospital delivery or as a result of subsequent infection. Interestingly, the death rate for home deliveries was appreciably lower. Today the death rate is only a small fraction of this, but it is still significant. In the United States, for white women the figure is 2.4 per 10,000 live births; for black women, the figure is four times higher. And for older women (age 40 to 45) the risk is 10 times higher than

---

10. This process is sometimes accelerated by an obstetrician through the deliberate infusion of oxytocin or through mechanical dilation of the cervix. However, recent statistics show an increased number of fetal complications following induced labor.

it is for women aged 20 to 25. Still higher risks are experienced by women with special health problems such as heart disease, kidney disease, or diabetes. These death rates approach those of young men in a war. As a species, humans owe much of their success to a complex nervous system which requires a longer gestation period for completion. The price for this success is paid by the females of the species, even in an age of modern medicine.

Thus far our attention has focused on the factors involved in successful fertilization and embryological development. But the human species differs from others such as rabbits and rats in that sexual pleasure is not invariably linked to pregnancy and birth. The two processes are separated by the techniques of contraception ("against conception").

Contraception

Though it is sometimes forgotten, contraceptive techniques may be used by males as well as females. Some of the more common methods in use today are listed in Table 1, along with an estimate of the failure rate.

TABLE 1
Methods of contraception and their failure rates[a]

MALES	HIGH	LOW	
Withdrawal	38	10	
Condom	28	7	
Vasectomy	0	0	
FEMALES	HIGH	LOW	AVERAGE
Rhythm	38	0	25
Lactation	26	24	25
Diaphragm and spermicidal jelly	35	4	12
IUD			
0 to 12 mos.			2.4
12 to 24 mos.			1.4
Pill	2.7	0	0.5
Tubal Ligation	0	0	0
Abortion	0	0	0

[a] Figures represent number of conceptions per year per 100 users.

Withdrawal is probably the oldest and still most widely used contraceptive method in the world. Unfortunately, it is unreliable. It requires a great deal of skill and practice on the part of the male, because only 3 to 5 seconds elapse from the time he becomes aware of imminent climax and the actual ejaculation. Furthermore, even ejaculation into the vaginal opening, given normal vaginal secretions, puts the sperm in a position to reach the oviduct.

The *condom* is a thin rubber sheath which covers the penis, and it reduces much of the sensation for the male. There is a possibility of failure due to breakage of the condom, especially if insufficient room is left in the tip for the collection of sperm.

Vasectomy involves the cutting and tying of the two sperm ducts (the vas deferens) (Fig. 6-18). Many misconceptions exist about the nature of this operation. The operation is so simple it can be done in a doctor's office under a local anesthetic. A vasectomy is *not* equivalent to castration, because the blood supply to the testes remains undisturbed and the testes continue to function in testosterone production. Therefore after vasectomy both sex drive and secondary sexual characteristics remain undisturbed. The man is fully capable of a sex act with ejaculation, but the ejaculate contains no sperm. At the present time, vasectomy should be considered an irreversible operation, since reuniting the sperm ducts is much more difficult than cutting them, and subsequent fertility cannot be guaranteed. If a man feels he might want children at a later time, it is possible to store sperm for 10 years or longer in a frozen state. He could therefore "bank" his sperm prior to the vasectomy and use it later for artificial insemination. In 1970, 750,000 men in the United States underwent vasectomies.

The *rhythm* method is simple in concept—it involves sexual abstinence by the woman during her fertile period. The difficulty comes in discovering when the fertile period occurs. It is believed (without much evidence) that an ovum must be fertilized within 24 hours after ovulation if a viable zygote is to be formed. It is further believed (again with little evidence) that sperm in the female reproductive tract retain their ability to fertilize an egg for approximately 2 days. Therefore the fertile period of a woman extends from 2 days before ovulation (lifetime of a sperm) to 1 day after (lifetime of an ovum). Therefore one should know the day of ovulation and know it 2 days in advance! Contrary to typical textbook charts, ovulation need not occur on day 14 of the menstrual cycle. One can calculate the date of ovulation retrospectively, by counting back 14 days from the onset of menstruation. Thus a woman with a 28-day cycle ovulates on day 14, while a woman with a 34-day cycle ovulates on day 20. Women in their teens and early twenties have notoriously irregular periods, and approximately one woman out of six continues to have such irregular periods throughout life.

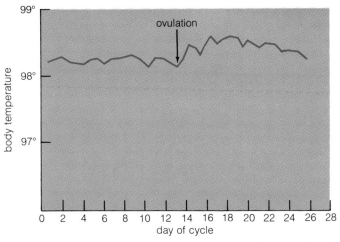

Figure 6-29 THE ELEVATION OF BODY TEMPERATURE AT OVULATION.

Another method of detecting the time of ovulation is to make a day-by-day chart of the basal body temperature (the temperature on first awakening in the morning). On the day of ovulation there is approximately a $\frac{1}{2}°F$ rise in body temperature (Fig. 6-29). Unfortunately, a similar rise in body temperature can occur as the result of a slight fever, mental upset, tension, or any number of other factors. There are other, complex medical tests to detect ovulation, but these require daily visits to a gynecologist's office. Until a simple and reliable method is developed to detect the time of ovulation, the rhythm method will remain one of the least effective methods of contraception available.

Lactation is another old method of birth control. For 7 to 9 months after birth, ovulation is suppressed in a nursing mother. However, milk production may continue much longer than this, with the result that a nursing mother may become pregnant before she stops nursing.

Spermicidal jellies or creams are relatively ineffective by themselves but provide considerable protection when combined with a *diaphragm* (Fig. 6-30). The diaphragm mechanically blocks the cervical opening, and any sperm that make their way under the edges are killed by the spermicidal jelly. Diaphragms come in five or six different sizes and must be fitted by a physician. They must also be refitted every 2 or 3 years prior to pregnancy, and immediately after delivery, when the cervical dimensions change. In use, the diaphragm cannot be felt by either partner, but optimal protection requires high motivation and intelligence. It must be seated properly, must remain in place for 6 hours after the last exposure, and cannot be used more than two exposures in a row.

The *intrauterine device,* or *IUD,* is a plastic or metal loop which is inserted through the cervix in a straightened-out form and then

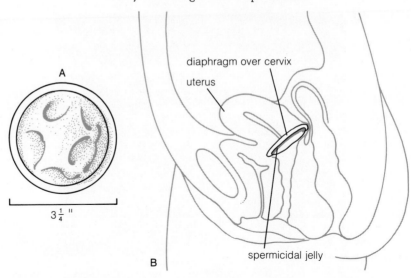

Figure 6-30 THE DIAPHRAGM. (A) External view. (B) Diaphragm inserted, with a coating of spermicidal jelly.

coils up inside the uterus so it cannot fall out (Fig. 6-31). It has a short length of nylon string which protrudes through the cervix and can be checked to confirm that the IUD is still in place.

No one is sure how the IUD works. One theory holds that it changes the rate of travel of the ovum through the oviduct, causing it to arrive in the uterus too soon. Another theory proposes that mechanical irritation of the uterine wall prevents implantation.

The advantages of the IUD are low failure rate, ease of insertion (2 minutes), low cost, and the fact that once it is in place it can be forgotten. The disadvantages are possible cramping and bleeding during the first few months, and outright expulsion in many women.[11] The IUD is expelled by 40 percent of women who have never been pregnant, but by only 25 percent of those who have.

A steroid *pill* is generally a combination of a synthetic estrogen and progesterone or a sequence of the two. It is taken once a day, irrespective of sexual activity, and any lapse of memory increases the chances of pregnancy. In fact, it is possible that essentially all failures that do occur can be explained by forgotten pills. The pill has a very low failure rate, allows for spontaneity in sex, and at the present time is known to produce few serious side effects. The one clearly estab-

---

11. In extreme cases, infection or perforation of the uterus may result. While data are incomplete, the overall mortality due to complications resulting from the usage of IUDs appears to fall in the range of 5 deaths per 1 million users. Thus the IUD is safer than the pill.

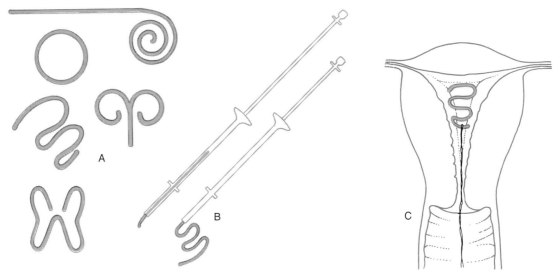

Figure 6-31 THE IUD. (A) A few of the common shapes in which IUD's are manufactured. (B) An IUD pulled into a device used for insertion (top) and expelled (bottom). (C) The IUD in place inside the uterus. Note nylon string passing through cervix.

lished side effect of the pill is an increased risk of blood clotting. Studies made in Great Britain in 1968 showed that the incidence of death due to blood clotting among users of the pill was 13 per 1 million women per year. This was eight times higher than among nonusers. However, this should be compared with the death rate due to pregnancy, which was 228 per 1 million pregnancies in the same age group. Newer versions of the pill, by cutting down on their estrogen content, have reduced by half the deaths due to blood clotting.

Another British study completed in 1975 and disseminated in the United States by the FDA implicates the pill as increasing the risk of heart attacks in older women. Among women in their 30's, nonusers suffered 19 fatal heart attacks per 1 million women per year. Users of the pill suffered 54 fatal heart attacks. For women over 40, the death rates were 117 and 547 respectively. In consequence, the gynecology advisory committee of the FDA has recommended that women over 40 adopt other forms of contraception.

The mode of action of the pill varies among different classes of the pill. The traditional progesterone-based pill acts on the hypothalamus by a feedback mechanism. There is no midcycle surge of LH, and ovulation is inhibited.

Tubal ligation involves the cutting and tying of the oviducts. It is thus equivalent to a vasectomy in males. The operation is a major one, involving abdominal surgery. However, unlike in the male, there is a significant possibility of reversal. The success of reversal has been about 30 percent in recent years, and has been going up.

Abortion is neither recommended nor desired as a routine means of contraception. However, it remains an important backup method when routine measures fail. Data from New York State (where 200,000 legal abortions were performed in the year following passage of the 1970 abortion law) show that abortion, when carried out under medical auspices, is many times safer than natural pregnancy and childbirth. This is especially true if the abortion is performed in the first 12 weeks of pregnancy. In 1973, the United States Supreme Court ruled that no state may restrict a woman's right to an abortion in the first 12 weeks of pregnancy.

Future Methods of Birth Control

The pill, used today by tens of millions of women across the world, has been clinically available only since 1960. At the time of its introduction, it represented a radical departure in birth control technology. There is every reason to think that similar new developments will occur in the future.

In the short term, there is likely the development of a once-a-month pill or once-a-year pill. These would be long-lasting steroids given by injection or silastic implantation. A "morning-after pill" has already been licensed by the FDA in the form of DES a synthetic estrogen analog. However, in its present form this pill has such unpleasant side effects that it is not suitable for regular use. It prevents implantation of the fertilized egg.

The prostaglandins are a class of nonsteroid hormones which can act as abortifacients (abortion inducers)—they induce menstruation or labor, depending on the stage of pregnancy. Much effort is now being invested in eliminating the side effects that attend the use of these compounds.

In the longer term, contraceptive advances will require a much better understanding of the reproductive process as a whole. What factors regulate sperm production in the male? What controls the travel of the ovum down the oviduct? Does the egg release an attracting hormone for the sperm? What factors regulate early cell divisions? The questions are endless, and there are no answers at present. In fact, reproductive physiology is possibly the least studied area of physiology as a whole. When we can describe in molecular terms the contraction of heart muscle or the propagation of a nerve impulse, it seems incredible that there still does not exist a simple, reliable test for the time of ovulation. Such a test could add as much to contraceptive art as any pill now under development.

Bibliography

Eric T. Pengelley, *Sex and Human Life,* Addison-Wesley, 1974. This small, profusely illustrated book, explains human reproductive physiology and contraceptive techniques in a comprehensive yet nontechnical way.

John Tyler Bonner, *On Development,* Harvard University Press, 1974. This is a book of ideas, not a detailed stage-by-stage observation of embryonic development. Bonner treats development in a much wider context than would the typical embryology text. His discussion ranges from the assembly of cell organelles to the development of insect architecture. A brief excerpt follows (*kinety* refers to a surface cover with cilia).

> Recently Sonneborn and his co-workers have done an interesting experiment in which, by microsurgery, they reversed a small portion of a kinety. The result is a *Paramecium* with part of one row of basal bodies in which the fine structure and details are all pointed 180° away from the rest of the surface. This anomalous kinety is now inherited; it appears to be a permanent fixture of the progeny (which have been carried through 800 generations). This again shows that the cortex is made up of macromolecules that assemble in a particular pattern, and that this pattern, even in a disturbed state, is directly inherited. As Sonneborn points out, this is nongenic inheritance, yet clearly the synthesis of the substances necessary for this replication of the surface macromolecules must be controlled by the genome. It is an exaggerated and beautiful case of what we have emphasized before: direct inheritance from one generation to the next is not restricted to the DNA of the genome, but many other substances and structures are built up from previous cell cycles. In this case we have a large and exceedingly complex cortex whose pattern of fitting together is a property of the macromolecules at the cortex and is not directly under nuclear control. Over what must have been a long time and a vast number of cell cycles, a surface structure evolved. The structure itself had properties such that its immediate form is independent of the nucleus; at the same time, it is totally dependent on the nucleus, we presume, for the synthesis of its specifically shaped building blocks.

Adolf Portmann, *Animal Forms and Patterns,* Schocken Books, 1967. This little book suffers grievously in its translation; nevertheless the originality and intellectual power of the author shine through. Portmann is a nonmolecular biologist, who speculates about the visual significance of animal form. A refreshing, beautiful book.

S. E. Luria, *Life, The Unfinished Experiment,* Charles Scribner's Sons, 1973. Winner of the 1974 National Book Award, Luria covers a lot of ground in this brief volume, from molecules to the evolution of the human mind. He looks at common processes, such as the healing of a knee cap or the growth of a nerve to its target organ, and reminds us how little we know about them.

D. I. Balinsky, *An Introduction to Embryology,* 4th ed., W. B. Saunders, 1975. A widely used, well-balanced text.

GENERAL READING

TEXTBOOKS

Part III

Populations

Figure 7-1 FAMILY.

chapter 7

Genetics: The Science of Inheritance

Genetics has often been seen as the heavy hand of fate that parcels out fortune and misfortune with no regard for fairness or compassion. Has not everyone wished for the beautiful body of the lifeguard or the movie starlet? Yet few of us are so blessed. To be sure, beauty is a subjective and dispensable quality and a rich, rewarding life can be led without it. Of much greater consequence are the lives blighted by crushing defects in body structure and function. Down's syndrome, cystic fibrosis, hemophilia, Tay-

Sachs disease, various forms of mental retardation are all examples of defective body plans decreed by an individual's genes. Each is a story of personal and family tragedy, and of unrealized human potential.

In the past, such genetic defects were accepted as acts of fate. Personal adjustments were made in a passive way, much as we still accept personal beauty or ugliness as unchangeable facts of life. However, recent developments in medicine and cell biology have been changing the picture for many gross birth defects, and for the developmental process itself. Genetic counseling can alert parents in high-risk situations. The technique of amniocentesis (Chapter 6) permits the examination of the fetus at an early stage of pregnancy. The presence or absence of a suspected birth defect can be determined with a great deal of confidence.

Genetic counseling, amniocentesis, and determination of fetal status all represent presently available and widely used technologies. Beyond these lie other technologies that await the solution of smaller or larger technical problems. These technologies of the near or more distant future will make it possible to decisively modify our physical (and mental) makeup. These techniques, variously called "genetic engineering" and "human engineering," have so far been used only rarely on human subjects. Bacteria, amphibians, and small mammals have been the chief experimental organisms. But a technique that has been successful in a vertebrate such as a frog or a mouse offers no theoretical obstacles to use in a human.

Out of countless incredible experiments, I will mention one as an illustration of genetic technology possibly just over the horizon. Cloning is a technique that can be used to produce an unlimited number of "Xerox copies" of an individual. The experimental organism used so far has been the frog, an animal convenient because of its relatively large egg and convenient external embryonic development. The 1N nucleus of an unfertilized egg is removed by means of a fine glass capillary and is replaced with a 2N nucleus taken from an intestinal or skin cell. The egg behaves as though it were fertilized and proceeds to develop into an adult individual genetically identical to the donor of the 2N nucleus.

What rationale might be offered for extending the technique to humans? Cloning permits a bypass of the chance mechanism of sex. It allows for the development of an individual whose genetic makeup is totally specified—perhaps a Nobel Prize winner, or a talented musician, or a great political leader. Self-cloning would also permit the production of spare body parts to replace organs lost to disease, aging, or accident. Even if the best possible eugenic or euphenic intentions are used to justify experiments of this type, they raise a host of questions of a political and ethical nature. Genetic engineering involves a complex technology that will be crushingly expensive if applied to humans. Who would receive such treatment? Would individuals with the greatest wealth also receive the most perfect bodies? What effect would such laboratory intervention have on the present biological unit of reproduction, the human family? And finally, since such manipulations involve a building of humans, what sort of society would the builders be building?

It is ironic that the father of modern genetics was a monk. His name was Gregor Mendel, and in 1865 he clarified the then-chaotic notions about inheritance by reporting on a few simple and brilliant experiments with garden peas. Mendel's reception was undoubtedly the greatest scientific fiasco of the past century. Whereas Darwin's conclusions were strongly resisted by some, they were equally strongly advocated by others. But Mendel could not find a single scientist who understood him in 1865. Thirty-five years later the world rediscovered Mendel, long after he had abandoned his experiments in discouragement and lay buried in a graveyard in Brünn.

Mendel had succeeded where all others had failed because of two innovations. Where others had tried to understand inheritance in its totality, Mendel followed one trait at a time. And where others sought absolute answers, Mendel sought statistical ones; he counted. In other words, Mendel treated the agents of inheritance (today we call them genes) as physical objects passed on from generation to generation by the laws of chance. Today we know that the physical objects are DNA fragments, and we know that they are indeed passed on from generation to generation by the probabilistic mechanism described by Mendel.

The firm foundation stones laid by Mendel in 1865 today support a mighty edifice of genetic knowledge. The field of genetics long ago became too vast for any one individual to encompass, and working geneticists today have such titles as molecular geneticist, phage (bacterial virus) geneticist, human geneticist, and population geneticist. However, in their broadest outlines, all genetic processes follow similar principles. The principles discovered in garden peas are equally valid for fruit flies, land snails, or human beings. Since we have a certain proprietary interest in the last-mentioned species, we shall choose most of our illustrations from the field of human genetics. We shall examine genes from three different perspectives: the mechanism of *gene expression* in single cells or organisms, the process of *gene transmission* from one generation to the next, and the distribution of *genes in populations*.

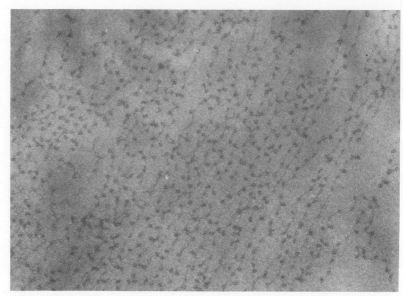

Figure 7-2 ELECTRON MICROSCOPE PHOTO OF CHROMOSOMES ISOLATED FROM NUCLEI OF MOUSE CELLS. These chromosomes are fully extended.

Gene Expression

A *structural gene* is a stretch of DNA long enough to code for a single protein. In addition, certain regions of DNA (such as promoters and operators) function not in coding for proteins but in the control of other genes (Chapter 3). These are called *control genes*. For the purposes of the present discussion, we will concentrate on the behavior of structural genes. Whether or not a given structural gene contributes to the makeup and behavior of a cell depends on whether or not its potential for protein synthesis has been realized. As explained in Chapter 2, it is the mix of proteins in a cell that ultimately defines its structure and capabilities. Genes that are present in a cell but do not contribute to protein synthesis remain silent—they are not expressed in that cell. It is estimated that, in a typical human cell, no more than 5 to 10 percent of the genes present participate in protein synthesis and are therefore expressed.

Genes and Chromosomes

The individual genes in a simple organism like a bacterium are strung together in a single continuous "naked" strand of DNA called a *chromosome* ("colored body"). The chromosomes of higher organisms are more complex: besides DNA, they also contain a type of basic protein called *histone,* as well as other kinds of nuclear proteins.

GENES AND CHROMOSOMES / 239

There are also variable amounts of RNA (as mRNA). Details of the molecular architecture of the chromosome are still obscure, but high-resolution electron microscope photographs show chromosomes of higher organisms as a series of connected beads. The connecting string is DNA, present as a continuous strand snaking around repeating cores of chromosomal protein (see Fig. 7-2). Presumably the chromosomal structure shown in Figure 7-2 represents DNA in an inactive state. Activation requires unwinding of the DNA from the protein spools (Fig. 7-3).

One of the more unusual attributes of these higher chromosomes is their ability to change length as needed, by contracting by a factor of approximately 1,000. This is accomplished by a process of coiling, supercoiling, and super-supercoiling of the chromosomes. The reason such drastic changes in length are necessary is that the chromosomes periodically replicate themselves and the replicas are parceled out equally to the two daughter cells during the normal division of the cells (mitosis). It is only in their shortened form that the chromosomes are compact enough to be pushed off to opposite sides of a dividing cell. This is also the only time when chromosomes are compact enough to be seen by the ordinary light microscope and followed as individual entities.

A typical mammalian chromosome contains tens of thousands of genes. Genes on the same chromosome tend to be passed on as a unit from generation to generation. However, as we will see shortly, genes linked on a single chromosome need not be frozen into permanent associations but can be reshuffled into new chromosomal arrangements.

The Diploid State

The body cells of human beings (and of most other higher organisms) contain two copies of each kind of chromosome. Such cells are said to be *diploid,* or 2N. Thus a human body cell contains a total of 46 chromosomes, consisting of 23 different pairs. Two chromosomes paired in this manner are said to be *homologous.*

Let us now look at the gene complements of two homologous chromosomes. The sequence of genes on one chromosome controls the same set of cellular functions as the sequence of genes on the other homologous chromosome. For example, if gene 1 on one chromosome codes for the respiratory protein hemoglobin, then gene 1 on the homologous chromosome will also code for hemoglobin. Most of the time the two genes are identical in every respect. In this case, the cell is said to be *homozygous* for the gene in question. However, about 30 percent of the time in humans there is a detectable difference between the two genes. Thus, if gene 1 codes for hemoglobin, gene 1'

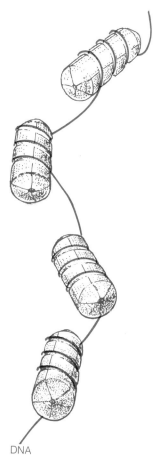

DNA

Figure 7-3
PROPOSAL FOR DNA PACKING IN CHROMOSOMES OF HIGHER ORGANISMS. Presumably the structure shown here and in Figure 7-2 represents DNA in the inactive state. Transcription would require uncoiling from the histone spools.

may code for hemoglobin but with a slightly different amino acid sequence. In this case the cell is said to be *heterozygous* for the gene in question. Two genes concerned with the same function but which carry measurably different information are called *alleles* ("other" or "different").

If a cell carries conflicting instructions for the synthesis of a given protein, which set of instructions will be expressed in the living cell? Two kinds of limiting situations may be observed. In one case, both alleles contribute equally to the functional makeup of the cell. A common example involves the genes that code for blood group antigens (Chapter 4). A series of gene loci exist that code for surface antigens on human red blood cells. Each such gene locus may be expressed in at least two allelic forms. One example is the gene locus coding for the M and N blood group antigens. Gene L^M codes for M antigen, and gene L^N codes for N antigen. It is found that three kinds of individuals exist: those with type M blood, those with type N, and those with type MN. The first group is homozygous for the L^M gene, the second group is homozygous for the L^N gene, and the third is heterozygous, with both alleles contributing equally to the observed antigen production. Alleles of this type are said to be *co-dominant*.

More commonly, the heterozygous state results in the apparent full expression of one allele and the submergence of the other allele. The allele whose presence can be overtly observed is termed the *dominant* allele. The allele whose presence in the heterozygote is masked is called the *recessive* allele. To illustrate the behavior of dominant and recessive alleles, let us choose hemoglobin synthesis as an example. Hemoglobin is the red protein used to transport oxygen in the red blood cell (Chapter 4). The hemoglobin molecule is assembled from four smaller protein subunits. Two are identical and are called *alpha* chains. They are proteins containing 141 amino acids each. Alpha chains are made by a specific gene. Two others are similarly identical and are called *beta* chains. They contain 146 amino acids each, made by a different gene, the beta gene. The amino acid sequence of both chains is known.

Many Americans, mostly of African origin, carry a modified form of the beta gene called the *sickle cell* gene. We will call the normal allele the S allele and the sickle cell allele the s allele. The s allele produces a protein made of 146 amino acids, wherein the amino acid in the position 6 has been changed from the normal glutamic acid to valine. This modified beta chain is then incorporated into the hemoglobin molecule, but the resultant hemoglobin behaves strangely. At low oxygen tension, such as occurs deep in the body tissues, it

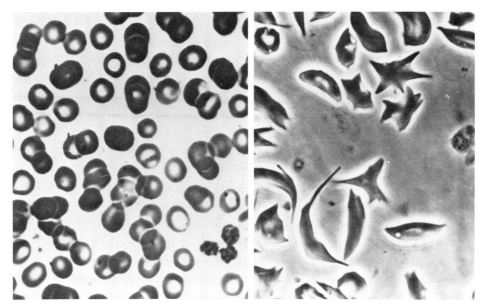

Figure 7-4 RED BLOOD CELLS FROM AN INDIVIDUAL SUFFERING FROM SICKLE CELL ANEMIA (right). Normal cells are shown at left.

tends to crystallize. The result is a strange twisting and wrinkling of the RBC carrying it (Fig. 7-4). Such RBCs are fragile. They rupture easily, have a much shorter lifetime than normal RBCs, and may start clots and occlude blood vessels. This can result in a cascade of catastrophes reaching into every region of the body. Damage to the heart may lead to heart failure. Damage to the bones may lead to osteomyelitis and rheumatism. There may be mental impairment and paralysis due to ministrokes in the brain. Other complications include kidney failure, pneumonia, and abdominal pain. In addition, the resultant anemia leads to general lassitude, poor physical development, and mental impairment. Even with modern medical treatment, half the afflicted individuals do not survive beyond age 20. A person homozygous for the sickle cell gene is said to carry *sickle cell disease.*

But what of the individual who is heterozygous for the sickle cell gene? The individual's genetic makeup is *Ss,* and he or she produces both kinds of beta chains for incorporation into hemoglobin. The resultant hemoglobin has properties more like those of normal than sickle cell hemoglobin. Sickling can still occur, but only under conditions of extreme anoxia, and the individual, who is said to carry the *sickle cell trait,* can look forward to a relatively normal life. Thus, in the case of the hemoglobin system, the presence of the normal *S* allele turns out to have greater importance for RBC function in the heterozygous individual than does the presence of the *s* (sickle cell) allele. We therefore call the *S* allele the *dominant allele,* and the *s* allele the *recessive allele*. Dominance and recessiveness seems to be the rule rather than the exception in genic interactions. As a consequence, we cannot deduce the genetic makeup of an individual (the *genotype*) by simple observation of external appearance the (*phenotype*). An individual with two normal S alleles has the same phenotype as an individual who is heterozygous for s.[1]

Genes and the Environment

Genes do not act in a vacuum. The genetic makeup of an individual reflects only a potential which becomes actualized in a specific environment. The genes of the penguin seem to have produced a slow-moving, clumsy animal—if the penguin is observed on land. But the same penguin in water becomes a powerful, handsome diver and swimmer. To take a human example, a newborn child with PKU (Fig. 2-13) suffers catastrophic consequences in a normal environment. But if the amino acid phenylalanine is removed from his or her diet, the child has a chance of achieving normal development. Or consider a healthy young male growing up in London in the year 1700. He grows up lively and strong and is pressed into service by the Royal Navy. He goes on a long sea voyage during which his body falls apart as he develops scurvy: bleeding gums, swollen joints, bones that break easily, and general exhaustion. Death follows. The reason for the symptoms is vitamin C deficiency. His body (like all human bodies) lacks the enzymes needed to synthesize the vitamin, although, for example, the rats traveling on the same ship thrive because they have no such enzyme deficiency. This genetic defect of the sailor's body was no detriment in his normal environment in London, because his diet there was likely to include fresh fruits and vegetables, good sources of vitamin C. A change in environment from home to ship thus converted an unessential missing gene into a lethal birth defect.

Not all cases of genetic-environmental interaction can be analyzed as simply as PKU or vitamin C synthesis. Does IQ have a genetic basis? How about diabetes? Schizophrenia? Cancer? One of the most useful tools in the study of such questions has been the comparison of identical and nonidentical twins. Identical twins are derived from a single fertilized egg. Their genetic makeup is therefore identical. Nonidentical or fraternal twins result when two eggs are produced at about the same time and both are fertilized. They

---

1. This statement is true for the normal North American environment. As we shall see later, there exist certain environments where the heterozygote does not behave like the homozygous normal genotype.

are therefore no more genetically similar than any two siblings are. They may even have different fathers. Comparison of identical and fraternal twins with respect to a given trait should give us some idea of the relative contributions of genetics and environment to the expression of the trait. Table 1 lists some qualitative traits.

TABLE 1
Differences (averages) between identical and fraternal twins with regard to some qualitative traits

TRAIT	IDENTICAL REARED TOGETHER	IDENTICAL REARED APART	FRATERNAL REARED TOGETHER
Stature (diff. in cm.)	1.7	1.8	4.4
Weight (diff. in kg.)	1.9	4.5	4.5
IQ (diff. in Binet points)	3	6	8.5

Here, any differences between two identical twins should reflect environmental differences alone. Differences between fraternal twins are accounted for by a combination of environmental and genetic influences. Thus all traits listed suggest a genetic component. However, a comparison of this sort does not give us a good idea of the *range* of variation possible from different environments. The variation in environment for two twins is likely to be small.[2] Of special interest is the small number of identical twins who have been reared apart since birth. A 24-point difference in IQ was recorded for one such pair. The two children were adopted into very different environments. One dropped out of school after second grade, and the other went on to finish college. This is an extreme example, and most identical twins reared apart show less variation in IQ than fraternal twins reared together.

Nonqualitative traits may be compared as to degree of concordance; for example, if one twin is mentally retarded, what percentage of the time will the other twin also be retarded? Table 2 shows

---

2. Geneticist L. S. Penrose has said that, if one looked at twin data uncritically, one might conclude that clothing was inherited biologically.

TABLE 2
Concordance (percent of similarity) for two types of twins with respect to certain human characteristics

TRAIT	IDENTICAL	FRATERNAL
Measles	95%	87%
Tuberculosis	65	25
Diabetes mellitus	84	37
Cancer	61	44
Site of cancer (where both have cancer)	95	58
Down's syndrome (mongolism)	100	6
Epilepsy	72	15
Schizophrenia	38	12
Manic-depressive psychosis	77	19
Mental retardation	97	37

the degree of concordance for certain human characteristics. What can we learn from such a comparison? In the case of measles, when one identical twin catches it, the other is almost sure to follow. But since fraternal twins show much the same result, we assume that the effect is largely an environmental one. The measles virus that struck down one twin is likely to be lurking in the same places that the second twin frequents.

The result with tuberculosis may be a little more surprising. The cause of tuberculosis is a well-known bacterium, *Mycobacterium tuberculosis*. While the great majority of city dwellers are exposed to the bacillus, only a small fraction succumbs, and here susceptibility to infection seems to have a large genetic component. Down's syndrome shows most clearly the importance of the genetic factor. Here the individual's cells carry an extra copy of gene 21. The excess gene product unbalances normal metabolism enough to produce a series of physiological changes along with severe mental retardation. There is very little in the way of environmental factors that can counteract these effects. The other traits on the list fall somewhere in between measles and Down's syndrome on the environmental-genetic axis. For these both environment and heredity are important in producing the observed effect.

Polygenic and Pleiotropic Effects

The implied model of gene expression we have assumed thus far is that one gene produces one polypeptide, which in turn produces one character trait. Polygenic and pleiotropic effects introduce an element of complication in this simple model. The prefix "poly-" means "many," and the prefix "pleio-" means "more." A *polygenic* character is one that is controlled by many genes. A *pleiotropic* gene is a gene that affects many characters.

Many of the most obvious human characters are polygenic in nature. Consider skin color, a trait that children obviously inherit from their parents. If skin color were controlled by a single gene existing in two allelic forms (A_D = dark and A_d = light), we would expect all humans to fall into one of three pigmentation classes: dark (A_DA_D), intermediate (A_DA_d), and light (A_dA_d). In fact, human skin pigmentation shows continuous variation from very dark to very light. Such continuous variation can be explained by assuming a minimum of two, and more probably three to six, gene loci that contribute to skin pigmentation. Thus, in a five-gene model, the darkest color is given by the genotype $A_DA_DB_DB_DC_DC_DD_DD_DE_DE_D$ and the lightest by the genotype $A_dA_dB_dB_dC_dC_dD_dD_dE_dE_d$; a genotype such as $A_DA_dB_dB_dC_DC_DD_DD_dE_dE_d$ represents one of the many intermediate shades. If it is assumed that the genes act additively and with equivalent force, then a single gene for either light or dark pigmentation would make only an imperceptible change in the phenotype. This means that, if North American blacks and whites were to freely interbreed, the skin color of the resulting population would be very close to that of the present-day white population. American blacks at present constitute about 10 percent of the total population, and a 1:10 dilution of their genes for skin color would produce a mix of phenotypes resembling those of present-day Spain or Italy. Both of these countries absorbed their Moorish invaders several centuries ago.

Other traits that show polygenic control include body height, mean arterial blood pressure, IQ, length of life, degree of resistance to disease, and dimensions of body structures such as length of arm or weight of the heart. The great majority of such polygenic characters show a continuous range of variation in phenotype in the general population. When the phenotypes are plotted against their frequency in general population, a characteristic bell-shaped curve is observed (see Fig. 7-5).

The converse of the polygenic effect is the pleiotropic effect—

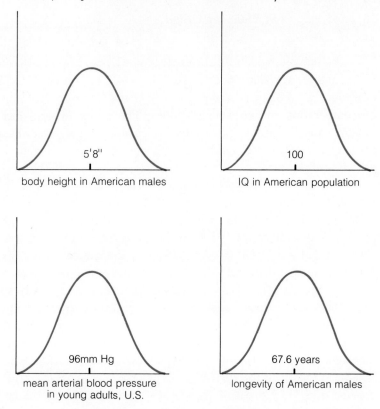

Figure 7-5 FREQUENCY DISTRIBUTION OF PHENOTYPES IN POLYGENIC INHERITANCE. The greatest number of individuals in the populations studied show a mean value for the trait measured. As the measured value of the trait diverges above and below the mean, progressively smaller numbers of individuals are found possessing the trait.

in which a gene affects several character traits. We have already examined one such gene—the sickle cell gene which affects such diverse body organs as bones and joints, the kidneys, and the brain. Sickle cell anemia represents perhaps the best-understood example of gene action in all of human genetics at this time. We understand in molecular detail the events in DNA, RNA, and protein that lead to the effects described. We also understand the environmental reasons for the high incidence and persistence of this gene (page 263).

In most other cases of pleiotropic gene action, the nature of the gene defect is not so well understood. Medical science often recognizes a *syndrome:* a group of specific and abnormal traits present in an individual, which are transmitted to the offspring as a unit. Such a syndrome may be reasonably assumed to depend on a single gene. A famous example is Marfan's syndrome, which may have affected Abraham Lincoln. Individuals with this syndrome have abnormally long legs, fingers, and toes, and are "pigeon-chested." There may also be accompanying heart defects and an abnormal positioning of the eye lens. Though it is hard to understand how such diverse symptoms may result from a single gene defect, it is possible that the primary defect involves an error in connective tissue synthesis.

Other syndromes have been traced to a specific enzyme defect. The Lesch–Nyhan syndrome condemns its victims to severe cerebral palsy, mental defects, self-mutilation, and aggressiveness toward others. It leads to death in childhood. This brutal syndrome has been traced to a defect in uric acid metabolism—the afflicted individuals lack the enzyme hypoxanthine-guanine phosphoribosyltransferase. This is an extreme example of antisocial behavior that depends on a single gene. It is possible that other, lesser degress of deviant behavior may be ultimately traced to gene defects of the type discovered in the Lesch–Nyhan syndrome.

Meiosis and Gene Transmission

Gene transmission is a simple process in organisms, such as bacteria, that typically reproduce vegetatively. A copy is made of all the genes in the parental cell. The copy is then passed on to one daughter cell, while the original is passed on to the other daughter cell. The process involved is mitosis (Fig. 1-21), and this is also the process used to pass genes on from body cell to body cell during vegetative growth. But humans are sexually reproducing organisms, and gene transfer from one generation to another involves interaction between two individuals—the father and the mother.

Maternal and Paternal Contributions

The male and female contribute very nearly equal amounts of genetic information to the future offspring.[3] This was certainly not obvious to most observers before the end of the nineteenth century, because of the confusing effects of dominant and recessive alleles. For example, a brown-eyed mother and a blue-eyed father produce a brown-eyed child. Has the father's gene for blue eyes been lost? Not at all—it is present in the cells of the child, but since the blue-eyes allele is recessive to the brown-eyes allele, it is not expressed in the phenotype. That same invisible blue-eyes gene will ultimately be

---

3. The female contributes a little bit more than the male, because she contributes all the cytoplasm for the fertilized egg, and she may contribute more genes in her sex chromosome (see page 251).

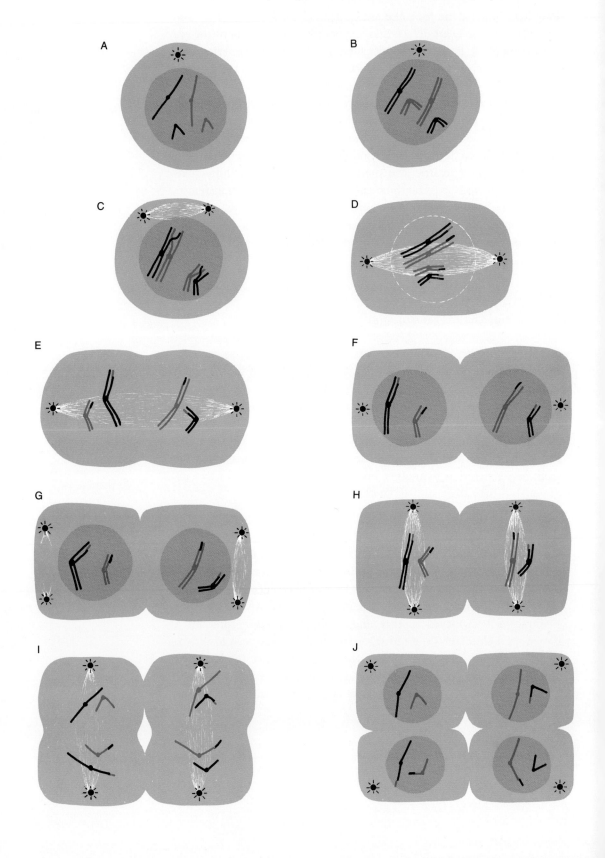

Figure 7-6 MEIOSIS. (A) A cell before the onset of meiosis. The hypothetical cell contains two kinds of chromosomes, a long type and a hooked type. The brown chromosomes have come originally from the mother, the black ones from the father. (B) As the chromosomes enter meiosis, each doubles. The resulting two *chromatids* are held together by a centromere. (C) Homologous chromosomes pair with each other. A close juxtaposition results in occasional interchange of chromosomal material (*crossing over*). (D) Chromosomes complete exchange of material, nuclear membrane dissolves, spindle fibers form across cell, and chromosomes attach themselves to spindle fibers by means of centromeres. (E) Chromosomes migrating to opposite sides of the cell. (F) End of first meiotic division. Each daughter cell receives only one copy of each homologous chromosome, hence the chromosome count of each cell has been reduced by one-half. (G) Centrioles divide again as cells prepare for second meiotic division. (H) Chromosomes attach themselves to spindle fibers by means of centromeres. (I) Separation of chromatids and migration to opposite poles. (J) Products of the second meiotic division.

packed into the gametes of the child and may emerge again in visible form in some later descendant.

The male and female genetic contributions are carried by the male and female gametes—the sperm and egg. In order to produce a sperm or egg, the mitotic mode of cell division cannot be used in humans. An entirely different mode of cell division must be used. It is called *meiosis*.

The root "meio-" means "less," and a meiotic cell division yields fewer chromosomes in the daughter cells than were present in the parental cell. In fact, the resultant number of chromosomes is less by a factor of exactly $\frac{1}{2}$. But each daughter cell receives one copy of each *kind* of chromosome present (Fig. 7-6). Such cells are called *haploid* or $1N$, as opposed to the diploid or $2N$ parental cells.

The sperm and egg are living cells that have become specialized for fusion with each other.[4] The result of the fusion (fertilization) process is a $2N$ cell, the *zygote*, which is the first cell of the future offspring. The zygote is transformed into a new individual through a process of repeated mitotic cell division accompanied by differentiation. It is obvious that, without the reduction in chromosome number during meiosis, repeated fertilization in each generation would lead to a doubling and redoubling of the chromosome number, an impossibility.

---

4. Exceptions include animals such as bees, aphids, some mites, and rotifers. In these organisms, a haploid egg can develop without fertilization by sperm. This "virgin birth" produces a haploid adult.

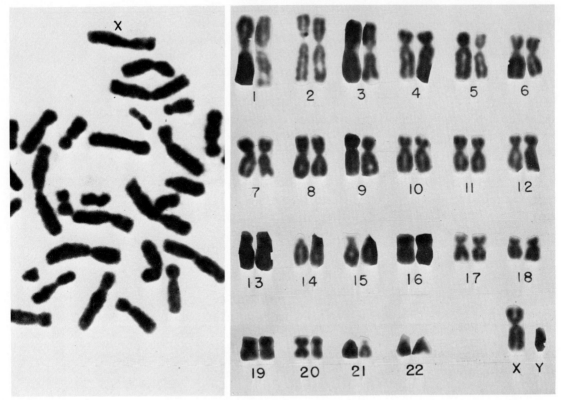

Figure 7-7 THE NORMAL MALE KARYOGRAM (CHROMOSOME DISPLAY). The chromosomes have been isolated during metaphase and therefore each contains two chromatids. Left photo shows typical appearance of chromosomes during metaphase stage of cell division. Right photo shows arangement of chromosomes into homologous pairs.

Sex Chromosomes. Sperm and eggs thus act as the links between generations in sexually reproducing organisms. They are the bearers of the chromosomal cargo destined for the new individual. Because of the nature of meiotic cell division, the sperm and egg each carry one copy of each kind of chromosome present in the parent organism. They therefore carry equal chromosomal cargoes. There is one important exception to this statement. All chromosomes in the diploid cell exist as duplicate copies, except for the sex chromosomes. Two forms of this chromosome exist in humans—an X and a smaller Y. The presence of two X chromosomes makes the bearer a female, hence all eggs contain an X chromosome. A paired X and Y makes the bearer a male. Half the sperm carry the X and half the Y copy. Thus it is the father's sperm that determines the sex of the child (Fig. 7-8). When left to their own devices, the X-bearing and Y-bearing sperm have an approximately equal chance of reaching an egg first

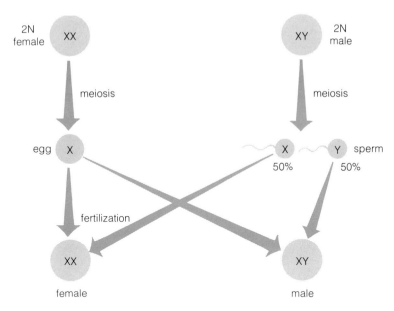

Figure 7-8 SEX DETERMINATIONS IN HUMANS.

and fertilizing it. Thus roughly equal numbers of boys and girls are born. But since the X chromosome is larger than the Y chromosome, slight physical differences exist between X-bearing and Y-bearing sperm. In recent years scientists have taken advantage of these differences to enrich sperm with respect to either X or Y content. The enriched sperm can then be used for artificial insemination. While this technique is still far from being able to guarantee either a male or female offspring, the odds are much improved that the resultant sex will be the desired one.

The X chromosome, besides being involved in sex determination, also carries a number of genes along its length which are quite unrelated to sex.

Some examples include the genes for hemophilia (bleeder's disease), red-green color blindness, and the Lesch–Nyhan syndrome (p. 247). All these traits are recessive to the normal alleles. However, since a man carries only one X chromosome, a recessive gene is fully expressed. A woman, on the other hand, is usually able to suppress such deleterious effects because her other X chromosome is likely to carry a healthy dominant gene.

Thus about 8 percent of the boys in the United States are red-green color blind, while only about 0.4 of the girls show this defect. Hemophilia is found in about 1 male in 10,000 in the general population and is almost unknown in females.

Crossing-over

We have seen that meiosis is the mechanism that operates to maintain a constant chromosome number in sexually reproducing organisms and provides for roughly equal chromosomal contributions by male and female to the next generation. Beyond this, the meiotic process leads to another, very far-reaching result. While meiosis acts to include one copy of each kind of parental chromosome in the gamete, the chromosomes packaged into the gamete need not have the same constitution they had in the parental cell. During meiosis, homologous chromosomes may be torn into fragments and may interchange such fragments with each other through the process of *crossing-over*.

The result is a *recombination* of genes. Genes which had previously occupied different chromosomes might find themselves suddenly thrown together and linked on the same chromosome. This process has been extremely important to the story of biological evolution, and may in fact constitute the chief raison d'être of sex. After all, sex is not necessary for reproduction per se. Both individual cells and entire organisms can reproduce asexually. The cuttings of trees, the runners of strawberries, the rhizomes of crabgrass, and the summer hatch of aphids and water fleas are all examples of successful reproduction without sex. And yet, sex is such a universal and important urge that we must seek an explanation for it. What sex contributes is a kind of juggling process which continually juxtaposes genes into new combinations, and it does this via the process of recombination during meiosis. We must remember that the million or so genes that inhabit our bodies are not an indifferent assortment of protein synthesizers. Eons of evolution have selected those genes that work best together.

Let us illustrate this with an example from Europe, the common land snail, *Cepaea nemoralis*. Two genes control the appearance of the shell. A gene for color exists in two allelic forms, brown and yellow.[5] A second gene for banding also exists in two allelic forms, banded and nonbanded. One would therefore expect to find four types of shells: yellow banded, yellow unbanded, brown banded, and brown unbanded. In fact, only yellow banded and brown unbanded are the forms commonly found. The brown unbanded form is most common in forest habitats, where its shell blends in with the brown leaf litter. The yellow banded form is most common in field habitats, where it blends in with the stalks of grass and green vegetation. Apparently the two genes are tightly linked to each other and travel as a unit. Though recombination can still tear them apart, the frequency of recombination is directly proportioned to the distance that separates

---

5. The brown shades are variable and range into pink.

the two genes. The genes for color and banding pattern are located so close to each other on a chromosome that recombination between them is rare. In fact, this reasoning is used in the "mapping" of relative gene positions on chromosomes. Genes that are usually inherited as a unit are assumed to be on the same chromosome. And once genes have been assigned to a given chromosome, their relative positions on the chromosome can be deduced from a study of recombination frequencies. For example, if linked genes A and B recombine 2 percent of the time and the recombination frequency for both A-C and B-C is 1 percent, we can say that the order of genes on the chromosome is A-C-B (Fig. 7-9).

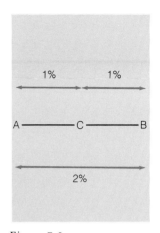

Figure 7-9
MAPPING OF GENE SEQUENCES BY MEASUREMENT OF RECOMBINATION FREQUENCIES. The closer two genes lie to each other on a chromosome, the more rarely will they be split by recombination during meiosis.

The ultimate in close linkage is exemplified by genes making up a single operon (Fig. 3-17). For instance, in the *lac* operon of *Escherichia coli*, the three structural genes necessary for lactose utilization lie immediately next to each other. A major new advantage results: genes that lie next to each other can be turned on an off as a unit (via operator and regulator genes). The probability that recombination will rip apart such adjacent genes is vanishingly small. The assembly of originally unrelated genes into a single operon is perhaps the ultimate triumph of recombinational gene shuffling. We see here the enormous evolutionary payoff of the sexual process. Three things have happened:

1. Favorably interacting genes have been brought together in the same individual.

2. These genes have been placed on the same chromosome so that they will be passed on as a unit to future generations.[6]

3. Since the genes have been positioned immediately next to each other, they can be controlled by the subtle and economical agency of operator and regulator genes.

Most biologists feel that the meiotic reshuffling of genes implied in the sexual process is a more important source of evolutionary variation than is direct mutation.

Independent Assortment and the Laws of Chance

There is one more result of the meiotic process that has important consequences for the process of gene transmission. This is the fact that the chromosomes assort randomly—a given sperm or egg winds up with an assortment of maternal, paternal, and recombined chromosomes that depends completely on chance. This means of

---

6. This is a moot point for the bacterium, since it carries only a single chromosome. However, for the organism with multiple chromosomes, this result has important consequences.

course that, if two different genes are located on two different chromosomes, they will also assort randomly. This was one of the original conclusions of Mendel.

Events that take place according to chance are not necessarily unpredictable. But the kind of predictions we can make depend on probabilities of occurrence, not absolute assertions. For instance, consider a lump of radioactive thorium-234 undergoing decay. If we could isolate a single atom of thorium-234 and observe it, we would have no way of predicting whether it would decay in the next hour, next day, or next month. However, by observing a large number of thorium-234 atoms, we can predict with great accuracy that exactly half of them will decay in 24.1 days.

Again, consider your own body. The water molecules in your body are in a continual state of agitated motion. These movements are all out of phase with each other, like a million clocks that produce a common buzz instead of a synchronized ticking. If by some working of chance all these water molecules were suddenly to move upward at the same time, you might rise into the sky like a rocket. Yet the probability of such a coincidence ever coming to pass is so low that we can safely say it has never happened. Thus, precisely because our constituent molecules move by chance, we can walk with absolute predictability on the ground.

Let us return to genetics. The reason again why we can accurately predict the behavior of chromosomes and genes in the bulk is because we know that chromosomes individually move by pure chance. However, our predictions will be valid only if we talk about the summated results of many, many genetic events. Genetic theory has no predictive value for small sample sizes. Thus a couple has no idea whether their next child will be a boy or a girl, yet they know that out of 120 million babies due to be born next year close to 60 million will be boys. With this important qualification, we can now look at the fate of a typical gene through several generations.

Genetic studies tend to avoid human subjects for several reasons. Their genetic makeup is seldom known, their generation time is long, and they mate according to the dictates of their hearts, not their geneticists. Let us therefore do an imaginary experiment with two large, imaginary populations. Population A carries two normal S alleles for the beta chain of hemoglobin, while population B carries two sickle cell (s) alleles and therefore suffers from sickle cell anemia. What would happen if individuals from population A mated with individuals from population B? (We will ignore the medical problems involved.) All the offspring (the F_1 or first filial generation) would be free of sickle cell anemia and would carry the sickle cell gene as a recessive. Now, what would happen if members of the F_1 genera-

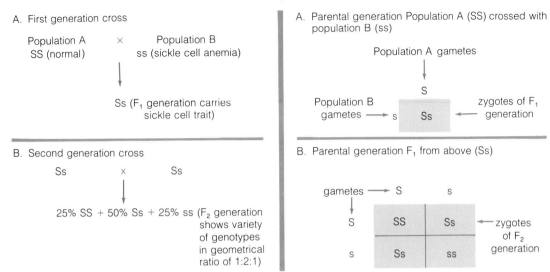

Figure 7-10 (left) HYPOTHETICAL CROSS BETWEEN TWO HUMAN POPULATIONS. (A) First generation cross. The F_1 generation carries the sickle cell trait. (B) Second generation cross (random mating between members of the F_1 generation). The F_2 generation carries a variety of genotypes in the geometrical ratio of 1:2:1.

Figure 7-11 (right) EXPLANATION OF RESULTS OF FIG. 7-10. (A) First generation cross: each parental population incorporates a single kind of allele in its gametes. (B) Second generation cross: each parental population may produce gametes bearing either of two alleles.

tion mated with each other? In the F_2 (second filial) generation, three-fourths would be apparently normal, although two-thirds of the apparently normal would still carry the sickle cell gene. One-fourth of the total would carry two copies of the sickle cell gene and thus be exposed to the full force of the disease (Fig. 7-10). For simplicity, the expected ratios of *SS* to *Ss* to *ss* would be 1:2:1. The explanation for these results is given in Figure 7-11. The two alleles are incorporated with equal frequency into sperm and eggs. The various gametes can fertilize each other with equal probability.

Random Assortment of Two Characters. Now let us complicate our imaginary experiment by following two traits at a time: hemoglobin synthesis and eye color. Let us assume that population A (*SS*) is blue-eyed and population B (*ss*) is brown-eyed and contains no genes for blue-eyedness. Eye color is controlled by a single gene, with brown eye color (*B*) dominant to blue (*b*). We assume that the genes

A. Parental generations Population A (SSbb) crossed with population B (ssBB)

Population A gametes

	Sb	
Population B gametes sB	SsBb	← zygotes of F₁ generation

B. Parental generations F₁ from above (SsBb)

gametes →	SB	Sb	sB	sb
SB	SSBB	SSBb	SsBB	SsBb
Sb	SSBb	SSbb	SsBb	Ssbb
sB	SsBB	SsBb	ssBB	ssBb
sb	SsBb	Ssbb	ssBb	ssbb

← zygotes of F₂ generation

Figure 7-12 GENETIC TRANSMISSION OF TWO INDEPENDENTLY SEGREGATING GENES (BETA CHAIN GENE OF HEMOGLOBIN AND GENE FOR EYE COLOR). (A) Population A (SSbb) crossed with population B (ssBB). (B) F₁ from above (SsBb) crossed with itself.

for eye color and hemoglobin synthesis are located on different chromosomes. Thus population A may be represented as *SSbb,* and population B as *ssBB.* If the two populations are crossed, the F₁ generation will all be *SsBb* genotypically and brown-eyed normals phenotypically (Figure 7-12). However, the F₂ generation will show a range of new combinations of traits: normal brown-eyed, normal blue-eyed, sickle cell brown-eyed, and sickle cell blue-eyed in phenotypic ratios of 9:3:3:1. The explanation is given in Figure 7-12, and the new assortments of traits are due to the independent assortment of chromosomes into gametes as a result of meiosis. Such new assortments of traits resulting from the independent assortment of chromosomes once more serve to generate the new variants used as the raw material in evolutionary selection.

To summarize this discussion of gene transmission generated by meiosis, the overall mechanism therefore includes the following features.

   1. Male and female gametes carry equivalent genetic information.

   2. Chromosomes do not maintain their physical integrity but exchange material between homologous pairs, leading to new juxtapositions of genes.

   3. Maternal and paternal chromosomes assort at random in the formation of gametes.

Population Genetics

Genesis (from which the word genetics is derived) has a beautiful explanation for the origin of human races. After the great flood, all people lived together and spoke one language. They decided to build a huge city for themselves, with a tower on top which would reach into the clouds. Everyone worked together and made good progress, but when God saw the tower pushing higher and higher, he became jealous. He came down to the city and caused everyone to speak in a different tongue, whereupon the town was named Babel and all the people scattered across the world and lived in separate groups.

To the biologist, this story makes a certain amount of sense. Certainly human races could neither originate nor maintain themselves if all people lived together and communicated without hindrance. But what do biologists mean by the term "race"? To answer, we will have to look at some other concepts first.

Gene Frequencies

It has been estimated that perhaps 30 percent of the million or so genes found in the human organism occur in alternative forms. That is, two or more allelic forms of each of these genes may coexist in the human species. For the sake of simplicity, let us assume that only two allelic forms of a particular gene exist and that one is dominant to the other. How can we discover the relative frequencies of the two alleles in a given population? To do this, we must learn some simple laws of probability. The probability of any event can vary from 0 to 1. A probability of 1 means the event will always occur; a probability of 0 means it will never occur. For example, the probability of rolling a three with a die is $\frac{1}{6}$. Dice have six sides, and any side is equally likely to be rolled. The probability of rolling a four is also $\frac{1}{6}$. The probability of throwing heads with a coin is $\frac{1}{2}$, since of the two possible orientations of a coin, heads and tails are equally probable (Fig. 7-13).

Figure 7-13 THE PROBABILITIES OF THREE CHANCE EVENTS. (A) Probability of throwing "heads" with a coin–$\frac{1}{2}$. (B) Probability of rolling "three" with a die–$\frac{1}{6}$. (C) Probability of picking "five of hearts" from a deck of cards–$\frac{1}{52}$.

258 / *Populations: Genetics—The Science of Inheritance*

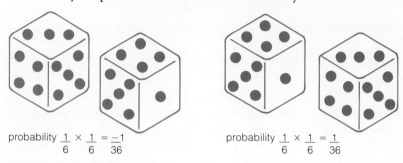

Figure 7-14 THE TWO POSSIBLE WAYS OF ROLLING A "THREE" AND A "FOUR" IN TWO TRIES.

Let us now ask ourselves what is the probability of rolling a three on a die and then rolling a four. The probability of both of these events occurring in sequence is $\frac{1}{6} \times \frac{1}{6} = \frac{1}{36}$. Every time we roll a three we have only one chance in six of also rolling a four. To state this rule formally: *If a particular double event is composed of two independent single events, then the probability of the double event is the product of the probabilities of the single events.*

Now let us look at another kind of situation. What is the probability of our rolling a die twice in such a way that it will come up three once and four once? Obviously there are two ways of doing this (Fig. 7-14). We can either roll a three followed by a four (probability $\frac{1}{6} \times \frac{1}{6} = \frac{1}{36}$), or we can roll a four followed by a three (probability $\frac{1}{6} \times \frac{1}{6} = \frac{1}{36}$). Since both situations satisfy our requirement, the overall probability will be the sum of the individual probabilities, or $\frac{1}{36} + \frac{1}{36} = \frac{2}{36} = \frac{1}{18}$. To state this formally: *If a type of double event may occur in two mutually exclusive ways, then the probability of the type is the sum of the probabilities of the different ways.*

Let us now use these techniques to look at the gene frequencies in a population. Let us look at a gene that exists in two allelic forms, A and a, with A dominant to a. Let us further say that the frequency of the A allele is proportional to p, and that the frequency of the a allele is proportional to q. Here p and q are the probabilities of picking out A or a in a random sample of the population. Of course, the sum of p and q must equal 1. (This is analogous to saying that the probability of throwing heads *or* tails with a coin is 1.) We can now calculate the relative frequencies of the various genotypes:

Genotype	Probability
AA	$p \times p = p^2$
Aa ($Aa + aA$)	$2 \times p \times q = 2pq$
aa	$q \times q = q^2$

Since the above genotypes are the only ones possible, the sum of the

individual probabilities must once again be equal to 1, or
$$p^2 + 2pq + q^2 = 1$$
This is known as the *Hardy–Weinberg* equation, and it allows us to calculate the frequency of a given allele in a population if we know the frequency of q^2, or homozygous recessive.

Let us use the example of diabetes. The genetics of this disease is not terribly "clean," in part because of the large environmental component. Diabetes is not overtly expressed in many individuals who may have the necessary genetic makeup. For instance, diabetes is extremely rare in China. This probably does not mean that the Chinese are lucky with their genes, but that they are lucky with their diet which does not feature the high carbohydrate load common to the United States. In spite of these complications, many experts feel that diabetes is inherited as a one-gene recessive trait.[7] The estimated frequency of diabetics in the United States is 6.8 percent, of whom approximately one-third show overt diabetes. If D is the normal allele and d is the diabetic,

$$d^2 = 6.8\% = .068$$
$$\text{and } d = \sqrt{.068} = .26$$
$$D = 1 - .26 = .74$$

Genotype	Frequency
DD (unaffected)	$(.74)^2 = .55$
Dd (carriers)	$2(.26)(.74) = .38$
dd (diabetics)	$(.26)(.26) = .07$

These figures are shocking, but they illustrate the power of the Hardy-Weinberg equation. The gene for diabetics is carried by 45 percent of the United States population (38 percent carriers plus 7 percent diabetics), a result scarcely guessed by casual observation. (For practice, work out the gene frequency for PKU if PKU is inherited as a single-gene recessive and the incidence is 1 per 10,000 births.) Note that the dominance or recessiveness of a gene has nothing to do with its relative frequency. For example, the blood group O gene, which is recessive, is the most abundant allele in the American population. Likewise, having five fingers on a hand is recessive to having six fingers. It is only when an allele reduces the reproductive potential of an organism that its frequency is lowered by natural selection[8]

---

7. In addition, recent evidence suggests that some cases of diabetes are the result of viral damage and not of genetic inheritance.

8. Other factors that may affect relative gene frequencies are mass movements of people, statistical fluctuation due to small numbers, nonrandom mating, and mutation.

Human Races

Anthropologists make frequent studies of gene frequencies in human populations. Like all busy people, they prefer tests that are quick and unambiguous. Among the all-time favorites are human blood groups. At least 16 different gene loci are required to specify antigens on the surface of the human RBC. Each of these gene loci exists in a variety of allelic forms. The best known is the A, B, O system, a system consisting of three alleles. A and B are codominant and both are dominant to O. This yields six possible genotypes and four possible phenotypes:

Genotype	Phenotype (blood group)
AA, AO	A
AB	AB
BB, BO	B
OO	O

Any number of interesting trips to distant and exotic places have been financed with the ostensible purpose of taking a few drops of blood from natives' fingertips for the study of gene frequencies. As might perhaps be expected, different populations have differing gene frequencies. Table 3 lists frequencies for two blood groups, the A-B-O system and the Rh system:

TABLE 3
Blood group frequencies

POPULATION	O	A	B	AB	RH−	RH+
American Indian (Utes)	.97	.03	.00	.00	.00	1.00
Australian aborigine	.43	.57	.00	.00	.00	1.00
Basque (early European)	.57	.42	.01	.00	.30	.70
English	.48	.42	.08	.01	.15	.85
French	.40	.42	.12	.06	.17	.83
German	.36	.42	.14	.06		
Italian	.46	.33	.17	.03		
Russian	.32	.34	.25	.08		
Chinese	.34	.31	.28	.07	.02	.98
Japanese	.30	.38	.22	.10	.01	.99
Hawaiian	.36	.61	.02	.00		

POPULATION GENETICS / 261

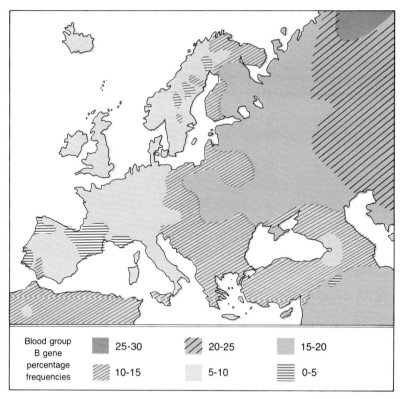

Figure 7-15 FREQUENCY OF BLOOD GROUP B GENE ON EURASIAN CONTINENT.

Let us now go back to the question we asked earlier: What do biologists mean by the term "race"? A *race* is any population with aggregate gene frequencies significantly different from other populations. Admittedly this is a very "soft" definition. How different do the gene frequencies have to be before the difference becomes significant enough? How many genes have to be considered? Depending on where one draws the boundaries, the human species may be made up of 1,000 races or 12, or 3, or 1. Asking how many races exist in the human species is like asking how many lumps exist in a vanilla pudding. Differences in gene frequencies most certainly exist between populations, but different populations tend to be separated by gradients in gene frequencies, not sharp breaks. A good example is the Eurasian distribution of the gene for blood group B (Fig. 7-15). This gene was spread westward from its point of origin in Central Asia by means of the successive Mongol, Tartar, and Turkish invasions of A.D. 500–1500. Only isolated groups such as the Basques of France and Spain and the Lapps of Sweden and Finland remained untouched, because mountains, inaccessible terrain, or determined resistance prevented interbreeding. The gene frequency for the blood group B gene remains highest in the regions of Eastern Europe that remained longest under Mongol and Tartar domination. Eastern Europeans lack the oriental facial features one might expect, because presumably such characters are polygenically determined, and the complex was broken up by dilution with genes from the surrounding populations.

Such mass movements of people have occurred repeatedly in human history, throwing previously genetically isolated populations into contact with each other. While evidence from polygenic traits such as facial features and skin color is quickly lost, the alleles of single-gene characters such as blood groups can remain as silent markers. On the basis of such blood group data, it is estimated that about 25 percent of the genes of American blacks have originated from the surrounding white population.

During the 1930s, the Nazis gave rather wide currency to the concept of a "pure race." A pure race is a genetic impossibility, because a race can only be defined in terms of gene frequencies, and every known human group shows great internal variability of gene composition. To say that two populations differ in gene frequencies is not the same as saying that each population is homogeneous for a given set of genes. In fact, there is no single gene (other than very rare mutations) that is the exclusive property of one population only—all genes are shared by the human species as a whole.

Changes in Gene Frequencies

If genes do fluctuate in frequency from one population to another, what is the causative agent? Genes continually undergo changes because of *mutation,* and mutations are almost always deleterious. Mutation rates are extremely low for most genes (about 1 per 100,000 gametes per generation). Nevertheless, considering the large number of genes involved, it is probable that everyone carries from 5 to 10 such devastating mutations. These alleles are almost always recessive and are present in the heterozygous state, which is why they are not expressed. This rationalizes the almost universal taboo against incest; many more such deleterious genes would be expressed as homozygous recessives if fathers were allowed to marry their daughters or brothers their sisters.

Selection is another major force affecting the gene frequencies of different populations. An obvious example of variability is human skin color, a trait controlled polygenically. The skin, in the presence of ultraviolet light, synthesizes vitamin D which is necessary for deposition of calcium in the bones. The recommended daily dose of vitamin D is 400 IU per day. If this dose is increased to 1,800 IU per day or higher, toxic effects may appear. The body deposits so much calcium that calcium winds up not only in the bones, but also in soft tissues such as the heart, the aorta, and the kidneys. Irreversible damage results. Thus the darker skin pigmentation of people living in the tropics protects them against a vitamin D overdose resulting from excessive ultraviolet light. Genes for light skin would be strongly selected against in such an environment.

The gene for sickle cell anemia represents a curious case. In its homozygous form, the gene is deadly and would be very strongly selected against. Yet in tropical and subtropical regions such as parts of Africa, the Mediterranean, and India, the gene is present in surprisingly high frequency. For example, its frequency in Nigeria is 31 percent. Why does it persist in the face of such strong negative selection? The distribution of the sickle cell gene parallels the distribution of malaria and, in its heterozygous form (as the sickle cell *trait*), this gene offers partial protection against malaria. Since a far larger fraction of the population carries the heterozygous form of the gene than carries the homozygous form, the population as a whole benefits from the gene. The price paid is in the form of the unfortunate homozygous recessives. When this gene is transported to a nonmalarial environment, it becomes wholly deleterious, and its frequency would be expected to drop. The frequency of the gene in American blacks is about 9 percent.

Diabetes seems to represent an analogous situation. Does the individual heterozygous for diabetes enjoy a reproductive advantage over the homozygous dominant? We do not know, but it would be hard to explain such a high frequency of a deleterious gene in any other way.

Another factor that may affect gene frequencies is *genetic drift*. The phenomenon of genetic drift is basically a phenomenon of sampling error. The Gallup Organization calls about 1,700 representative Americans in order to find out how 205 million Americans feel about a question. If only 17 people were called, the results would be far less representative of the country as a whole. Likewise, if we looked at the gene frequencies of a small sample of a population, they would not accurately represent the larger surrounding population from which the individuals were drawn. If this small starting population were then to breed exclusively within themselves, their descendants would likewise reflect the altered gene frequencies, modified by intervening statistical fluctuations in gamete production.

Such events are not unknown in human history. It is said that the tribes of Israel originated from the descendants of Abraham, who

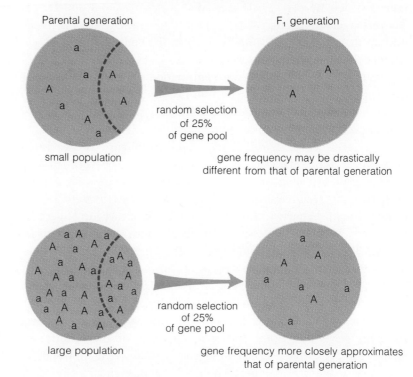

Figure 7-16 GENETIC DRIFT. The circles represent two gene pools (populations) with identical gene frequencies. If it is assumed that in both cases only 25 percent of the individuals will contribute genes to the gene pool of the next generation, then sampling can be expected to lead to larger divergence from the parental mean in the smaller population.

left the city of Ur. The Hawaiian Islands were originally settled by a small band from Polynesia. The American Indians, who spread across the North and South American continents, were descendants of small hunting bands that crossed the Bering Strait. Gene frequencies in these groups are very different today from those in their presumed ancestors (compare the Ute, Hawaiian, and Chinese frequencies for blood groups O, A, and B—Table 3). This aspect of genetic drift is called the *founder effect*.

However, the phenomenon of drift need not involve migration by small samples of the population. There is another sense in which a gene pool may undergo a sampling process. A population is sampled every time it reproduces, because not all individuals of a population take part in the reproductive process. Among humans some remain single, some are sterile, some marry but do not wish children, and some die in wars or accidents or lose their children at an early age. Thus only a fraction of the population contributes its genes to the next generation. Even among those individuals engaging in sexual reproduction only one-half the genes find their way into each gamete.

If one samples an equal fraction of a small and a large population, the sample of the smaller population will show greater sampling error (see Fig. 7-16). Therefore genetic drift will play a role even in a population that shows no migration, providing the population is a small one. While genetic drift may not play an important role in human populations today, it probably was highly significant in earlier times when the population of the earth was small and people lived in smaller groups. The result has been to make the human species an extraordinarily diverse species genetically. In the past, genetic diversity was always a species' most valuable asset in the drama of survival. This should give us some encouragement about the future of the human species.

Bibliography

Garrett Hardin, *Nature and Man's Fate,* New American Library, 1959. A prolific writer, Hardin has not hesitated to comment on social issues of the day from a perspective of genetic and evolutionary theory. If Hardin steps where angels fear to tread, he does it with a combination of grace and intellectual force.

TEXTBOOKS

A. Srb, R. Owen, and R. Edgar, *General Genetics,* 2nd ed., W. H. Freeman and Co., 1965. Though the text needs updating, it remains a clear and well-balanced treatment of organismic and molecular genetics.

Curt Stern, *Principles of Human Genetics,* 3rd ed., W. H. Freeman and Co., 1973. A thorough, authoritative, yet readable treatment of human genetics.

Michael Lerner, *Heredity, Evolution, Society,* W. H. Freeman, 1968. Genetic processes, with their human implications.

ARTICLES

Bernard D. Davis, "Prospects for Genetic Intervention in Man," *Science* 170, 1279–1283 (1970). Davis suggests that control of polygenic traits is much less probable than the cure of monogenic birth defects.

Leon R. Kass, "The New Biology: What Price Relieving Man's Estate," *Science* 174, 779–788 (1971). Kass raises the ethical issues behind such techniques as cloning.

V. A. McKusick and R. Claiborne, eds. *Medical Genetics,* H. P. Publishing Co., 1973. A collection of essays by leading researchers in the field of human genetics.

K. Z. Morgan, "Never Do Harm," *Environment* 13, 28–38 (1971). Presents the hazards of radiation, especially from misuse of medical X-rays.

Figure 8-1 CHARLES DARWIN.

chapter 8

On a hillside in the Catskill Mountains of New York State there sits a huge boulder. The boulder is made up of sedimentary Devonian sandstone, and it has been here for 400 million years. Today it overlooks a valley overgrown with sugar maple and ash. In the distance lies a chain of gently sloping mountains which disappear in a blue haze on the horizon. As I sit on the gray boulder and look out into the distance, I try to imagine what it has seen since it has stood here. The annual march of seasons across the hillside must seem like a faint vibration superimposed on the larger cycles of marching glaciers, continental drift and separation,

Evolution

and crustal upthrust. When the gray boulder was young, the dinosaurs were still 150 million years in the future. The first mammals would not appear for 220 million years, and it would be 260 million years to the first flower.

But now the gray sandstone mass is old. The exposed rock face is eroding rapidly as fingers of frost and plant roots pry between sedimentary plates. The relentless forces of water and wind chisel away the surface, slowly uncovering the record of life of the boulder's youth. Interpositioned between the sand grains are the imprints of long-extinct ammonites, and perhaps inside are the traces of an armored fish. But even as the rock undergoes dissolution, it can look forward to at least another million years of contraction before all its substance has traveled down the hillside and into the ocean where it was born. A million years! An immeasurable time in human affairs, but not so in an evolutionary reckoning.

To the biblical creationists of the first half of the nineteenth century, the world was a well-defined place that was at most a few tens of thousands of years old. To Charles Darwin, the earth was a hoary planet with a history reaching back almost without limit. Limitless time—this perhaps more than any other factor is essential for the evolutionary thinking of the past century. So long as the age of the earth was reckoned in thousands instead of billions of years, all rocks seemed ordinary and all speculation along the lines proposed by Charles Darwin and Alfred Russell Wallace seemed futile. The magnificent sweep of Darwin's thought is not the expression of some personal world view. The nineteenth century was a century of adventure and receding horizons. The atom, electromagnetism, x-rays, the electron, the laws of thermodynamics—these were nineteenth-century discoveries, and they did not long retain their academic character.

To this list of nineteenth-century discoveries we have to add one more: the sense of time. When Charles Darwin set out on his voyage around the world as a naturalist aboard the H.M.S. *Beagle,* he carried with him a book entitled *Principles of Geology* by a then unknown young man, Charles Lyell. It was to Lyell that Darwin owed his conception of time. Lyell held that geological phenomena were the result of natural forces operating over enormous periods of time, and operating in a manner no different from their operation today. Sedimentation, upthrust, and erosion, though working a grain of sand at a time, could wear down old continents and create new ones. Lyell's view of time differed sharply from the view prevailing in the early part of the century. His work did much to prepare public attitudes to accept Darwin's ideas 30 years later, and it was to Charles Lyell that Darwin dedicated his *Origin of Species.*

Toward the end of his life, the question of the earth's antiquity returned to haunt Darwin. Lord Kelvin, widely regarded as the leading physicist of his day, set out to calculate the age of the earth by a study of its internal temperature and rate of cooling. He concluded that the earth must have cooled to its present temperature in only 10 to 30 million years, a span of time clearly too short to allow free play for Darwin's evolutionary mechanism. Darwin's mechanism proposed random variations arising over lengthy periods of time and becoming fixed in natural populations by surviving the winnowing process of natural selection. But how could such an inefficient mechanism generate the present diversity of life in the short time span calculated by Kelvin? Arrogant and relentless, Kelvin pressed his attack, and Darwin retreated and began to doubt himself. He died deeply troubled about the value of his life's work. Had he lived a decade longer, he would have found Lyell's time table vindicated by the discovery of a completely unsuspected phenomenon: the natural radioactivity of the earth's rocks. It is this process of radioactive decay that has kept the core of the earth hot, in spite of the rapid natural cooling stressed by Kelvin. Measurement of the abundance of natural decay products of uranium and thorium indicates that the earth is in excess of 5 billion years old.

The Process of Evolution

Darwin's theory of evolution, first published in 1859, has been vastly enriched in detail, but its large outlines remain unchanged. His statement was approximately as follows.

    1. All species produce offspring in numbers greater than can be supported by the finite resources of their environment.

    2. Natural variability arises randomly among members of each species.

    3. Therefore a struggle for survival results. Variants best adapted to their environment survive through a process of natural selection, and they form the next generation in their own image.

The Species

Both Darwin and his successors assigned a central role to the concept of the *species*. The species has been defined as the evolving unit, but it has proved to be a surprisingly slippery concept to pin down in practice.[1] The species is a defined concept, and the difficulties that arise in using the concept operationally may be illustrated by an example offered by G. G. Simpson.

---

1. In fact, there is good reason to think that many of the important evolutionary events take place at a level below that of the species, the population. We will touch on this later.

Simpson's example is the concept of identical twins—twins derived from a single fertilized egg. We believe that millions of such people have been born, yet in not a single case have we been able to prove this directly—by observing a fertilized egg or early embryonic stage dividing in two and then going on to produce two separate individuals. We accept people as identical twins because of other evidence we expect to follow—evidence such as the similarity of protein structure and character traits. To put it another way, the statement that identical twins are derived from a single egg is not a verified fact but an assumption. But to explain twin similarities by any other mechanism would require far more shaky assumptions.

We face a comparable problem in evolution. Like the splitting of a fertilized egg, the evolution of a species has never been observed.[2] Yet the practicing taxonomist is often called on to decide whether two organisms are members of the same species or not. The answer is often an educated guess.

The question of whether two individuals are identical twins or not is answered by protein analysis, and especially by the ability to tolerate reciprocal skin grafts. Let us see what kind of evidence is sought in deciding whether or not two organisms belong to the same species.

1. The definition of a species is applied only to sexually reproducing organisms. This immediately removes from consideration a large number of plants, bacteria, and other organisms that reproduce by vegetative rather than sexual mechanisms. Common examples among plants include hawkweeds, the prickly-pear cactus, many blackberries, and many members of the sunflower, rose, citrus, and grass families. Thus a thicket of blackberry plants may all be derived from the same parent. Since no gene transfer occurs with any other blackberry plants, do we call the thicket a separate species? To most biologists the question presents a headache which is avoided by restricting the definition of species to sexually reproducing organisms.

2. A species consists of one or more populations that may be distributed in space as well as in time (that is, the population of all human great-great-grandparents is largely dead but is still part of the human species). These populations may be physically distinct from each other. Populations that can be distinguished by physical or other means are called *races* or *subspecies* (see Chapter 7).

---

2. At least not in a natural situation. A species may be artificially created in the laboratory.

3. All members of a population interbreed (exchange genes). This is more commonly a potentiality rather than an actual practice. Likewise, distinct populations may interbreed. For populations separated by space or other geographical barriers, interbreeding again becomes a potentiality rather than a normal practice. To put it another way, a man from Vancouver, B.C., could without difficulty marry a woman from Adelaide, Australia, and raise a family of four, if the opportunity arose for a meeting and a love affair. The fact that they are not likely to do so does not place them in separate species.

4. All populations of a species are reproductively isolated from all other populations outside the species. This is a key aspect of the species definition, and one that may be hard to verify in practice. The fact is that the great majority of species are named after the examination of a few dead specimens in a museum. In these cases nothing is known about the breeding behavior of the organism. Breeding isolation is inferred on the basis of distinctive physical characteristics. That is, it is assumed that two organisms that do not interbreed will have accumulated sufficient differences in physical characteristics to make the fact of genetic separation obvious.

This may not always be true. For example, there are two species of western fruit flies, *Drosophila pseudoobscura* and *D. persimilis,* that share a common range in the west (the range of *D. pseudoobscura* is more extensive than that of *D. persimilis*). Physical differences between the two species are so minute (differences in male genitalia, differences in bristles on front legs, differences in chromosomes) that for many years the two were considered a single species. Yet of thousands of flies examined in the wild not one example of hybridization (interbreeding) between the two has been found. Hybridization between the two is possible under laboratory conditions, but in this case the male hybrids are sterile. Thus the evidence for separate species for the two fly populations is excellent.

The evidence is not always so neat. Two species may not interbreed in nature because their geographical ranges are widely separated, or because their mating behaviors are incompatible. Both these factors operate to prevent gene flow between the lion of Africa (*Panthera leo*) and the tiger of India (*Panthera tigris*). Yet when lions and tigers of opposite sex are kept together under the unnatural conditions of the zoo, interbreeding may occur. The hybrids ("ligers" or "tiglons") show traits intermediate between those of the two parents. Nevertheless, the lion and the tiger are still considered good species, because no ligers or tiglons are born under natural conditions.

272 / Population: Evolution

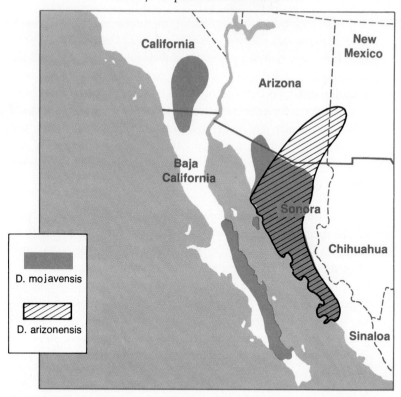

Figure 8-2 GEOGRAPHICAL RANGES OF TWO FRUIT FLIES, DROSOPHILA MOJAVENSIS AND D. ARIZONENSIS.

A still more perplexing situation involves two desert species of fruit flies, *Drosophila mojavensis* and *D. arizonensis*. The *mojavensis* species exists as three geographically separated populations (Fig. 8-2). The range of the Sonora population of *D. mojavensis* overlaps with the range of *D. arizonensis*, yet no hybridization between the two takes place. This indicates that the two belong to different species. Furthermore, although the three populations of *D. mojavensis* are geographically separated and therefore do not interbreed, they interbreed easily under laboratory conditions to form healthy, fertile offspring. This provides indirect evidence for placing them in the same species.

The entire edifice begins to crumble, however, when one performs laboratory mating tests between *D. arizonensis* and either the California or Baja California populations of *D. mojavensis*. Such crossings produce small numbers of healthy, fertile offspring. Does this mean that *D. mojavensis* and *D. arizonensis* are not deserving of species rank after all? Ernst Mayr has proposed the term *semispecies* for forms such as the three races of *D. mojavensis,* and *superspecies* for the *D. arizonensis–D. mojavensis* complex. That is semispecies refers to incipient species while superspecies refers to closely related and typically geographically separated species that were once races of a single species.

We have spent (some would say wasted) this much time on a series of increasingly confusing examples to illustrate one important point: the species is an evolving unit. As such, we must expect to catch species in occasional acts of evolutionary transition from poorly differentiated populations to genetically fully separate entities. We seem to have done this in the case of the *D. mojavensis–D. arizonensis* complex, and countless other examples have been offered by biologists from Darwin onward. It has been said that if we could devise a foolproof way to fix species identities, then evolutionary theory would surely be in trouble.

Evolutionary Change in the Species

We have said that the species is a population or group of populations that undergoes evolutionary change. But what sort of change is evolutionary change? Briefly, to evolve is to undergo a change in the genetic makeup of the species. This involves a change in gene frequencies of the sort described in Chapter 7; some alleles show an increase in frequency, while others decrease. In addition, individual genes undergo more subtle mutational changes, as we shall describe later in this chapter. But perhaps the most important kind of genetic change cannot be described as a simple accumulation of mutations or an increase or decrease in the frequency of a particular gene. For genetic change to produce a significant evolutionary effect, there must be an orchestrated change in a large number of genes. Whether this result is achieved by coordinated change in structural genes or by a change in regulatory genes is unimportant. What is important is that all the genes carried in the gene pool of a population work together in a synergistic way to produce not just a random assortment of phenotypic changes but a meaningful whole of physiological capabilities and behavioral drives that serves to adapt the organism to a specific niche in its environment (see Chapter 13 for a discussion of the ecological niche).

To illustrate, let us consider an animal such as the modern horse (Fig. 8-3). The horse is an animal of the open plains, where it grazes on grass. Why does the horse have such a long head instead of a short one like the lion? Because a long head is necessary to accommodate the large grinding teeth necessary for macerating tough stalks of grass. Since such a grass diet always contains large amounts of sand

Figure 8-3 HORSES.

and other abrasive material, the teeth are worn down and a horse would quickly become toothless if it were not for the fact that its teeth have very long crowns which extend deep into the jaw and keep pushing upward as the grinding surfaces become worn. This is why the head must be deep as well as long.

Why is the neck so long? It has to be long enough to reach to the ground when the horse is feeding. Why is it so thick? It contains the massive musculature necessary to keep the heavy head aloft. These muscles are anchored on long vertebral processes in the shoulder region.

Why are the legs so long? Since the horse's teeth have been modified for grazing, they are useless as either an offensive or defensive weapon. The horse's defense is based on speed and alertness, with a backup in the form of kicking hoofs. Speed is achieved by the elongation of ankle bones and the lifting of the foot off the ground until only the very tip of the toe touches the ground (the horse's "ankle" is halfway up the leg, at the hock). This toe is capped by a heavy nail called a hoof. The horse has lost four of its ancestral five toes, because for a heavy animal a single toe offers a structurally sounder support—there are no ligaments between the toes to give way under heavy impact.

THE PROCESS OF EVOLUTION / 275

Why is the horse so large-bodied? Because a diet of cellulose (the main food constituent of grass) requires bacterial help during the digestion process. No mammal has evolved a digestive enzyme able to break down cellulose, while bacteria have. The solution used by grazers such as horses and cows is to house the required bacteria in their digestive tracts. These must be proportionately very long and complicated in order to give the bacteria time to complete the breakdown process, and thus the grazing animals themselves tend to be large and wide-bellied.

And why does the horse have a tail? To help drive off the horseflies that land on its sparse-haired sides. We could go on with finer and finer points, but it should be clear by now that all parts of the horse's body are interrelated and function as a unit to make the horse succeed in a life on the plains.

We can now begin to see the importance of the need for reproductive isolation of each species. If a species possesses a coadapted set of genes that allow it to succeed in a specific ecological niche, then hybridization with another species might break up this genetic complement and cost the species its fineness of ecological adaptation. To take a ridiculous example, what would happen if a horse were able to hybridize with a tapir? A tapir belongs to the same mammalian order as the horse. It is much smaller than the horse (about 3 feet high), lives in the dense tropical jungles of Malaysia and South America, subsists on succulent vegetation, and takes to water when threatened. The tapir lacks both the dental and digestive apparatus to feed on an all-grass diet and lacks the speed to escape predators on an open prairie. The horse, in contrast, is physically too large for a life in the forest and would be unable to find its grassy diet there. Any hybrid between the two would therefore be unsuccessful as either a

Figure 8-4
SOUTH AMERICAN TAPIRS, DISTANT RELATIVES OF THE HORSE AND RHINOCEROS.

jungle or a plains animal. In general, evolution could produce none but the most generalized organisms if gene flow between species were not restricted.

Isolating Mechanisms. If reproductive isolation between species is so important to evolutionary success, then what mechanisms do organisms use to maintain their reproductive isolation? To use a human analogy, two species may avoid unwanted hybrid offspring by practicing either contraception or abortion. As in the case of humans, contraceptive mechanisms are more widely used, are less expensive biologically, and are generally more effective in maintaining reproductive isolation. Contraceptive isolating mechanisms act to prevent fertilization and zygote formation between individuals of different species. But unlike in the case of humans, the most effective form of contraception between organisms in different species is abstinence. The following are all examples of isolating mechanisms of the contraceptive type.

1. There may be differences in habitat between the two species. We have already mentioned the horse and the tapir, animals that prefer the plains and the jungle, respectively, and which therefore are unlikely to meet in order to mate.

2. There may be seasonal or temporal differences in breeding periods. For example, of the two common American toads, *Bufo americanus* (the American toad) breeds early in the spring and has finished its breeding program by the time the closely related *B. fowleri* (Fowler's toad) is ready to begin.

3. There may be differences in behavior that make it impossible for the two animals to complete their courtship rituals. Of course, this mechanism is not possible for organisms other than animals. Among the animals, however, it is probably the single most important isolating mechanism in operation. Just as among humans a budding romance may be terminated by a thoughtless gesture or comment, so among animals there must be correct reciprocal cues between mating partners if the mating act is to be consummated. These reciprocal cues will be lacking if the wrong species is present. The examples are endless, but one that is perhaps most easily observed is the mating behavior of fireflies. Observe an open field at night in the early summer as fireflies dot-and-dash the darkness with little points of light. The males fly overhead while the females wait in the grass below. Follow the signals of a single male for a while to establish the pattern of flashing. If you repeat this for a series of males, you will discover that several kinds of flashing patterns may be present in the same field, generated by males of different species. The

THE PROCESS OF EVOLUTION / 277

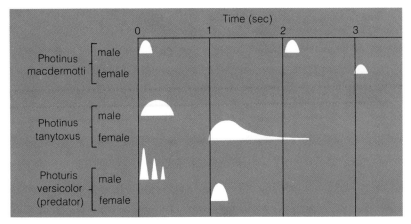

Figure 8-5 SIGNAL PATTERNS IN FIREFLIES. The male signals while in flight, and the female answers from the ground. The *Photuris* female is a *femme fatale* who is able to mimic the signal responses of *Photinus* females. When the deluded *Photinus* male approaches, he is devoured. She does not practice this behavior with males of her own species.

females on the ground respond by emitting a characteristic light pattern when a male of the correct species is overhead. Thus the fireflies are guided to mates of the correct species (Fig. 8-5).

Such behavioral synchronization appears to be almost universally present among members of the animal kingdom. The absence of such mechanisms among the plants may account for the much higher incidence of hybridization between different plant species. However, even among the animals, hybridization is much enhanced when the organisms are removed from their natural environment. The behavioral isolation mechanisms are thus not foolproof.

4. There may be mechanical incompatibility between different species. This appears to be an especially important factor among insects, in which the positioning of genitalia must be so precise that even a slight mismatch makes mating impossible.

5. There may be gamete incompatibility. The sperm of a male of the wrong species may not survive in the reproductive tract of the female, or a pollen tube of the wrong species may not be able to develop after deposition on the stigma (Chapter 10).

In contrast to the foregoing mechanisms of isolation, which prevent interspecific fertilization, aborting or postzygotic isolating mechanisms allow fertilization and development to take place. However, the hybrid formed may be weak or sterile. An example of this kind of isolating mechanism is the crossing of a horse and a donkey to produce a mule. The mule is a physiologically healthy animal that suffers evolutionary abortion because it cannot reproduce. These are obviously more expensive processes biologically and appear to be less effective as agents of isolation.

To summarize, we have seen that evolution consists of genetic change in a species. This change is of a coordinated sort and acts to generate a closer fit between a species and its ecological niche or way of life. Such an improved fit is possible only if the species maintains its genetic isolation from other species adapted to other ways of life. The isolating mechanisms may be contraceptive or abortionlike in nature.

Let us ask next what processes are responsible for the genetic changes in question. For instance, how could such an incredible animal as a horse originate?

In theory at least, the genetic processes of mutation, genetic drift, and migration could produce such an animal if given an infinite amount of time. But these processes are random. While they produce genetic change, there is on adaptive focus to the change. Only natural selection can focus such random changes into ecologically successful channels. In this process it is greatly aided by the power of sexual recombination in more rapidly assembling favorable gene combinations. An analogy that has been used is that of asking a monkey to compose "A Midsummer Night's Dream" by randomly hitting the keys on a typewriter. It is conceivable that the play could be written after many billions of years of random tapping of the keys. This would be analogous to dependence on the mechanisms of mutation, drift, and migration. The play could be produced much more rapidly if the monkey were given a pair of scissors and a copy of all the other plays of Shakespeare. By cutting up the words and reshuffling them into new though still random arrangements, progress would be more rapid. This would be analogous to adding the genetic mechanism of chromosomal recombination to the processes of mutation, drift, and migration.

But progress would be still more rapid if the monkey with scissors were also provided with an editor. The editor would cull from the stream of nonsense those arrangements that showed successively closer approximations to "A Midsummer Night's Dream" and hand only these back to the monkey for further refinement. This would be analogous to adding to the preceding random genetic mechanisms the mechanism of natural selection. However, even with the editing hand of natural selection, evolution remains an extremely slow process. The earliest recognizable ancestor of the modern horse was *Eohippus*, a dog-sized, hoofed mammal with four toes on each foot and a diet of shrubs and leafy vegetation. The transition from *Eohippus* to the modern horse took 50 million years.

THE PROCESS OF EVOLUTION / 279

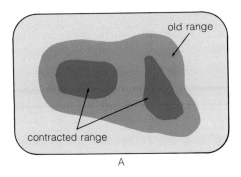
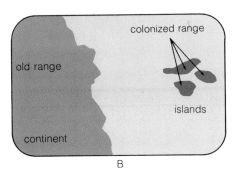
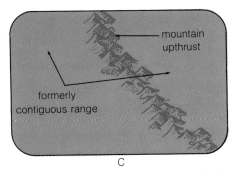

Figure 8-6 ESTABLISHMENT OF GEOGRAPHICAL ISOLATION. (A) Shrinkage of species range. Examples: tapirs, lungfish. (B) Outmigration. Example: finches of Galapagos Islands. (C) Splitting of range by geological forces. Example: separation of India and Tibet by Himalayas.

The Formation of New Species

Natural selection may produce two kinds of evolutionary change in a species. If a species exists as a single interbreeding population and natural selection acts uniformly on the entire population, then each generation will differ by an infinitesimal amount from its predecessor. But after 1,000 or 100,000 generations of persistent selection, sufficient differences may have accumulated that the earlier and later populations would have to be ranked as separate species. This kind of evolution is called *phyletic evolution* and, while it generates new species, it does not change the total number of species in existence.

However, if a species exists as a cluster of separate populations that do not interbreed, then natural selection, acting differentially on the separate populations, might accumulate enough genetic changes to convert such populations into separate species. The selection process would act in a differential way if the environmental pressures on the separate populations were different. This type of evolution is called *primary speciation*. This is the process responsible for the generation of most of the enormous diversity of life on earth today.

The barrier that normally acts to bar new species formation is sexual reproduction. For instance, if a human being were born with a series of mutations giving the owner a bushy tail, would the individual be likely to establish a new, bushy-tailed species of *Homo?* Not at all—the genes for tail-bearing would be diluted in the general gene pool and probably disappear with as little visible trace as would the genes for dark skin color when diluted in a much larger pool for light skin color (see Chapter 7). For genetic differences to be established between two populations, the populations must be reproductively isolated. Such reproductive isolation is ultimately maintained by the various kinds of isolating mechanisms we have discussed. But in the vulnerable early stages of speciation, geographical isolation is probably the only reliable way to protect the genetic individuality of two populations. Such geographical barriers may be as insubstantial as a river or a desert for some species, or as formidable as the Himalayas or the oceans for others (Fig. 8-6).

Once the separated populations have established strong isolating mechanisms and achieved species status, they can be brought back into contact with each other without amalgamation into a hybrid population. Such recently established species show close similarities of body form and behavior, and are called *sister species*. Examples include *D. mojavensis* and *D. arizonensis* mentioned earlier, the American toad and Fowler's toad, the northern and the southern flying squirrel, the lion and the tiger, the wolf and the coyote, and countless others. *Homo sapiens* lacks a sister species.

There exists another important mechanism of species formation which requires no period of geographical separation and which produces a new species in a single generation. This mechanism is restricted almost entirely to plants and involves the establishment of *polyploidy* in the plant. A polyploid organism contains more than the normal diploid ($2N$) number of chromosomes in its cells. The polyploid state can be achieved in a number of ways, but the method of evolutionary interest requires that two dissimilar organisms hybridize. The hybrid receives a $1N$ complement of the chromosomes from one parent and a second $1N$ complement from the other parent. This hybrid is sterile, because during meiosis it is unable to generate gametes with a consistent chromosomal content (Fig. 8-7).

During the normal meiotic process, homologous doubled chromosomes pair to produce assemblies of four chromosomes (quadrivalents) which undergo crossing-over and are then partitioned out in an equal way during the two meiotic divisions. The result is identical $1N$ gametes. However, when the hybrid undergoes meiosis, homologous pairing cannot occur, since the chromosomes contributed by the two parents are too dissimilar. In the absence of pairing, chromosomes

THE PROCESS OF EVOLUTION / 281

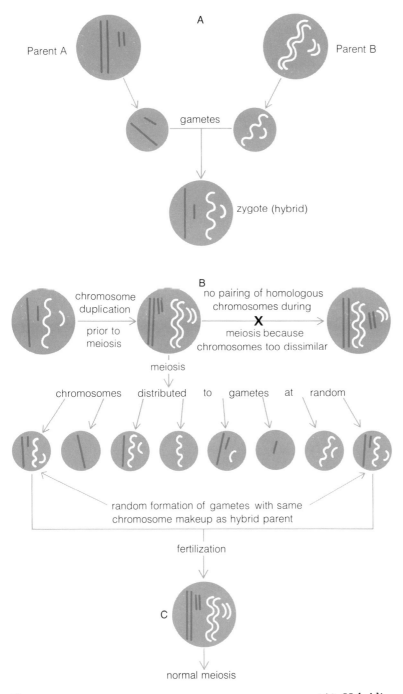

Figure 8-7 FORMATION OF NEW SPECIES BY POLYPLOIDY. (A) Hybridization between two dissimilar species. (B) Formation of gametes by hybrid during meiosis. (C) Formation of zygote with 4N (polyploid) chromosome count.

are distributed at random during the first meiotic division. The resultant gametes contain a jumble of chromosomes and, when the gametes attempt to fuse, they fail to reconstitute the chromosomal arrangement of either the hybrid or of its parents. In very rare cases (the greater the total number of chromosomes, the rarer) *all* the chromosomes present in the hybrid somatic cell wind up in a single gamete.

In the example shown in Figure 8-7, the number of chromosomes present in the 2N hybrid is 4, and the chance of such an event occurring is $(1/2)^4$ or $1/16$. With 10 chromosomes present, the probability drops to $(1/2)^{10}$, or approximately $1/1,000$. With 20 chromosomes, the probability is $(1/2)^{20}$, or approximately $1/1,000,000$, and with 40 chromosomes the probability is approximately one chance out of a trillion. The human 2N chromosome number is 46, a fairly typical number among mammals. Typical chromosome counts among higher plants are in the range of 6 to 14. This is one reason why animal hybrids such as the mule, formed by crossing a horse and a donkey, are sterile.

If two gametes with a full chromosome count should meet, they would establish a zygote with double the chromosome count of the hybrid parent. Such a polyploid organism would be able to undergo normal meiosis and would therefore be fully fertile.[3] It would be reproductively isolated from both its hybrid parent and from its normal grandparents, because of difficulties with chromosome pairing and separation during gamete formation. Such a polyploid plant would immediately embark on a separate evolutionary adventure. Because such polyploid species incorporate the genes from two related parental species, they generally show a hybrid vigor and compete effectively with the parental species. A surprising number of our common plants are polyploid, including important food crops such as wheat and the potato. Other examples include cotton, tobacco, many blackberries, the redwood tree, and almost all ferns (Table 1).

Thus we see that plants, which lack the important behavioral isolating mechanisms common among animals, are able to establish reproductive isolation by the novel method of polyploid formation. This gives them a mechanism of speciation that animals lack. The three common mechanisms of speciation we have discussed are summarized in Figure 8-8.

---

3. The requirement for two such gametes means that, in practical terms, only organisms capable of self-fertilization can establish the polyploid state. Self-fertilization is almost unknown in animals but is very common among plants. This is another reason why this mechanism of species formation is restricted to the plants.

THE PROCESS OF EVOLUTION / 283

TABLE 1
Percentage of polyploid species among higher plants

REGION	PERCENTAGE POLYPLOIDY
West Africa	26
Northern Sahara	38
Great Britain	53
Greenland	71
Worldwide average	50

Faunal Regions and Continental Drift

Thus far we have looked at the factors that tend to create new species. Let us ask ourselves now why the various species of the world show the kind of distribution they do. There is a story by Jules Verne (very exciting to an 11-year old boy!) concerning the hijacking of a nineteenth-century clipper ship. The ship sets sail for South America, but a sinister and mysterious man on board sabotages the compass so that the ship misses South America and drops anchor off southern Africa. The passengers and crew disembark and head inland, heading for what they expect to be a ranch on the pampas. The sinister man wants to encourage this belief. Suddenly one of the crew sights an animal running quickly in the distance. It is an ostrich. Like a

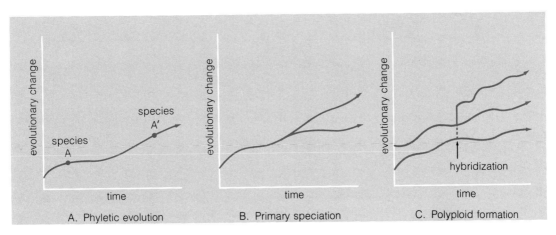

Figure 8-8 THREE COMMON MECHANISMS OF SPECIATION.

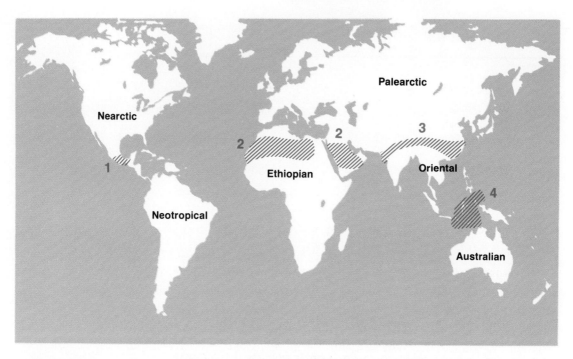

1 and 3 Mountains
2 Deserts
4 Water

Figure 8-9 FAUNAL REGIONS OF THE WORLD. The geographical and ecological barriers between different regions are shown in color.

Figure 8-10 REPRESENTATIVE MAMMALS FROM DIFFERENT FAUNAL REGIONS. (A) Raccoon—Nearctic. (B) Vicuna—Neotropical. (C) Hedgehog—Palearctic. (D) African rhino—Ethiopian. (E) Gibbon—Oriental. (F) Echidna—Australian. The echidna is an egg-laying mammal not related to the hedgehog.

good European the crewman raises his rifle to shoot, but the hijacker deflects the shot, letting the ostrich escape. He assures the crewman that the animal was a rhea, a large bird somewhat resembling the ostrich. The hijacker knew, as everyone else does, that once you find an ostrich you know you are in Africa!

We find the same principle applying to almost all of the million or so other species of plants and animals on earth. Very few species are cosmopolitan—most are restricted to more-or-less well-defined territories. Interestingly, the individual ranges of these organisms do not form a patternless mosaic but are grouped into large aggregates. When we talk about the distribution of animals, we call these aggregates *faunal regions* or zoogeographical realms. See Figures 8-9 and 8-10.

The boundaries of the individual faunal regions are not as sharp as the boundaries of individual species. Rather, there is a tapering effect at the edges. For instance, the border between the oriental and Australian regions cannot be drawn as a fixed line. Instead, the islands of Indonesia which separate the two regions show a gradual decline in oriental fauna and an increase in Australian fauna going from west to east (Fig. 8-11).

This separation of the world into faunal regions is not a reflection of some intrinsic "best fit" of the varied animals to their own regions. The European rabbit has been a spectacular success in Australia, so much so that it threatens the survival of much of the native fauna. The mongoose of India has had a similar history in Hawaii. African honeybees are swarming across South and Central America. And the Norway rat has overrun the whole world and thrives wherever humans do.

The restriction of organisms to specific faunal regions has come about as the result of a historical process rather than some intrinsic determinism. Because of the filtering effect of the large barriers between the different regions, we can think of each region as a large island where, at least for the past 70 million years or so, evolutionary processes have worked in relative isolation from each other. What process accounts for the fragmentation of the world into separate faunal regions? We are deceived by maps which show a static earth. The map of the world that is true for today will not be true a million years from now, nor was it true a million years ago. The Atlantic Ocean is slowly spreading; Brussels is drifting away from New York, and Conakry away from São Paulo. Southwestern California is heading north toward Alaska. Iceland is adding new land in its interior. None of these changes will amount to much in your lifetime, but they may have amounted to a great deal in the lifetime of a species, and this includes many species still in existence today.

THE PROCESS OF EVOLUTION / 287

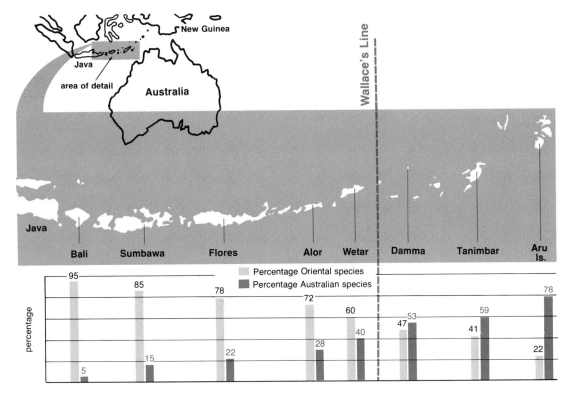

Figure 8-11 FAUNAL BOUNDARY BETWEEN ORIENTAL AND AUSTRALIAN REGIONS. The map shows the islands of Indonesia. Vertical bars below are a measure of relative abundance of Oriental and Australian forms of reptiles. Percentages of former are shown in color. For organisms other than reptiles, the tipping point between Oriental and Australian regions will come at different points. For instance, the transition point for insects is between New Guinea and the Australian mainland.

The present distribution of many species must be explained not in terms of the geography of the world as it exists today but as it existed 200 or 400 or 600 million years ago (Fig. 8-12). In the Paleozoic era, 400 million years ago, the world contained a single land mass, called Pangea ("whole earth") by geologists.[4] Successive fragmentations of Pangea resulted from movements of the earth's mantle, on which both the continents and the oceans ride. Seventy million years ago, the continents began to assume their present orientations. North America wrenched free of Europe, Australia tore free of the Antarctic, while India moved rapidly north, where it was to slam

---

4. Pangea was itself assembled through the aggregation of previously separate continents. The seams of these earlier aggregations are still visible in the form of mountain ranges such as the Urals and the Appalachians.

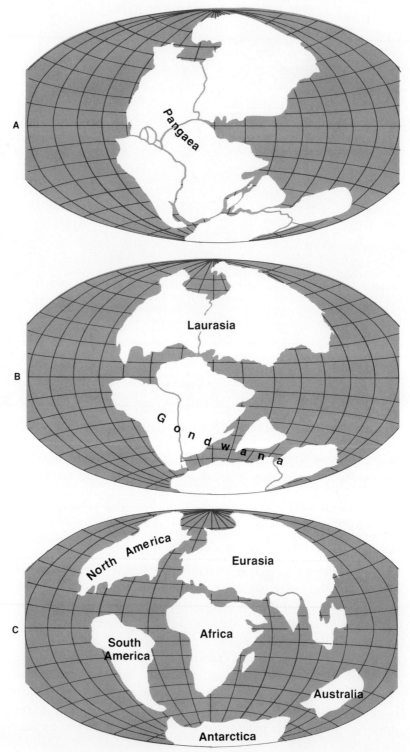

Figure 8-12 CONTINENTAL DRIFT. (A) Paleozoic era, 450 million years ago. Fishes and invertebrates dominant. (B) Beginning of Mesozoic era, 230 million years ago. Reptiles rising to dominance. (C) Beginning of Cenozoic era, 60 million years ago. Rise of mammals.

into southern Asia and throw up the high Himalayas. These mountains reached their final elevations only a million years ago. A species of goose still flies over these mountains in its annual migration from its summer range in Tibet to the winter range in India. The migration path was established when the Himalayas were lower and, as the mountains rose below them, the geese pressed higher and higher, until they were flying through an atmosphere where humans cannot survive without an oxygen mask. For these geese, the barrier between the Palearctic and oriental regions does not yet exist.

There are other examples of organisms whose present range can be understood only in terms of past geography. The lungfish are freshwater fish which evolved during the Devonian period, about 400 million years ago. Traveling along ponds and freshwater streams, they spread over the entire earth. Today lungfish are found only in South America, Africa, and a small patch of Australia. Fossil lungfish are found on all continents. Had the lungfish evolved more recently, it would have shown more restricted distribution, as do, for instance, the great apes (Pongidae). There is no American great ape, because by the time the Pongidae appeared in Africa about 25 million years ago the only way to reach the Americas was by boat!

Adaptive Radiation

What happens when a new species enters a given faunal region, either by migration or by speciation? It proceeds to diverge from all other species, and the divergence increases with time. What direction does this evolutionary movement take? As discussed before, the question of which gene combinations survive and produce the most offspring is decided by natural selection. Certainly an important component of this natural selection is the physical environment of the species. An arctic fox born without fur does not live long enough to propagate its forgetful genes. But of far greater importance is the *biological environment* of the species. In simplest terms, this has to do with its feeding relationships, since every living organism feeds on other life, or is itself eaten, or both.

Thus an insect cannot evolve into a honeybee if there are no flowers in its environment. But now let us suppose that an area already has a species of honeybee in residence. Let us assume that it feeds very efficiently on clover and less efficiently on a deep-throated flower such as phlox. Now a second insect species arrives on the scene, able to feed about equally well on both flowers. Those individuals of the second species that feed on clover will be in competition with the honeybee, while those specializing on phlox will have a less restricted food source and will be able to leave more descendants. Thus the second insect, in

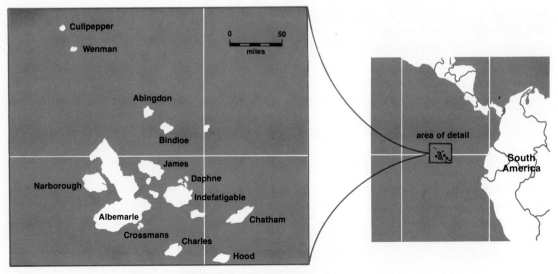

Figure 8-13 THE GALAPAGOS ISLANDS.

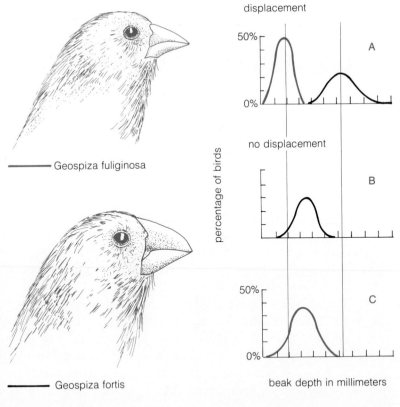

Figure 8-14 CHARACTER DISPLACEMENT IN DARWIN'S FINCHES. Two of the three species of ground-feeding finches are shown. They differ in beak depth, a fact that permits feeding on seeds of different sizes. (A) Character displacement on islands of Charles and Chatham, where *fortis* and *fuliginosa* coexist. The beak sizes of both species have been displaced away from the means shown in (B) and (C). The resultant specialization in feeding minimizes competition. (B) No character displacement on the island of Daphne, where *fortis* exists as the sole, all-purpose ground finch. (C) No displacement on the island of Crossman, where *fuliginosa* is the sole ground-feeder.

order to reduce competition with the honeybee, is strongly pushed toward specializations that encourage phlox feeding. The honeybee in turn tends to become a greater specialist on clover. This is an example of *character displacement*. The net result is that any new species in an area succeeds only if it can exploit a way of life that is not already being exploited by another resident species. Perhaps more rarely, a new species may succeed by outcompeting the resident species. New species continue to succeed until all conceivable ways of life (ecological niches) are filled. This process is known as *adaptive radiation.*

Many examples of this process can be demonstrated, and the first of these was described by Darwin on the Galapagos Islands (Fig. 8-13). These islands are volcanic caps sticking up from the Pacific Ocean. When they were formed millions of years ago, they must have been sterile. Essentially all forms of plants and animals found there today must have arrived by accident, carried by winds or ocean currents from far away. Most were related to the species found on the South American mainland 600 miles to the east, but a few had apparently arrived from the Polynesian islands even further away to the west.

Darwin was well aware of the geological history of the islands, and he appreciated the fact that their present plant and animal life must have descended from a few original colonizing individuals. But he was struck by the fact that such a diversity of species was found on islands close enough to each other to have a similar climate. If a new species had in fact been created on the Galapagos, the mechanism responsible would have had to be something other than a response to climate differences. We know today that a cluster of islands such as the Galapagos is unexcelled as a stimulus to evolutionary diversification. If *Homo sapiens* had evolved, for example, on Borneo instead of in Africa, we might well be cohabiting the earth today with one or two other species of humans. (This presumes, of course, the delayed invention of the boat.) In such an area, a neighboring island is close enough to be colonized by occasional accident, yet far enough away to provide the geographical separation two populations need to speciate.

When Darwin visited these islands, located on the equator about 600 miles west of Ecuador, he found a number of strange species of animals, including 14 species of finches, small seed-eating birds with massive beaks. On the South American mainland, finches have remained finches, small birds feeding on the ground and specialized for a diet of seeds. On the Galapagos, no other small land birds were present, and the invading finches moved systematically to exploit all the available ways of life. One group remained on the ground and

Figure 8-15 CAMARHYNCHUS PALLIDUS USING A CACTUS SPINE TO PROBE FOR INSECTS IN TREE BARK.

fed on seeds. However, the three species in this group (*Geospiza magnirostris, G. fortis,* and *G. fuliginosa*) became subspecialized for eating seeds of different sizes, a fact that is reflected in their beak sizes (Fig. 8-14). The three species provide a particularly clear example of character displacement. On islands where a given species exists as the sole ground feeder, its beak is of intermediate size. When more than one species inhabits the same island, the beak sizes of each species are displaced away from the common mean. This reduces competition between the closely related species.

Other groups of finches turned to alternate food sources and feeding strategies. (1) One group of two species feeds on the abundant cactus seeds of the islands. (2) One species alternates between cactus seeds and seeds on the ground. (3) One species feeds on tree seeds. (4) Seven species have become specialized for a diet of insects. One of the seven (*Camarhynchus pallidus*) has become essentially a woodpecker, probing for insects in the bark of trees. But instead of the long tongue and air-hammer-like beak of the traditional woodpecker, *C. pallidus* has evolved a behavioral substitute: it picks up a sharp cactus spine in its beak and uses it to probe for insects (Fig. 8-15). Two species among the insect-feeding group feed on insects in trees but behave more like wood warblers than finches; they have small, needlelike beaks, bright colors, and are aerial acrobats.

On a much larger scale, the same process of adaptive radiation has taken place in each of the faunal regions of the world. Fossil evidence shows that on each continent such radiations have occurred

THE PROCESS OF EVOLUTION / 293

several times. During the Triassic period 230 million years ago, reptiles rose to dominance on land. They eventually produced many of the types of animals common today. There were flying reptiles, the pterosaurs. (One of these had a wingspan of over 60 feet, larger than a Phantom jet. Appropriately, it was discovered in Texas.) There were large herbivores such as *Triceratops*. There were large carnivores such as the tyrannosaurs. And there were swimming forms such as the ichthyosaurs (Fig. 8-16). The large majority of these had disappeared by the Cenozoic era, 63 million years ago. The causes for the extinction

Figure 8-16 CONVERGENT EVOLUTION PRODUCED BY TWO WAVES OF ADAPTIVE RADIATION. The reptilian species at left in each pair are all extinct.

of the large reptiles remain shrouded in mystery. But the disappearance of so many important forms meant that their former ecological niches were left unoccupied. These empty niches were soon filled by another class of vertebrates, the mammals.

The animal world of today might not seem too drastically changed to a visiting reptile from the Triassic. Mammals have replaced the reptilian world with one that bears a strong resemblance to it. While no one would mistake a lion for a tyrannosaur, an ichthyosaur swimming among a school of killer whales could pass. This is an example of *convergent evolution*. More important than the physical resemblances between the two groups are the functional similarities; both mammals and reptiles produced herbivores, carnivores, flying forms, swimming forms, and so on.

It is important to appreciate that evolution in these cases has not been running backward. The evolutionary process is always unidirectional, because of the nature of mutations and recombinations, the source of genetic variability. In theory every mutation and recombination is reversible, but the probability of an exact reversal is so minute

that it can be neglected. To take the chromosomes of a human being as an example, there are more recombinants possible than there are atoms in the universe. The probability that simple chance will produce a preexisting arrangement is ridiculously small. This is why, once two species start to diverge evolutionarily, they grow progressively different and never revert to their original starting point. In the convergent evolution of reptiles and mammals we see a shared response to the presence of similar ecological niches. Internally, however, the killer whale and ichthyosaur are as different as a human and a box turtle.

Much closer evolutionary convergence is possible between contemporaneous groups in two faunal regions. For example, we would not expect to find a reptilian analog of the horse in the Triassic period, since no grasses existed then. Nor would a reptilian monkey find any bananas. But comparison of the Australian and North American faunas of today, for example, shows some remarkable similarities indeed (Fig. 8-17). The two groups are quite distantly related—the Australian mammals are marsupials, while the North American ones are placentals. But since external conditions on the two continents

Figure 8-17
Convergent evolution between placental mammals (left) and marsupial mammals of Australia (right).

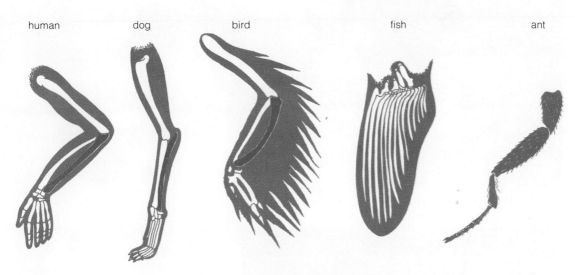

Figure 8-18 FRONT LIMBS OF FIVE ANIMALS.

today are much more similar than they were for fossil reptiles versus modern mammals, physical as well as behavioral resemblances are amazing. Apparently the most efficient tunneling, insect-eating, underground mammal has weak eyes, short paddlelike legs with claws, a fat body, no earlobes and, in short, looks like a mole. The most efficient medium-size mammal that lives underground but feeds on aboveground vegetation looks very much like a woodchuck, and so on.

To summarize, we have seen that the same process that has operated on a microscale on the Galapagos Islands has operated on a much vaster scale in the various faunal regions of the world. This is the process of adaptive radiation. Beginning with a single species or group of species, adaptive radiation stimulates the proliferation of forms until all possible ways of life in the region are filled.

The Construction of Evolutionary Trees

Thus far we have looked at the evolutionary process from an operational point of view. We have examined the nature of genetic change in species, the modes of speciation, and the logic behind the diversification of life into empty habitats. Let us now take a look at the techniques used to deduce evolutionary relationships between organisms. This is a matter of deducing historical relationships by looking at the

THE CONSTRUCTION OF EVOLUTIONARY TREES / 297

physical evidence of body structure. The classic technique makes use of the principle of homology: two structures that are similar have been inherited from a common ancestor. The more modern approach extends this principle to the molecular level and compares related proteins and nucleic acids in the species under study. We shall examine each of these techniques in turn.

Let us begin by trying to deduce the evolutionary relationships between five organisms: an ant, a human, a bird, a dog, and a fish. We simplify the problem of comparison by looking at a little bit of each organism at a time. The structure of the forelimbs is an easy place to begin. Look at the skeletal structures in Figure 8-18. Obviously there is a basic similarity which links the first four appendages but not the fifth. The limb of the ant is constructed along principles entirely different from those of the other four. Its muscles are located inside the skeleton instead of outside, and the skeleton itself is made of chitin, not calcium phosphate as the other four. We assume that similarity of structure implies relationship by descent. We therefore say that the human, bird, dog, and fish share a more recent common ancestor than any of them do with the ant. A diagram of that statement might look as follows:

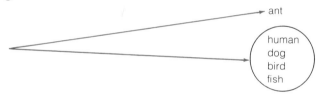

If we now look at the remaining group of four, it is apparent that the front appendage of the fish looks least like that of the other three, again indicating a more distant common ancestor:

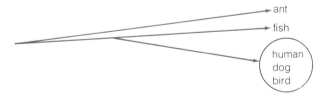

Finally, in the remaining group of three, the wing of the bird looks like the others, and again we indicate this in our diagram as follows:

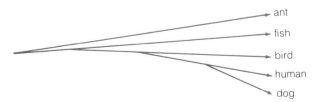

Such a diagram has a time dimension. From fossil evidence, we know that the human and the dog shared a common ancestor perhaps 70 million years ago, the human and the bird about 300 million years ago, the human and the fish perhaps 400 million years ago, and the human and the ant more than 500 million years ago.

We can make some further observations. The limb bones of the human, bird, dog, and fish are all made of the same material (calcium phosphate), they all originate in the same way during embryological development, and to a greater or lesser extent they all follow the same architectural plan. If we extend our look at the muscles, blood vessels, and nerves associated with the front limb, we find that these are likewise arranged in more-or-less similar ways in the different limbs. We therefore say that not only have the vertebrates descended from a single common ancestor, but their front limbs are also derived from a single ancestral structure. Thus we say that the limbs of the human, bird, dog, and fish are *homologous* (from the Greek for "agreeing"). Structures that are not derived from a common ancestral structure but which nevertheless serve the same function are said to be *analogous* (from the Greek for "proportional"). Thus the leg of an ant and the limb of a dog are analogous but not homologous.

Homologous structures are not always easy to recognize as such. The three little bones inside the mammalian ear are partly derived from the jawbone of a reptile. The jawbone in turn is derived from the first gill arch of a primitive group of fish, the jawless fishes such as lampreys. The jawless fish used their gill arches to give mechanical support to the intervening gill slits, through which water had to flow freely.

What we can learn from such comparisons? Perhaps the most impressive lesson is the economy of the evolutionary process. There is no such thing as a brand-new organism on earth, or a brand-new organ. The flower of the water lily is just a made-over cone of a coniferous plant. The wing of a bird is a made-over front leg of a reptile. Our teeth are the made-over scales of fish. The same holds true for the internal tissues; the uterus is a made-over oviduct, and the lungs arose as an outpocketing from the back of the throat of a fish. Thus all evolutionary processes are conservative ones, building on the raw materials bequeathed by previous generations. Each living organism, from the tiniest duckweed to the highest-flying hawk, represents an infinity of small changes and creative adaptations which have added up to something complex and marvelous.

By patient comparison of a variety of organ structures in both living and fossil forms, taxonomists have reconstructed the historical process leading to the present diversity of life on earth. These rela-

THE CONSTRUCTION OF EVOLUTIONARY TREES / 299

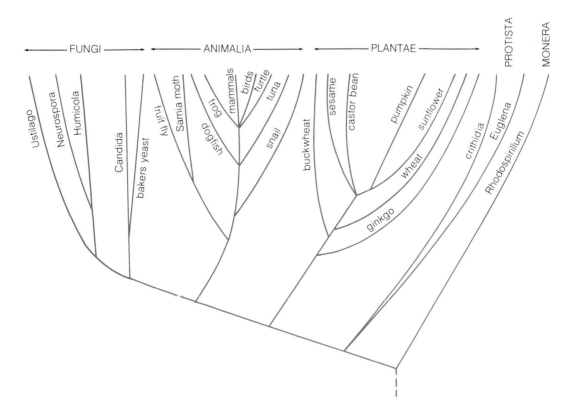

Figure 8-19 EVOLUTIONARY RELATIONSHIPS BETWEEN LIVING ORGANISMS AS DEDUCED BY DIFFERENCES IN STRUCTURE OF CYTOCHROME C. The progression of time is toward the top of the tree, and branch points represent extinct common ancestors. Line length to nearest branch point is a measure of evolutionary distance between two organisms. Longer lines indicate a larger number of amino acid residue differences between corresponding cytochrome molecules. The lines are not all drawn to the same scale.

tionships are represented as a tree. Living species are drawn as branches reaching up to a common level on top, and extinct species as branches terminating lower down. It is important to remember that such "primitive" forms as reptiles and amphibians have an ancestry as old as our own. The mammals did not evolve from present-day reptiles or amphibians or fish. Rather, all share a common ancestor in time. A simplified version of the evolutionary tree is given in Fig. 8-19, which omits all extinct species. Much important information for the construction of such a tree of life has come from the study of evolution at the molecular level.

Evolution at the Molecular Level

Evolution and entropy have always been two great antagonists. While galaxies recede and stars burn out, a small corner of the universe (or perhaps several such small corners besides our own biosphere) has been growing more complex and ordered through the process of evolution. The fact that this process occurs at the cost of an entropy increase in the rest of the universe does not diminish its strangeness. The step-by-step buildup of complexity is perhaps best illustrated by looking at the evolution of protein molecules, since they are the agents that finally determine all the gross characters of an organism, such as the shape of the appendages, the method of feeding, and the location of the nervous system to name a few.

While in theory every protein molecule making up an organism has been evolving and is suitable for study, in practice only a few have been chosen. They are all proteins that are easy to isolate and purify and which have relatively short chain lengths. If they are to serve for comparative purposes, they must also be proteins found in many different groups of organisms.

The protein must first be isolated in crystalline form. This is a difficult procedure for any but the most abundant proteins. Next the amino acid sequence of the protein is determined. This used to be a far more difficult project, requiring years of work for even the smallest proteins. With modern instrumentation, the job can now be done in a matter of months. The method used is as follows. (1) The protein is split into overlapping smaller fragments with the use of two or more different protein-splitting enzymes (trypsin and chymotrypsin are commonly used); (2) the fragments are purified, and their amino acid sequence determined by chemical or enzymic means. Let us take as an example a polypeptide only 17 amino acids long and symbolize each amino acid by means of a capital letter. Let us then say that trypsin hydrolysis gives us the following fragments (for our model, we assume that trypsin cleaves peptide links to the right of vowels): A, I, T, TTSXU, MBLKSSTBO. And chymotrypsin hydrolysis gives the following (we assume that chymotrypsin cleaves peptide links to the right of the letters S and B): S, OT, TB, LKS, XUMB, AITTS. If you know that the two methods of hydrolysis produce fragments with regions of overlap, can you reassemble the fragments into their proper sequence? (Answer: AITTSXUMBLKSSTBOT.)

Given such difficulties, it is not surprising that only relatively few proteins have been studied in a comparative way at this time.

These include the following: cytochrome c (mitochondrial protein), hemoglobin and myoglobin (respiratory proteins), fibrinopeptides (involved in blood clotting), histone (chromosome structure), and trypsin, chymotrypsin, and elastase (proteolytic enzymes).

To see how changes in amino acid sequence may arise, let us take as an example position 100 in cytochrome c (**Fig. 8-20**).

In dogfish cytochrome, this position is occupied by the basic amino acid lysine; in the rattlesnake we find glutamic acid, and in the pigeon glutamine. Such changes may be explained as single-base replacements in DNA, or point mutations. Thus, the codon AAA, coding for lysine, can be transformed into nine other codons by single-base substitutions:

1. AAA-lysine (basic amino acid)
2. GAA-glutamic acid (acidic amino acid)
3. UAA-chain termination
4. CAA-glutamine
5. AGA-arginine (basic amino acid)
6. AUA-isoleucine
7. ACA-threonine
8. AAG-lysine
9. AAU-asparagine
10. AAC-asparagine

Some of the new codons, such as AAG, are conservative and code for the same amino acid. Such mutations would not be detected by the present technique. Others, such as AGA, are semiconservative in that the new amino acid substituted has properties similar to the one replaced. The rest are radical substitutions, with UAA (chain termination) being the most radical of all. This mutation destroys the function of the protein, and the protein is therefore eliminated when it does occur. Position 100 of cytochrome c turns out to be one of the least essential and therefore most variable positions in the molecule, hence we are not surprised that a change as drastic as the substitution of an acidic residue (in the rattlesnake) for the basic lysine does not lead to loss of function. Position 79, however, is absolutely crucial, and the cytochrome of every organism tested, from human to bacterium to sesame seed, contains lysine in this position. We have no reason to believe that mutations in this position have occurred less frequently than elsewhere. They have simply been aborted in every case. There

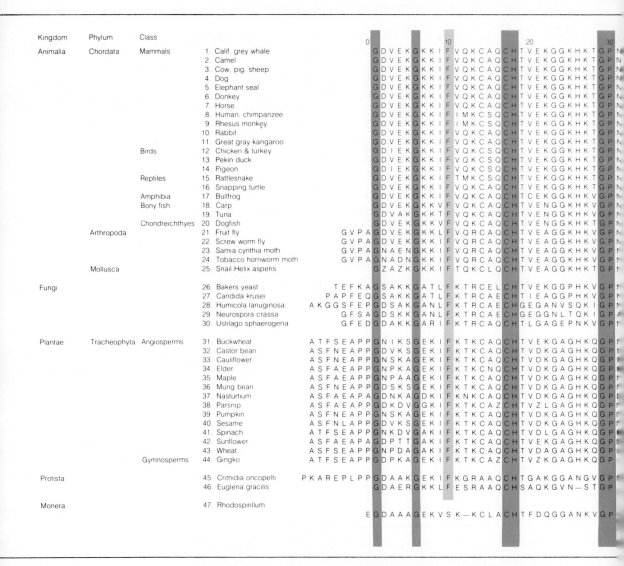

Figure 8-20
AMINO ACID SEQUENCES OF CYTOCHROME C FROM 47 SPECIES.
Symbols used:
alanine—A
valine—V
leucine—L
isoleucine—I
methionine—M
phenylalanine—F
tyrosine—Y
tryptophan—W
lysine—K
arginine—R
histidine—H
trimethyllysine—X
aspartic acid—D

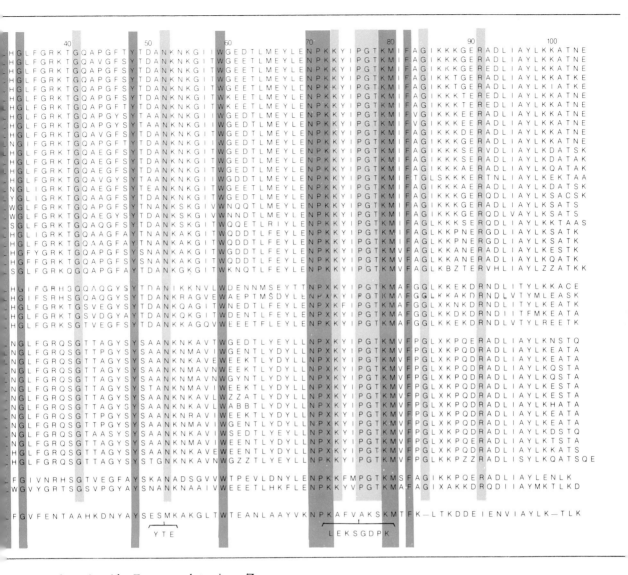

glutamic acid—E
cysteine—C
glycine—G
proline—P
serine—S
threonine—T
asparagine—N
glutamine—Q
glutamic acid *or*
glutamine—Z
deletion (—)
Light color indicates invariant residues among eukaryotes.
Dark color indicates absolutely invariant residues.

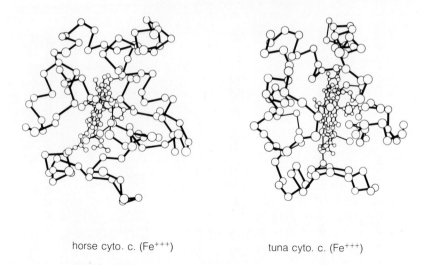

horse cyto. c. (Fe^{+++})   tuna cyto. c. (Fe^{+++})

Figure 8-21 TERTIARY STRUCTURES OF HORSE AND FISH CYTOCHROME C.

are many such unchangeable positions in the cytochrome c molecule. For instance, the cysteine in position 17, the histidine in 18, and the methionine in 80 seem to have never varied in the entire history of living organisms. These groups are responsible for the binding of the heme group to the protein, and any change here is fatal (Fig. 8-21).

The simplest explanation for such regions of invariance is that selection works not on the amino acid sequence but on the folded molecule, and that this folding has not changed significantly since the origin of mitochondria. X-ray diffraction studies on horse and fish cytochromes show no important change in their three-dimensional structures (Fig. 8-21).

Moreover, chemical tests show that the cytochromes of all eukaryotic species tested are compatible with the electron donors or acceptors of all other eukaryote species.[5] Again, such chemical compatibility would not be imaginable without a three-dimensional fit. While fish, fungi, snakes, and insects have gone on separate evolutionary journeys in the exploration of alternate environments and lifestyles, their mitochondria have changed only in nonessential ways. This is an expression of the principle we have stated before: all eukaryotic cells are fundamentally similar—the life of a cell inside a cactus is not very different from the life of a cell inside a salamander. The first and most important task they all face is the conversion of

---

5. The bacterial cytochrome is so radically different that it is totally incompatible with eukaryote electron donors or acceptors. Protistan cytochrome c shows only partial activity.

food molecules into ATP, and this has remained the function of the mitochondrion and its component proteins. Therefore the selection pressures that serve to adapt a species to its environment are only secondarily related to the selection pressures that adapt the cytochromes to their oxidases and reductases.

However, it is perhaps too early to say that *no* functional differences have arisen among eukaryote cytochromes. Perhaps more sensitive methods of measurement will reveal such selective advantages, as they have been revealed among seemingly neutral characters at the organ level. Even a very minute selective advantage can become overwhelming when measured over spans of tens or hundreds of millions of years.

There are two striking things about the cytochrome sequences in Figure 8-20. The first is the relative conservatism of this molecule, which has endured in functionally constant form over a span of more than a billion years. The second is that, where changes do occur, they cluster themselves in a way that reflects the evolutionary history of their host organism. That is to say, sequences among mammals are more or less similar, as are sequences among birds, plants, insects, and so on. And the differences between mammals and reptiles are smaller than the differences between mammals and fish, which in turn are smaller than the differences between mammals and insects, and so on. Cytochrome c therefore offers a molecular clock for measuring the slow beat of evolution. It is a clock that can be erratic at times, but the erratic behavior can be rationalized and compensated for. Most important, it is a clock that has been ticking longer and more systematically than any other measure of evolutionary time. The fossil record has given us vague and unsatisfactory answers to some of the most basic evolutionary questions. What is the relationship of vertebrates to the invertebrate phyla? What is the relationship of eukaryotes to prokaryotes? How are prokaryotes related to each other? The answers to these questions are written more clearly in the molecules of living organisms than in the fossil remnants of dead ones.

We can begin by calibrating our clock. By consulting Table 2, we find that the average mammal cytochrome c differs from the average reptile cytochrome c in 11 amino acid residues (expressed as differences per 100 residues). We correct this figure for repeated mutations, since we are interested in total evolutionary change. This gives us an average difference of 15 per 100 residues. From the fossil record, we know that the lines leading to modern mammals and to modern reptiles diverged 300 million years ago. We can therefore calculate that it takes an average of 20 million years for cytochrome c to mutate in one amino acid position. We can make similar correlations for other groups (see Table 2).

When these and other comparisons are plotted, we see the relationships in Figures 8-19 and 8-22. The results are not unambiguous There is considerable variability among the members of a given group. But this is to be expected, since we have based our comparisons on only a single homologous character. To expect a complete answer from such a comparison would be as unreasonable as expecting all of vertebrate phylogeny to be sorted out on the basis of limb structure alone. Once the cytochrome c comparisons are correlated with comparisons based on other proteins (especially such universal ones as enzymes of the glycolytic pathway), protein phylogeny will be on a much surer footing. But even this preliminary result shows the power of the method in reaching back into time periods so distant that the fossil record has been obliterated.

After cytochrome c, perhaps the second most widely studied proteins have been the oxygen-binding proteins hemoglobin and myoglobin. Like cytochrome c, these proteins contain complex metallo-organic functional groups called *heme groups* which serve as the active sites, or working parts, of the molecules. In myoglobin, the heme group forms a relatively stronger bond with an oxygen molecule than it does in hemoglobin. Myoglobin is thus used as an oxygen-storage molecule, while hemoglobin serves for oxygen transport. As discussed in Chapter 4, the hemoglobin molecule is a tetramer with each subunit containing both a heme group and a protein moiety. There are two kinds of protein chains in the hemoglobin of an adult mammal,

TABLE 2
Evolutionary divergence and sequence changes in cytochrome c

GROUPS	CHANGES PER 100 RESIDUES CORRECTED FOR REPEAT MUTATIONS	MILLIONS OF YEARS SINCE DIVERGENCE	
		FROM FOSSIL RECORD	EXTRAPOLATED
Mammals vs. reptiles	15 ± 4	300	
Birds vs. reptiles	13 ± 6	240	
Fish vs. amphibians	25 ± 2	400	500
Insects vs. vertebrates	47 ± 4		940
Plants vs. animals	103 ± 4		2,060

THE CONSTRUCTION OF EVOLUTIONARY TREES / 307

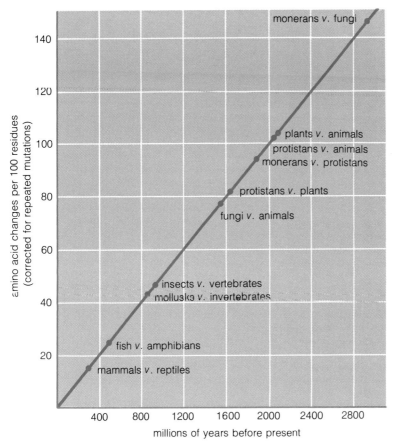

Figure 8-22 ESTIMATION OF EVOLUTIONARY AFFINITIES BY CYTOCHROME C MUTATION RATES. The line is drawn by comparing mammal–reptile amino acid differences with known time of divergence as shown by the fossil record. Other dates of divergence are extrapolated on the basis of amino acid differences.

alpha (α) and *beta* (β).[6] After sequence analyses and tertiary structure determinations were carried out on Hb_α, Hb_β, and myoglobin, it became obvious that the three proteins were variations on the same basic pattern of sequence and folding (Fig. 8-23). The same was true of the minor chains of Hb, Hb_γ, and Hb_δ. The hemoglobins and myoglobin are therefore examples of *homologous proteins*—proteins derived from a common ancestral protein. But the perplexing thing is that all of these molecules may coexist in the same animal—analogous to a situation in which an animal has not just a set of arms but also wings and flippers. It is easy to explain

---

6. Many mammals also carry a special form of Hb during fetal life, called *gamma* (γ), in place of β. In addition, primates have a small fraction of Hb in which β is replaced by *delta* (δ).

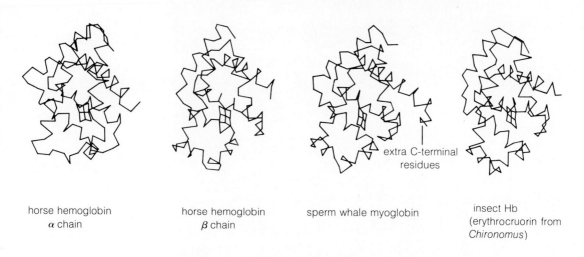

horse hemoglobin α chain

horse hemoglobin β chain

sperm whale myoglobin

insect Hb (erythrocruorin from *Chironomus*)

Figure 8-23 SIMILARITIES IN TERTIARY STRUCTURES OF HEMOGLOBIN ALPHA CHAIN, HEMOGLOBIN BETA CHAIN, AND MYOGLOBIN.

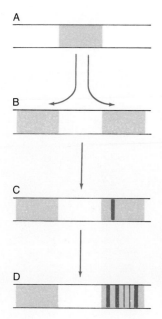

evolutionary divergence between two proteins in separate lines of descent, but how can we explain such divergence within a single organism? The simplest postulate would involve an accidental doubling of an ancestral gene, followed by mutation in one of the duplicate genes (Fig. 8-24). Such gene multiplication is known to occur all the time, and it would provide a buffering mechanism enabling the organism to absorb major mutations in the alternate gene copies. Such continued mutation might eventually result in a gene with a new function. In the case of hemoglobin and myoglobin, these func-

Figure 8-24 EVOLUTION OF A NEW PROTEIN FOLLOWING GENE DUPLICATION. (A) Ancestral gene. (B) Accidental doubling of gene. (C) Mutation destroys original gene function in right-hand gene. A comparable mutation in a single gene would be lethal if an important protein function was lost. (D) Continued mutations and natural selection produce a gene with a new function.

tions are closely related. A much higher degree of functional divergence is represented by the enzymes lysozyme and α-lactalbumin. Lysozyme hydrolyzes bacterial cell walls, while α-lactalbumin is a component enzyme used to synthesize the milk sugar lactose. Sequence analysis of the two proteins reveal a recent evolutionary divergence.

As for cytochrome c, the differences in primary structure between myoglobin and the hemoglobins can be used as calendars of evolutionary time. The hemoglobin–myoglobin system shows a more rapid rate of change than does cytochrome c; the time needed to replace 1 percent of the amino acid is estimated as 5.8 million years (instead of 20 million years). The more rapid rate of change is probably due to the fact that the globins react with a smaller molecule than does cytochrome c.[7] Mutations probably appear at comparable rates in the two systems, but acceptable mutations are rarer in cytochrome c than in the globins. The more rapid clock of the globins makes them more useful as chronometers of recent events. Thus hemoglobin data suggest that the human and gorilla together diverged from other primates about 28 million years ago, while the human and gorilla diverged from each other as recently as 5 million years ago. A date this recent is challenged by those who set a greater store by the fossil record, which places the human-gorilla split at 15 to 20 million years ago. Such apparent conflicts should lead ultimately to more sophisticated protein- as well as fossil-based phylogenies. But perhaps it is the very existence of such paradox and conflict that provides the students of evolution with a rich and inexhaustible goal of study.

Unanswered Questions in Evolution

Though launched scientifically over a century ago, evolution remains today a young science, full of excitement and surprise. The major questions of evolution have been far from answered, either with respect to specific events or with respect to the processes responsible for these events. We finish this chapter with an example of the latter. One of the foundation stones of Darwin's thinking was the concept of the struggle for survival. Too many individuals are born in relation to available resources, and those best adapted to external conditions

---

7. The globins react with the relatively small oxygen molecule. Since only a small portion of the globin is in contact with its substrate, the remaining portions of the molecule are relatively free to undergo variation. Cytochrome c transfers electrons from one bulky protein (cytochrome c reductase) to another (cytochrome c oxidase). Thus much of the surface of cytochrome c is in contact with its substrates, resulting in more stringent spatial and chemical requirements.

leave the most descendents. Implicit in this theory is the assumption that the best adapted individuals elbow their less favored brothers and sisters out of the way.

But students of animal behavior can point to many examples of behavior that run counter to this expectation. Bees rush any potential hive intruder and attack it. But a bee that stings an intruder cannot withdraw the sting and therefore loses its own life. A crow gives a warning cry at the sight of a cat, allowing its less observant neighbors to escape but drawing attention to itself. Yearlings of the Florida scrub jay do not breed for 1 to 3 years after maturation but stay with their parents, helping them feed the hatchlings and guarding the nest against predators. There is an obvious selective advantage to such behavior. By helping close relatives who are likely to share their own genes, altruistic individuals increase the likelihood that their genes will be represented in future populations. Most examples of cooperative behavior appear to involve close relatives. Thus all the worker bees in a hive or all the worker ants in an ant colony are sisters, since all have been born of the same queen. And scrub jay yearlings do not help birds other than their own parents. But it may be too early to say that altruistic behavior among unrelated individuals does not play an important role in the evolutionary process.

Jane Goodall, in a beautiful book, *Innocent Killers,* describes the great grazing herds of the Serengeti: the wildebeest, zebra, Thomson's gazelle, and Grant's gazelle; as well as some of their predators: wild dogs, hyenas, and golden jackals. Both the hunters and the hunted cooperate, the former to kill, the latter to escape. Sometimes the cooperation extends to amazing lengths. When a herd of zebra sleep at night, a few zebras always remain awake and alert as sentinels. As soon as a predator is detected, the sentinels sound an alarm and the group is instantly on its feet.

During a chase, the young zebras remain close to the mares. The stallions actively defend the mares and foals against such predators as hyenas and wild dogs. Furthermore, when one family group is chased, it tries to join up with other family groups, until a mass of 200 zebras or more is moving as a unit. The stallions remain in the rear, viciously kicking and biting their attackers. Thus often the intended victim escapes with its life.

Goodall vividly describes one particularly harrowing hunt. A pack of wild dogs was making a slow, stalking approach to a group of about 20 zebras which included a mare and her young foal. When the dogs had come within 20 yards, the nervous zebras turned and began to trot away. The dogs then raced in pursuit, and the zebras increased their pace to a canter. They ran in a tight group, with the mare and foal in the middle. As they ran across the plains, other zebra groups joined them. The zebras were running just fast enough to keep pace with the slowest of their group—in this case probably the foal. The mare and foal lagged behind, but whenever one of the dogs drew near, a stallion lunged at it, ears back and teeth bared. The dogs managed to avoid the stallions every time but kept pressing in on the mare and foal. Ultimately, in the confusion the mare and foal became separated from the rest of the herd, along with a yearling that stayed with the mare and may have been her own offspring. The pursuers immediately surrounded the three, while the rest of the herd vanished over a slight rise.

Left on their own, the three zebras stopped running and grouped into a tight defensive formation, with the foal in the middle. The dogs made successive rushes at the foal, but were always repelled by the mother or yearling who moved no more than a step or two away no matter what the dogs did. As the minutes sped by and the dogs became bolder, the situation grew desperate for the zebras. Suddenly the dogs changed strategy. One of them jumped for the mare's head, its jaws aimed for the upper lip. Once a dog has a zebra in such a grip, he pulls hard and does not let go until the rest of the pack has disemboweled the victim. The mare jerked her head away at the last second, and the dog's jaws snapped shut on empty air. But now the dogs sensed blood, and they leapt forward again. The end seemed imminent.

Suddenly, the ground was vibrating with pounding hoofs. As Goodall looked up, she saw 10 zebras bearing down quickly on the scene. A moment later the 10 closed ranks around the mother and her two offspring. Wheeling around, the group galloped off in the direction of the main herd. The dogs pursued for about 50 yards but were unable to penetrate the close group and gave up the chase.

Though Goodall had never before observed such a dramatic rescue, it is clear that zebras show significantly more social cooperation than do wildebeests or gazelles. Zebras lack the fleetness of gazelles or the impressive horns of wildebeests, yet they make up one of the most successful groups on the Serengeti. The conclusion seems inescapable that this success depends in large measure on their cooperative, caring behavior. But this brings us back to the question of the mechanism involved. Clearly, altruistic behavior within a family group leads to perpetuation of the genes responsible for such behavior, even if the caring individuals themselves lose their lives in the process. But what about caring toward strangers? Would not the most altruistic individuals expose themselves to the greatest danger and therefore see their genetic contribution eliminated?

Among present-day evolutionists, the genetic basis of such co-operative behavior represents an unanswered question, and a hotly debated one. But at a time when the world is being increasingly compared to a jungle, it is perhaps reassuring to know that the real jungle is not without its quota of caring and concern.

Bibliography

GENERAL
READING

Loren Eiseley, *The Immense Journey,* Vintage Books, 1956. A collection of essays by a master storyteller on the general theme of evolution. Eiseley beautifully conveys the richness of the evolutional process in this excerpt, taken from his essay "How Flowers Changed the World."

A few nights ago it was brought home vividly to me that the world has changed since that far epoch. I was awakened out of sleep by an unknown sound in my living room. Not a small sound—not a creaking timber or a mouse's scurry—but a sharp, rending explosion as though an unwary foot had been put down upon a wine glass. I had come instantly out of sleep and lay tense, unbreathing. I listened for another step. There was none.

Unable to stand the suspense any longer, I turned on the light and passed from room to room glancing uneasily behind chairs and into closets. Nothing seemed disturbed, and I stood puzzled in the center of the living room floor. Then a small button-shaped object upon the rug caught my eye. It was hard and polished and glistening. Scattered over the length of the room were several more shining up at me like wary little eyes. A pine cone that had been lying in a dish had been blown the length of the coffee table. The dish itself could hardly have been the source of the explosion. Beside it I found two ribbon-like strips of a velvety-green. I tried to place the two strips together to make a pod. They twisted resolutely away from each other and would no longer fit.

I relaxed in a chair, then, for I had reached a solution of the midnight disturbance. The twisted strips were wistaria pods that I had brought in a day or two previously and placed in the dish. They had chosen midnight to explode and distribute their multiplying fund of life down the length of the room. A plant, a fixed, rooted thing, immobilized in a single spot, had devised a way of propelling its offspring across open space. Immediately there passed before my eyes the million airy troopers of the milkweed pod and the clutching hooks of the sandburs. Seeds on the coyote's tail, seeds on the hunter's coat, thistledown mounting on the winds—all were somehow triumphing over life's limitations. Yet the ability to do this had not been with them at the beginning. It was the product of endless effort and experiment.

Arthur Koestler, *The Case of the Midwife Toad,* Vintage Books, 1971. An "intellectual thriller"—the true story of the Viennese scientist Paul Kammerer. His researches into evolutionary theory conflicted with the predictions of Mendelian genetics, and the resultant controversy led to Kammerer's suicide in 1926.

Charles Darwin, *The Origin of Species,* Modern Library. Like *Das Kapital, The Interpretation of Dreams* and the Bible, *The Origin of Species,* is a book more commonly referred to than read. But even a superficial browsing through Darwin's work will illustrate its intellectual power and depth of study. In the following excerpt, Darwin considers the dispersal mechanisms of living organisms.

> Some species of fresh-water shells have very wide ranges, and allied species which, on our theory, are descended from a common parent, and must have proceeded from a single source, prevail throughout the world. Their distribution at first perplexed me much, as their ova are not likely to be transported by birds; and the ova, as well as the adults, are immediately killed by sea-water. I could not even understand how some naturalised species have spread rapidly throughout the same country. But two facts, which I have observed—and many others no doubt will be discovered—throw some light on this subject. When ducks suddenly emerge from a pond covered with duckweed, I have twice seen these little plants adhering to their backs; and it has happened to me, in removing a little duck-weed from one aquarium to another, that I have unintentionally stocked the one with fresh-water shells from the other. But another agency is perhaps more effectual: I suspended the feet of a duck in an aquarium, where many ova of fresh-water shells were hatching; and I found that numbers of the extremely minute and just-hatched shells crawled on the feet, and clung to them so firmly that when taken out of the water they could not be jarred off, though at a somewhat more advanced age they would voluntarily drop off. These just-hatched molluscs, though aquatic in their nature, survived on the duck's feet, in damp air, from twelve to twenty-hours; and in this length of time a duck or heron might fly at least six or seven hundred miles, and if blown across the sea to an oceanic island, or to any other distant point, would be sure to alight on a pool or rivulet.

George Gaylord Simpson, *Horses,* Doubleday, 1961. Because the fossil record of the horse is more complete than that of almost any other mammal, its evolutionary story can be told in a way approaching its true complexity. As Simpson shows, the path from the "dawn horse" to *Equus* was a tortuous one, with many blind ends and dramatic changes in life styles.

Theodosius Dobzhansky, *Mankind Evolving,* Bantam Books, 1970. A great geneticist looks at man.

Edward O. Wilson, *Sociobiology: The New Synthesis,* Harvard University Press, 1975. On the biological importance of cooperation.

TEXTBOOKS

L. E. Mettler and T. G. Gregg, *Population Genetics and Evolution,* Prentice-Hall, 1969. A brief, scholarly, and not overly mathematical treatment of the genetic basis of evolution.

G. L. Stebbins, "Processes of Organic Evolution," 2nd ed. Prentice-Hall, 1971. A short text comparable to the above, with a greater emphasis on the evolution of plants.

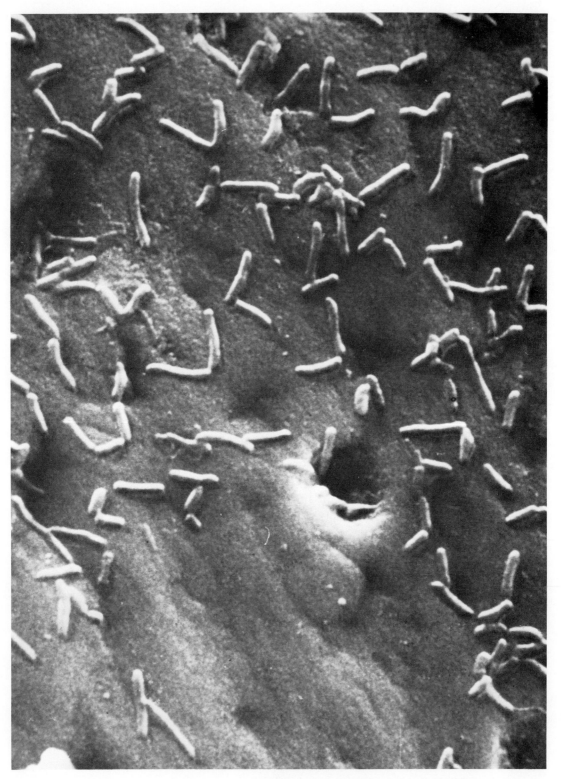

Figure 9-1 BACTERIA ON THE SURFACE OF A TOOTH.

chapter 9

The Variety of Life

Adam, the biblical first man, led a very busy life in the Garden of Eden. Whenever God created a new creature, he brought it to Adam to be named, and whatever Adam called the living creature, that was its name. While we cannot be sure of this, it seems probable that, after naming the three-hundred thousandth beetle, Adam must have felt there must be a better way. The better way of naming living creatures is an organized system called *taxonomy,* or the laws of classification of living creatures. That is to say, we do not treat every creature as completely independent and unrelated to every other creature. Instead, we see living organisms

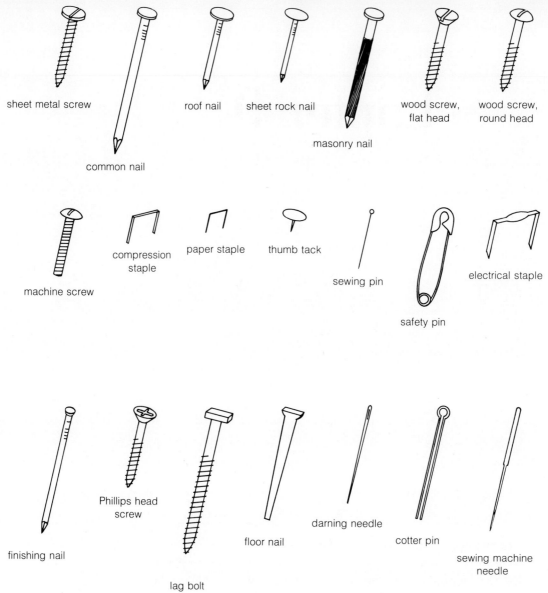

Figure 9-2 AN ASSORTMENT OF FASTENERS.

as having stronger or weaker affinities with each other, as discussed in the previous chapter. The taxonomic system we use today is a "natural" system. That is, two people working independently of each other should be able to look at the same organisms and arrange them in identical groupings dictated by the decipherable evolutionary family trees. Unfortunately, the practice does not always meet the expectations of theory. To see why, let us take a simple mechanical analogy. Take a look at the pile of objects shown in Figure 9-2.

Now, what is the natural way of grouping these objects? Taxonomist A takes a look and says there are two large groups: (1) the screws, which includes all objects that must be twisted; and (2) everything else, which includes all objects that are pushed. But then what about the masonry nail, which has threads and twists as it is being pounded? Taxonomist B looks at the assemblage and says, "Twisting versus pushing is a superficial difference. A screw and a nail are both held in place by friction. A more natural division would put the screws and the nails together, along with the staples and tacks. The other division would contain the needles and pins, since these are frictionless devices." But a critic could then question the machine screw, which is also frictionless and fits into a threaded hole. Taxonomist C, seeing more basic diversity, might propose four natural groupings: the nails, the screws, the staples, and the pins. But again his critics could point out that the paper staple and the cotter pin have more in common with each other than with their neighbors, since both are bent after insertion.

Arguments of this type abound in the taxonomy of living organisms, and it is therefore no wonder that taxonomists keep changing their minds and disagreeing with each other. But perhaps the areas of agreement are more important than the areas of disagreement. No one questions the close relationship between a crustacean and an insect, just as no one would question the close relationship between a wood screw and a sheet-metal screw. Agreement most often breaks down in determining both the largest and smallest natural units. These represent the two ends of the evolutionary spectrum— the events of the dim past, where differences are so large between groups as to obscure relationships, and the events of the immediate past, where differences between two groups are so slight that it becomes a question whether they may not in fact be branches of the same group. We are not concerned here with such subtleties, but it is important to realize that the natural taxonomy in use today can lead to different results in the hands of different taxonomists. The nomenclature in use today classifies all living organisms according to the following categories:

> Kingdom
> Phylum
> Class
> Order
> Family
> Genus
> Species

Kingdom is the largest, most inclusive grouping, and species the smallest. As an illustration, Table 1 shows how four familiar organisms are classified.

TABLE 1
Classification of four familiar organisms

	HUMAN	HORSE	EASTERN DIAMONDBACK RATTLESNAKE	COMMON SUNFLOWER
Kingdom	Animalia	Animalia	Animalia	Plantae
Phylum	Chordata	Chordata	Chordata	Tracheophyta
Class	Mammalia	Mammalia	Reptilia	Angiospermae
Order	Primata	Perissodactyla	Squamata	Dicotyledonae
Family	Hominidae	Equidae	Viperidae	Compositae
Genus	*Homo*	*Equus*	*Crotalus*	*Helianthus*
Species	*Homo sapiens*	*Equus caballus*	*Crotalus adamanteus*	*Helianthus annuus*

This system of classification tells us a great deal about an organism: it tells us who its relatives are, and it tells us much about its basic mode of life. If you have never seen a sunflower, this is what you could learn from its taxonomic description:

1. The kingdom of Plantae includes photo-synthetic organisms of a certain degree of complexity. Excluded are the blue-green algae.

2. The phylum Tracheophyta includes plants with a vascular system—basically all the important land plants. Excluded are the mosses.

3. The class Angiospermae includes plants that have flowers. Excluded are coniferous trees, ferns, and certain primitive land plants.

4. The order Dicotyledonae includes plants with two seed leaves and broad photosynthetic leaves, such as pea, maple, and apple. Excluded are monocotyledons, plants with a single seed leaf, such as corn, grass, palm, and orchids.

5. The family Compositae contains plants with compound flowers, such as the daisy, the dandelion, and the goldenrod. It is the largest family of flowering plants in North America. Excluded are other families such as mustards, peas, roses, and violets.

6. The genus *Helianthus* includes all sunflowers, such as the prairie sunflower, the western sunflower, and the giant sunflower. Excluded are other composites such as daisies, goldenrods, thistles, and hawkweeds.

7. The species *H. annuus* is a unique classification—the common sunflower. It refers to a single interbreeding and coevolving assemblage of organisms.

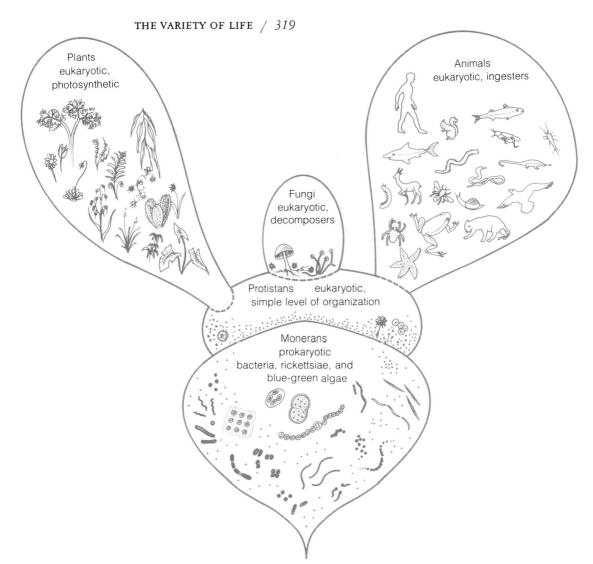

Figure 9-3 THE FIVE KINGDOMS OF LIVING ORGANISMS AND THEIR EVOLUTIONARY RELATIONSHIPS. Note the indeterminate boundaries between protistans and higher kingdoms. The viruses have not been included because they are not alive in the same sense as the other organisms listed.

In terms of everyday experience, only seldom do we identify a living organism this completely—down to the species level. When we say that a tree is a maple or birch or oak, we are identifying it only to the genus level. When we say that an insect is a bug or a beetle or a dragonfly, we are doing less than that—we are identifying it only down to the level of order. And when we say an animal is a sponge or a rotifer or a roundworm, we have identified it only in terms of phylum. Even so, the identification conveys a great deal of information, since all sponges, rotifers, and roundworms follow predictable life-styles.

There was a time, less than 20 years ago, when biologists recognized only two kingdoms—plant and animal—to which all living organisms belonged. This raised troublesome questions about where to place, for instance, the bacteria. Were they animals with cell walls, or plants without photosynthetic ability? The consensus today is that there is sufficient diversity in the living world to recognize at least five separate kingdoms (Fig. 9-3). In terms of either living mass or functional importance, there are vast differences among the five kingdoms. Protistans today are definitely a minor kingdom, yet they represent a necessary evolutionary bridge between the monerans and the plants, fungi, and animals. In this chapter we will look at three kingdoms: the monerans, the protistans, and the fungi. The next chapter is devoted to the plants, and in Chapter 11 we will look at the animal kingdom. A brief summary of the major groups within each kingdom is offered at the end of the book.

The Monerans

The kingdom Monera includes the smallest living organisms on earth—bacteria and the blue-green algae. Since human beings are among the very largest organisms alive, we have, not unexpectedly, directed the bulk of our attention to other organisms in our own size range.[1] Monerans have been so neglected that we have no idea how many species the kingdom contains. It was not until the 1940s that biologists even recognized the fact that bacteria have sex.

The Bacteria

This neglect has not discouraged the bacteria, however. They exist in numbers like the stars, on our bodies, in the soil, floating through the air, and suspended in water. One of my beautiful memories involves swimming at night in the waters off Woods Hole, Massachusetts. There rich coastal waters are inhabited by a variety of photoluminescent organisms, including the bacterium *Photobacterium fisherii*. When this bacterium receives a pulse of oxygen, it emits light like a firefly. In the daytime, the light is too weak to be seen. But at night, as one's body moves through the water, the turbulence created brings little bubbles of oxygen below the surface, causing the bacteria to luminesce. One's arms, legs, and body give off

---

1. The one exception to this statement is the bacterium of the human intestine, *E. coli*. This organism has replaced the fruit fly as the chief experimental organism of genetics and has been studied exhaustively biochemically. It is on its way to becoming the first entirely defined living organism.

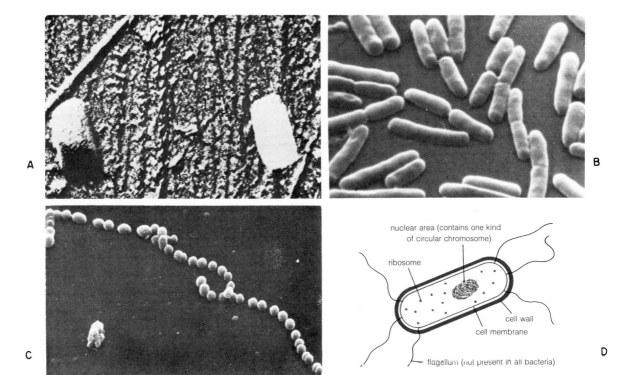

Figure 9-4 FOSSIL AND MODERN BACTERIA. (A) A fossil bacterium (*Eobacterium isolatum*) thought to be 3.1 billion years old. It was found in precambrian rock in South Africa. The imprint in the rock at left shows that the bacterium came originally from the rock and is not a recent contaminant. (B) Modern bacteria (*Pseudomonas aeruginosa*) as seen by scanning electron microscope. (C) Modern bacteria (*Streptococcus pyogenes*). These bacteria cause a number of serious infections in humans, including scarlet fever, strep sore throat, and gangrene. (D) Diagrammatic view of a modern bacterium.

continuous sparks of white light. Now, imagine a world where all bacteria have the same ability to luminesce. It would be a world of light. Our skins would glow, and the intestinal tract would be a coiled beacon. Large rivers, such as the Ohio, the Mississippi, the Colorado, the Delaware, and the Hudson, would shine brilliantly from bacteria feeding on their organic material. So would estuaries, small lakes, and even some larger lakes such as Lake Erie. The soil would glow as bacteria cooperated with fungi in recycling the dead remnants of plants and animals. All animals would glow as special hot spots. Even the air would have a faint glow, growing dimmer at higher altitudes. But since we can see them only with powerful microscopes, the bacteria carry on their lives around us invisibly.

There is little doubt now that bacteria represent the most primitive form of life on earth. The first cell that emerged as a living force from the inanimate soup of the young earth must have looked very much like a present-day bacterium (Fig. 9-4). On the outside was a tough, rigid cell wall. In modern bacteria, the cell wall is very

322 / Populations: The Variety of Life

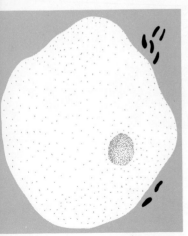

Figure 9-5
SIZE RELATIONSHIPS BETWEEN BACTERIA AND A HUMAN CHEEK CELL.

different chemically from the cell walls of plants, which are made of cellulose. The bacterial cell wall is stitched together in three dimensions by long chains of amino sugars and shorter chains of amino acids. This strange type of cell wall is also found in blue-green algae but in no other kingdom in the living world.

Inside the cell wall is a typical cell membrane. The cell membrane encloses cytoplasm with none of the familiar organelles found in higher cells. For instance, there is a nuclear area but no well-defined nucleus bounded by a nuclear membrane. This is why the monerans are called *prokaryotes* ("before nucleus"). Likewise respiratory enzymes are present but they are not organized into mitochondria. There are no Golgi bodies, centrioles, mitotic apparatus, or endoplasmic reticulum. This does not mean that bacteria cannot carry out the functions mediated by these organelles in higher organisms or *eukaryotes* ("true nucleus"). They synthesize DNA and protein, they can carry on aerobic respiration of the same type used by eukaryotes (as in the Krebs cycle), they can secrete substances into the external medium, they can undergo cell division, and so on. Bacteria use one all-purpose organelle to accomplish all these tasks—their cell membrane with associated enzymes and nucleic acids. This was the basic formula for life on earth, and all higher forms are elaborations of this primordial theme. If other fundamental modes of life ever existed on earth, they were outcompeted and driven into extinction by the bacteria. But it is noteworthy that the bacteria themselves were never driven into extinction by their more advanced descendants. They are formidable competitors.

Perhaps more than anything else, the success of the bacteria rests in their small size (Fig. 9-5). Small size means a high surface/volume ratio (see Chapter 1, page 6), and this in turn allows rapid movement of foods and respiratory gases across the cell membrane. The result is a potential for very high metabolic rates. When all conditions are optimal, bacteria can divide and double their mass every 20 minutes. In comparison, the most actively growing human cells divide approximately once every 20 hours. Such rapid cell division also implies a high mutation rate, since each cycle of DNA replication introduces the possibility of variation. The result is that bacteria are tremendous opportunists—they take instant advantage of a favorable environment by multiplication, and they can quickly combat an unfavorable environment (such as the presence of an antibiotic) by suitable mutation. This is why pharmaceutical chemists are engaged in a continual race to produce new antibiotics as old ones become ineffective because of genetic changes in the target bacteria.

Another important survival mechanism among the bacteria is the ability of some species to form *spores* ("seeds"). Bacteria form spores when faced with temporary environmental inclemency: exhaustion of food supply, drying, rise or fall in temperature. The spore contains all the DNA of the bacterial chromosome, a few enzymes, one or more spore membranes, and a thick cell wall (Fig. 9-6). Importantly, it contains only a very small amount of water. Since the chemical processes of life take place in water, a spore represents a state of suspended animation. How long a spore can survive is not known, but dried spores at the British Museum have germinated after being stored for 120 years. A bacterial spore can withstand several hours of boiling in water and represents the most resistant form of life known. Presumably spores could also survive a flight in space, and there has been widespread scientific concern that "dirty" space vehicles might contaminate the moon and outer planets with terrestrial bacteria. This would be a particularly serious problem in the exploration of Mars, which today appears to be the most likely candidate for possible extraterrestrial life.

Another bacterial winning card is the substantial cell wall. The cell wall forms an armor protecting the bacterium against osmotic changes. A number of antibiotics destroy bacteria by preventing synthesis of the cell wall. Included on this list is penicillin, perhaps the most phenomenally successful antibiotic of all time.

The food habits of bacteria cover the entire spectrum of possibilities. A small number are photosynthetic though only under anaerobic conditions. The great majority of bacteria feed on ready-made carbon compounds (as do humans). If a bacterium is lucky, the carbon compounds are available in water solution and are readily transported into the cell. In most cases, however, the food source is a macromolecule which must be digested outside the cell before absorption can take place. Thus a food such as cellulose is broken down into glucose outside the body of the bacterium, by release of the bacterial enzyme cellulase. Bacteria exist that can utilize every known natural carbon compound, even such unusual ones as oil slicks (providing oxygen is present). The problem comes with unnatural carbon compounds such as plastics, chlorinated hydrocarbon insecticides (for example, DDT), and a mixed bag of other compounds such as PCBs and herbicides. In their billions of years of evolution, bacteria have never had cause to evolve enzymes to break down these molecules. These compounds are therefore piling up in the environment faster than they can be decomposed. In the long run, there is little doubt that bacteria will evolve enzymes to cope with such com-

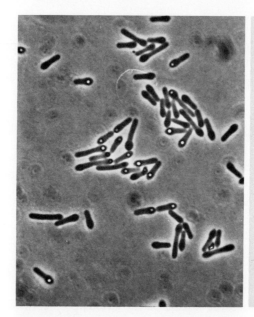
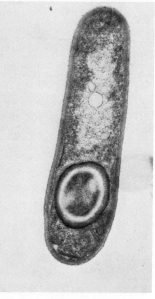

Figure 9-6 SPORES IN BACILLUS SPHAERICUS. The view at left is seen via a phase contrast light microscope. The view at right is seen in an electron microscope and shows a single bacillus with its internal spore.

pounds, but the evolutionary long run must be reckoned in millennia, not in generations. The evolution of an enzyme to decompose DDT is a far more formidable bacterial task than the evolution of antibiotic resistance.

A few species of bacteria are *parasites,* meaning they feed on living organisms. Perhaps it is better to call these bacteria *pathogenic* (disease-causing). Bacteria such as those responsible for syphilis, rheumatic fever, and cholera do not produce their devastation by outright feeding on the body cells of their host. Typically, they release toxins which kill or disturb the function of target cells. Other bacteria may disturb internal homeostasis by more indirect methods such as siphoning off food or provoking a strong immune response.

Bacterial pathogens do not plague the human species alone—there are specific microorganisms that infect birds, other mammals, insects, and even plants. Once the current romance with chemical insecticides wears off, some of these microorganisms may develop into more ecologically responsible agents of insect control.

One bacterial species is already widely used as an insect control agent. The Japanese beetle, a small, metallic beetle accidentally imported from the Orient, feeds on plants (Fig. 9-7). The beetle is effectively controlled by the milky disease bacterium, a soil bacterium that infects the beetle grubs and multiplies so abundantly that the insect's blood turns a milky white. The bacterium, once established, can remain in the soil for many years in the form of spores. It

Figure 9-7 A JAPANESE BEETLE WITH THE REMAINS OF ITS MEAL—A ROSE.

specifically infects only beetles of the family to which the Japanese beetle belongs—the scarabids—and is harmless to any other group of insects, soil animals, birds, mammals, or plants. Although the milky disease bacterium spreads slowly by natural means, its spread has been accelerated through deliberate dispersal by federal, state, and local groups. Hundreds of thousands of acres in the eastern states now enjoy permanent protection against the Japanese beetle as a consequence.

But there is a dark side to the story. Just as bacteria represent relatively unexploited opportunities for insect control, they also represent unexploited opportunities for biological warfare against humans. A class of enzymes discovered in 1973 (the restriction endonucleases) makes it possible to introduce the genes of other species into bacteria. Some of the possible experiments include (1) introduction of bacterial genes from another species that confer either resistance to antibiotics or ability to form bacterial toxin, (2) introduction of the genes of viruses, and (3) insertion of animal genes into bacteria. Whether by deliberate design or by the accidental escape of such genetically engineered bacteria, the consequences of such experiments could be horrendous. Plagues such as swept through Europe in the Middle Ages or through non-Western societies after exposure to Europeans would again march through crowded cities. Bacteria carrying genes of tumor viruses could cause widespread cancers. As a result, in 1974 a group of distinguished scientists from the United States National Academy of Sciences called for a complete moratorium on experiments of the first and second types, and called for extreme caution in undertaking experiments of the third type. This is the first time that scientists have voluntarily agreed to refrain from a trail of experiments that could produce great danger to the human species. It remains to be seen whether or not this call for abstention will be respected.

But we should not leave with a jaundiced view of the bacteria. On balance, the human species has more to gain than to fear from them. They synthesize vitamin K inside the intestinal tract. They are responsible for the production of such food products as vinegar, buttermilk, sauerkraut, salami, and many cheeses. Much of the mass of Swiss cheese consists of the bodies of dead bacteria. Even many antibiotics (such as streptomycin) are of bacterial origin. They play a role in the formation of fossil fuels and many mineral ores. And in the larger sense, they carry out crucial ecological tasks—recycling of carbon compounds, fixation of nitrogen, and helping in soil formation.

The Rickettsiae

The rickettsiae are a shadowy group living on the border between the bacteria and the viruses. Like bacteria, these very small parasites contain both DNA and RNA and the unique moneran type of cell wall. But like viruses, which we will discuss shortly, they appear incapable of life outside a living cell. They cause various serious diseases in humans, including Rocky Mountain spotted fever and typhus. The man who discovered the causative agents for spotted fever and for typhus was Howard Taylor Ricketts (1871–1910). Ricketts died at the age of 39 from typhus accidentally contracted during his laboratory studies, and it is for him that the group is named. It is felt today that the rickettsiae represent degenerate bacteria.

The Blue-Green Algae

The green scum that appears in stagnant waters and rich ponds resolves itself into a collection of tangled filaments when viewed under the microscope. Each filament is a necklace of individual cells of blue-green algae. Unlike true algae (see Chapter 10), the blue-greens lack an internal nucleus and chloroplasts. Their chlorophyll is bound instead to a complex system of internal membranes (Fig. 9-8). In fact, a single cell of a blue-green alga strongly resembles a single chloroplast. Besides photosynthesis, the blue-greens have the important ability to fix nitrogen (convert atmospheric nitrogen gas into ammonia). All living things require nitrogen for life,

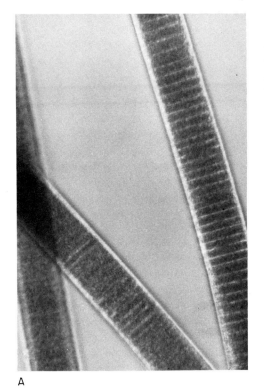
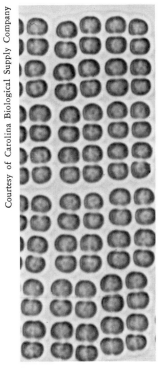
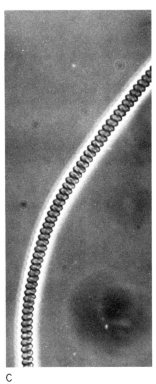

A B C

Figure 9-8 BLUE-GREEN ALGAE. (A) *Oscillatoria,* a filamentous form. The filaments move by a mysterious side-to-side motion. (B) *Merismopedia,* a multicellular form that grows in flat sheets one cell layer thick. (C) *Spirulina,* a helical form without visible internal subdivisions. It moves in a manner similar to that of *Oscillatoria.* (D) Internal structure of a blue-green algal cell. The photosynthetic lamellae fill the entire cell and are not localized in chloroplasts as in eukaryotes.

D

but nitrogen gas is chemically inert, hence unusable directly. Both blue-green algae and certain species of bacteria can reduce nitrogen gas to ammonia, which can then be incorporated by other living organisms. Without a source of fixed nitrogen, all other life on earth would grind to a halt. In a spectacular displacement of an essential natural process, the bacteria and blue-green algae have today yielded to humans their primary position as nitrogen fixers. The nitrogen fertilizers we use to squeeze ever-higher yields of crops from the land are industrially produced. The amounts applied to farms in the industrialized world are so excessive that fixation of nitrogen by soil microorganisms is repressed. Farmers become technologically hooked—they cannot stop using artificial fertilizers even if they want to. Much of the excess is leached by rain and carried into bodies of water, where it causes further problems (Chapter 12).

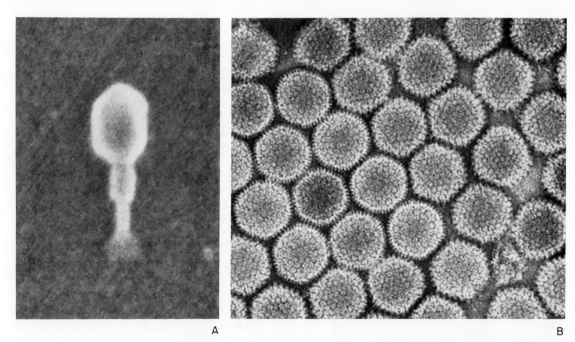

Figure 9-9 VIRUSES. (A) A bacterial virus. The tail structures are used to attach to bacterial cell walls. (B) An array of adenoviruses. This is a group of viruses associated with human respiratory ailments, including the common cold. Individual protein subunits in the protein coat can be distinguished.

The Viruses

Viruses are biological structures, but they cannot be called living organisms in the sense in which we have used the word. That is, a virus carries out none of the repair and maintenance functions necessary to counter second-law disintegration. Any living cell must work to stay alive, and the evidence of such work comes from its ongoing respiration. A virus, whether dissolved in water or sitting dry in a bottle, never respires. It therefore has no better chance of preserving its complex chemical structure than any other complex chemical compound (Fig. 9-9).

Then how does a virus replicate? By subversion. It captures for its own ends the cellular machinery of a living cell, forcing it to channel all its resources into the manufacture of viral particles instead of normal cellular components. The result of course is death of the cell. This type of virus is called *virulent*.

Viruses can exploit another strategy for multiplication. After a virus enters a living cell, its DNA can become incorporated into the genome of the host cell. Thereafter the incorporated viral genes replicate in synchrony with replication of the host genome. The viral genes do not otherwise harm the host cell. Such a virus is called *temperate* (Fig. 9-10). A temperate virus can be occasionally transformed into a virulent one by environmental stimuli such as ultraviolet radiation or exposure to a variety of chemicals. When such a formerly temperate virus becomes virulent and leaves its host, it may carry with it a segment of the host's DNA. After attacking a new host, it again assumes a temperate life style and incorporates its nuclear cargo (both viral and host DNA) into the DNA of the new host. The process is called *transduction* ("to lead across") (Fig. 9-11).

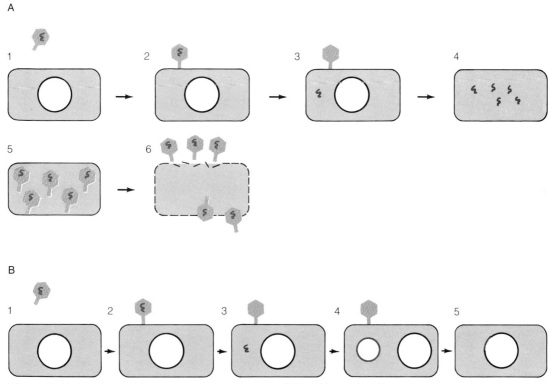

Figure 9-10 VIRULENT AND TEMPERATE VIRUSES. (A) Life cycle of a virulent virus. The virus illustrated belongs to the bacteriophages, or viruses that infect bacteria. The virus attaches to the bacterial cell wall (1, 2), makes a hole in the wall, and injects its DNA into the bacterium (3). There follows the destruction of host DNA (4), synthesis (4), and encapsulation (5) of viral DNA, culminating in destruction of the bacterium and escape of new virus particles (6). (B) Life cycle of a temperate virus. Attack and injection of viral DNA occurs as before (1–3). The injected DNA then becomes circular (4) and integrates into the host DNA at a specific site (5). The integrated viral DNA replicates in synchrony with bacterial DNA.

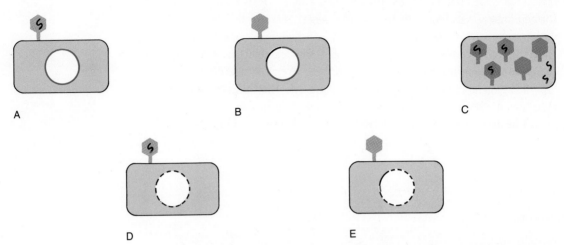

Figure 9-11 TRANSDUCTION. (A) Attack on bacterium #1. (B) Integration of viral DNA into host genome. (C) Turning virulent, the virus multiplies and destroys the bacterium. The viral DNA retains a fragment of host DNA. (D) Attack on bacterium #2. (E) Integration of viral DNA plus cargo of bacterial DNA into genome of host #2.

Until recently, the process of transduction had been described only in bacteria. But in 1974 George Todaro and co-workers at the National Cancer Institute announced an example of transduction in a mammalian system. Perhaps most surprising was the evolutionary distance between the two hosts involved: the domestic cat and primates.

In general, viruses of either the virulent or temperate kind are highly host-specific. A virus that attacks a cucumber plant does not attack a human being who plants the cucumber or eats it in a salad. But Todaro was able to show that all breeds of domestic cat carry genetic material incorporated from an unknown primate donor ancestral to humans, the great apes, and some Old World monkeys. The incorporated DNA is quite stable and only very rarely betrays its presence by erupting as a C-type virus. The transfer took place 5 to 10 million years ago, as estimated from mutational differences between the DNA segment found in primates and in cats. It is impossible to guess at the present time how general such interspecific DNA transfer will prove to be. But if the process proves to be more than an accidental quirk, then the viruses will have to be seen in a new light as potential agents of evolutionary change.

Structurally, a virus is far simpler than even the simplest bacterial cell. The coat consists of one or at most a few kinds of proteins. Packed inside is nucleic acid, either DNA or RNA but never both, as in living cells. It is the job of the outer coat to protect the cargo of nucleic acid and allow it to penetrate its host cell. The coat is often left behind on the outside.

RNA viruses form an interesting group. Before their nucleic acid can integrate into the host genome, it must be transcribed into DNA. The host cell produces a viral enzyme, RNA reverse transcriptase, which reverses the normal transcription process; viral RNA becomes a template for DNA synthesis. Many RNA viruses appear to be tumor-producing.

Viral infections are common among vertebrates, higher plants, arthropods, and bacteria. They are quite rare among fungi, nonflowering plants, and most invertebrate animals such as mollusks and annelids. There is some evidence to suggest that the present pattern of viral infections reflects the multiple origins of viruses. In fact, some biologists believe that the creation of viruses is an ongoing process that results from the separation and encapsulation of host nuclear material.

The Protistans

The word "protistan" has the same root as "protein," or "prototype," meaning "first." The protistans constitute the first eukaryotes—organisms with true nuclei, mitochondria, chloroplasts, and so on. How did they originate? A great deal of chemical and cytological evidence suggests that one moneran organism assumed a mutualistic existence inside another to give rise to mitochondria, chloroplasts, and even the eukaryote cilium or flagellum. That is, a mutual interdependence evolved between the host organism and its internal guest.

The protozoan *Myxotricha paradoxa* offers a hint of how such an interdependence may have originated. *M. paradoxa* lives in the gut of Australian termites. These termites, like their American cousins, possess mouthparts for chewing wood but lack the digestive enzymes needed to degrade the wood to an edible form. This is the function carried out by *M. paradoxa*. The protistan swims vigorously in the gut of the termite, engulfing and digesting tiny bits of wood chewed up by its host. At first sight, it appears to be propelled by masses of ordinary cilia scattered over the surface pellicle. But a closer look reveals that these are not cilia at all but spirochetes—flagellated bacteria like the ones that cause syphilis. The spirochetes

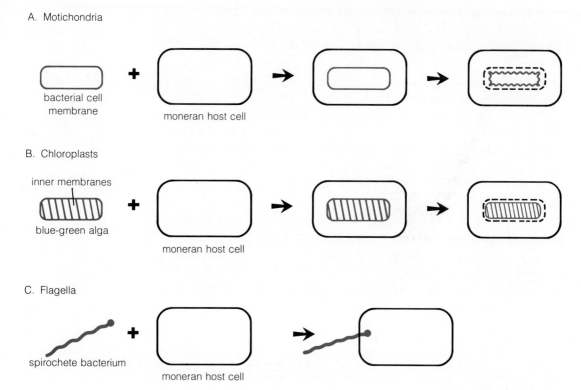

Figure 9-12 PROPOSED ORIGINS OF EUKARYOTIC CELL ORGANELLES.

are attached to the surface of the protistan at regular intervals and drive it forward at high speed. Floating in the cytoplasm of the protistan are oval bodies that once more turn out to be bacteria and that may function as a source of the enzymes needed for the digestion of wood.

Myxotricha paradoxa thus stands at the interface of two sets of mutualistic associations. It lives in the shelter of the termite's gut and provides the termite with an essential digestive capacity. But *Myxotricha* in turn serves as host to its bacterial symbionts, from which it derives propulsion and a necessary set of enzymes. It seems not unreasonable to imagine that similar associations, hundreds of millions of years older, resulted in the present cellular forms of eukaryotes.

Mitochondria presumably evolved from an internal bacterium. Both mitochondria and bacteria have DNA that is circular and naked. Nuclear DNA of eukaryotes is linear and bound to protein. Both mitochondrial and bacterial ribosomes are 15 percent smaller than cytoplasmic ribosomes of eukaryotes. Protein synthesis in bacteria and mitochondria is inhibited by the same chemical agents. The inner membrane of a mitochondrion resembles the cell membrane of an aerobic bacterium in its capacity for electron transport. The outer membrane of the mitochondrion appears to be contributed by the host cell. In fact, many internal enzymes of mitochondria are also specified by genes in the host cell nucleus. This is why mitochondria cannot lead an independent existence outside the host cell, as could their microbial ancestors (Fig. 9-12).

Similar arguments link the chloroplast to a prehistoric blue-green alga. There are similarities in DNA, ribosomes, protein synthesis, structure of internal membranes, and chlorophyll. (Blue-green algae contain chlorophyll a, while eukaryote photosynthesizers contain chlorophyll a plus a second kind of chlorophyll.) Even the mechanism of photosynthesis follows a similar pattern and is unlike the mechanism used by photosynthetic bacteria.

The protistans of today are all either single-celled organisms or contain a small number of cells bound in a loose association. These colonial protistans show no differentiation into specialized tissues. Instead, there is frequent specialization of function at the subcellular level (Fig. 9-13). Since protistan cells are often thousands of times larger than cells of other eukaryotes, they may be functionally equivalent to small animals, complete with legs, digestive tract, eyespots, and an internal coordination system, all built into a single cell. Other protistans have affinities with plants and fungi. In fact, it is very difficult to draw a well-defined boundary between protistans on the one hand and plants, fungi, and animals on the other. The protistans broke important new ground, evolutionarily speaking. Their achievement of eukaryotic cell structure put them within striking distance of the three major modes of life on earth: that of plants, animals, and **fungi.**

The Fungi

No gardener or homeowner ever planted a fungus in the garden or along the driveway. Yet along with bacteria, soil animals, and the humble grasses, the fungi are all around us. Given a bit of moisture and some organic material to feed on, they can work wonders; they can consume an acre of fallen leaves, an uprooted tree, or even a house with a leaky roof.

The working part of a fungus is called a *hypha* ("thread") (Fig. 9-15). The hypha is a long string of cells which permeates the material to be consumed. Unlike an animal, a fungal hypha lacks mouthparts or a digestive system. Hence it can feed only on foods in water solution, which it absorbs through its cell membrane. In order to solu-

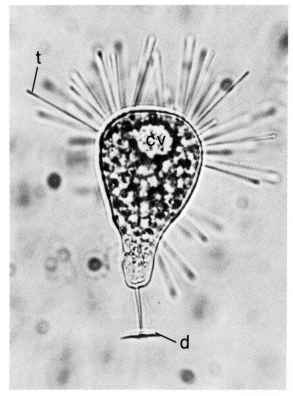
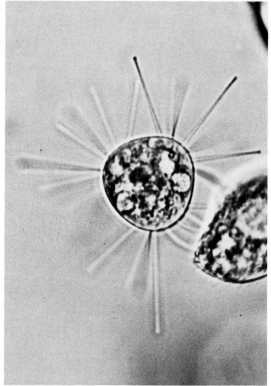
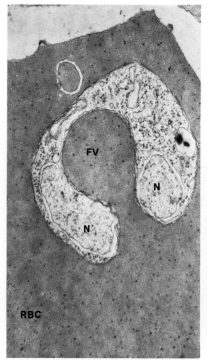
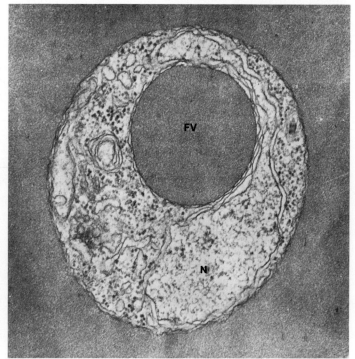

Figure 9-13 TOKOPHRYA INFUSIONUM, A CARNIVOROUS PROTISTAN FROM FRESHWATER PONDS (p. 334, top). (A) Side view. Young Tokophrya are covered with cilia and swim vigorously. The adult shown here loses its cilia and becomes permanently attached to a firm substrate by means of a basal disk (d). It feeds on other protistans by means of feeding tentacles (t). Also visible is a contractile vacuole (cv), used to bail out excess water (× 350). (B) *Tokophrya infusionum* feeding on *Tetrahymena,* another protistan. The organism is seen here from the top and its basal disk is hidden beneath. *Tokophrya* belongs to a group called the suctorians. It does not feed by suction, but ingests prey cytoplasm via a system of microtubules in the feeding tentacles.

Figure 9-14 PLASMODIUM COATNEYI, A PROTISTAN THAT CAUSES MALARIA IN THE RHESUS MONKEY (p. 334, below). (A) *P. coatneyi* inside an RBC ingesting a large "mouthful" of hemoglobin from the host cytoplasm. The ingested cytoplasm is forming a food vacuole (fv). The nucleus (n) is enveloped in double membranes and appears double because of its lobular nature. (B) *P. coatneyi* with fully formed food vacuole (fv). The elongated nucleus (n) fills much of the cell.

Figure 9-15 THE WORKING PORTIONS OF FUNGI. (A) Fungal hyphae. (B) Digestion of food by a hypha. The mechanism used is similar to that used by the bacteria. (C) Feeding by a fungal haustorium.

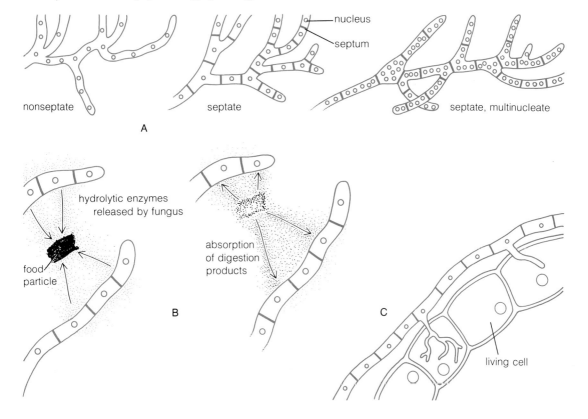

bilize a solid chip of wood, a piece of leather, or a loaf of bread, the hypha secretes external enzymes which depolymerize the food source into smaller, water-soluble molecules. These are then absorbed by the hypha and typically oxidized by Krebs cycle–electron transport enzymes to generate energy. Some fungi do not require oxygen for energy-generating processes. The most famous of these are the yeasts, which produce alcohol and carbon dioxide as metabolic end products under anaerobic conditions. The process is called *fermentation,* and has been used by people since prehistoric times in the brewing of beer and malt. An obvious and important limitation on fungal feeding is that the food source be biodegradable. Plastics, for example, are carbon compounds with a high energy content, but no fungal enzyme exists with the ability to solubilize them.

Hyphal feeding is restricted largely to dead materials. Living organisms, including plants, defend themselves against fungal attack by a variety of immunological and chemical means. However, some fungi do have the ability to penetrate living cells. Probably most of us will at some time be annoyed by ringworm or athlete's foot, both fungal infections. Fungal infection of plants can reach spectacular proportions. The Irish potato famine of 1845–1848, caused by the fungus *Phytophthora infestans,* killed a million people by starvation and forced another million to emigrate. More recent examples of fungal attack include chestnut blight and Dutch elm disease. The chestnut blight of the early part of this century essentially eliminated the American chestnut from North America. Prior to this time, the chestnut had been a major species of eastern oak forests. Dutch elm disease appears to be doing the same to the American elm. Caused by

Figure 9-16
A STREET IN WINNETKA, ILLINOIS, BEFORE (LEFT) AND AFTER (RIGHT) AN ATTACK OF DUTCH ELM DISEASE.

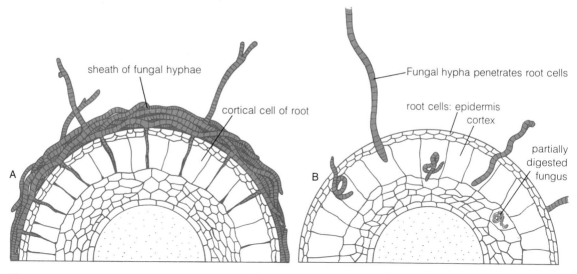

Figure 9-17 MYCORRHIZAE AND TREE ROOTS (CROSS SECTIONAL VIEWS). (A) Fungal sheath completely surrounds the root but does not enter cells of the cortex. (B) Fungal hyphae penetrate root cells, with some being digested by the plant cells. Both types of mycorrhizal associations benefit the plant.

the fungus *Ceratocystis ulmi,* which plugs the conductive vessels of the tree, it is spread by two species of bark beetles. Control attempts have focused on the beetles and have so far been unsuccessful. Many midwestern towns are undergoing drastic alterations as their elm-shaded streets are transformed into long stretches of dead stumps (Fig. 9-16).

The most common method by which such parasitic fungi attack their hosts is by means of a *haustorium* ("device to draw water") (Fig. 9-15). The haustorium penetrates a living cell and draws out nourishment from it. But in the living world, there is seldom a last word. The same haustoria that have been used to stunt and kill have in some cases evolved into essential plant structures.

Many plant roots are covered by a thick meshwork of fungi called *mycorrhizae* ("fungal roots") (Fig. 9-17). Fungi of the mycorrhizal layer may either penetrate root cells by means of haustoria or may simply send hyphae into spaces between cells. Such mycorrhizae are particularly common among tropical species. They apparently function in the absorption of minerals from the soil and in return receive organic materials from the root cells. In the tropics, they serve an important recycling function. The tropical soils that support rank jungles and lush rain forests are extraordinarily thin. When cleared of their forests and planted for farming, such soils give good yields for two or three years, followed by crop failure. This is

338 / Populations: The Variety of Life

because the high temperatures and moisture encourage the growth of fungi, which decompose the small amount of humus in the soil. The soil then loses most of its minerals and its water-holding capacity. The end result may be a hard, concretelike surface where nothing can grow. In an intact rain forest, soil mycorrhizae intercept all nutrients released from decomposing vegetation and return them to the growing trees. Crops planted in place of the rain forest lack such mycorrhizal associations.

Fungi are classified taxonomically on the basis of the visible part, the fruiting body. Mushrooms, morels, bracket fungi, and puffballs are all examples of fruiting bodies (Fig. 9-18). Of course, the fruiting body makes up only a small fraction of the total mass of the fungus, hidden below the surface as microscopic hyphae. The fruiting body itself is made up of a tangled mass of hyphae. The two largest groups of fungi are the *ascomycetes* or sac fungi (including morels, yeasts, cup fungi, some bread molds, and powdery mildews) and the *basidiomycetes* or club fungi (including most mushrooms, puffballs, bracket fungi, smuts, and rusts).

The ascus or the basidium is a fungal structure where $2N$ nuclei undergo meiosis to produce $1N$ nuclei. Thick walls are then laid down around the $1N$ nuclei to create fungal spores, which are then disseminated in staggering numbers. If all the spores from a single giant puffball germinated to produce adult puffballs, the weight of the puffballs would exceed the weight of the earth itself (Fig. 9-19).

Current taxonomic opinion recognizes five evolutionarily interrelated kingdoms of living organisms. In this chapter, we have considered three of them (with a brief digression to the viruses, which are not living organisms). The most primitive life pattern is that of the monerans, which include bacteria, rickettsiae, and blue-green algae. Monerans are prokaryotic organisms, lacking the true nuclei or advanced cellular organelles found in all other kingdoms. The protistans, our second kingdom, are eukaryotes, follow a variety of life styles, and are typically unicellular in body form. They represent a logical second step in evolution and merge indistinctly into the remaining kingdoms. The fungi, our third kingdom, are organisms evolutionarily derived from protistans. They have become specialized to feed by decomposition.

The organisms we have discussed in this chapter are the humble of the earth. Largely invisible and inaudible, they are nevertheless present in overwhelming abundance in every cranny of possible living space.

Figure 9-18
A GIANT PUFFBALL BEING USED AS A STOOL. The puffball is a fruiting body of a basidiomycete fungus.

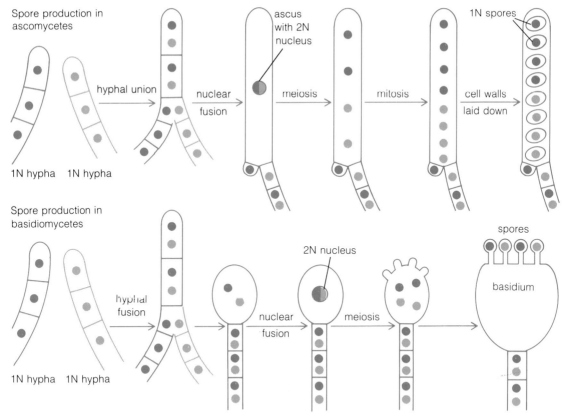

Figure 9-19 SEXUAL PROCESSES IN FUNGI. Note that in both the ascomycetes and the basidiomycetes the 1N generation dominates the life cycle. Only one 2N nucleus exists which immediately undergoes meiosis to return to the 1N state.

Bibliography

Helena Curtis, *The Marvelous Animals,* Natural History Press, 1968. About the protistans. A marvelous book.

R. G. Kessel and C. Y. Shih, *Scanning Electron Microscopy in Biology,* Springer-Verlag, 1974. The scanning electron microscope has finally given a visual personality to the tiny organisms of the world that they never had under the light microscope. This is a handsome book.

W. G. Walter, R. A. McBee, and K. L. Tempe, *Introduction to Microbiology,* D. Van Nostrand, 1973. A microbiology text for the non-science major.

C. J. Alexopoulos, "Introductory Mycology," 2nd ed., Wiley 1962. An introduction to the fungi.

GENERAL READING

TEXTBOOKS

Figure 10-1 A HONEY LOCUST IN SPRING.

chapter 10

In terms of living mass, plants are the most important form of life on earth. When biologists describe a natural community, they typically describe it in terms of its major vegetation, and this is done for good reason. Well over 90 percent of the living mass of such communities is made up of plants; plants not only provide food for all the animals of the community, but they also modify the local environment in a decisive way.

When we form a mental image of a plant, it is the tracheophytes we think of. The grasses of the prairie, the flowers of the garden, the trees of the forest, and the grains and vegetables of the farm—all

The Plants

these are tracheophytes. Before turning to this important phylum, however, we will first briefly look at the algae, a group of five phyla that some biologists do not even consider plants, and the bryophytes, or mosslike plants.

The Algae

We include the algae within the plant kingdom because, like higher plants, they function in photosynthesis. In addition, the green algae and higher plants have strong evolutionary affinities. But many biologists treat the algae as protistans, which they resemble because of their typically extremely simple body forms. Many are single-celled.

Though in a general sense the algae share a common life-style, five distinct phyla are recognized among them on the basis of fundamental biochemical differences such as types of chlorophyll, types of food storage, and types of cell wall material (see Table 1).

TABLE 1
The algal phyla[a]

PHYLUM	CHLOROPHYLL	FOOD STORAGE PRODUCT	CELL WALL COMPOSITION
Blue-green algae (moneran)	a	Cyanophyte starch	Moneran cell wall plus cellulose, pectin
Pyrrophyta (dinoflagellates)	a, c	Starch, fats, oils	Cellulose, pectin
Chrysophyta (diatoms)	a, c	Chrysolaminarin, oils	Cellulose, pectin, silica
Phaeophyta (brown algae)	a, c	Chrysolaminarin, oils	Cellulose, pectin, alginic acids
Rhodophyta (red algae)	a, d	Floridian starch	Cellulose, pectin, agar, calcium carbonate in some species
Chlorophyta (green algae)	a, b	Starch	Cellulose, pectin, calcium carbonate in some species

[a] Here the blue-green algae have been included for comparison, though of course they are not members of the algal group at all.

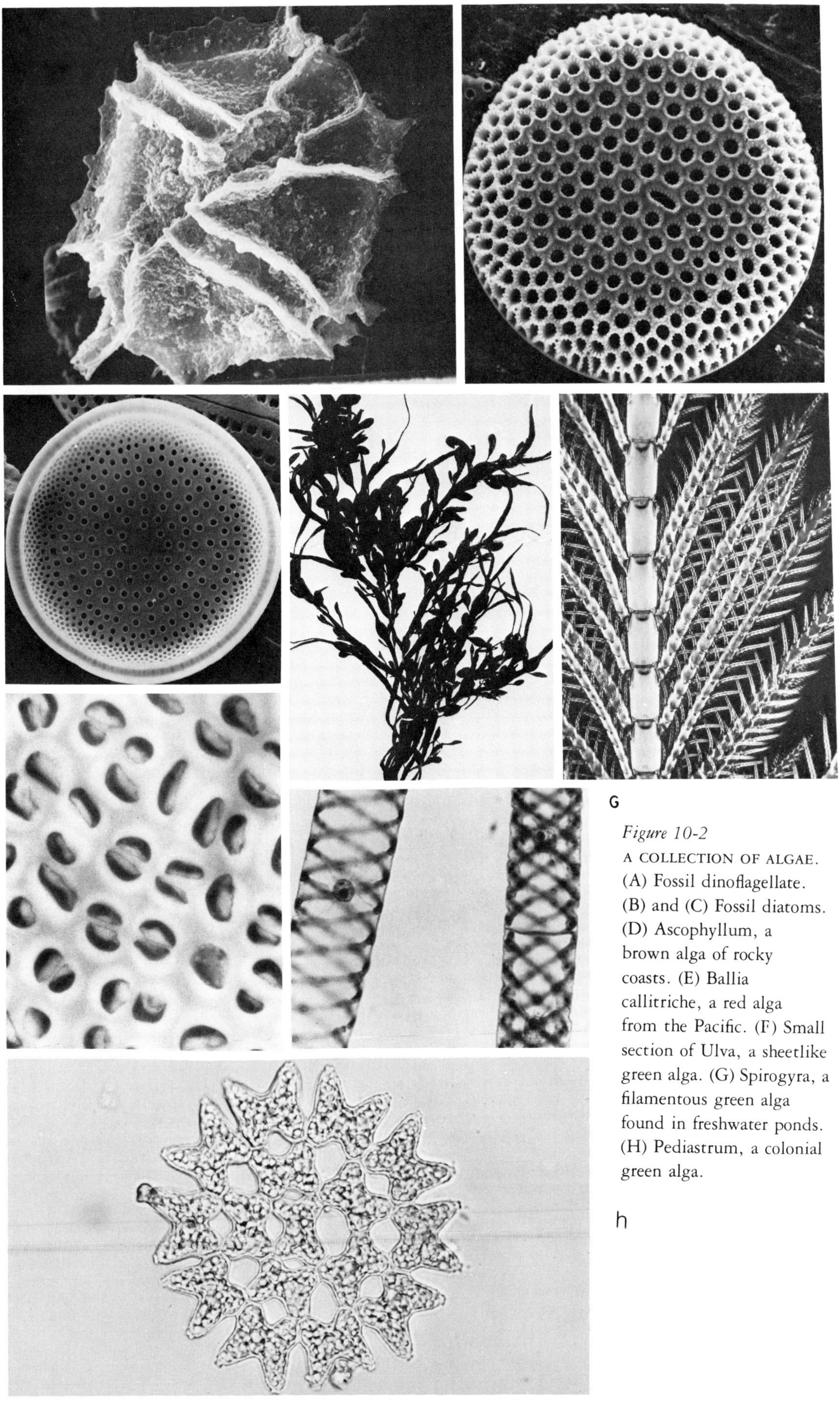

Courtesy of Carolina Biological Supply Company

Figure 10-2
A COLLECTION OF ALGAE. (A) Fossil dinoflagellate. (B) and (C) Fossil diatoms. (D) Ascophyllum, a brown alga of rocky coasts. (E) Ballia callitriche, a red alga from the Pacific. (F) Small section of Ulva, a sheetlike green alga. (G) Spirogyra, a filamentous green alga found in freshwater ponds. (H) Pediastrum, a colonial green alga.

Dinoflagellates ("terrible flagellates") are mostly single-celled organisms and mostly marine in distribution. Their contribution to total photosynthesis is second only to that of diatoms. They carry two flagella, one encircling the cell equatorially and the other running longitudinally. Some species produce a potent nerve toxin. When such species multiply in large numbers, they produce a "red tide" which can kill millions of offshore fish. This is why these tiny organisms are called the "terrible flagellates."

Diatoms ("cut in two") have two-part cell walls, one half fitting into the other like a shirt box. The cell walls are typically made of silica (glass) and carry species-specific, incredibly complex surface markings. They have traditionally served as visual test objects for light microscopes—the better the microscope, the more detail revealed. They are probably the most important photosynthesizers in the open oceans but are found in fresh waters as well.

Brown algae reach the largest size in the algal group. Some of the giant kelps of tropical waters may extend 200 feet in length, and the total length is limited only by the shearing stress of wave action. They also achieve the most complex body form. Many species (such as the common *Fucus* or *Ascophyllum* of rocky coasts) have a holdfast to anchor themselves to a firm foundation, as well as numerous gas-filled floats to maintain upright posture at high tide. The floats contain significant amounts of carbon monoxide. Many giant kelps even contain conductive tissue which carries photosynthetic products from the outer synthesizing regions to central regions where no photosynthesis takes place. The conductive tissue resembles the sieve tubes found in the phloem of higher plants.

Red algae are almost entirely multicellular. The typical red color of these algae is due not to chlorophyll but to phycoerythrin, an accessory photosynthetic pigment. Phycoerythrin absorbs light in the blue-green region of the visible spectrum. Blue-green light penetrates deepest into water, hence red algae can grow at greater depths than other plants. This phylum produces no flagellated cells at any stage of the life cycle. The male gametes are ameboid, while the female gametes are totally immobile. To direct the male gametes, the eggs release attracting hormones.

Green algae have shown exceptional evolutionary adventurousness. This phylum includes species ranging from unicellular (*Chlorella*) to those forming complex colonies (*Volvox*). Some have a marine habitat (*Ulva*), and others prefer fresh water (*Spirogyra*). Some are flagellated (*Volvox*), while others are immobile (*Pediastrum*). The plants may be haploid, with the zygote the only 2N cell (*Spirogyra*), or diploid, with the sperm and eggs the only

1N cells (*Bryopsis*); or the haploid and diploid stages may be equally important (*Ulva*). Some form cells of giant size (*Acetabularia* and *Valonia*).

It is therefore no wonder that this adventurous phylum was the one that gave rise to higher plants, the bryophytes and the tracheophytes. Much of the basic biochemistry of the higher plants was inherited from the green algae. This is why trees are green and not red or brown, why wood is made of cellulose and not glass, and why potatoes contain starch and not other polymers.

As can be seen from this brief discussion, the algae form a very diverse group. But they are more than biochemical curiosities—they play a crucial role in the living affairs of the world. It is estimated that they account for 50 percent of the world's total photosynthetic activity. Blue-green algae make only a negligible contribution in this respect.

The Bryophytes

The bryophytes include mosses, liverworts, and related forms. Physically speaking, it is a modest phylum. A field trip centered on the bryophytes is traveled on hands and knees, with a strong magnifying glass in one hand. The best places to look are damp woods, stable streamsides, and wind-sheltered places that receive plenty of rain. Even in their favorite habitats, mosses and liverworts appear almost as an afterthought, living in the shadow of the vascular plants. The only exception is a peat bog, where the peat moss *Sphagnum* creates a world to which all other species must adapt. The center of an old peat bog is a green, mysterious place. Almost all the flowering plants have disappeared, as have the mammals, and one can almost imagine oneself in a world of the Silurian period, 425 million years ago, when the first plants and animals were establishing themselves in a land existence. These pioneering plants were the bryophytes, and they hold the same position in the plant world that amphibians do among the vertebrates; they have one foot on land and the other still back in the water.

To understand what makes bryophytes able to live on land, we must look in some detail at their sex life. The sexual pattern of bryophytes was inherited from the green algae, their evolutionary ancestors. And the pattern is followed in modified form by all the higher plants that dominate the earth today. This pattern is called the *alternation of generations*. The concept is not familiar to us because it occurs only in plants, not in animals. A typical animal body, such as

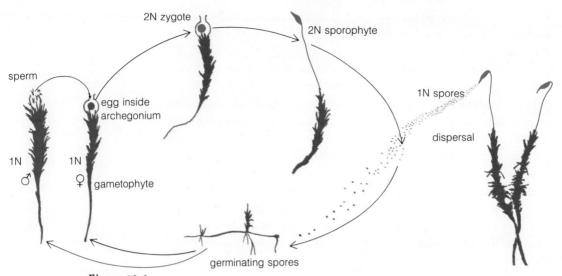

Figure 10-3
ALTERNATION OF GENERATIONS IN A MOSS. The sporophyte generation is shown in color.

our own, is made up of diploid ($2N$) cells. The only $1N$ cells are the sperm and the eggs, whose sole function is to find each other and fuse to form a $2N$ zygote.

However, $2N$ plants can produce another kind of $1N$ cell called a *spore*. Spore production from a $2N$ plant always occurs by meiosis. Plant spores are neither functionally nor structurally equivalent to the spores of bacteria, which are produced by asexual means and are designated for survival in harsh environments. The spores produced by either a multicellular green alga or a bryophyte then proceed to divide mitotically and develop into a complete $1N$ alga or bryophyte. This $1N$ plant can now produce gametes (sperm or eggs) by mitosis. It is therefore called the *gametophyte* ("gamete-plant") generation. The sperm wastes no time in finding the egg and fuses to form a $2N$ zygote. The zygote then divides mitotically to form an entire $2N$ plant. This $2N$ plant again undergoes meiosis in specific tissues to produce $1N$ spores, and is therefore called the *sporophyte* ("spore-plant") generation. Thus the life cycle of the plant involves an alternation between a $1N$ or gametophyte generation and a $2N$ or sporophyte generation (Fig. 10-3).

The links between the generations are the spores on the one hand and the gametes on the other. For any land plant, these cells represent the most vulnerable stages in its life cycle, simply because they are single cells and must face the harshness of the land environment alone. Bryophytes chose two very different kinds of mechanisms to protect the two kinds of cells.

Spores of bryophytes incorporate the protective features of bacterial spores—thick walls and low water content. Thus bryophyte

THE TRACHEOPHYTES / 347

spores have little to fear from the drying atmosphere of the land and can ride for considerable distances on air currents. The spores act as the dispersal stage.

The gametes and early embryonic stages of the bryophytes are protected by a different mechanism. The flagellated sperm can only swim in a watery medium. Hence it is released only when a rainfall or other water source provides a liquid environment. The egg is nonmotile and is closely surrounded by a layer of gametophyte cells (Fig. 10-3). This jacket of cells protects and nourishes the egg and remains to protect and nourish the developing embryo after the egg has been fertilized and begins to develop into a sporophyte.

While the dry spores and the jacket-enclosed egg and embryo represent successful adaptations to a land environment, the bryophytes have not entirely broken their close links to a watery environment. Fertilization must still take place in a water medium, and the egg attracts the sperm by means of water-soluble hormones. (Common table sugar is the sex attractant in some species.) Moreover, adult plants lack an efficient mechanism for water retention and lack true roots and a water conduction system. Therefore they must remain in close contact with groundwater and are restricted to relatively humid environments.

The Tracheophytes

The word "tracheophyte" means "plant with a windpipe." The group is named for its conductive xylem elements (see page 360) which superficially resemble the windpipes of animals. In addition to the prairie grasses, garden flowers, and forest trees, tracheophytes include more primitive plants such as ferns, horsetails, club mosses, and an extinct subphylum called the *psilopsids*. It is undoubtedly the flowering plants, or *angiosperms* ("seeds inside vessel") that form the evolutionary apex of the tracheophytes. Only the coniferous trees,

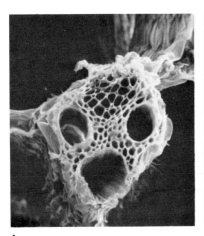

A   B

Figure 10-4
XYLEM VESSELS IN THE BULLRUSH SCIRPUS LACUSTRIS. (A) A vascular bundle showing three large xylem vessels and smaller phloem cells. (B) Closeup of xylem vessel at left.

members of the class *Gymnospermae* ("naked seeds"), approach them in importance on the earth. But there are only about 750 species of gymnosperms living today, as opposed to perhaps 250,000 living species of angiosperms. The internal and external taxonomic relationships of the tracheophytes are given in Table 2.

TABLE 2
Taxonomic relationships of the tracheophytes

Phylum Chlorophyta: green algae
Phylum Bryophyta: mosses and liverworts
Phylum Tracheophyta: plants with conductive tissues
    Subphylum Psilopsida (extinct)
    Subphylum Lycopsida: club mosses
    Subphylum Sphenopsida: horsetails
    Subphylum Pteropsida: plants with true leaves
        Class Filicinae: ferns
        Class Gymnospermae: plants with naked seeds (mostly conifers)
        Class Angiospermae: flowering plants

Given their overwhelming dominance of the land, it is surprising that essentially no flowering plants are found in the oceans, and only a small number in fresh waters. Here the primitive algae remain supreme, facing little threat from their more sophisticated descendants. This is because almost all the evolutionary advances of angiosperms have been concerned with adaptations to a land environment. Algae, though structurally simple, are perfectly adapted to their own medium. The transition from water to land was so drastic that successful adaptation required an evolutionary preparation stretching out for more than a billion years. The oldest living fossils date from about 3.1 billion years ago, and the green algae from about 1.4 billion years ago. The first tracheophytes appeared in the Silurian period, 425 million years ago, marking the conquest of land by plants. But it was "only" in the Cretaceous period, 135 million years ago, that flowering plants first appeared. That is a long wait even on an evolutionary time scale.

But consider the staggering problems that had to be overcome:

1. The algal method of reproduction had to be scrapped. Green algae use flagellated sperm which dry out and become immobile outside water. Angiosperms invented a completely new reproductive structure—the *flower*.

2. Algae, when removed from water, quickly dry out because they lack an outer covering that is waterproof. Angiosperms are

covered by a *cuticle* made of wax, which controls water loss.

3. A land plant must stay above ground in order to photosynthesize, but must reach below the ground to obtain water. The water must then be conducted to the aboveground portions. The solutions are *roots* and a *conductive system* (xylem).

4. Since roots are alive but cannot photosynthesize, they must be fed by the aboveground portions. The food must be conducted downward by a separate conductive system—the *phloem*.

5. A land plant experiences much more severe gravitational stress than a water plant and must therefore acquire a skeletal system. The tracheophyte skeleton consists of thick walls of solutions are *roots* and a *conductive system* (xylem).

6. Because a land plant must simultaneously respond to a variety of environments and because it contains many different kinds of tissues, it needs a communications system to send messages from cell to cell and integrate the development process. The communications system of land plants is a hormonal one, and in complexity it rivals the hormonal systems of animals.

Let us now take a closer look at some of these determinants of the life-style of an angiosperm.

The Flower

Flowers have always symbolized life and growth. We bring them to the sick as a symbol of health, we give them to those we love as a token of joy, and they are the natural playthings of happy children. This is entirely appropriate, because the flower is the generative organ of angiosperm plants. The reproductive processes in angiosperm plants have evolved in such an original way that it is worthwhile comparing them to the processes in animals.

In a water environment, there is little difference between the reproduction of animals and of plants. Both typically release their gametes directly into water, where a motile sperm seeks out a non-motile egg. The main difference between the two processes is that plants undergo an alternation of generations, producing sperm and eggs by mitosis from the gametophyte generation.

On land, animals as diverse as insects, birds, and mammals practice internal fertilization. The male introduces sperm directly into the body of the female, through the act of copulation. In a very roundabout way, the same is true of flowering plants, but again two generations are involved—a sporophyte (spore-producing generation) and a gametophyte (gamete-producing generation). There is an exquisite partition of function between the two generations. Both generations coexist within the flower, as do both sexes (in most flowers).

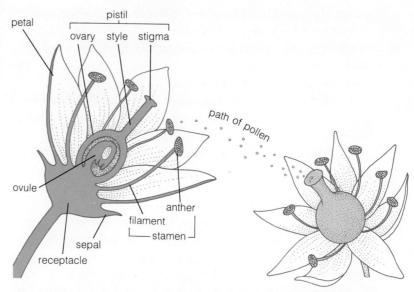

Figure 10-5
PARTS OF THE FLOWER.
The dotted line shows the path of pollen between two flowers of the same species.

But in terms of function, the flower accomplishes the same end results as the internal fertilization system of animals; it protects the gametes and developing embryo against drying, and it transfers sperm to the egg. But it goes beyond these basics to offer reproductive advantages that are enjoyed by no member of the animal kingdom.

An idealized flower (Fig. 10-5) consists of a series of modified leaves called *sepals, petals, stamens,* and *pistil(s)*. Stamens are the male reproductive organs and produce *pollen*. Pistils are the female reproductive organs and receive the pollen of another flower.

What is pollen? It is *not* comparable to the sperm produced by animals. An individual pollen grain is in fact a microscopic plant in its own right—it is the male gametophyte generation of the flowering plant. The pollen grain grows up from a $1N$ spore produced meiotically by the spore-producing tissue of the *anther*. But the pollen grain does not have much growing to do! The pollen grain shed by the anther contains only two cell nuclei, enclosed in a tough, species-specific pollen coat (Fig. 10-6). But if such a pollen grain is dropped onto a filter paper impregnated with a sugar solution, a strange thing happens. The pollen grain bursts open and sprouts a long tube. The tube grows so rapidly that its progress can be followed almost perceptably under a microscope. As the pollen tube elongates its growth is controlled by the tube nucleus. The other nucleus divides to yield two sperm. Incidentally, the same process of germination will take place if a pollen grain lands on the moist membrane of the nasal mucosa, much to the discomfort of hay fever sufferers. In this case, the pollen has made an understandable mistake in its choice of sexual object.

THE TRACHEOPHYTES / *351*

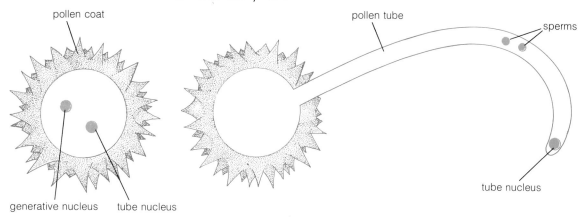

Figure 10-6 RESTING POLLEN GRAIN (LEFT) AND GERMINATING POLLEN (RIGHT).

The proper sexual object of the pollen grain is the *stigma* of another flower of the same species (Fig. 10-5). The stigma produces a sticky secretion which holds the pollen grain and stimulates germination of the pollen tube. The pollen tube actually digests its way down the style and into the ovary, guided by a chemical gradient of some sort, presumably calcium ions. Finally the pollen tube makes contact with the cell membrane of the female gametophyte which sits inside a structure called the *ovule* located inside the ovary.

A female gametophyte, though larger than the male, is hardly an imposing structure. It consists of a single cell membrane enclosing eight nuclei. One of these is the egg, destined to be fertilized by one of the two sperm. The other sperm is not superfluous—it fuses with the two *polar nuclei,* located in the center of the gametophyte cytoplasm, to produce a $3N$ tissue called the *endosperm.* The endosperm tissue, because of its $3N$ constitution, grows rapidly and balloons out the ovule to many times its original size. However, its ultimate fate is to serve as a source of stored food for the more slowly developing $2N$ embryo that resulted from the fusion of sperm and egg. (Fig. 10-7).

The female gametophyte, like its male counterpart, grew up from a $1N$ spore produced meiotically by spore-forming tissue (Fig. 10-7). The spore destined to become the female gametophyte is produced inside the ovule and is larger in size than its male counterpart, hence it is called a *megaspore* ("giant spore"). Though four such spores are produced per ovule, three degenerate and the remaining megaspore forms a single female gametophyte.

This gametophyte, after the characteristic double fertilization, develops into a *seed*. The seed contains at least two tough seed coats (derived from the integuments or outer coats of the ovule), internal

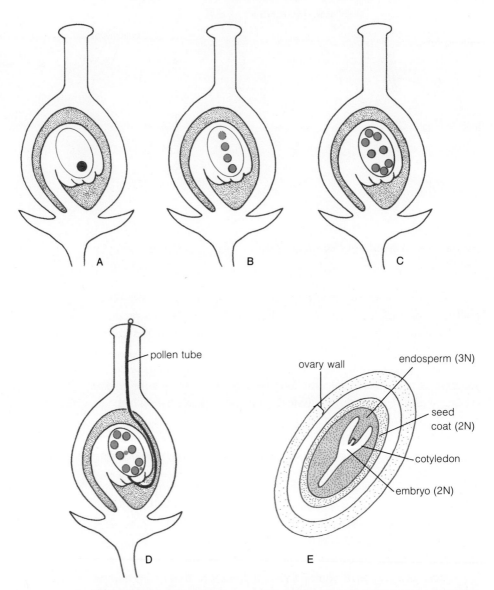

Figure 10-7 DEVELOPMENT OF THE SEED AND FRUIT IN A FLOWERING PLANT. $1N$ structures are shown in color. (A) View of the ovule inside the ovary. The ovule contains a megaspore mother cell, destined to give rise to the female gametophyte. (B) Meiosis of the megaspore mother cell produces four megaspores, of which three degenerate. (C) The surviving megaspore divides mitotically to yield a female gametophyte containing eight cell nuclei. (D) Fertilization. One of the sperms in the pollen tube fertilizes the egg to give a $2N$ zygote. The other sperm fertilizes the two polar nuclei to give rise to a $3N$ nucleus. The $2N$ zygote divides mitotically to produce the future embryo. The $3N$ nucleus divides mitotically to produce the future endosperm (nutritive tissue for the embryo). (E) Seed inside the fruit. The fleshy portion of the fruit has been derived from the wall of the ovary.

3N endosperm tissue, and a 2N *embryo*. The embryo, as its name implies, is a very young plant, complete with a rudimentary root, shoot, and one or two seed leaves or cotyledons. Nothing like a seed exists in the animal kingdom. If humans could produce seeds, they would consist of an embryo delivered while still inside its fetal membranes, plus an ovary, plus a food store sufficient for several weeks of growth. That may not sound like an advantage. The advantage of producing a seed is that the embryo inside is in a dormant state and remains so until conditions are suitable for growth. In the human situation, a woman might keep her seed in storage until she has established a career, or can provide a suitable physical and emotional environment for the growth of a child.

In the case of a plant, the seed typically remains dormant only until the next growing season. The alternative could be disastrous. A seed that germinated in the fall would reach the sensitive seedling stage by winter and would be killed by the cold. Then how does a seed "know" when to germinate? Mechanisms vary from species to species. Some seeds must pass through an obligatory cold period before they can germinate. For instance, a peach seed must spend 2 to 10 weeks at close to freezing temperature before its dormancy is broken. Other seeds contain natural inhibitors which must be removed before germination. Seeds of the salt bush, a desert plant from Utah, contain internal salt. This internal salt must be leached out by a heavy rain. Thus they germinate only when the ground has been thoroughly wetted. Still other seeds are encased in very tough seed coats which must be scarified before germination is possible. One of the most common mechanisms of scarification is to pass through the digestive tract of an animal, as is necessary with an apple seed.

Most seeds are viable for about 10 years if stored under cool, dry conditions. Grasses, such as domestic wheat and corn, last longer—from 30 to 50 years. Perhaps the record is held by the lotus seed. Several lotus seeds were recovered from a peat deposit in Japan, and radiocarbon dating established their age as 2,000 years. After scarification of the seed coat, the seeds were planted. The result was a breathtaking purple blossom as fresh as the morning, whose parent had died 2,000 years ago!

One of the most stunning evolutionary developments of angiosperms has been their entrapment of other species in their reproductive processes. This is not true among the gymnosperms, such as pine trees, where only wind is necessary to carry pollen from a male cone to a female cone and to disseminate the seeds afterward. Wind pollination works well if plants of a given species grow close together in massed assemblies, as in the case of a spruce forest, a bluegrass prairie, or a field of wheat. But wind pollination would be totally impractical for a single columbine growing in a spring meadow, or a trillium flowering in a rich wood. Such plants have enlisted the services of animal species in their pollination processes: insects, hummingbirds, and even bats and monkeys.

Probably the most important of these pollinators are the insects. Insects are an evolutionarily older group than are the flowering plants. But it is significant that the evolutionary record shows a great "flowering" of insect species around 150 million years ago, coinciding with the appearance of the first flowering plants. Today the connection between flowering plants and many insect species is so close that if one were to become extinct the other would also disappear. The most important insect pollinators are long-tongued insects—bees, moths, and butterflies. To attract such insects, flowers offer showy petals (insects see color), attractive fragrances, and a food supply in the form of nectar and excess pollen.

Moreover, it is to the advantage of the flower to attract only a single species of insect. An unselective insect that visits, say, mountain laurel and then lilac and honeysuckle, wastes its load of pollen on the wrong species. Consequently flower shapes have evolved so as to limit the possible species of pollinators. A snapdragon must be forced open by a strong insect such as a bumblebee. Bladder campion is white to attract moths and has a long corolla requiring the long tongue of a moth to reach the nectar at the bottom. Jack-in-the-pulpit has a rotten-meat smell to attract carrion beetles. Perhaps the apex of specialization is reached in the orchid family, where the flower-insect match is so specific that certain male wasps are regularly deluded into thinking the orchid is a female wasp. As they copulate with the flower, their movement serve to pollinate it. Similar adaptations are made by hummingbird-pollinated flowers of the tropics. The flower tube is shaped to fit the bird's beak, and the flowers are scentless but bright.

It is paradoxical that, while angiosperms have gone to great evolutionary lengths to attract insects and other animals, they have often found it necessary to protect themselves against these very same insects and birds. An insect feeding on pollen or nectar may inadvertently chomp on the ovary of the flower, destroying the potential seed. A very common protective device among the plants is to recess the ovary into the receptacle (Fig. 10-8), where it is much harder to get at. An *inferior ovary* is a strategy used by plants as diverse as dogwood, apple, and parsley. Another approach is to locate the source of nectar in a side pocket among the petals, away from the ovary. Such a *side nectary* is a very conspicuous feature of the columbine. A third

Figure 10-8 PROTECTIVE DEVICES IN FLOWERS. (A) Inferior ovary (narcissus in cross section). (B) Side nectary (columbine). (C) Composite flower (New England aster).

approach has been exploited by the composite family, including asters, daisies, and sunflowers. Here many individual flowers are closely packed into a single showy flower head. A pollinating insect must walk on the surface and cannot reach the ovaries down below. The success of the method is shown by the fact that composites make up the largest flowering family in North America.

There is a second stage in the reproductive cycle in which angiosperms have managed to enlist the aid of animals, and this is in the dispersal of seeds. The seeds of angiosperms mature inside the ovary of the pistil, and in fact this is the derivation of the word angiosperm ("seed inside a vessel"). A seed plus its matured ovary wall is called a *fruit*. Examples of fruits include the cherry, squash, pea pod, tomato, cucumber, and melon, but not the potato, cabbage, lettuce, or turnip. The fleshy portion of the fruit, made up of the ripened ovary wall, is often both delicious and nutritious. This is true for one reason: it is meant to be eaten. Even when the rest of the plant has evolved physical or chemical defenses to discourage feeding by animals, its fruit may be delicious. Such is the case of thorn-covered hawthorns, blackberries, and wild roses, or of the cyanide-filled black cherry. In fact, very few fruits are poisonous. The deadly nightshade, the loco weed, and the baneberry are the exceptions, not the rule. Once a fruit is consumed by a larger animal, the seeds contained inside may pass through the digestive tract unscathed and may be deposited far from the parental plant, along with a modest amount of fertilizer.

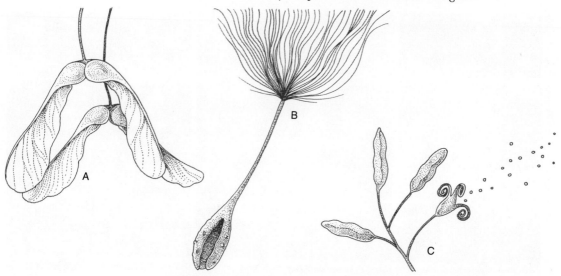

Figure 10-9 FRUITS AS AIDS TO SEED DISPERSAL. (A) Maple keys—aerodynamic flight. (B) Dandelion seed with parachutes—riding on air currents. (C) Touch-me-not—explosive disintegration of fruit pod and scattering of seeds.

Of course, not all fruit are designed to be eaten by animals. The parachute-borne dandelion fruit and the propellerlike fruit of maple or linden depend on wind and aerodynamics for dispersal (Fig. 10-9). The fruit of witch hazel and jewel weed shoot out their seeds by a ballistic mechanism. But even nonedible fruit may depend on animals for dispersal. The beggar ticks of old fields (fields in the process of returning to woodland) carry dozens of tiny barbs which cling to the fur of passing animals or to the clothing of strolling nature lovers. When they are finally picked off they are far from their parental plant and ready to colonize new spaces.

How did the flower evolve? There is little doubt that the ancestral structure resembled a pine cone carrying both male and female scales. If one peels away the petals of a primitive flower like the magnolia or the tulip tree, the central cylinder of pistils and stamens still looks very much like a tiny, green pine cone (Fig. 10-10). Further changes involved a reduction in the number of stamens and pistils.

One important difference between a pine cone and a flower is that the seeds of the pine cone are borne naked on the surface of the scales, while the seeds of angiosperms are wrapped inside the basal portion of the pistil. It is thought that the pistil evolved as an open leaf bearing spore-producing tissue along its margins. Such a leaf is still present in ferns. The leaf then curved inward and fused to form the ovary of modern angiosperms (Fig. 10-11).

Figure 10-10 A MAGNOLIA FLOWER. This primitive flower resembles a bisexual pine cone with petals. The central core consists of many pistils spirally arranged. Below these are the more broadly spreading stamens.

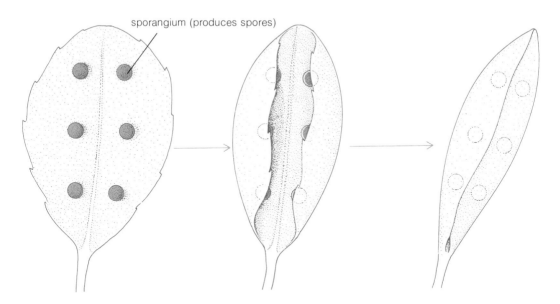

Figure 10-11 EVOLUTION OF THE CARPEL. The carpel probably evolved as a modified leaf bearing sporangia (left). Such leaves are still found in present-day ferns. With time, the edge of the leaf curled inward and fused, producing a closed chamber called the carpel. The carpel shown here contains a single chamber. Fusion of several carpels produces a multichambered ovary.

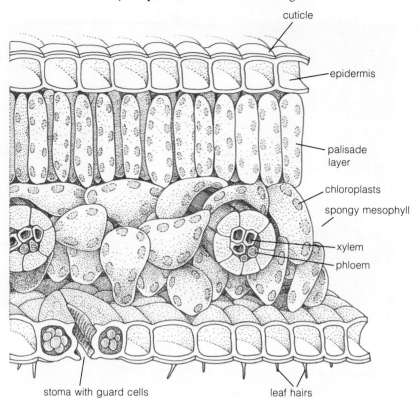

Figure 10-12 CROSS SECTION OF A LEAF. Note the large internal surface area to facilitate gas exchange.

The Leaf

The leaf of a modern tracheophyte drives the rest of the plant. Here a careful architecture of chloroplast-bearing cells is used to convert the energy of photons from the sun into the chemical energy of sugar molecules (Fig. 10-12). (See Chapter 2.) This reaction consumes carbon dioxide from the atmosphere and releases back molecular oxygen. The other necessary raw material, water, is piped into the leaf by means of conductive tissue continuous with the stem and the roots. The difficulty is that water evaporates quickly under prevailing conditions of photosynthesis, and a plant loses much more water due to evaporation than it uses up to make sugar. A single corn plant transpires 2 quarts of water a day, and an acre of corn loses 300,000 gallons per growing season.

These transpiration losses would be much higher if special water-conservation measures were not used by the leaf. First, a leaf is covered by a thin, waxy cuticle which acts as a water barrier. If the cuticle were absent, all terrestrial leaves would shrivel and dry up in a matter of hours, much as rock weeds, brown algae, shrivel up when exposed at high tide.

Figure 10-13
STOMATA ON THE SURFACE OF A LEAF OF THE BULLRUSH (SCIRPUS LACUSTRIS). Each stoma is lined by two guard cells. Also visible is a single leaf hair projecting from the surface.

Unfortunately, higher plants have not managed to invent a cuticle that can stop water but not carbon dioxide and oxygen. Hence the cuticle must be pierced by openings to permit gas exchange. These openings are called *stomata* ("mouths"), and a plant can maintain its stomata in either an open or closed state by means of a pair of guard cells. The exact mechanism responsible for guard cell opening and closing is by no means clear. Factors such as light and dark, carbon dioxide and oxygen tension, and potassium ion concentration exert a controlling effect. However, the ultimate factor controlling the opening or closing of each stoma is the entrance or exit of water from its pair of guard cells. As a simple mechanical analogy, imagine a pair of tubular balloons lying side by side, with a strip of stiff tape on their facing sides. When the balloons are inflated, each one bulges outward to create an opening in the middle. Guard cells have a similar cell wall thickening along their facing sides, and the entrance of water causes the cells to open. Loss of water, through wilting or osmotic pressure changes, causes the cells to close.

A typical leaf carries most of its stomata on the underside. Often the underside of a leaf has a velvety texture, caused by microscopic leaf hairs which create an area of still air just outside the leaf. This again serves to limit water loss.

Xylem and Water Transport

As can be seen, even the most advanced land plants, the angiosperms, continually leak water into the atmosphere. The process is unavoidable, since only a total closing of the stomata and a consequent shutdown of photosynthesis would prevent water loss by transpiration. As the parenchyma cells (the thin-walled, photosynthesizing cells) of the leaf dry out, they lose turgor pressure and the familiar phenomenon of wilting results. Prolonged wilting may lead to death. This kind of disaster is avoided by continual replacement of the water lost by transpiration. The only large source of such water is below the ground, and a land plant must send down roots deep enough to reach this source. For a beach plum growing atop a coastal sand dune, the roots may have to travel 200 feet down to reach groundwater. The above-ground portion of a beach plum is seldom more than 10 feet tall.

To move the water from the root tip to the leaf, a vascular plant uses a conducting system called *xylem* ("wood"). Xylem makes up the great bulk of a tree. Any piece of lumber is likely to be almost pure xylem tissue, but xylem is also prominent in herbaceous plants. The chief difference between the two is that herbaceous plants make only one set of xylem vessels in their lifetime, while woody plants add new xylem every year. Xylem tissue is dead. The cells that originally formed it have disintegrated, leaving behind only their elaborate cell walls. These cell walls typically form hollow cylinders. The end plates have either disappeared completely or contain larger pores which allow free movement of water (see Chapter 1). The result is that long tubular elements stretch the length of a vascular plant, from root tip to stem to the innermost chambers of a leaf. These tubular elements carry water and minerals from the soil to the leaf and stem.

What is the force that pumps groundwater hundreds of feet into the air, to the topmost leaf of a tulip tree or a redwood? This force resides in the leaf itself. As moist parenchymal cells lose water by evaporation through the stomatal openings (transpiration), their osmotic pressure rises, and adjacent fresh water enters the cells. This fresh water comes from a neighboring xylem vessel which terminates in close vicinity (Fig. 10-12).

The xylem vessel carries an essentially unbroken column of water which reaches down to the root. The walls of the xylem vessels are strong and thick and do not collapse even under extreme pressure. Therefore any loss of water from the leaf end produces a negative pressure at the root end. The column of water connecting the leaf to the root is under enormous tension and, if one were dealing with an equivalent thickness of steel cable, the cable would snap. The column

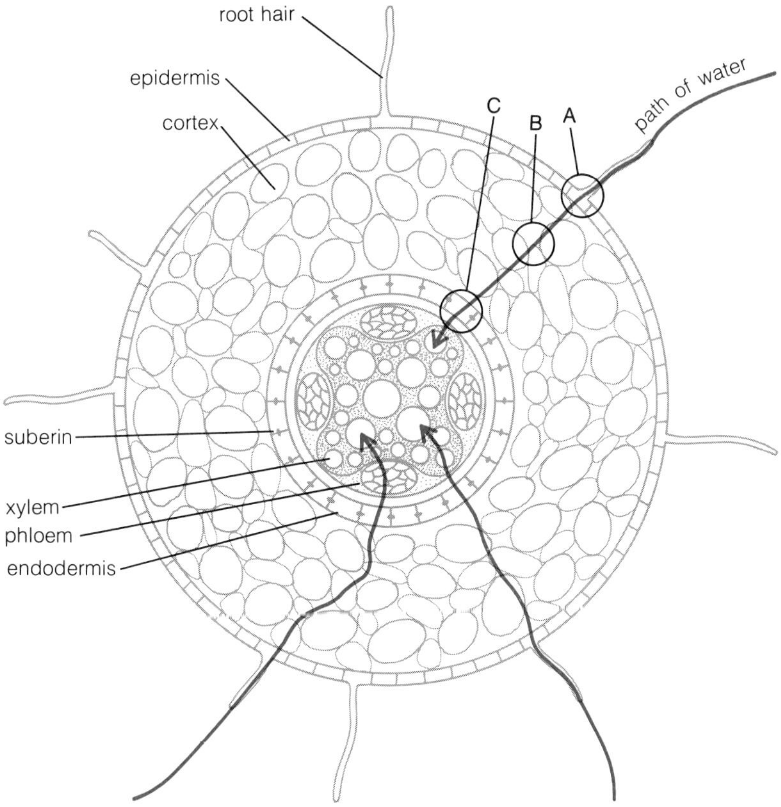

Figure 10-14 CROSS SECTION OF A ROOT. The path of water is shown in color. (A) Epidermal cells with root hair. (B) Cortical cells. (C) Cells of the endodermis with radial cell walls plugged by the waxy substance suberin.

of water does not snap for two reasons: water molecules have an enormous affinity for each other, and they have an enormous affinity for the protein and cellulose constituents of the xylem cell wall. Tension from above pulls water to the top of a tree, not pressure from below. Once a deciduous tree loses its leaves in the fall, the internal train of water is arrested in place. It must be there next spring, for it could not be pushed or sucked into place if it were not.

The Root and Water Transport

The next question is, how is water replaced at the bottom of the xylem train as it is lost at the top? When a root is viewed in cross section (Fig. 10-14), it is seen that the xylem is not in direct contact with groundwater. It is separated from groundwater by three intervening layers of cells: the epidermis with its root hairs, the cortex, and the endodermis.

The root absorbs water largely through its *root hairs*. These are cellular projections of the epidermis and are found only near the growing tip of the root. New root hairs are continually generated by a living plant in its shifting search for water. Water enters a root hair because of an osmotic pressure difference between the inside and outside of the cell. If the osmotic pressure of the groundwater rises, water no longer enters the root hairs. In the fall of 1972, an employee of the parks department, either through malice or ignorance, dumped a half truckful of rock salt in Cunningham Park in New York City. Although the supervisors were immediately notified and the salt removed within one day, six large trees in the immediate area died. The small amount of salt that had remained on the surface had worked its way into the groundwater and raised its osmotic pressure sufficiently so that the trees were literally starved of water, wilted, and died. The use of rock salt to melt snow in the winter takes a similar toll of trees lining roads and highways. But since deciduous trees have lost their leaves in the winter and have only minimal water requirements, the effect is much smaller than if the salt had been applied during the growing season.

From the root hair cells, water makes its way across the *cortex* by a further osmotic pressure gradient. The root hair cells, swollen with fresh water from the ground, now have a lower osmotic pressure than the adjacent cortex cells. As the latter take up water, they in turn lose it to their more centrally located neighbors, and so on. Much of the water also travels between the cells of the cortex, by a similar osmotic principle.

But just before this water reaches the xylem, it meets a barrier—a single layer of cells called the *endodermis*. (Having nothing to do with the endoderm of embryological tissues in animals.) These cells are tightly joined, unlike the cells of the cortex, and their radial cell walls are impregnated with a waxy, waterproof substance *suberin*. Therefore all soil water and minerals must cross through the living cytoplasm of these cells. The endodermis thus acts as a gatekeeper, controlling access of water and minerals to the xylem.

Aerobic metabolism and ATP production are needed for the sustained flow of water through the endodermis. One thus sees the paradoxical situation of trees dying after their roots have been drowned by water from a beaver dam. The trees die because the flow of water stops, even though their roots are drowning in water. The submerged roots lack sufficient oxygen for endodermal cells to continue their function. But the endodermal cells do not pump water directly. They need ATP for the pumping of ions into the area adjacent to the xylem. The pumped ions raise the osmotic pressure in the xylem and allow water to enter across the endodermis. The ions are removed again by a pumping mechanism when water reaches the

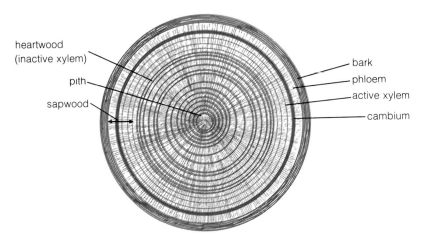

Figure 10-15
CROSS SECTION OF A WOODY STEM. Only the outermost layers of a tree trunk are active in conduction. The inner layers of xylem become plugged with resins as the tree ages and are called heartwood. The heartwood may be lost to rot, insect destruction, or fire, with no impairment of conductive function. But such hollow trees have lost much mechanical support and are vulnerable to ice and windstorms.

leaves. In a sense, the root from the endodermis inward may be thought of as a semipermeable sausage casing enclosing a fluid of high osmotic pressure. The semipermeable casing is made up of the cell membranes of the endodermal cells, and the same cells are responsible for creation of the osmotic pressure difference. When a plant stem is cut with the roots still in the ground, fluid oozes out of the cut stem. In extreme cases this root pressure may push the fluid 50 to 60 feet into the air. However, even this high pressure is not sufficient to explain the rise of water to the top of a 300-foot tree. Only the tension developed by transpiration from the leaves can do this. In summary, two mechanisms are seen to drive the water in plants. One is a pull from above, generated in the leaves. The other is a push from below, generated by osmotic phenomena in the root hairs and cortex and by endodermal pumping.

The Phloem and Carbohydrate Transport

The roots, stems and leaves of a land plant are mutually interdependent. No segment can survive by itself, yet thrives because of the contributions of the others. We have seen the role of the roots and the stem in supplying water for the leaves. The problem faced by the roots is their lack of photosynthetic tissue. While they typically extract necessary oxygen directly from the soil, their supply of carbohydrate must come ultimately from the leaves. This carbohydrate is transported by a separate conduction system, the *phloem* ("the bark"). In the stem of a mature tree, the bark and the phloem form the outer shells of the tree. A layer of embryonic cells, the *cambium* ("to change"), separates the phloem from the xylem (Fig. 10-15). The cambium generates both the xylem and the phloem and increases the girth of the tree in annual spurts. In the spring, when the cambium is very active, the bark and phloem are easily stripped.

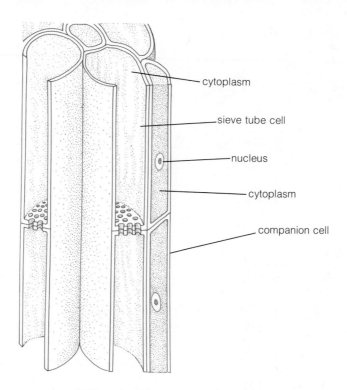

Figure 10-16
SIEVE TUBE CELLS FROM THE PHLOEM AND THEIR COMPANION CELLS. The sieve tube cell contains no nucleus but does contain ytoplasm. The companion cell contains both.

If the stripped area extends completely around the tree, the tree dies because its phloem conduction system has been interrupted. The North American Indians, who lacked metal axes, used this as the standard method of clearing land for cultivation.

A phloem cell is alive, unlike a typical xylem cell in which only the cell wall remains. The functional phloem cell in flowering plants is the sieve tube cell (see Fig. 10-16). The cell is cylindrical, with a simple cell wall perforated at the top and bottom like a sieve. Sieve tube cells are stacked on top of each other to form long tubes. Inside is a cell membrane enclosing a cytoplasm devoid of a nucleus or cell vacuole. The cytoplasm also contains very few mitochondria. Tightly joined to the sieve plate cell is a small *companion cell,* which contains a nucleus, endoplasmic reticulum, many mitochondria, and other organelles. It is thought that the companion cell supplies the sieve tube cell both with energy and any needed nuclear information.

There does not exist a totally satisfactory theory to explain conduction in phloem cells. It has been hard to observe conditions inside a phloem cell without interfering with its function. One organism that manages to enter a phloem cell without interfering greatly with its function is the aphid. Aphids are plant-feeding insects whose mouthparts are structured as delicate syringes or stylets. The stylet penetrates the outer layers of a young stem and is inserted directly into the cytoplasm of the sieve tube cell. If the stylet is cut, it will continue to ooze out cell sap for hours or days. The most remarkable aspect of the cell sap taken from a sieve tube cell is its very high concentration of sucrose, or table sugar. The sucrose concentration ranges from 10 to 25 percent, far higher than in any other part of

THE TRACHEOPHYTES / 365

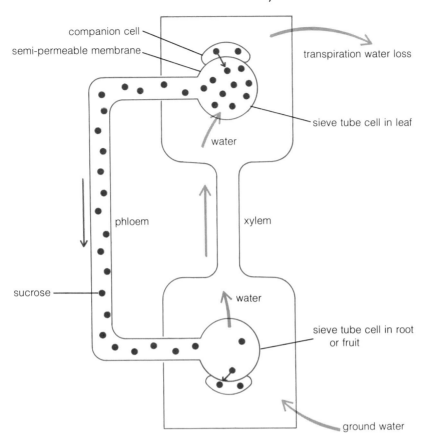

Figure 10-17
THEORY OF BULK FLOW IN PHLOEM. The diagram represents phloem conduction in the summer, when carbohydrates are made in the leaves and stored in the roots and fruits. In the spring, the carbohydrate flow is reversed as sap rises from roots to growing buds.

the plant. Such a high sucrose concentration would result in high osmotic pressure in the phloem cells.

The bulk flow theory of phloem conduction is based on this osmotic principle. The theory states the following:

1. Glucose made by photosynthetic cells in the leaf is transported to the companion cells of phloem.

2. The companion cells convert glucose into the double sugar sucrose and actively transport it into the sieve tube cells.

3. As a result of the high osmotic pressure, water enters the sieve tube cells. This water comes from adjacent xylem or parenchymal cells.

4. Because of its rigid cell walls, the sieve tube cell cannot swell. Instead, there is bulk flow of cytoplasm into adjacent sieve tube cells. (This is the part of the theory that causes the most controversy.)

5. When the phloem finally reaches a storage organ, the sugar is removed, by a reversal of the active transport described before. The storage organ is typically a root, but may be a growing fruit, stem, or other part of the plant.

Figure 10-18 CALIFORNIA REDWOODS.

A schematic diagram of the bulk flow theory is shown in Figure 10-17. This theory explains the reversal of sap flow in the spring, when the tree sap rises from the roots up to the developing buds high in the tree. At this time, osmotically inactive starch stored in the roots is hydrolyzed and pumped into the root phloem cells as sugar.

Anchoring and Support against Gravity

A giant redwood tree is by far the largest organism that has ever lived. It is larger than a whale, a dinosaur, an elephant, or a prehistoric monster. No animal of this size could possibly survive—it would be crushed to death by its own weight. The redwood does more than survive—it can withstand thousands of years of windstorms, blizzards, and changing weather, growing ever taller and more majestic. It falls only to the chainsaws that cut it into picnic tables and patio furniture. Anchoring is provided by the same roots that snake out into the earth in search of water. The roots of even much smaller trees have such a tenacious hold on the earth that tree stumps often

The Hormonal System of Plants

The study of plant hormones began with Charles Darwin. This amazing biologist, whose scientific interests ranged from geology to the facial expressions of animals, showed in 1880 that the phototropic (turning toward light) response of a growing grass seedling is controlled by the growing tip of the seedling. It was left to Frits Went in 1926 to prove that the growing tip produced a hormone he called *auxin* ("to grow"). It took several more decades to prove that naturally occurring auxin was indoleacetic acid, a small nitrogen-containing molecule related to an amino acid. Many synthetic compounds also have strong auxinlike activity, of which more later. A growing plant turns toward the light because the illuminated side contains less auxin than the dark side, causing faster cell elongation on the dark side. An inner translocation mechanism moves the auxin from the light side of the stem to the dark side (Fig. 10-19).

Auxin also controls the geotropic response of a plant (turning with respect to the earth's gravitational field). If a seed falls on the ground sideways, the mature plant does not also grow up sideways. Instead, the stem curves upward and the roots downward. This is

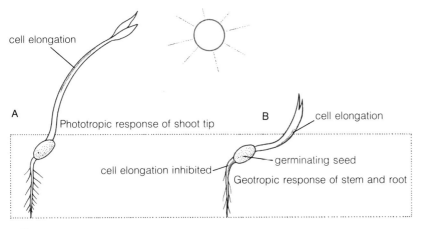

Figure 10-19 THREE HORMONAL EFFECTS OF AUXIN IN YOUNG SEEDLINGS. Areas with elevated auxin concentration are shown in color. (A) Phototropic response of the shoot tip. (B) Positive geotropism of root and negative geotropism of the stem. High auxin concentrations inhibit growth of root cells but stimulate growth of stem cells.

because more auxin accumulates at the bottom of a horizontal stem or root than at the top (Fig. 10-19). The stem cells are stimulated to elongate, producing upward flexion. The elongation of root cells is inhibited by high auxin concentrations, producing downward flexion of the root. The auxin molecule itself is much too small to sink to the bottom of a cell or plant tissue in response to gravity (just like dissolved sugar does not sink to the bottom of a coffee cup). However, large inclusion bodies such as starch grains do sink to the bottom of a cell. It is thought that these inclusion bodies direct the gravitational movement of auxin.

Auxin stimulates the development of fruit. A growing pollen tube releases auxin which stimulates cell enlargement in the ovary or receptacle. In fact, fertilization is not necessary for the development of the ovary wall. The result has been the artificially produced seedless orange and seedless grape. Pollination in these plants is prevented, and the immature ovaries are sprayed with synthetic auxins. However, all attempts to produce a seedless cherry, peach, or plum by similar techniques have failed.

Auxin inhibits lateral bud development. Growing tips secrete auxin which moves down the stem and inhibits the growth of lateral branches. If the tip is clipped, the inhibition is removed and lateral buds suddenly sprout into branches. These lateral branches then form new growing tips and inhibit lower buds until clipped in turn. Hedges of privet, forsythia, or yew are maintained in their thick and bushy state by annual clipping of the growing tips. The budworm performs this operation where it is not wanted—on the growing tips of lumber trees such as spruces and white pine. The result is a crooked tree whose value as lumber is much reduced.

Auxin also prevents leaf and fruit drop. Leaves drop in the fall because of the growth of a layer of cells at the base of the petiole (leaf stalk) called the *abscission layer* ("cutoff layer"). The abscission layer functions in the winter to seal the branch against water loss. In the growing months of spring and summer, the development of the abscission layer is inhibited by auxin produced in the growing tip of the branch. In the fall, this growth activity declines. Diminished auxin production results, and auxin inhibition of the abscission layer is lost. The mucilage that holds adjacent cells together becomes weak and gelatinous, and any slight wind movement breaks the leaf off. This principle is important to commercial fruit growers, who spray their crops with auxins to prevent fruit from dropping before it is picked. Only picked fruit is commercially valuable. The opposite practice is followed by cotton farmers. A cotton field can be sprayed with an auxin inhibitor about a week before harvesting. The leaves drop off, allowing mechanical cultivators to remove the cotton bolls more easily.

Figure 10-20 A TREE THAT GROWS IN QUEENS. The stem of this honey locust was buried for a year during a major construction project. When the soil was removed at the end of the year, a large adventitious root had developed. This root saved the life of the tree, since the original roots could not function due to a lack of oxygen deep under the soil.

Auxin is also responsible for lateral root formation. Any portion of a plant that is in contact with the ground can develop lateral roots. While typically only the roots themselves are in contact with the ground, a plant stem can also respond with root production under proper conditions (Fig. 10-20). For riverside plants such as willows, this is an important survival mechanism. These plants are so often uprooted by floods that the ability of the uprooted stem to put out new roots makes the difference between survival and disappearance. Again, the principle has been exploited by horticulturists. By use of synthetic auxins, almost any cutting from a tree or herb can be induced to form roots and establish itself as an independent plant. Species such as holly, widely used in ornamental plantings, grow very slowly from seeds. They are propagated by tree farms through the use of auxin-treated cuttings.

It is obvious that the auxins have come to play an important role in agriculture and horticulture. These are industries of life. But auxins have also come to play an important role in the industry of death. During the Indochina war, the United States Air Force sprayed millions of acres of forest, cropland, and mangrove swamp with synthetic

Figure 10-21 PLANES ON A DEFOLIATION MISSION IN INDOCHINA.

auxins, arsenic compounds, and other plant killers. The synthetic auxins used were 2,4-D and 2,4,5-T (Fig. 10-21). These compounds have been widely used in this country as weed killers. In Indochina (mostly South Vietnam), they were used in concentrations ranging from 13 to 130 times greater than those recommended for weed control. The compounds caused feverish growth of the sprayed plants, followed by exhaustion and death. According to estimates of the United States National Academy of Sciences Committee on the Effects of Herbicides in Vietnam, 1.25 million cubic meters of merchantable timber were lost as a result. The most serious damage occurred to coastal mangrove forests. These trees were extremely sensitive to the herbicides, and a single spraying was sufficient to kill. Photographs of the area show the death of all vegetation as far as the eye can see. According to the committee, recovery will take over a century.

Other classes of plant hormones have not yet been exploited with the same degreee of ingenuity as the auxins, in part because less is known about their physiological effects and their mode of action.

Cytokinins ("cell dividers") play the major role in promoting cell division. While auxins can stimulate cell division under some conditions, such as lateral root formation, they must work in conjunction with other plant hormones to accomplish this.

Gibberellins (isolated initially from the fungus *Gibberella fujikuroi*) stimulate stem elongation. Cabbage treated with gibberellin shoots up into a tall-stemmed plant many feet high. The same effect is observed with dwarf strains of corn, rice, and fruit trees. Unlike the auxins, the gibberellins are not involved in tropic responses, do not inhibit lateral bud growth, and do not prevent leaf abscission.

Growth inhibitors are a varied assemblage of compounds that counteract the effects of auxins, gibberellins, and cytokinins. These are the hormones that maintain the seed in a dormant state until growth conditions are suitable and maintain dormancy in buds throughout the winter. We have already mentioned the role of sodium chloride in maintaining dormancy in the seeds of the desert salt bush. An important agent responsible for seed dormancy is abscissic acid. The compound has been misnamed; initially it was believed to function in leaf abscission. But as with the terms prostaglandin, pituitary, and vitamin ("vital amine"), the initial term has stuck, even though later studies showed it to be inappropriate.

Undoubtedly many *ripeners* exist in the plant world, but so far the only one isolated has been a simple hydrocarbon, ethylene. When you buy apples or oranges prepackaged in plastic bags in the supermarket, the bags always come with large perforations in the sides. They are there for good reason: fruit packaged in an airtight bag would ripen so quickly it might rot before it could be sold. A ripening fruit releases ethylene. Ethylene is a gas and stimulates other fruit to ripen. As they ripen, more ethylene is released, and thus a self-accelerating cycle is established.

Flowering hormones or florigens are relative latecomers to the known list of plant hormones. They are responsible for the development of flowers and angiosperms. Florigens of May flowers such as trillium or narcissus are produced in the spring. Florigens of summer-flowering flowers such as black-eyed susan or bedstraw are produced in the summer. And the fall-flowering flowers, such as goldenrod or New England aster, produce their florigen in the fall.

The study of plant hormones is a very active field of research today. There is no question that the near future will see a "flowering" of knowledge in this field, as new hormones and entirely new classes of hormones are discovered. As many plant growers have observed, the more one learns of plants, the more they seem like humans.

In this brief survey of the plant world we began with a look at five algal phyla. These simple photosynthetic organisms have strong affinities with the protistans and are all adapted to an aquatic existence. Their sexual reproductive cycles involve an alternation of generations. One of these phyla, the green algae, was ancestral to the bryophytes, the first plant phylum to invade the land. Bryophyte adaptations to a land existence remain imperfect; fertilization requires a water medium because of flagellated sperm, they lack a waterproofing cuticle, and they lack conductive tissue to bring water, food, and hormones to all plant tissues. As a consequence, bryophytes remain physically small and restricted to moist habitats.

All these defects were remedied by the tracheophytes, especially the most advanced class of the phylum, the angiosperms. This is the group that overwhelmingly dominates the earth today, and to which we devote the rest of our discussion. A waterproofing cuticle covers the photosynthetic surfaces, and both water loss and gas exchange are controlled by stomata. The reproductive organ is the flower. Alternation of generations still occurs, with the gametophyte generation greatly reduced at the expense of the sporophyte generation. The flower has totally freed angiosperms of a water dependence and in addition leads to the production of seeds and fruit. The seeds and the fruit both serve to disperse the species and to synchronize growth with conditions that are environmentally optimal.

The conductive tissues of angiosperms serve several interrelated functions: transport of water, transport of foods, transport of hormones which coordinate development, and anchoring and support. During the period of active growth, xylem vessels move water upward, while the phloem moves food and hormones downward. The two systems interact to form a circulatory system whose driving force is based on osmotic pressure differentials. Unlike vertebrate circulatory systems, however, the plants can reverse the direction of nutrient flow as required, can use their conducting vessels as internal skeletal systems, and can generate new conducting vessels in annual increments to stay perpetually young.

Bibliography

GENERAL READING

Donald Culross Peattie, *A Natural History of Trees,* 2nd ed., Houghton Mifflin, 1950. Written by a man who loves them.

R. T. Peterson and M. McKenney, *A Field Guide to Wildflowers of Northeastern and North Central America,* Houghton Mifflin Co., 1968. The best of the plant field guides, this book will tell you at once and without ambiguity the difference between Solomon's seal, false Solomon's seal, and the starry false Solomon's seal. A separate volume is available for Rocky Mountain wildflowers.

Edgar Anderson, *Plants, Man, and Life,* University of California Press, 1969. A delightful book, as fresh and colorful as the wildflowers that Anderson describes. The author discusses perhaps the most far-reaching scientific advance of the human race—the domestication of wild plants. A sample excerpt:

It is quite likely that a good many of our crops were not originally used for the purposes which we would now suppose the only reason for growing them. Considered merely as vegetables, the history of squashes and pumpkins is a curious puzzle. All the cultivated species are large and have pleasantly flavored flesh. The wild species are uniformly small and repulsively bitter. How could any such plant have been tolerated even as a crude food, until it developed modifications permitting its use at least as a famine food? Well, in the first place, one has only to live in a Mexican village for a few months to realize that the pumpkin flesh is not the only portion of the plant which makes delicious food. The seeds of many varieties are excellent and on the whole the seeds are almost as important in Mexico as the flesh. So many Americans have appreciated them in the bars of Mexico City that they are now beginning to appear in our specialty shops with other salted nuts. Primitive man, however, had still different needs for pumpkins and squashes. We already know enough about their long and complicated career of domestication to be fairly certain that they started as rattles in ceremonies and dances, and as primitive dishes for eating and for storage.

C. L. Duddington, "Evolution and Design in the Plant Kingdom," Thomas Y. Crowell Co., 1974. The writing style is clear and free of jargon, and all aspects of the plant experience are discussed. Duddington's examples are taken from all over the world, and there is an encyclopedic fullness about this book that one does not expect from its little size.

C. L. Wilson, W. E. Loomis, and T. A. Steeves, *Botany,* 5th ed., Holt, Rinehart, and Winston, 1971. A widely used, well-written text.

F. C. Salisbury and C. Ross, *Plant Physiology,* Wadsworth Publishing Co., Plant physiology at a more advanced level.

M. C. Ledbetter and K. R. Porter, *Introduction to the Fine Structure of Plant Cells,* Springer-Verlag, 1970. Electron micrographs by two masters of the art. The pictures are accompanied by detailed explanations.

TEXTBOOKS

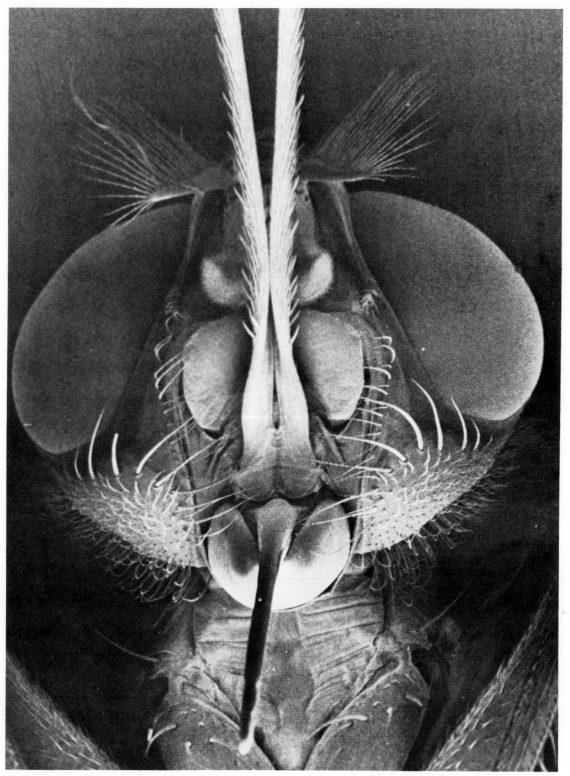

Figure 11-1 HEAD OF A TSETSE FLY.

chapter 11

The Animals

To a biology watcher with a social conscience, the animals are an unnecessary part of the world. They are unproductive—in a sense all animals are parasitic on plants. But they are very appealing parasites; in them, living organization has reached its greatest complexity. It is therefore no surprise that our perception of life tends to be most acute within the animal kingdom. The stroller who cannot tell the difference between a black birch and a black cherry can easily distinguish between a chipmunk and a squirrel, two animals with a much closer taxonomic relationship than a cherry and a birch. But even within the animal kingdom, our everyday lore tends to center on only one group, the vertebrates. How many people can distinguish between a lacewing and a lace bug, both common insects of the suburban backyard? Or between a sand flea and a sow bug, not insects at all but land-dwelling relatives of lobsters and crabs? A treatment as brief as this can hardly be expected to describe all the varieties of animal life. I shall instead set two modest goals. On the one hand, I shall present a very general view of the evolutionary relationships within this group. On the other hand, I shall try to give a little of the flavor of day-to-day life of some of our common neighbors in the animal world. Since the most common wild animals any of us are likely to meet are insects and vertebrates, we shall pay special attention to these groups.

First, what characteristics set animals off from members of the other kingdoms? Fundamentally, all animals use the same method of feeding. Unlike plants, animals cannot manufacture their own food but must obtain it from plants, from other animals, or from decomposing material. Their method of feeding is ingestion; they swallow food whole, in smaller or larger chunks, and digest it internally. In contrast, the fungi break down their food externally by means of hydrolytic enzymes and then absorb the products of decomposition. But it is hard to make categorical statements in biology. There are some animals, such as starfish, that feed on clams and oysters exactly as a fungus might, if the fungus had arms to pry open the shell of the victim. The same kind of exceptions exist to all other generalizations about the animal kingdom.

Another specialty of animals is that they either move through their environment or work to bring their environment to themselves. Even the apparently motionless mussel generates water currents to sweep small bits of food inside its shell. Such movement is essential, because an animal rooted to one spot would quickly starve to death. For example, a cow's high energy needs and low efficiency of energy conversion dictate that it must be supported by a wider area of the earth's surface than an equivalent biomass of plant life. Further, movement and multicellularity imply the need for rapid coordination, and this need is served by a nervous system. No other living kingdom has developed one. And finally, the fact that animals are multicellular (as opposed to their unicellular protistan ancestors) has made possible a tremendous division of labor among cells. The cell that is a specialist always outperforms the cell that is a jack-of-all trades.

The Coelenterate Phylum

The *coelenterates* ("hollow intestine") are the simplest animals that can be placed in the evolutionary mainstream of animal development. There is a simpler form, the sponges, but these represent an evolu-

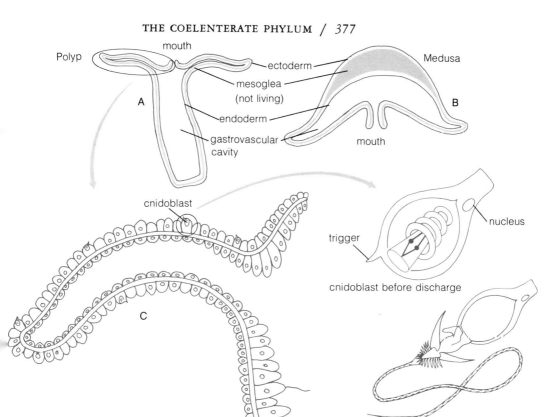

Figure 11-2 THE COELENTERATES. (A) Polyp form. (B) Medusa form. In organisms where both forms alternate during a life cycle, the polyp form is the feeding form, the medusa form the dispersal and sexually reproducing form. (C) Cellular detail of *Hydra,* a typical polyp form. The inset shows a cnidoblast before and after discharge of its nematocyst.

tionary dead end which seems to have generated none of the other animal phyla living today. The coelenterates include jellyfish, hydroids, and corals. The phylum shows two basic body forms (Fig. 11-2). The polyp stage is sessile and primarily a feeding stage, while the medusa stage is motile and primarily concerned with sexual reproduction. Both stages contain only two cell layers, an *ectoderm* and an *endoderm,* separated by a nonliving *mesoglea* ("glue in the middle"). Often the ectoderm and endoderm may be only one or two cell layers thick, to facilitate gas exchange and diffusion of food molecules and wastes. Nevertheless, coelenterates often achieve large sizes, either by containing large amounts of mesoglea, or by living as colonial forms. The most spectacular of these colonial forms are the corals. Each individual is microscopic and lives inside a shell made of calcium carbonate. Trillions of corals may aggregate into a single colony and, when this happens, they can become a geological force extending boundaries of islands and building atolls and reefs in warm tropical waters. The most spectacular of these is the Great Barrier Reef of Australia, which is several miles wide and extends for over 1,000 miles along the eastern edge of Australia. Like all living organisms, such a barrier reef may be poisoned, parasitized, and destroyed by the forces of nature or by human activity. A section of Bikini atoll was blown away by a hydrogen bomb, and the Great Barrier Reef is threatened today by offshore oil drilling and by the crown-of-thorns starfish, a voracious feeder on living corals.

Each individual coelenterate is a carnivore, though most corals and some hydroids contain intracellular algae which supplement the animal's diet by contributing photosynthetic products. Coelenterates capture other small animals by means of their tentacles, which distinguish between living and nonliving prey and reject the latter. On living prey, there is a furious attack. The tentacles lash out, lassoing the victim and riddling it with venomous darts called *nematocysts* ("thread pouches"). A nematocyst is enclosed within a unique cell called a *cnidoblast* ("stinging cell") (Fig. 11-2). A nematocyst behaves somewhat like a rubber glove with a finger inverted. When air is blown into the glove, the finger pops out. Of course, it is not air but water that rushes into the nematocyst. Some nematocysts function only to bind the victim, while others penetrate its tissues and inject a paralyzing venom. The attack of the large colonial jellyfish, the Portuguese man-of-war, can be fatal to a swimmer.

Food captured by a coelenterate's tentacles is passed through the mouth into the *gastrovascular cavity* ("cavity serving as stomach and circulatory system"). Cells of the endodermis release enzymes which digest the prey. The digestion products are absorbed by endodermal cells and some are transmitted to cells of the ectoderm. Undigested material is expelled through the mouth.

Structurally the coelenterates are so simple that one must marvel at their relatively complex behavior and convoluted reproductive cycles. Yet their structure is already too complex to have evolved directly from single-celled protistan precursors. As in embryological development where the gastrula develops from the blastula and morula, so in evolutionary development the coelenterates probably evolved either from a small blackberrylike mass of cells or from a hollow-sphere multicellular ancestor. But no such possible ancestral form survives today, and neither have any traces been discovered in the fossil record.

The Flatworm Phylum

The word "worm" does not convey a great deal of biological information, because so many different kinds of animals look like worms. I remember a walk through the woods with our daughter when she was

THE FLATWORM PHYLUM / 379

three years old. It was spring, and I had been lifting up flat rocks to show her snakes warming themselves underneath. For a moment I turned my attention to something else, and my daughter wandered off. Then I heard an excited shout, "I found a snake!" There was my daughter, crouching down, staring intently at something on the ground. I hurried over and followed her pointing finger just underneath. There, moving slowly, and waving its head from side to side, was an alarmed brown caterpillar! Since my daughter had never seen a caterpillar before, it met all of her criteria for a small snake.

Similarly, the term "worm" covers a tremendous diversity of biological organisms that may have much less in common with each other than a human being and a sparrow. The simplest of all the wormlike phyla is the phylum *Platyhelminthes* ("flat worms"). Flatworms are not exactly objects of common everyday experience, though they may intervene destructively in human life. The liver fluke lives in the human body and may wreck it by burrowing through the liver and other organs. The disease is called schistosomiasis and infects tens of millions of people in tropical and semitropical areas in the Middle and Far East, Africa, and South America. Tapeworms are members of another class of flatworms.

But perhaps the most typical flatworms are the free-living forms. Most are marine, but they are also easy to find in rich ponds and streams, where they feed on decaying materials and on "slow game," including each other. Here they are easily found by dropping in a piece of liver tied to a string. When the string is pulled out an hour later, several flatworms may be clinging to the liver. A free-living flatworm distantly resembles a medusa form of a coelenterate which has replaced its nonliving mesoglea with a true tissue layer called the *mesoderm* (Fig. 11-3). But whereas a medusa swims mainly up and down and

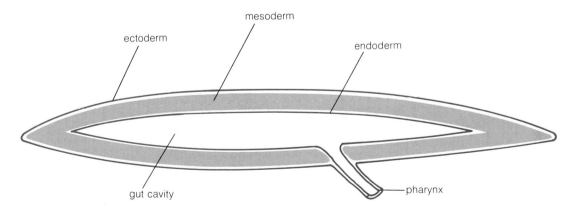

Figure 11-3 A FLATWORM (POSTULATED ANCESTRAL FORM). Note the resemblance to the medusa form of the coelenterate (Fig. 11-2). However, the flatworm contains three tissue layers and is bilaterally symmetrical.

is radially symmetric,[1] a flatworm moves horizontally. Therefore one end has become specialized as a head, with receptors for light and for chemical senses and a local concentration of nervous tissue. The animal is bilaterally symmetrical (as are humans and almost all other animals). Food still enters and leaves by the same opening but, unlike that of a coelenterate, a flatworm's digestive cavity branches through the mesoderm in a complex way so that no body cell need be very far from its food source. Food is digested and distributed in the same manner as in coelenterates, and this is the factor that dictates the flat body shape of the flatworms. Both food and respiratory gases must be transported internally by diffusion, hence the mesoderm cannot develop very extensively in three dimensions without starving itself. The lack of a circulatory system has placed firm limits on the flatworm body plan.

Free-living flatworms have proved attractive to researchers. They are the simplest animals to show learning ability. No whiz kids, they require hundreds of trials to learn to take one arm and not the other of a T maze. They also show legendary powers of regeneration. A flatworm chopped into four or five sections does not die. Each piece regenerates any missing parts, such as head, mouth, or tail, and continues as an independent adult. If one sections a trained flatworm, each regenerated adult retains the learning of the original trainee. It thus appears that learning is not stored in the brain alone, but throughout the animal's nervous system.

The Annelid Phylum

While the *annelids* include leeches and a variety of marine worms, perhaps the best known example is the garden earthworm, the archetypal worm (Fig. 11-4). The earthworm beautifully illustrates the derivation of the name of the phylum, Annelida. The word means "little ring," and the body of an earthworm is encircled by dozens of these rings. Each ring in fact represents a complete partitioning of the body into a semiautonomous segment. Each body segment carries its own set of appendages, its own nerve ganglion, its own excretory organs, and its own separate loop of the circulatory system.

One of the few body systems that fails to reflect this segmentation is the digestive tract, which runs the length of the earthworm. Unlike the digestive tract of flatworms, the annelid digestive system

---

1. A radially symmetrical object (such as a vase or a cylinder) is symmetrical around a central axis of symmetry. A bilaterally symmetrical object (such as a hammer or a book or a human being) is symmetrical about a single central plane of symmetry.

THE ANNELID PHYLUM / 381

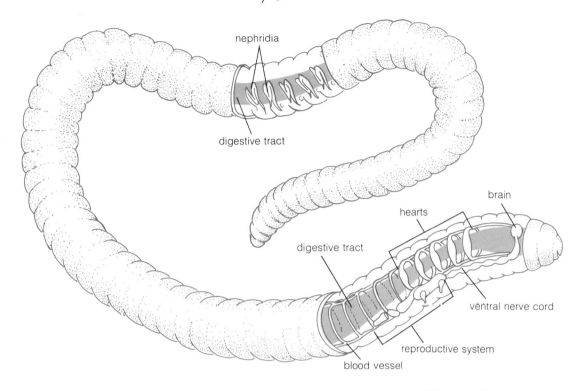

Figure 11-4
AN EARTHWORM IN
CUTAWAY VIEW.

forms a tube which begins with the mouth and ends with the anus. Food therefore moves through in a single direction, allowing digestive organs to carry out a logical sequence of enzymic hydrolyses. In terms of food, the earthworm is the gardener's favorite animal because it ignores green leaves until they die and start to rot. The earthworm, like most higher animals, does not have enzymes to decompose the cellulose of plant materials. Instead it feeds on bacteria and hydrolyzed food compounds found in rotting leaves.

In addition to its tubularity an earthworm's digestive tract shows another important advance over the flatworm plan. This is the fact that an earthworm's digestive tract is not embedded in a solid matrix of mesoderm. Instead, the digestive tract, along with its mesodermal components (muscles, blood vessels) runs through a spacious body cavity called the *coelom* ("hollow"). The muscles and blood vessels of the outer body wall can therefore act completely independently of those of the digestive tract. If a flatworm requires some mechanical mixing of its intestinal contents, it must twist and gyrate its entire body. However, an earthworm can completely separate the mechanical aspects of digestion and locomotion.

For an earthworm, the processes of digestion and locomotion are intertwined in a different sense. The medium through which an earthworm moves is the earth, and it literally eats its way through. Its mouth and throat have powerful muscles which allow them to act as a suction cup. An earthworm sucks in a bit of the earth and then inches forward into the newly created space by means of peristaltic movements of its body. It anchors itself against the sides of its burrow by means of several hundred tiny eversible bristles. At the same time, the earth that has been sucked in is passed through the digestive tract. The excreted castings contain largely inorganic components of the soil. Earthworms play an important ecological role in making the soil more hospitable to plants. Their burrows improve aeration and water penetration for plant roots, and the castings provide a plowing and overturning action which turns over up to 5 tons per acre per year in a field, and up to 20 tons per acre per year in the woods.

For purposes of feeding, earthworms remain largely in the upper inch or two of soil, where the greatest concentration of rotting plant material is found. But in late fall these submarines of the soil move deeper underground to escape the frost. I remember once digging a trench about 5 feet deep, going through solid clay. As the trench progressed slowly, I thought how shallow the inhabited layer of the earth really is, since at this level there was no obvious sign of life, not even plant roots. Suddenly my pickax exposed an earthworm that had settled down for its winter sleep, in a deep-diving feat comparable to that of a man-made submarine descending into the abyssal depths of the sea.

In contrast most other annelids are water dwellers. This habitat is dictated by their method of respiration: they breathe through their skin. Some annelids increase their respiratory surface by means of feathery projections called *gills* (Fig. 11-5). Others, such as the clam worm, accomplish the same thing by breathing through dozens of flat, papery legs called *parapodia* ("pseudolegs"). Parapodia also function as swimming organs. An earthworm does not deviate greatly from this pattern. Since gills and parapodia would interfere with burrowing, an earthworm breathes directly through its skin. This means that the surface of the skin must remain moist. Only moist cells remain alive, and only living cells exchange oxygen and carbon dioxide efficiently. This is why an earthworm emerges from its burrow only briefly and at night, when the dew protects it against drying out. (Earthworms are colloquially called night crawlers.) Paradoxically, a heavy rain can be as destructive as a drying wind. The rain-clogged earth contains little oxygen, forcing earthworms to the surface for air. Once above ground, they die because their pink skins provide no protection against the ultraviolet radiation of the sun.

Figure 11-5
MARINE ANNELIDS. These organisms withdraw into protective tubes when danger threatens. Here the head areas with flower-like gills have been extended. The gills are used for breathing and as an aid in feeding.

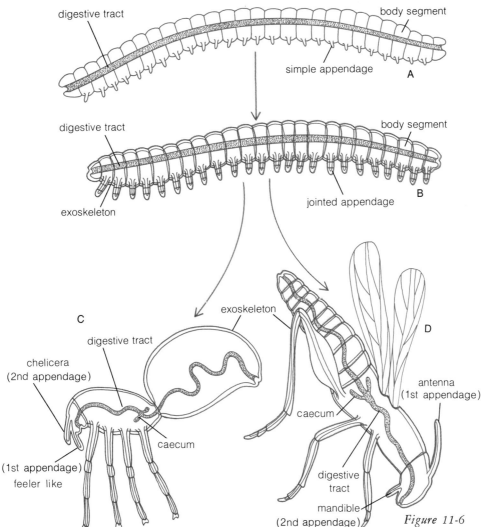

Figure 11-6
THE EVOLUTIONARY TRANSITION FROM ANNELID TO ARTHROPOD.
(A) Annelid ancestor.
(B) Primitive arthropod.
(C) Advanced arthropod of the chelicerate line.
(D) Advanced arthropod of the mandibulate line.
In both advanced forms, note the shortening of the body, specialization of appendages, and internal branching of the digestive tract.

The Arthropod Phylum

This giant phylum represents one of the two pinnacles of evolutionary development in the animal kingdom (the other is our own group, the chordates). If evolutionary success is measured in terms of numbers of species, then the arthropods have no competitors. Just one class of arthropods, the insects, contains about 750,000 species (the total for the chordates as a group is about 45,000 species). And approximately 15 new insect species are described and classified every day, with authoritative projection of the total pegged at about the 3 million mark. It is obvious that only mammalian taxonomists could have named the present evolutionary era the age of mammals. To any impartial observer, this is the age of arthropods.

Arthropods have close evolutionary links to the annelids, though at first glance a butterfly, a highly specialized arthropod, has little in common with an earthworm. The earliest arthropods most probably resembled present-day centipedes. To turn an annelid into an arthropod required only a few basic structural changes, yet these changes had the most profound consequences (Fig. 11-6). An arthropod is, first of all, an armored animal. Its external covering provides both protection and support. This covering or *exoskeleton* ("external skeleton") occurs in many animal phyla, such as the mollusks (clams, oysters, and so on), brachiopods, and bryozoans. But the exoskeleton of arthropods is *articulated;* it contains hinges allowing the animal within to move. The exoskeleton covers the appendages as well and, unlike the simple bristles of an earthworm, arthropod appendages are jointed (the word "arthropod" means "jointed legs"). This makes the appendages more efficient in locomotion and useful for other purposes as well, such as the manipulation of food. Through time, arthropod appendages also evolved into offensive weapons, organs of copulation, sensory receptors, pollen carriers, hypodermic needles, sound generators, and almost every bizarre tool one can think of. Just as humans rose to prominence because of the dexterity of their hands, the arthropods rose because of the dexterity of their feet (Fig. 11-7).

Important as the exoskeleton was, the success of arthropods also had much to do with their annelid inheritance. Whereas annelids all have the long, thin wormlike body plan, many arthropods are compact, agile animals. The annelid body plan is dictated at least in part by consideration of the digestive system. One reason why the annelid body is long is because its digestive tract must be long to do an effective job of digestion and absorption (Fig. 11-6). However, a thin, wormlike animal is likely to be slow and clumsy when moving around. Advanced arthropods, such as spiders and insects, have resolved the dilemma by adding internal blind ends to their digestive tracts, an option made possible because of the surrounding coelom.

Another undeveloped option in the annelid body plan is the segmented body plan. Each annelid segment has body parts much like those of neighboring segments: paired appendages, nerve ganglia, excretory organs, and so on. Arthropods have followed the tactic used by every successful higher organism: specialization and division of labor. Body segments of advanced arthropods are not carbon copies of the neighboring segment (Fig. 11-8). Segments in the head region are specialized for sensory reception and food manipulation; the nerve ganglia are fused, the appendages form mouth parts. Thus these arthropods chew with their feet. Appendages of the thoracic segments function as legs. A total of six or eight legs allows much more rapid

Figure 11-7 SCANNING ELECTRON MICROSCOPE VIEW OF AN ARTHROPOD EXOSKELETON. (A) Overall view of the tip of the leg of a tse-tse fly. The bristles have a sensory function. The foot contains a set of claws as well as a fringed area. (B) Enlargement of the fringed area shows that each tentacle terminates in a tiny suction cup. This allows the fly to walk on a pane of glass, or walk upside down on a smooth ceiling.

Figure 11-8 SCANNING ELECTRON MICROSCOPE VIEW OF THE HEAD OF AN ANT (CARDIOCONDYLA WROUGHTONI). The mandibles are closed in front. A blackberry-like compound eye is visible at left. The terminal knob of an antenna is projecting forward at right ($\times$ 500).

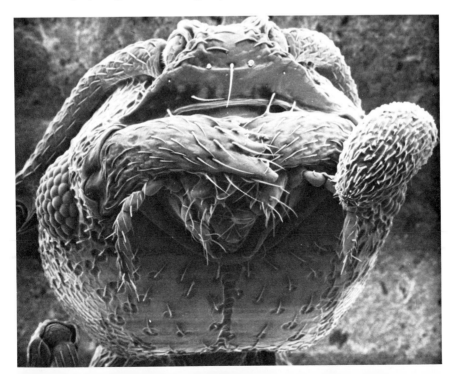

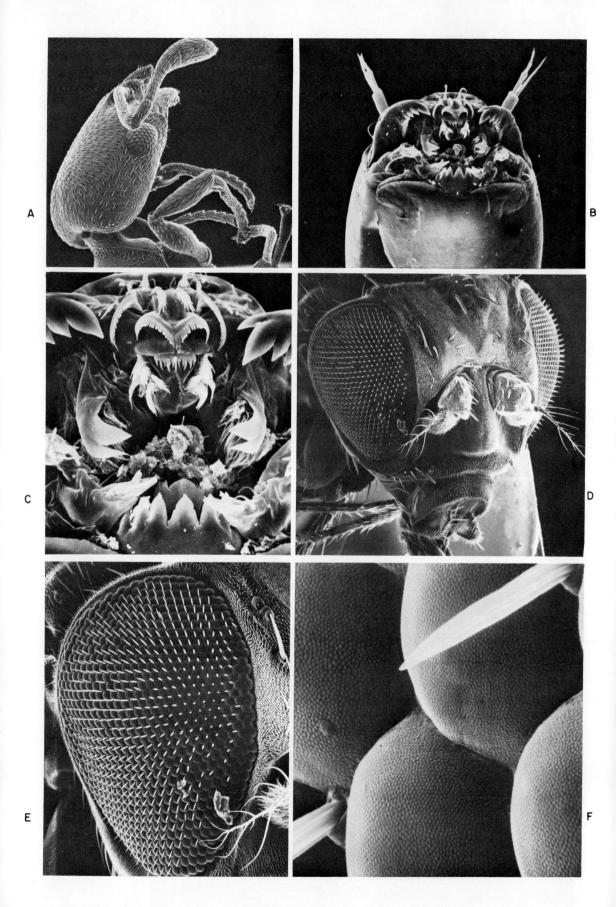

Figure 11-9 PORTRAITS OF ARTHROPODS. (A) Head and front legs of an ant (*Cardiocondyla*). (B) Head of a chironomid insect larva (*Glyptotendipes*), an inhabitant of polluted waters, where it feeds by filtering suspended matter. (C) Close-up of mouth area of *Glyptotendipes*. (D) Head of fruitfly. (E) Close up of compound eye of the fruitfly. (F) Close up of individual subunits of the compound eye (ommatidia). The bristles have a sensory function.

movement than the several dozen carried by a millipede. The posterior segments carry no appendages at all in insects or in chelicerates but are specialized for digestive and reproductive functions.

With these basic strategies, the arthropods have achieved dominance of every living medium on earth: water, land, and air. They have invaded the land more than once, and each new invasion has produced highly successful adaptations. The insects and the arachnids represent two such widely different classes (Fig. 11-6). The arrival of the insects on land, about 350 million year ago, coincided with the development of the first extensive land vegetation. In this strange new world, insects had an unlimited food supply and essentially no competitors. Even today, the most serious enemies of insects are other arthropods.

The same kind of endless vista opened up with the conquest of air by the insects. Three-hundred-million-year-old fossils show giant dragonflies, with a wingspan of three feet. These insects were in the air for 150 million years before the first birds appeared, and for 250 million years before the first bats.[2] It is perhaps this long head start on land and in the air that has made insects the most diverse of terrestrial life forms. Insects are not necessarily the most abundant animals in terms of numbers of individuals—that record is probably held by another group of arthropods, the minute copepods (Crustacea) that feed on marine plankton. But even so, it is estimated that the world's insect population numbers about 10^{18}, or a billion billion. This works out to about 300 million insects for every human being. This inventive and numerous life form deserves a better response than that of our popular culture, which gags at a worm in an apple, or a *Parcoblatta pennsylvanica* under the kitchen sink. The best remedy for this kind of ignorance is a 10 or 15× pocket magnifier that will turn every walk in a weedy lot into a safari (Fig. 11-9).

---

2. Though we generally do not think of spiders as flying animals, young spiders can travel for hundreds of miles on currents of air by extruding a length of spider silk. Even the gentlest updraft picks up the silk and carries the baby spider for great distances. This allows for rapid dispersal of these animals.

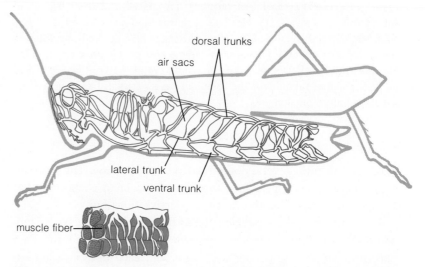

Figure 11-10 THE TRACHEAL SYSTEM OF A GRASSHOPPER. The inset shows the junction between a terminal tracheole and a body muscle.

The Life of the Insects

What is it like to live as an insect? For one thing, the body parts may be in strange places. A fly tastes with its legs, a moth smells with its antennae, and a grasshopper hears with the side of its chest. Insects do not breathe through the mouth, nor does their blood carry oxygen. Insects respire by means of *tracheae* ("windpipes")—ducts that open to the sides of the body and carry air in a branching system to every muscle of the body (Fig. 11-10). The tracheal system replaces the gills of more primitive arthropods and has made insects at home on land.

But there is one factor that shapes and directs the insect life-style more than any other, and this is its *exoskeleton*. Like our own internal skeleton, the insect exoskeleton provides attachments for body muscles and therefore permits rapid movement. But unlike our skeleton, the insect exoskeleton is not living tissue. It contains a celluloselike material called *chitin* ("coat of mail"), which may be impregnated with hardening substances such as calcium carbonate. It therefore cannot repair itself or increase in size like mammalian bones. In order to grow, an insect must literally climb out of its exoskeleton, inflate itself by swallowing air or water, and then lay on a new and larger exoskeleton. This process is called *molting,* and it is certain that, if insects could converse, their conversations would often come back to these crucial milestones in their lives. Molting may be accompanied by more than an increase in size—the entire form and behavior pat-

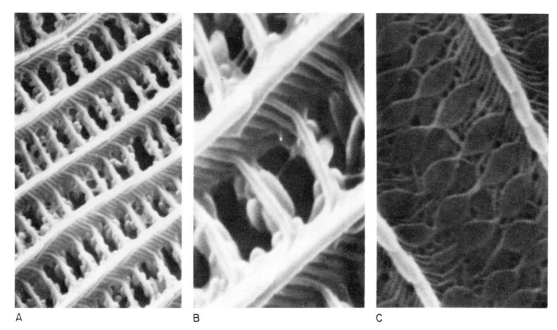

A B C

tern may be changed, in a process called *metamorphosis* ("transformation"). Who would suspect that the stolid, wormlike caterpillar would molt into the delicate, fluttering butterfly? Likewise, the juvenile form of the dragonfly spends its life underwater in a pond, stalking mosquito larvae and pupae, before predator and prey both metamorphose into aerial forms and leave the scene. Metamorphosis allows the juvenile and adult forms of the insect to coexist without competition. Juvenile forms are typically concerned with feeding and increase in body size. Adult forms instead have become specialized for sexual reproduction and may not feed at all. Thus a mayfly larva may spend 2 years feeding in a pond or stream and then emerge as a winged adult which mates, lays eggs, and dies in 24 hours. The winged form of an insect, no matter how short-lived, is the final adult form.

Figure 11-11
SURFACE DETAIL OF THE SCALES OF THE WING OF THE ORANGE SULFUR BUTTERFLY. Such scales give the butterflies and moths their bright colors. The scales rub off easily, allowing the animals to escape from spider webs. (A) Surface of an orange-colored scale. (B) Higher magnification of (A). (C) Surface of a black-colored scale.

The wings of a flying insect are an outgrowth of its exoskeleton, and they are never molted. The most primitive flying insects, such as dragonflies, have two pairs of wings which cannot be folded. Dragonflies are very skilled fliers, but their large, fixed wings make them more visible, hence vulnerable, on the ground. This probably explains the extinction of the giant dragonflies of the past. More modern flying insects, such as beetles, use only their hind wings for flying. Their front wings have been modified into a stiff pair of covering wings. On landing, the hind wings are immediately folded and tucked under the front wings, and the beetle can run into a crevice and hide, or perhaps blend in with the background.

The insect exoskeleton has another important function: it acts as a water barrier. The body of a common housefly contains no more than a tiny drop of water, and without protection such a drop of water would evaporate in a minute. Yet a fly can remain active for many hours without drinking. It is protected by a waxy cuticle over the exoskeleton analogous to the cuticle on the leaves of higher plants. If a fly is shaken in a bottle filled with charcoal, the charcoal removes the cuticle without damaging the underlying exoskeleton. Such a fly dies in a few minutes in normal dry room air, but can survive in a chamber maintained at 100 percent humidity.

The exoskeleton has served the insects well, and they have taken this arthropod inheritance and modified it to their own ends, as with the techniques of camouflage and the invention of wings. But the exoskeleton has imposed one serious limitation on insects: it has kept them small in size. The giant insects of science fiction would pose a threat to no one, simply because their massive exoskeleton would be too heavy to carry and too difficult to replace in molting. A factor of equal importance in limiting body size is the tracheal system of breathing. An increased volume would place impossible demands on the duct system.

The small body size of insects necessarily requires a small brain size and a shortened life span. Two opposing forces are at work in the insect nervous system. The processes of advanced segmentation result in the fusion of ganglia into larger and more complex assemblies. This increase in nervous complexity is opposed by the small body size, which limits complexity. The result has often been described as stereotyped behavior. While the behavioral pattern may be quite intricate, it is sometimes described as being totally predictable because it has been "wired in" inside the insects' nervous system. But many students of insect behavior have demonstrated that this view is too simple. Perhaps most convincing are studies by Niko Tinbergen of Oxford University and his co-workers such as Aubrey Manning. In 1952 Manning began a study of bumblebee behavior. He showed clearly that a bumblebee is capable of learning and depends on this ability in its everyday experience. Manning set up an observation post near some houndstongue plants, large plants with abundant nectar. The flowers studied began to produce nectar on May 20th, yet the first bumblebees did not visit the cluster until May 26th. Prior to this time, they had been carrying pollen of a variety of other flowers. By their second or third visit the bumblebees (which Manning had marked with colored dots so that individual bees could be recognized), showed signs of recognizing houndstongue and specializing on this species. They made occasional visits to plants resembling houndstongue, such as ragwort and thistle, even though these were not in flower at the time. And on ragwort or thistle they searched for flowers in exactly the same spots where houndstongue would be expected to carry them.

Soon the bees specializing on houndstongue developed a fixed routine. Each bee worked out a specific order in which it visited the plants. After all the plants had been visited, the bee flew to its nest to unload the nectar, and then returned ½ hour later. Now Manning pulled out a single plant from the series. When one of the regulars returned, it flew to the spot where the plant had stood and circled for 40 seconds, searching for its accustomed landing spot; then it gave up and flew on to the next plant. On its second circuit, the bee spent only 7 seconds searching. The third time, it took 1 second, and on the fourth visit it flew to the empty site but went on at once. On its fifth visit it skipped the site altogether. By any standard except a human one, this must be considered a very fast learning response.

Even more impressive are digger wasps of the genus *Ammophila*. Like many other wasps, *Ammophila* preys on other insects. *Ammophila adriaansei* preys on the caterpillars of butterflies and moths. Before it captures a caterpillar, it digs a combination grave and cradle which extends to a depth of about 3 inches into sandy soil. The grave is for the caterpillar, and the cradle for the *Ammophila* egg that will feed on the caterpillar after hatching. When the nest is finished, the *Ammophila* seals the entrance with little pebbles or bits of wood and flies off to hunt a caterpillar of the correct type.

The wasp stings the caterpillar and paralyzes it without killing it. It then makes its way back to the nest, mostly dragging the heavy caterpillar along the ground or making short hopping flights with the caterpillar in its claws. When it reaches the nest, it opens the door, drags the caterpillar to the bottom, and deposits a single egg on top of it. Then it closes the nest again, doing a more careful job than before. It rakes sand over the pebbles, so that the entrance is completely camouflaged.

Then the wasp moves off to start a second nest, repeating the process as before. But 2 days later it returns to the first nest, digs out the entrance, and descends to the bottom for an inspection. By then the egg has hatched. The larva has consumed the caterpillar and is hungry for more. Seeing that things have indeed progressed to this point, the wasp leaves, closing and camouflaging the door, and flies off to bring more caterpillars. It may return 5 to 10 times with a caterpillar, never letting the larva go hungry, and this provisioning may continue for another 2 days. While doing this, the wasp may start yet a third nest. Finally, on returning for a provisioning trip, the wasp discovers that the larva has entered a dormant stage and formed a *pupa* ("puppet"), from which the adult will emerge. Then the wasp leaves the nest for the last time, covers it with pebbles, rakes in sand grains, and pounds them in place with its head. In the meantime, it has been making inspection and provision trips to its other two nests. Thus it has been operating up to three nests at the same time. It has remembered the exact location of each nest despite the careful camouflaging of the entrances. And it has kept track of the exact stage of development in each nest, responding in the appropriate way. If a given nest does not receive sufficient provisions by nightfall, the *Ammophila* returns to its task in the morning. This is constancy and dedication we usually associate only with mammalian mothers.

But the limits of insect intelligence are quickly apparent. There has been no convincing demonstration of reasoning in any insect. Thus while ants can learn complex mazes to find food, they do not think of building a platform out of leaves or pebbles to reach food suspended just above them. There are simply too few neurons in the insect nervous system to allow for this sort of flexibility of response. Moreover, very few insects are born while their parents are still alive. A newly hatched insect must be able to function on its own in the world and must be able to do so at once. It is not the beneficiary of any learned information passed on from the previous generation. Thus there is no time for the slow processes of testing by imitation and experiment that more creative behavior requires. This kind of behavior was left for the chordates, and more specifically for members of two classes within this group, the birds and the mammals.

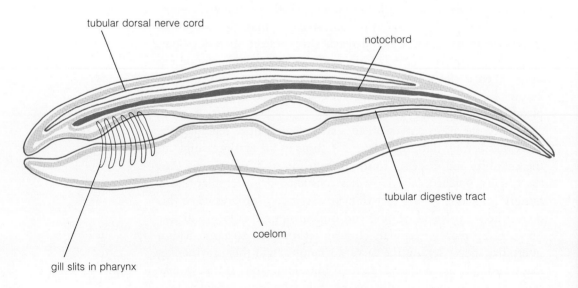

Figure 11-12 A PRIMITIVE CHORDATE (SCHEMATIC).

The Chordate Phylum

Along with the arthropods, the chordates dominate the land of today. Within the animal kingdom, both groups have an ancient history as evidenced by fossil remnants. The oldest arthropod fossils date from 600 million years B.P.,[3] while the oldest chordate fossils extend to 500 million years B.P. As with all fossils, these represent minimum ages. And in both cases, it is not the original forms that are important today, but their distant descendants. For the chordates, these include modern fish, birds, and mammals.

TABLE 2
Body plans of five animal phyla

TRAIT	COELEN-TERATES	FLAT-WORMS	ANNE-LIDS	ARTHRO-PODS	CHOR-DATES
Symmetry	Radial	Bilateral	Bilateral	Bilateral	Bilateral
Tissue layers	2	3	3	3	3
Digestive system	Pouch	Pouch	Tubular	Tubular	Tubular
Coelom	—	—	+	+	+
Circulatory system	—	—	+	+	+
Segmentation	—	—	+	+	+
Blastopore	—	—	Forms mouth	Forms mouth	Forms anus

The primitive chordate showed little evidence of a future major role within the animal kingdom (Fig. 11-12). Like annelids and arthropods, the earliest chordates had a tubular digestive tract, a true coelom, and a segmented body form (Table 2). However, all these structures had a separate origin in the chordate line, by a process of convergent evolution. The evidence for this is largely embryological. There exist several fundamental differences between the way a chordate embryo develops and the way an annelid or arthropod embryo develops. Cleavage in chordates is indeterminate. In annelids and arthropods, it is determinate. What this means is that identical twins or identical quadruplets cannot develop in the annelid-arthropod line but can in the chordate line. In the annelid-arthropod line, once the fertilized egg starts to divide, each daughter cell is destined to form a specific portion of the future embryo, such as the skin or digestive tract. Its fate has been determined. If a cell is removed from an early stage, the remaining cells will not be able to form the body part in

---

3. Before the present.

question. However, a cell removed from a two-cell or four-cell stage in a chordate can form a genetically identical twin or quadruplet.

Another fundamental difference between the two lines concerns the fate of the blastopore in the gastrula (Chapter 6). In the annelid-arthropod line, the blastopore becomes the future mouth. Annelids, arthropods, and their relatives among animals with a tubular digestive tract are called *protostomes* ("mouth first"). Chordates and their relatives, however, convert the blastopore into the future anus, and the mouth is formed secondarily. Chordates are thus deuterostomes ("mouth second"). Protein sequence analyses (Chapter 8) confirm that the two lines have followed separate evolutionary histories for a very long time. In fact, the evolutionary history of the chordates was a major mystery until N. J. Berrill of McGill University pointed out similarities in embryological development between chordates and echinoderms (starfish, sea urchins, brittle stars, and so on). However, there is little doubt that present-day echinoderms such as the starfish have undergone drastic changes in body form and bear little resemblance to the ancestral forms that once shared a common history with the chordates.

The primitive chordate possessed a *tubular dorsal nerve cord*, a stiff internal rod called a *notochord* ("chord in the back"), and *gill slits* in the throat or pharynx. The animal's body form reflects an oceanic life-style. The notochord provides internal stiffening and allows the animal to swim more efficiently, or perhaps to burrow in the bottom sediments. Strong muscles alone do not make for rapid movement unless they are anchored to rigid supports which can push against the surroundings (imagine a boneless runner or weight lifter). The notochord was made of cartilage, and it became the basis of the internal skeleton of the vertebrates. Even today, the human embryo in the course of its development first forms a notochord, which is then surrounded by bone and incorporated into the backbone (Chapter 6, page 217). The movable internal skeleton of the vertebrates played a key role in their later success.

The primitive chordate fed much like a clam or mussel does today; it sucked in a stream of water through its mouth and then strained out small food particles on the surface of gills, passing out the waste water through gill slits in the side of the pharynx. A developing human embryo still forms ridges along the sides of the throat where gill slits would have been present. Besides performing a feeding function, gills also served to extract oxygen from the circulating water and to give off carbon dioxide. While this body plan served the chordates well in the water, it was totally unsuited for survival on land, and at least 150 million years elapsed before they could step on land and call it home.

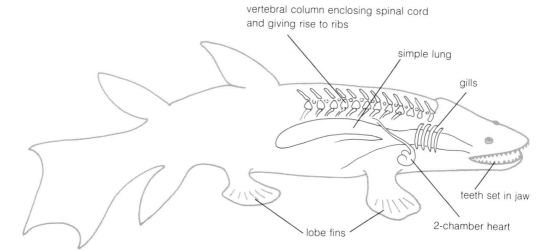

Figure 11-13 A LOBE-FINNED FISH. Though it is a water-dwelling animal, this fish is preadapted for a land existence.

We pick up our story of chordate evolution in the Devonian period, about 100 million years after the appearance of the earliest chordates. This period has been called the age of the fishes, and by this time a new class of chordates dominated both the fresh waters and oceans of the world. This was the class *Osteichthyes* ("bony fish"), and it is to this class that almost all modern fish belong.

We tend to think of a fish as being totally incompatible with a land existence (witness the expression "a fish out of water"). In fact, bony fish were already preadapted for a land existence. The groundwork had been laid, so to speak, and relatively minor changes were necessary to complete the transition from water to land. The specific type of bony fish thought to have made the transition is the group known as the *lobe-finned fishes* (Fig. 11-13). Like all bony fish, they had jaws with teeth. This meant that they could seize and chew larger animals. There is no place on land for a filter feeder. Again in common with all other bony fish, the spinal cord of these fish was enclosed in a bony, flexible vertebral column to which were attached large numbers of ribs. This strong internal skeleton would be a necessity on land, where the stresses of gravity are much stronger than in water.

Lobe-finned fish still breathed primarily by means of gills, but they also had a primitive pair of lungs. In many modern fish, this primitive lung has been modified into a swim bladder, which helps regulate the internal buoyancy of the fish. But even today fish with lungs survive in scattered regions of the world, from Australia to

South America. They are freshwater species and live in warm, stagnant waters where dissolved oxygen levels may be too low to allow the use of gills. Lungfish therefore supplement their oxygen supply by gulping air.

Finally, the fins of lobe-finned fish were attached to the body by means of bony stalks. Such lobe-finned fish were thought to have disappeared forever about 75 million years ago, but during the last 30 years several have turned up in the nets of fishermen off the African coast. This generated excitement comparable to that which would be created by the trapping of a live dinosaur during a routine hunting expedition!

A Devonian lobe-finned fish never had to worry much about becoming a fish out of water. Its simple lungs allowed it to gulp air, and lobed fins were sturdy enough to allow belly-dragging progress over land. It should not be imagined that these fish felt any strong urge to leave the water and settle on land. It is much more reasonable to suppose that the initial sojourn on land was a forced one. Great climatic changes took place during the Devonian period, including the drying up of many lakes and swamps. A lobe-finned lungfish that found itself trapped in a shrinking body of water ventured out across land not in order to stay there but in order to reach another body of water. Evolutionary selection then worked to enhance this ability, even though the initial goal of such evolutionary change would be a conservative one: to allow the fish to return to water. But once this initial step had been taken, the stage was set for the emergence of a new class of vertebrates: the amphibians.

The Amphibians

The term "amphibian" means "life on both sides," referring to life on both land and water. The amphibian one is most likely to encounter is a frog or a toad. However, these jumping, tailless animals represent highly specialized forms and are as different from the ancestral amphibians as bats are different from ancestral mammals. The first amphibians resembled present-day salamanders (Fig. 11-14). The external body form is still fishlike, but a variety of structural changes announce this as a land form. The fins are gone, replaced by two pairs of limbs. Adult forms have lost their gills and depend for oxygen on their lungs and skin. In order to make the skin a respiratory surface, it must remain moist. This is why the skin of a frog or salamander has a slimy feel. The skin of a toad, having lost much of its respiratory function, is much drier.

There is also an important difference between the circulatory systems of the fish and of the amphibian. The heart of the fish has

THE CHORDATE PHYLUM / 397

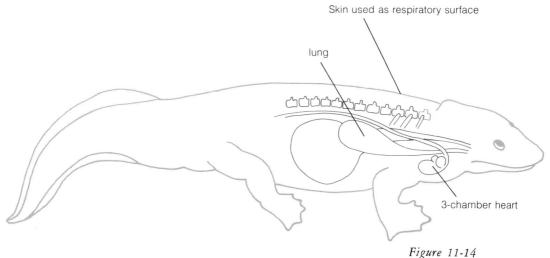

Figure 11-14
A PRIMITIVE AMPHIBIAN

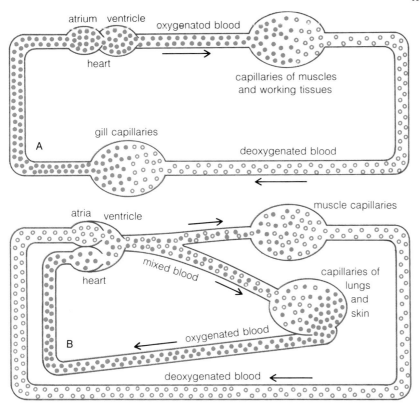

Figure 11-15 THE CIRCULATORY SYSTEMS OF AN AMPHIBIAN AND A FISH. In the fish (above), the systemic and respiratory capillary nets are arranged in series. This creates a greater frictional resistance to blood flow and puts a greater load on the heart. In the amphibian, respiratory and systemic capillary beds are in parallel, but there is a mixing of oxygenated and deoxygenated blood in the single ventricle.

two chambers: a collecting chamber (the atrium) and a pumping chamber (the ventricle). A single ventricle pumps blood through two capillary systems in series. Blood first courses through working tissues, such as muscles and the digestive tract, where oxygen is lost by the blood; then the entire flow is passed through the gills, where oxygen is picked up from the water. Then it returns to the heart to be pumped out again (Fig. 11-15). As a result, the muscles receive well-oxygenated blood. But the two capillary systems in series present a high resistance to blood flow. The higher energy demands of an amphibian terrestrial existence would overtax a heart working in this manner.

The heart of an amphibian consists of three chambers, two atria plus one ventricle. The ventricle pumps blood to muscle capillaries and to skin and lung capillaries in parallel, not in series. The load on the heart is much reduced, making higher flow rates possible. But a new problem is introduced. The oxygenated blood from the lungs and skin and the deoxygenated blood from the muscles enter the two atria of the heart separately but must leave through the single ventricle. A mixing of the oxygenated and deoxygenated streams occurs. In fact, the mixing is only slight, because the ventricle contains internal baffles which function to keep the two streams separate.

The limbs, respiratory organs, and circulatory system of an amphibian represent major steps in the conquest of land. But much of the amphibian's body conspires against these steps and still ties it to a watery existence. The limbs are generally weak and make the amphibian clumsy on land. The moist skin requires a humid environment or continual wetting to function. A frog left overnight in a little boy's bedroom dies because of drying of the skin. Fortunately, most little boys catch toads! Even the lungs function poorly. A frog does not draw in air with its chest as higher vertebrates do—it must swallow its air. And amphibians have made no changes in the reproductive system of fish. Eggs and sperm are still released into water, where fertilization occurs. The larva that hatches out looks more like a fish than an amphibian and is totally dependent on water. The result is that most amphibians can spend only part of their lives on land. They are still tied to water as birds are tied to land. These remaining ties to water were broken by direct descendants of the amphibians, the reptiles.

The Reptiles

The reptiles have had a bad press. The serpent with its forked tongue misled a weak woman. The dragon had an unfortunate taste for virgins. The monsters that haunt fairy tales and children's dreams are all recognizably reptilian. The explanation for this state of affairs

THE CHORDATE PHYLUM / 399

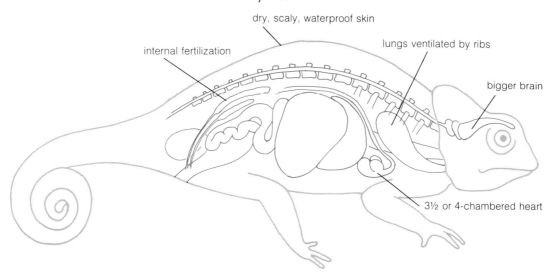

Figure 11-16 CUTAWAY VIEW OF A REPTILE, THE CHAMELEON.

must be sought more in the theories of Freud than in biological fact. One can hardly imagine animals more shy and harmless than our two most common reptiles, the box turtle and the garter snake. And when one speaks of the evolutionary achievement of the reptiles, one must recognize them as a milestone in vertebrate evolution. Reptiles hold the same position among vertebrates that tracheophytes do among plants—they were the first true colonizers of the land (Fig. 11-16).

The reptilian body feels dry to the touch. The skin is no longer used as a respiratory surface and is waterproofed by a covering of close-fitting scales. To compensate for the loss of oxygen from the skin, the lungs have become more efficient. The internal surface of the lungs is more highly divided than in amphibians and provides a larger surface area. More important, the lungs are ventilated bellows-fashion by movement of the ribs.

The limbs of a reptile are stronger than those of amphibians. The word "reptile" is derived from the Latin word for "crawl." Nevertheless, several lizards achieve very respectable running speeds. The basilisk lizards of South America have flat hind feet and are so speedy that they can run across the surface of water. Some have been observed to run a quarter mile in this fashion.

In order to provide for the increased energy needs of a land existence, the reptilian heart has a partial (sometimes complete) separation of the ventricle. Thus mixing of oxygenated and deoxygenated bloodstreams is reduced, and working tissues receive higher concentrations of oxygen.

The reptilian brain, while small by mammalian standards, shows an increase in complexity over the amphibian model. It contains the first small cerebral hemispheres, used for higher functions in the mammals. This is in keeping with the increased complexity of land existence.

But perhaps the most original contribution of the reptiles was in the field of reproductive physiology. They abandoned the ancient practice of external fertilization. The male introduced sperm directly into the body of the female, where it fertilized an enormously large egg. The fertilized egg then received a waterproof shell and was laid somewhere on land. This was the *amniote egg,* which had its own food supply, oxygen supply, and waste disposal (Chapter 6). Even the principle used in the disposal of nitrogen wastes was original. The nitrogen wastes of fish and amphibians are moderately to severely toxic and would kill the baby reptile in the developing egg. Therefore reptiles disposed of wastes by converting them into uric acid, a compound so insoluble that there was no danger of poisoning inside the tight confines of the shell.

The amniote egg was as essential to success on land as was the development of limbs. It completely freed reptiles of dependence on water. In fact, it sometimes made them dependent on land. The green turtle is a reptile that spends its entire life in the open ocean. Yet in order to reproduce, it must return to a specific sand beach to lay its eggs.[4] This represents the most vulnerable stage in the life of the green turtle. The eggs and the adults are taken by humans, and the newly hatched young are decimated by frigate birds during their run to the open ocean. During the past 400 years the green turtle population has been reduced from tens of millions to a few thousand individuals. Yet green turtle soup is still available in the "better" restaurants of New York City.

The approximately 6,000 species of reptiles known today represent only a remnant of the mighty host that ruled the earth during the Mesozoic era (250-70 million years B.P.). This class is divided into three major orders: Chelonia (turtles), Crocodilia (crocodiles and alligators, the closest living relatives of both the birds and the dinosaurs), and Squamata (snakes and lizards). In addition, there is the minor order Rhynchocephalia, exemplified by a living fossil, the tuatara of New Zealand.

---

4. For example, all green turtles of the western Caribbean and Gulf of Mexico return to a single beach at Tortuguero, Costa Rica, at breeding time. Other populations have other, equally restricted nesting areas.

Figure 11-17 AMPHIBIANS (A—B) AND REPTILES (C—E). (A) Fowler's toad. (B) A tree frog from South America. The toes end in adhesive disks for easy movement over leaves and branches. (C) Head of an iguana. (D) A newly hatched turtle. The site of attachment to the yolk sac is still visible on the shell below. (E) A chameleon at the beginning of its strike. When fully extended, the tongue can reach as far as the rest of the body.

The Birds

The class Aves, or birds, is a direct offshoot of the reptiles. In fact, the relationship between birds and reptiles is so close that Robert T. Bakker of Harvard University has proposed stripping the birds of their status as a separate class of vertebrates and including them in a new class, the Archosauria. The archosaurs would include present-day crocodilians, as well as the dinosaurs. According to Bakker, the dinosaurs never became extinct but survive today as birds. One of the most intriguing elements of this theory is the possibility that, unlike other reptiles, the dinosaurs were warm-blooded. Evidence for this comes from dinosaur bone structure, geographical distribution, and predator/prey ratios. Whether or not birds are simply feathered, flying dinosaurs, the fact is that almost all the differences between birds and classic reptiles can be related to the needs of flight.

The process of flight requires an increased energy supply, and this is achieved by raising the body temperature. A bird may have a normal body temperature of 112°F (as opposed to 98°F in humans). At this temperature, about twice as much energy is generated per unit of body weight as is at 98°F. This high body temperature is kept at a constant level, and birds fly as fast on cold days as they do on hot ones. Feathers are used to insulate against heat loss. These are highly modified reptilian scales.

In keeping with the increased energy needs, the heart contains four chambers and essentially functions as two separate hearts. One half pumps oxygen-rich blood to muscles and working organs, and the other half pumps oxygen-depleted blood to the lungs. The lungs of birds are extremely efficient in oxygen uptake because, unlike other vertebrate lungs, they contain no dead space. Air passes completely through the lungs and into air sacs which are tucked into the long bones and other body cavities.

Skeletal changes include modification of the front limbs into wings (Fig. 11-18). The strong flight muscles are anchored to an enlarged breastbone or keel. The skeleton from the shoulder to the hip region is rigid, a necessity in flight. The skeleton itself is very light; the feathers of a large bird may weigh more than the bones do. The loss of the front limbs to flight has created the need for a new organ of manipulation, and this function has been assumed by the head and neck. The neck of a bird contains twice as many vertebrae as the neck of a mammal, and a bird can do things with its neck that no mammal can hope to duplicate. For example, try rubbing the small of your back with your lips. Birds do it all the time.

Throughout the body, steps have been taken to cut down weight. There is no urinary bladder as in other land vertebrates. Wastes are

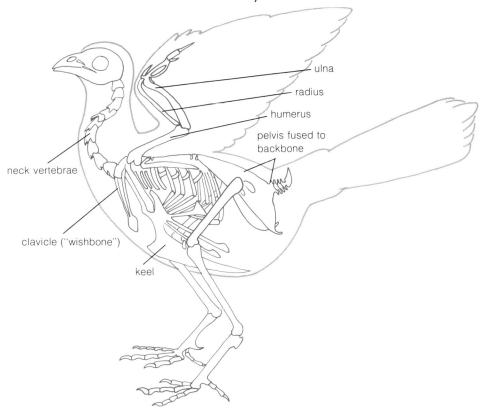

Figure 11-18 A BIRD IN CUTAWAY VIEW, SHOWING SKELETAL STRUCTURE. Note the long, flexible neck, the large breastbone (keel) used as an anchor for flight muscles, and the rigid, fused skeletal structure of the body which makes for stability in flight.

excreted as they are formed, in the form of uric acid. Digestion is very rapid: an average of 1½ hours instead of the 6 to 12 hours typical of mammals.

Finally, there has been a great increase in the size of the nervous system. Life in the air demands accurate perception and instant reaction. A slow mulling over of the situation will not do. Much of a bird's increased brain size therefore is concerned with muscular coordination and sensory reception. However, the term "birdbrain" is an unfair one. Careful studies have shown that blue jays fashion simple tools out of scrap paper in order to reach food. Konrad Lorenz described his experiences with pet jackdaws, which fluttered behind him on bicycle rides through the countryside. They always returned home, even when they lost their master along the way.

The complex nervous system of a bird takes time to develop. When a baby reptile hatches out of its egg, the parent may not be

Figure 11-19 VARIATION IN BEAK FORM IN BIRDS. (A) A woodcock. The long bill is used to spear insects and small animals in wetlands such as woodland swamps. (B) D'Arnaud's barbet. This ground-feeding bird of Africa has a thick beak powerful enough to crush dried fruits and seeds. (C) A brown pelican. The large bill pouch can comfortably hold a full load of fish. (D) Red-tailed hawk. The beak of the hawk is shaped for the tearing of flesh. (E) A timberline wren from South America. The thin beak is used for catching insects in flight.

present at all, but the young animal can survive on its own. When a little chick hatches out of its egg, one parent is always present and often both parents are present. The chick is assured of parental support and protection. Without that support it would have as little chance of survival as a newborn mouse or rabbit. It is not the development of bones and muscles that requires this long time of dependence —it is the brain and nervous system.

The Mammals

Like birds, the mammals owe much to their reptilian ancestry. By an accident of geology, examples of the earliest mammals have been preserved on the islands of New Guinea and Australia. The most primitive of these, such as the echidna and the duck-billed platypus, lay reptilian eggs, have a reptilian cloaca (single opening for reproductive and excretory tracts) and, like the reptiles, maintain a fluctuating body temperature. They can justifiably be called reptiles with hair, and that is what they would have been called if the line had ended here. But it didn't, and the line produced you and me, two organisms startlingly different from reptiles.

The technical definition of a mammal stresses points that border on the arcane: a single bone in the lower jaw, the presence of three inner ear bones, differentiated teeth, RBCs without nuclei, and so on. The functional definition is much simpler. Mammals are land vertebrates possessed of high metabolic activity and high intelligence. All their body architecture is oriented toward maintaining these characteristics.

The high metabolic rates of mammals are achieved in the same way as the high metabolic rates of birds: by a rise in body temperature. The same metabolic pathways are present as in the reptiles and in distant ancestors before them. The metabolic pathways are simply made to run faster. And as in birds, the mammalian elevated body temperature does not fluctuate with changes in external temperature. Just as a hot stove requires good insulation, so a hot mammalian body minimizes heat loss by means of hair. The legs of mammals are swung directly below the body, unlike the typically sprawling limbs of reptiles. Such a leg placement allows for greater running speed. Further, the belly is raised off the ground, and heat loss is reduced. In keeping with higher metabolic needs, the heart is truly four-chambered, as in the birds. The lungs are ventilated not only by means of ribs but by means of the diaphragm, a sheet of muscle that completely separates the chest cavity from the abdominal cavity. Contraction of the diaphragm creates a partial vacuum in the chest cavity, causing air to rush into the lungs.

Mammals have the largest and most complex brains in the vertebrate group, and their behavior is correspondingly more flexible than that of any living reptile or amphibian (Chapter 5). The large mammalian brain is biologically expensive. The embryological development of this brain is so long that no egg could contain enough internal nutrients to allow its full development. Consequently, the young of advanced mammals develop inside the mother's body, where they are nourished by the placenta. After birth, maternal care continues as the young are suckled by means of mammary glands.

A long period of maternal care allows the brain to reach its full structural complexity and also allows for external inputs—the processes we call learning. Learning in mammals proceeds by a variety of methods. There is the familiar conditioning process described by B. F. Skinner of Harvard University (and by many before him): an appropriate response is rewarded, while an inappropriate response is not. As a result, a pigeon can be taught to dance a figure eight or a rat can be taught to follow a path through a complex maze. Much parental teaching proceeds by a combination of imitation and conditioning. Such teaching is an essential part of the mammalian growth process—instinct alone is not an adequate guide to survival. Even the relationship between a cat and a mouse is not instinctive; anyone who has seen a kitten with its first mouse knows that the kitten must be *taught* to kill. So must a wolf cub, a mountain lion, or a human.

Mammals indulge in another, more creative, kind of learning called *play*. Very little of this kind of behavior is seen among other vertebrates. Even the relatively brainy birds lead an earnest, plodding kind of existence in comparison, and "as free as a bird" is seldom as free as a mammal. Of course, play has serious components. But, even if a little kitten is playing at stalking and attack, it is always ready for diversion and novelty: a pounce on its own tail, or a ball deliberately pushed only so it can be chased. There is little question that this kind of behavior makes for a more creative, wider ranging learning process than might be available to the more earnest pigeon or grass snake.

Mammals, along with insects, represent one of the pinnacles of animal development. Although ancestral mammals evolved as land animals, their descendants also include representatives who have become masters of the air (Order Chiroptera—the bats) and water (Order Cetaceae—the whales and dolphins, and Order Sirenia—the manatees). The life-styles represented within these mammalian orders stretch the imagination, and to do justice to a good sampling of these would require many, many books the size of this one. Here we look more closely at only one of these orders, the primates. It is the order of mammals to which human beings belong.

Figure 11-20 AN ASSORTMENT OF MAMMALS. (A) Dolphin. (B) African elephant. (C) Cheetah at a kill. (D) White rhinoceros. (E) Fox cub at its den. (F) Red kangaroo.

The Primates

What is a primate? First, the order is an ancient one among mammals, representing a stem group which has retained many of the characteristics of the first mammals. Three examples of these are as follows.

1. A typical primate has five fingers on each limb. Five is both the maximum number of fingers possible for mammals and the number bequeathed to them by the amphibians. There is no known mammal with six or seven fingers, though of course many with fewer than five are known. The modern horse carries only one finger per leg, capped by a big nail we call a hoof. To a taxonomist, the horse represents a specialized mammal, with toes built for carrying a heavy weight at high speeds. A horse can always outrun a human but of course cannot use a hammer or play a violin.

2. A typical primate possesses four types of teeth: incisors, canines, premolars, and molars. Reptile teeth show little variety, though there may be many more inside the mouth. Mammals more dentally specialized frequently lack some of the four types. For example, a squirrel and a woodchuck lack canines, and a cow lacks both canines and upper incisors. Although a squirrel can easily beat a monkey in a nut-opening contest, it lacks the versatility of dental performance the monkey is capable of.

3. One more example: primates have two bones in their lower forelimbs—the radius and the ulna. The ulna is the bone used in flexing the elbow, and an extension of it forms the "funnybone." The radius is the bone necessary for making twisting motions such as turning a key or a screwdriver. In certain mammals such as horses, sheep, and antelopes the two bones are fused. This increases the weight-bearing capacity of the lower forelimb and aids in running, but it forever disqualifies an antelope from a job as a mechanic or locksmith.

To summarize, evolutionary specialization, leading to enhanced performance at such tasks as running, burrowing, gnawing, or scratching, almost invariably brings with it a closing down of other available options for the animal in question. A better runner becomes a poorer manipulator, a better gnawer, a poorer hunter, and so on. While specialization can lead a species to increased success under well-defined circumstances, it can also lead the species to ruin if those circumstances should change. The road to extinction has been traveled by many an organism that adapted too rigorously to its environment and then saw that environment change. In this sense, primates have in many ways chosen to remain generalists rather than specialists. A generalized animal, besides being at home in a greater variety of environments, also keeps in reserve its options for future evolutionary change. Thus birds evolved from reptiles but certainly not from such a specialized reptile as a snake, which has totally lost the forelimbs that could be modified into wings. The spectacular changes leading to the emergence of human beings can be explained only by taking account of the generalized nature of the primate stock.

At the same time, primates have not entirely passed up the process of specialization—the purely generalized, omnipotential animal does not exist. The main thrust of primate specialization consists of adaptation to *arboreal* life (life in trees), and much of this has been passed on to us. Some of these arboreal adaptations are obvious ones, others are less so.

Among the obvious adaptations is the development of *prehensile limbs* ("capable of grasping"). In most primates, all four limbs are involved and, in New World monkeys, the tail as well. Of course, tree-dwelling animals do not need prehensile limbs in order to travel through branches; witness the nimble squirrel which depends on its sharp claws instead. Far from developing their claws, primates have reduced them to flat nails instead, which function mainly as reinforcements for the fingertips.

Dependence on grasping limbs instead of claws has given primates a bonus that squirrels do not enjoy; primate forelimbs are beautifully fitted for manipulation and exploration. The upper arm can flex and rotate to assume almost every conceivable position, and the fingers can perform the fleet and independent motions demonstrated by violinists, pianists, and typists.

Life in the trees puts a premium on good *vision*. If a monkey jumping for a branch misestimates the distance, it will not leave offspring to inherit its poor vision. Besides distance perception vision is significant in other ways to the primate life-style. Fruit is an important part of the primate diet, and there are few better ways to locate fruit than by its color. It is therefore no surprise that, alone among the mammals, the primates are possessed of color vision—surely one of the most delightful senses to have been bequeathed to human beings!

A word about the estimation of distances is in order. The next time you are a passenger in a car, try closing one eye for a moment and then estimating the distance of the car ahead of you. It becomes bewilderingly difficult! There is very little loss in the sharpness of the image, but much of the sense of distance is lost. This is because distance perception requires two eyes, with an overlapping field of vi-

Figure 11-21
THE BULGING EYES OF
THE RABBIT.

sion. All mammals have two eyes, but not all of them have a very large overlap between the two fields of view. Take the rabbit, for example. The two bulging eyes, located on the sides of the head, give the rabbit a 360° field of vision—equivalent to the human wish to "have eyes in the back of the head." But there is little in the way of an overlap in front, and so, while a rabbit can see an approaching (or pursuing) coyote from any direction, it has only an approximate idea of how far away it is.

There is one problem in moving the eyes to the front of the face: animals with a large snout find that the snout blocks part of the view. The primate solution has been to reduce the size of the snout. This result costs them dearly in the ability to *smell,* since the length of the snout largely controls the amount of nasal mucosa available for smelling. This is not a serious loss to most primates; smells are strongest along the humid, cool floor of the forest, not high up in the trees filled with breezes and sunlight.

Another fact of human biology we owe to the primates is the *small number of children* born at a time—generally no more than two. This reflects the problems a primate mother would encounter in tending more than one infant high up in the branches of a forest canopy. Even with one infant per mother, accidents do happen, and infants can plunge 100 feet or more down to the ground. The motherhood instinct is so strong that a howler monkey will rush down to the ground to save an infant, something she does not do under any other circumstances.

A species can follow two opposing strategies in maximizing the number of offspring. In one case a large number of infants may be produced per litter, with little parental care and heavy mortality. The other strategy is to produce only one or two offspring per litter, but to invest heavily in parental care to ensure survival. This is the strategy followed by primates and carried to the extreme of an 18-year-long dependency period in some humans. A 4-year-old chimp that loses its mother and is left on its own has almost as little chance of survival as would a 4-year-old human left alone. Among primates, heavy parental care is extended even to the period *in utero,* and primates have abnormally long gestation periods.

Another primate characteristic is complex *social behavior.* Apes and monkeys typically live in bands with involved social hierarchies, feeding behavior, displaying, grooming behavior, strategies for defense and offense, and a system of communication based on sounds and visual signals.

Again there is a strong human carryover. One of the strongest punishments imposed on a prisoner is solitary confinement—the withdrawal of human company. The emotional stress is so severe that mental derangement can occur in a matter of weeks. It is to be expected that all human societies manipulate this strong social drive. For example, Western societies have traditionally discouraged physical touching between adults. Legal exceptions have been touching between man and wife, and between lover and beloved. Touching satisfies the strong primate need for grooming, and the narrow channeling of this urge has contributed to the kind of social structure we accept as normal: a tight, self-sufficient nuclear family embedded in an otherwise atomized society.

The primate way of life has understandably put a premium on *brain development,* and primates, especially the monkeys and apes, possess larger brains in proportion to body weight than typical mammals do. This, too, has been passed on to the human species.

Homo Sapiens

We conclude this chapter by taking a brief look at the known facts of human body form and evolution. What is the secret of this large, successful, astonishing species? First, we are the only primates to abandon the use of arms for locomotion and to walk exclusively on our feet. This is a rather simple observation, but one that indicates a revolution in behavior and body anatomy as well. Perhaps the most comparable event in evolutionary history is the evolution of birds from reptiles, a process that likewise involved abandonment of the front limbs for locomotion and their evolution into organs of flight.

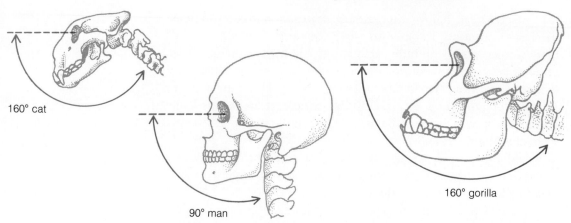

Figure 11-22 INSERTION OF THE FORAMEN MAGNUM IN THE HUMAN SKULL AS COMPARED TO THE SKULL OF A CAT AND A GORILLA.

To be sure, chimpanzees, orangutans, and gorillas can stand erect on their hind legs and even move a fair distance in this manner. But sooner or later they return to quadrupedal walking, since an ape's body is not built to withstand the resulting strain for long. Walking on the hind legs involves more than an act of will—it involves a basic rearrangement of muscle and bone.

Let us begin with the skull. Should a cat or gorilla stand up on its hind legs, it is faced with the problem of its face staring upward. This is because in a quadrupedal animal the spinal cord and vertebrae enter the skull at an angle of approximately 160° with respect to the line of sight. In humans, the foramen magnum (the opening in the skull through which the spinal cord enters) has migrated downward, forming an angle of approximately 90° with respect to the line of sight. Besides solving the problem of seeing ahead instead of upward, this downward migration of the foramen magnum makes possible the thin, graceful neck of the human compared with the massive, muscular neck of the gorilla. The human head rests on the vertebral column in a balanced position, while a gorilla's head is pulled down by the forces of gravity, requiring restraints in the form of massive muscles and large anchoring neck vertebrae (Fig. 11-22).

Next, changes had to be made in the shape of the spinal column. In all quadrupedal animals, the spinal cord is shaped in the form of an arch. Engineers as long ago as Roman times discovered that an arch can bear much more weight than a straight span, hence bridges are commonly built in this shape.

The problem with an arched backbone is that a gorilla would have a tendency to tip forward when standing up. The human solu-

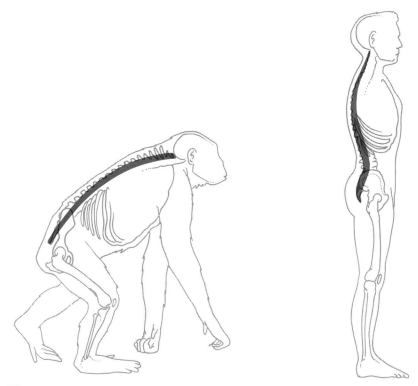

Figure 11-23 THE ARCHED BACKBONE OF THE GORILLA AS COMPARED TO THE S–SHAPED BACKBONE OF THE HUMAN.

tion is an S-shaped backbone which is stable in the upright position. The vertebrae in the human backbone become progressively larger as they approach the pelvic region, since they must bear the entire weight of the upper torso. Though a human vertebra can typically support up to 700 pounds of weight, it is not uncommon for the lubricating disks between vertebrae to slip sideways because of excessive weight. The result is pinching of nerves and internal pain in the lower back region. This clearly represents an unsolved problem in the adaptation to bipedality (Fig. 11-23).

The bones of the human pelvic girdle (the hip bones) have also undergone considerable modification. These bones must transmit the entire weight of the upper torso to the legs, instead of transmitting only half the weight as in a quadruped. The ileum is consequently more massive than in apes. Its bowled shape also serves to cradle the viscera.

The legs have also been "done over." There are strong new muscles in the buttock and calf region. The former are responsible for swinging the leg back and forth in a typical walking gait. The

Figure 11-24
A FAMILY OF WHITE-HANDED GIBBONS. The long, powerful arms of the gibbon are perfectly suited for brachiation through trees.

calf muscles connect by means of the Achilles tendon to the heel and are necessary to cushion the impact of the foot coming down on the ground. Neither of these shapely areas is well developed in apes.

The leg itself is proportionately longer than in the apes, allowing for increased speed in walking and running. In fact, a gibbon's legs are so short that it can move upright only by holding its arms aloft like a Greek dancer (Fig. 11-24).

And substantial work has been done on the human foot to improve its performance. The arch principle, abandoned in the backbone, reappears in the foot. An arched foot helps cushion the impact that results from the whole weight of the body coming down on the narrow surface of the foot. Flat feet are a guarantee of pain when walking. The design of the foot as a shock absorber has cost it one function that it retains in the apes. The big toe in the human foot is no longer opposable as it is in the apes, hence it is relatively useless in picking up objects and in grasping.

The freeing of the human hand and arm from the task of locomotion has allowed it to blossom forth as a superb organ of manipu-

Figure 11-25
A MALE BABOON IN A THREAT GESTURE. Note the long canines and the U–shaped dental pattern.

lation. The quick movements of the pianist, the caress of the lover, the iron grasp of the lumberjack illustrate the range of performance of this amazing organ. As we will see shortly, it is the human hand, coupled with the brain, that is largely responsible for the worldly success of our species. As a matter of fact, the general feeling among anthropologists is that it was the hand that provoked the human brain to its amazing development.

Finally, and presumably in a response to the change in habitat of our early ancestors from forest to plain, the dental pattern underwent a marked change. The jaw of a gorilla is elongated, with parallel sides and with large canines in front. These canines are so large that the jaws interlock permitting only up-and-down motion of the jaw. A gorilla's canines serve two purposes: they shred the foods of the forest—fruits, berries, and succulent leaves; and they are formidable weapons of offense. Observation of gorillas shows that this unchivalrous species is most likely to use its bared canines as a threat to intimidate females. However, a pack of male baboons has been observed to attack and dismember a threatening leopard.

Why did the human species throw away this useful weapon? Loss of the canines, coupled with a loosely hinged jaw opens up the possibility of a grain diet, without ruling out other fare as well: meat, fruits, and the like. Grass seeds are too tough to be chewed up without some sideways motion of the molars, and the human molars represent a fairly efficient grinding surface.

Humans are essentially hairless. Equipped with an abundance of sweat glands, the body has considerable cooling ability such as few other animals possess. Since we evolved in a tropical climate and probably evolved as hunters, this cooling of the body gives a decided advantage in the chase. In fact, *Homo sapiens* is the only carnivore that stirs forth in the noonday sun in the tropics. Subsistence hunters of the Kalahari hunt large game by selecting a victim and then driving it relentlessly. Without rest or food, the animal eventually exhausts itself and falls to spears and arrows.

Finally, and perhaps most uniquely human, is the use of tools and language, a complex we call culture. We will touch on this again.

Human Evolution

Let us now try to sort out the specific evidence that might show how a line of primates evolved into a line of primitive and then advanced hominids.[5] The evidence at hand is based on a few bone fragments. Unfortunately, the fossil record for the early evolution of humans is much less complete than, say, the fossil record for the evolution of the horse. There are two reasons for this. First, our ancestors lived in tropical forests, and this sort of habitat is the worst possible for the preservation of fossils. There is rapid decomposition by fungi and torrential rains, as well as the destructive activities of small ground animals. Second, it is thought that hominids evolved from small population groups with a great deal of internal variation, a fact that reduces the chances of finding fossils or fitting them unambiguously into a logical pattern. Given these limitations, here is a reasonable hypothesis regarding human evolution.

The story begins in Africa, perhaps 15 to 20 million years ago. At this time the climate was undergoing rapid change, with the result that much of the rain forest was drying up and changing into a savannah—grassland with occasional trees. This presented two choices to the tree-loving primates of the forest. They could try to become

---

5. Hominidae is the family that human beings belong to. There are no other hominids living today besides *Homo sapiens*. Other primate families include the great apes, the Old World monkeys, New World monkeys, and a number of more primitive families linked in the suborder prosimians: lemurs, tree shrews, lorises, and others.

HUMAN EVOLUTION / 417

even better adapted to the forest or they could migrate across the danger-filled savannah with its big cats, sparse shelter, and uncertain food sources. The first option was chosen by the ancestors of the present-day great apes. In a sense, they displaced our ancestors from the forest by competing more successfully for a shrinking resource. The second option—striking out across the savannah—was the option chosen by our ancestors. In all probability, they were simply trying to find a fresh, uncrowded forest, much as the Devonian lobe-finned fish was trying to crawl to another pond. Any adaptations to survival on the savannah these ancestral humans may have had initially served a conservative purpose—a return to the forest home from which they had been driven.

We have very few clues as to what these early creatures were like. At present the likeliest candidate for this ancestral position is a medium-sized primate called *Ramapithecus* (Rama for the Indian god who used an army of monkeys to cross from India to Ceylon; pithecus meaning "ape.") The only surviving fragment of *Ramapithecus* is a jaw fragment, but this is enough to establish its relationship to *Homo sapiens*. The jaw is paraboloid, not U-shaped as in typical apes, indicating different kinds of diet. The paraboloid jaw is more suited to a diet of grains and roots, which are found in the savannah.

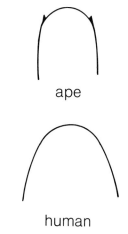

Figure 11-26
THE PARABOLOID JAW OF THE HUMAN AS COMPARED TO THE U-SHAPED JAW OF AN APE.

Ramapithecus lived 12 million years ago, and there is a great gap in the fossil record for the next 8 million years or so. Then the clues start coming more quickly.

Dating from 4 million years ago and stretching on for the next 3 millions years, there lived hominids whose remains were first discovered in South Africa. Hence the name *Australopithecus* (southern ape).[6] *Australopithecus africanus* was physically small and mild-looking, stood about 4½ feet tall, and weighed 50 to 80 pounds. They lacked sharp claws and teeth. Even the brain was unimpressive: a cranial capacity of about 500 cubic centimeters in the range typical of apes (a gorilla's brain capacity has a range of 420 to 750 cubic centimeters). They were fully bipedal. What makes the discovery of these bones so exciting is that *Australopithecus* was a user of tools. Scattered around their bones are pebbles with the edges chipped off. These may seem like rather ineffective tools; but testifying to their usefulness is the finding of bones of much small game in the vicinity of *Australopithecus* bones. The game includes antelope, an animal this hunter could not have consumed without the help of the chipped pebbles. Even if the kill were made with a club or long animal bone,

---

6. In fact, at least two species of *Australopithecus* appear to have coexisted. The small-bodied *A. africanus* is thought to have contributed to the future evolution of humans, while the more massive *A. robustus* died out without issue.

Figure 11-27
STONE-AGE TOOLS.
(A) An Australopithecine pebble tool. (B) A hand ax and scraper manufactured by *Homo erectus*. The tool was shaped from a single block of stone. (C) A scraper used by the Neanderthals. This tool was made by flaking off chips from a central core.

Australopithecus could not pierce the skin with teeth alone. As L. S. B. Leakey demonstrated, an antelope can be skinned and dismembered in just 20 minutes with the use of such a pebble tool. It is possible that pebble tools were also used to manufacture other tools, such as sharpened sticks for digging, but these of course have disappeared.

The relatively small brain size of *Australopithecus* was a most unexpected finding. Biologists had naturally assumed that a considerable brain capacity was required for a tool culture. What seems to be true instead is that the hunting and tool-using life-style of *Australopithecus* selected for an increasingly larger brain. Hunting in a band requires close cooperation and puts a premium on the ability to communicate. Further, the use of tools presupposes a certain foresight and anticipation, qualities we associate with higher mental functions. Thus there exists a positive feedback between hunting and tool-using and brain development.

The thread of the story is picked up again at about 700,000 B.P., with the discovery of fossil remnants of a much brainier hominid called *Homo erectus*. *Homo erectus* had a brain capacity of 800 to 1,200 cubic centimeters. Like *Australopithecus*, they walked fully erect, made tools, and hunted. But the tools of *H. erectus* show a great advance over those of *Australopithecus*. For instance, there are hand axes with a definite knifelike edge. But perhaps the most important tool used by *H. erectus* was fire. Hence the game they hunted includes the largest and most formidable of animals: elephants, horses, rhinoceri, and wild cattle. Bringing down game like this required a group effort and possibly a long expedition. There is evidence of permanent campsites which they returned to after their expeditions. If present-day subsistence hunters are a guide, the men returned not only because of sexual ties to the women (who, unlike other primates, are permanently receptive) but also because the women as gatherers of fruits, berries, edible roots, and the like, provided most of the calories for the diet. There is another advantage to the base camp. To a nomadic animal that travels with a group, even a slight injury such as a sprained ankle can lead to death, because it must fall behind the group and loses its protection. With a base camp, it can rest and recover and rejoin the wandering part of the group when healing is complete.

The species prospered, and spread far beyond the boundaries of ancestral Africa. Fossil remnants have been found as far away as Java, Peking, and Hungary. The reason for the demise of *H. erectus* is not known, but perhaps the most likely place to look is the appearance of a new kind of hominid, the Neanderthal (Fig. 11-27).

The Neanderthal or *Homo sapiens neanderthalensis* was a variant form of our own species. The brain was slightly larger than ours (1,400 versus 1350 cubic centimeters average in modern populations). If brain volume is a direct indication of intelligence, it is perhaps surprising that, during a 60,000-year tenure on earth, the Neanderthal invented none of the tools that modern man has invented in a much shorter time. The tool kit of the Neanderthal consisted of implements of stone that show little advance over those of *H. erectus*. In terms of physical appearance, the Neanderthal's large cranium was pushed back, unlike ours which rises upward. This gave the Neanderthal a low forehead, which was further accentuated by heavy ridges over the eyes. Even so, given a suit and a shave, a Neanderthal man would draw little attention today on a city street. These early humans lived during a time when glaciers were advancing and retreating across the Eurasian land mass (100,000 to 40,000 B.P.), and their preferred homesites seem to have been caves. They showed a high degree of social cohesiveness, protecting the infirm and the less advantaged. For example, one cave in Turkey yielded a Neanderthal skeleton with evidence that the individual had been blind in one eye since childhood—the entire eye socket was infiltrated with bone. As an adolescent he had lost one hand in an unknown accident. And yet this individual lived to an old age, sheltered by his companions and presumably serving some useful function within the group. This caring extended even beyond life—the Neanderthals buried their dead, and analysis of pollen found at the gravesites shows that the dead were buried with small, bright Alpine flowers.

Modern humans (*Homo sapiens sapiens*) appeared in Europe about 30,000 B.P. and in Borneo about 40,000 B.P. It was they who painted the breathtaking spectacles at Altamira and Lascaux and displaced the Neanderthals, either by interbreeding or by warfare. It was they who colonized the American continent and they who invented agriculture, with the resultant swelling of the earth with people. Today we are toolmakers par excellence; we are no longer limited to fire, stone, wood, and natural materials but have discovered nuclear energy, electromagnetism, and the world of chemicals. And yet we often act irrationally. Our tools have become a danger to us, and we may turn out to have the shortest chapter in the story of humanity.

The Future Evolution of Homo Sapiens

The stereotyped view of the human of the future conjures up the picture of Charlie Brown: a being with a small body, weak arms and legs, supporting an enormous head. This projection is in keeping with our view of ourselves as an intelligent species, driven by rational impulses and concerned primarily with affairs of the mind. Even the term *Homo sapiens* means "wise man." Unfortunately, such a view has little support in fact. The size of the human brain has not increased at least for the past 100,000 years—if anything, it has decreased slightly in cubic measurement. And the accounts of history are only rarely relieved by examples of Matthew Arnold's brand of sweetness and light.

This admission of human intellectual limitation must be linked to a second, very ominous admission: the human species has become dangerously overspecialized. Human overspecialization is not the typical kind encountered in the rest of the biological world: an insect feeding on a single kind of plant nectar, a flower depending on a single insect species for pollination, or a woodpecker building a nest in a single kind of tree in a single kind of habitat. Human overspecialization consists of the stunning overuse of tools. We travel with motor vehicles, communicate with electromagnetic radiation, and are sheltered inside controlled semitropical environments. And when it comes to warfare, we can put our enemies to death by the millions by blasting them, poisoning them, cooking them, irradiating them, drowning them, suffocating them, or laboriously filling their bodies with metal fragments. This grotesque overdevelopment of our military capacity now poses a real danger to our survival as a species.

In the past, evolutionary mechanisms operated in the human species in the same way they operated in all other species. The agencies of natural selection, genetic drift, mutation, and migration have shaped the human gene pool and continue to shape it. But the question of the future direction such processes may take becomes academic if no human species remains to evolve. It is unfortunately too late to wish we had never thought of the hydrogen bomb or the cobalt bomb. Tool use is a behavioral trait, and its potential for disaster can be remedied only by development of another behavioral trait: the great human potential for cooperativeness and social cohesion, on a worldwide scale.

If Charlie Brown is a poor model for our future appearance on physical grounds, he is as good a model as any on grounds of behavior. When Charlie Brown loses a baseball game, he does not throw a grenade after his opponents to demonstrate his undiminished determination to win.

Bibliography

Desmond Morris, *The Naked Ape,* Dell, 1967. Desmond Morris delights in speculation of a provocative, far-ranging sort. In hands less skilled than his, the results might be vaporous.

Konrad Z. Lorenz, "King Solomon's Ring," Thomas Y. Crowell, 1952. Lorenz is one of the towering figures in the study of animal behavior, and shares with Tinbergen and Von Frisch the only Nobel Prize awarded in this area of study. He nevertheless manages to communicate to the general reader his delight and enthusiasm for his studies. But perhaps communication with readers is the easier task, because Lorenz, like King Solomon, is able to communicate with animals. The sample excerpt deals with his pet jackdaw Jock.

GENERAL READING

When Jock reached maturity, he fell in love with our housemaid, who just then married and left our service. A few days later, Jock discovered her in the next village two miles away, and immediately moved into her cottage, returning only at night to his customary sleeping quarters. In the middle of June, when the mating season of jackdaws was over, he suddenly returned home to us and forthwith adopted one of the fourteen young jackdaws which I had reared that spring. Towards this protégé, Jock displayed exactly the same attitude as normal jackdaws show towards their young. The behaviour towards its offspring must, of necessity, be innate in any bird or animal, since its own young are the first with which it becomes acquainted. Did a jackdaw not respond to them with instinctively established, inherited reactions, it would not know how to take care of them and might even tear them to pieces and devour them, like any other living object of the same size.

I must now dispel in the reader an illusion which I myself harboured up to the time when Jock reached sexual maturity; the kind of advances which Jock made to our housemaid, slowly but surely divulged the fact that "he" was a female! She reacted to this young lady exactly as a normal female jackdaw would to her mate. In birds, even in parrots, of which the opposite is often maintained, there is no law of attraction of opposites, by which female animals are drawn towards men and males towards women. Another tame adult male jackdaw fell in love with me and treated me exactly as a female of his own kind. By the hour, this bird tried to make me creep into the nesting cavity of his choice, a few inches in width, and in just the same way a tame male house sparrow tried to entice me into my own waistcoat pocket. The male jackdaw became most importunate in that he continually wanted to feed me with what he considered the choicest delicacies. Remarkably enough, he recognized the human mouth in an anatomically correct way as the orifice of ingestion and he was overjoyed if I opened my lips to him, uttering at the same time an adequate begging note. This must be considered as an act of self-sacrifice on my part, since even I cannot pretend to like the taste of fine minced worm, generously mixed with jackdaw saliva. You will understand that I found it difficult to co-operate with the bird in this manner every few minutes! But if I did not, I had to guard my ears against him, otherwise, before I knew what was happening, the passage of one of these organs would be filled right up to the drum with warm worm pulp, for jackdaws, when feeding their female or their young, push the food mass, with the aid of their tongue, deep down into the partner's pharynx. However, this bird only made use of my ears when I refused him my mouth, on which the first attempt was always made.

Jane Goodall, *In the Shadow of Man,* Houghton Mifflin, 1971. Jane Goodall has studied the social behavior of chimpanzees in a way that few others have, before or since: she has spent eleven years living with them in the wild, getting to know each animal as an individual. After reading her account, no one should say with certainty that the line between human and nonhuman is an absolute one. This brief excerpt is about death.

Olly's new baby was four weeks old when he suddenly became ill. I had been excited when I heard of his birth: would his elder sister Gilka show the same fascination for him as Fifi had for Flint? And how would Olly react if she did? Though I was not at the Gombe when the baby was born, I was there a month later when one evening Olly walked slowly into camp supporting him with one hand. Each time she made a sudden movement, he uttered a loud squawk as though in pain, and he was gripping badly. First one hand or foot and then another slipped from Olly's hair and dangled down.

While Olly sat eating her bananas, Gilka groomed her mother. Often I had watched Gilka working her way closer to her small sibling's hands just as Fifi had done two years earlier when Flint was tiny. This time, however, Olly permitted her daughter to groom the baby's head and back without attempting, as she usually did, to push Gilka's hands away.

Next morning it was obvious that the baby was very ill. All his four limbs hung limply down and he screamed almost every time his mother took a step. When Olly sat down, carefully arranging his legs so as not to crush them, Gilka went over and sat close to her mother and stared at the infant; but she did not attempt to touch him then.

Olly ate two bananas and then set off along the valley, with Gilka and me following. Olly only moved a few yards at a time, and then, as though worried by the screams of her infant, sat down to cradle him close.

V. G. Dethier, *To Know a Fly,* Holden-Day, 1962. Dethier, like Lorenz and Goodall, is also a student of animal behavior, in this case the common house fly. However, his approach is diametrically opposed to that of Lorenz and Goodall, in that all of the answers are teased out of the reluctant fly by means of ingenious experiment. In the following excerpt, Dethier's study of feeding behavior required him to get his flies hungry. One of the methods used was "walking" the fly.

An even simpler way to fly a fly, but one which requires more courage, is that of walking him as one would walk a dog. As we walked our flies up and down the corridors of the laboratory, people stared and shied away. The fly was only a blur at the end of a thread that stuck out horizontally in apparent defiance of gravity. The real trouble came in attempting to turn the fly around at the end of the corridor for the return trip.

Edwin Way Teale, *Near Horizons,* Pyramid Books, 1966. Teale has studied the common insects of his own backyard, in this case a few acres of scrub and marsh on Long Island. A warm, beautiful book.

Niko Tinbergen, *Curious Naturalists,* Basic Books, 1958. An account of studies in the behavior of the black-headed gull, the digger wasp, the bumblebee, the eider duck, and many others.

Howard Ensign Evans, *Wasp Farm,* Doubleday Anchor Books, 1973. The lives of the parasitic wasps—the tiny assassins of spiders, grasshoppers, flies, even other wasps.

G. B. Schaller, *The Year of the Gorilla,* University of Chicago Press, 1964. Schaller has studied the behavior of the mountain gorilla. Like Goodall he has lived in the wild, following his quarry till they get used to his presence. The result is illuminating for human behavior as well as the behavior of the mountain gorilla.

E. O. Wilson, *Insect Societies,* Harvard University Press, 1971. Wide-ranging and scholarly, this is nevertheless a readable account of the societies of ants, bees, wasps, and termites.

TEXTBOOKS

R. D. Barnes, *Invertebrate Zoology,* 3rd ed. W. B. Saunders, 1974. A large text that takes a careful look at all of the animals without backbones. Contains many painstaking and beautiful drawings.

Martin Wells, *Lower Animals,* McGraw-Hill, 1968. This is a much slimmer volume than the above and deals with the invertebrates from a functional point of view, e.g., how animals walk, how they regulate salt concentration. The level is moderately advanced.

Part IV

Communities

Figure 12-1 A WALK THROUGH THE FOREST.

chapter 12

The earth has a thousand faces, and all of her faces have a poetry. Sometimes the poetry is dramatic: the mist-shrouded spires of the California redwood forest, the deep organ music of a storm over Cape Hatteras, the baroque overlayerings of winter ice in Ontario. More often, the poetry of the earth is workmanlike and understated. Small animals rush for shelter as one lifts up a rock. Life-and-death struggles unfold in the overarching weeds of a vacant lot. A sapling reaches toward light under the dark canopy of an ash–maple forest. In this chapter we shall look at a few such everyday aspects of the earth, with special emphasis on the factors that

Field Biology

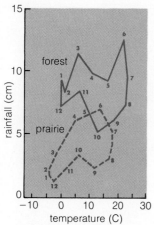

Figure 12-2

TWO CLIMATOGRAPHS: ANNUAL PLOTS OF TEMPERATURE AND RAINFALL. Top: Eastern deciduous forest. Bottom: Wyoming prairie. The numbers refer to months of the year.

428 / Communities: Field Biology

call forth the enormous differences we observe from one corner of the globe to the next. Or, to put it another way, why is the earth so various? Why are some forests redwood forests, while others are dominated by spruce and fir, by oak and hickory, by maple and beech, or by any of a hundred other species? And why are some areas of the earth not covered by forest at all?

The possible habitats for any living organism are limited by both physical and biological factors. By physical factors we mean such variables as climate, light intensity, nature of substrate (such as clay, sand, or solid rock), and the supply of water and minerals. By biological factors we mean the living organisms already present in a given habitat. For example, a chimpanzee, whose preferred habitat is a tropical forest, might do very well on the open plain if the plain did not contain the large cats and did contain more plants with edible fruits. In practice, physical and biological factors are so intertwined that it is hard to separate them.

No living species (including *Homo sapiens*) leads an independent existence. The fate of each species is inseparably linked to the destinies of others. When taken as large aggregates, such linked species constitute a *biome* ("living association"). On land areas, we speak of the tropical rain forest biome, the desert biome, the temperate grassland biome, the tundra biome, or several others. More specifically, a biome is a large association characterized on the basis of its dominant vegetation, because it is the vegetation that makes up the great bulk of living mass and that determines the kinds of animal life that can be present. The distribution of biomes is largely a function of temperature and rainfall, and each biome may be characterized by a specific climatograph—a plot of annual changes in temperature and rainfall (Fig. 12-2).

A biome is a large unit, often thousands of square miles in extent. For purposes of study and discussion, it is convenient to consider smaller living assemblages, such as a single pond or stream, or a single mountain slope covered with forest. One may even consider separately a single rotting log in a forest, because the living organisms inside differ sharply from those found outside. Such a smaller living assemblage is called a *community*. The boundaries of a community are always somewhat arbitrary (as are the boundaries of a biome). This is because no living assemblage is ever completely independent of its surroundings. Even ponds or marshes, which appear to have clear boundaries, receive an inflow of water and nutrients from the surrounding countryside and exchange migrating animals with their surroundings.

In spite of these difficulties of definition, one has no difficulty in recognizing when one has left a prairie or forest and is standing in a pond, or when one has left a high beach and is standing in a salt marsh. Biological communities have strong "personalities," and these personalities, like the personalities of the larger biomes, are shaped largely by external factors. The factors that determine the personality of a community are called the *limiting factors*—they are the external factors to which every member of a community must adapt. Often they represent the factor in least supply, such as water in a desert community or nitrate and dissolved oxygen in a peat bog. But the limiting factors may equally well be factors present in excess. Water plants of a salt marsh must adapt to a high salt content, organisms of a rocky shore must withstand the pounding surf, and bottom life in a fast stream must hold its own against the current. We shall now look at a few common kinds of communities, with a view to what makes each of them unique. In subsequent chapters we will try to see what makes them run.

The Soil

Life in the soil below forms as sharp a discontinuity with life above as does the life of the pond with its surroundings. Living soil, when packed into a covered aquarium, forms as lively a spectacle as would the same aquarium filled with fish from a coral reef. All that is needed for observation is patience and a good binocular microscope, because the wildlife of the soil is usually minute.

When we talk about the soil, we are generally talking about the upper 2 feet of the earth, and sometimes only about the upper few inches. This soil supports all the plant life growing on top, hence all animal life as well. Yet the soil itself is the product of living organisms. It takes more to create soil than the dissolving power of water melting down minerals or the fingers of frost cracking rocks. Soil scientists arrange soils into large groups, and it is interesting that these soil groups roughly parallel the boundaries of the large biomes of the earth. Prairie soil is different from the soil of a northern spruce forest, which in turn differs from the soil of an eastern deciduous forest. Soil has a stratified structure, and to see this one must take a sharp shovel and dig typically 3 or 4 feet down. One will expose a series of different-colored layers called soil *horizons*, as illustrated in Figure 12-3. The colors and thickness of the various horizons vary from place to place.

Figure 12-3
SOIL PROFILE OF A PRAIRIE SOIL (WYOMING). This is a fertile soil, easily tilled and well-suited for the raising of a variety of food crops. The parent material (C horizon) consists of water-worn gravel in a sandy matrix.

The uppermost or O horizon represents organic debris—partially decomposed leaves, branches, and grass stalks from the growing plants above. The A_1 horizon is dark in color and contains *humus*—finely divided organic material which has been worked into the earth by the activities of soil animals, especially the earthworms. Humus gives soil its water-retentive qualities and allows it to breathe. The A_1 and O layers combined are also called the *topsoil*. Humus, which originates from the remnants of plants above, is continuously destroyed through oxidation by fungi and soil bacteria. Therefore conditions that encourage the latter, such as heavy rainfall and high temperatures, make for a thin topsoil. The thinnest topsoils are found in the tropics, and the thickest ones in temperate prairies, where a modest rainfall is sufficient for luxuriant growth of grasses but not for fungal decomposition.

The bottom of the A_2 layer represents the average depth of penetration of rain water, and this rainwater leaches out minerals from the A_1 and A_2 horizons and deposits them in the B horizon. There is often a sharp visible boundary between these two layers. The bottommost layer is called the C horizon and represents relatively undegraded parent rock material.

The creation of topsoil is a very slow process, one that may call for centuries or millennia of work by soil organisms. Yet in its approximately 200 years of modern settlement, the United States has lost an average of one-third of its original topsoil through erosion resulting from poor farming practices. In essence this amounts to the mining of topsoil. Some Mediterranean countries such as Spain and Greece have lost essentially all their topsoil. These countries have kept their land under cultivation for thousands of years instead of for 2 centuries. This loss of the topsoil accounts for the red look of the Spanish landscape and its poor crop yields today.

Major Soil Animals

Organisms that live in the soil are all dependent for food on the plants above ground, and the great majority are found within the upper 1 or 2 inches. Here the soil texture is most porous and mobility of animals least restricted. The earthworm is the first citizen of the soil—it makes up 50 to 80 percent of the total biomass weight of living organisms, which is about 1,600 pounds per acre in a deciduous forest. On a weight basis, this is comparable to a herd of cattle grazing above ground. In terms of size, the earthworm is the largest of the soil animals, a veritable whale in relation to the other animals. Like whales, earthworms eat tiny organisms—the soil fungi that decompose organic material. Undigested material is excreted as casts, and in a deciduous forest the annual soil turnover may amount to

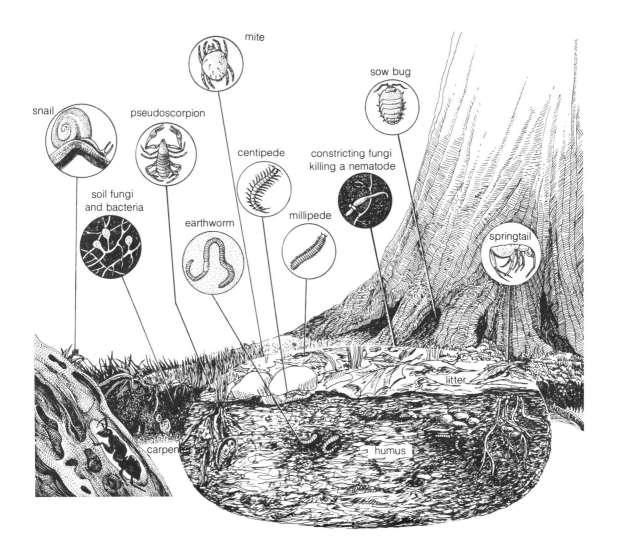

20 tons per acre. Earthworms thus play a key role in soil formation.

Millipedes are also among the largest soil organisms (Fig. 12-4). They eat the same things as earthworms but cannot burrow and are therefore restricted to the litter layer. They need moisture and are therefore scarce outside the forest.

Centipedes are comparable in size to millipedes. But while millipedes are retiring fungus feeders, the centipedes carry poisonous fangs and hunt other animals. They are the sharks of the soil community. Centipedes share their carnivore position with the ants of the soil. But ants feed by means of powerful crushing mandibles and are not restricted to a carnivorous existence as are centipedes. Ants are active and important burrowers, especially in prairie soils.

Figure 12-4
THE SOIL COMMUNITY. Only a small fraction of the soil organisms is shown here. It is these organisms that convert the inorganic matrix of soil minerals into a hospitable environment for life.

Moving gracefully on trails of slime are slugs and snails. They are unable to burrow and stay in the litter layer. But they are unusual in the animal world in their possession of cellulose-digesting enzymes. They can therefore feed on plant material directly instead of through a fungus intermediary.

Minor Soil Organisms

Much smaller and more numerous than the above animals are springtails, mites, and soil nematodes. Springtails are very primitive insects which are widespread from the Arctic to the tropics and have evolved a jumping device in their tails which allows these specks of living material to sail as much as 8 inches through the air. With mites (arthropods of the chelicerate line), their main diet consists of soil microorganisms: fungi, bacteria, and protozoa. Nematodes are far more primitive, wormlike organisms which move with a frantic, side-to-side wriggling motion. They are so numerous in the soil that a single spadeful of dirt may contain up to 10 million. They feed on almost any organic material available, including direct feeding on plant juices. Some plants, such as marigolds, fight back by releasing a nematode poison from their roots. Soil fungi also exist which prey on the nematodes in an almost animallike way. This involves strangling the unlucky nematode in a quickly tightening noose made of fungal hyphae (Fig. 12-4).

The bottom of the food pyramid (Chapter 14) is occupied by fungi, soil bacteria, and protozoa. These organisms feed largely on dead plant material from above and in turn form the food supply for larger soil organisms. Bacteria in particular are numerous beyond counting—a single gram of soil contains on the order of 1 billion of them. These organisms contribute to the bulk of the initial breakdown of organic materials from above. They are also responsible for that subtlest of aromas, the smell of freshly turned earth.

Feeding Relationships

In the end, the basic food resource in the soil community comes from dead plants and animals in the world above. Little of this is consumed directly by soil animals but is decomposed by soil bacteria, fungi, and protozoa. These then serve as food for the animals. Soil animals must all be adapted for movement through the soil. They do this by burrowing (earthworms), tunnel building (ants), or wriggling through tiny crevices (centipedes and millipedes—the many pairs of legs of these soil arthropods make for clumsy travel in the open but are ideally suited for the tight quarters of the soil litter).

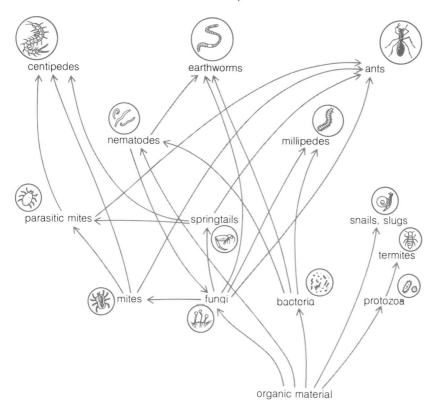

Figure 12-5
THE FOOD WEB OF THE SOIL (SIMPLIFIED).

The interaction of soil organisms is presented in simplified form in Figure 12-5. These tiny organisms, sometimes working over a time span of 10,000 years, have created a community below the ground which in some ways represents an ideal environment. There is a constant food supply, a relatively constant temperature, and a constant film of life-supporting moisture. Potential natural disasters are few. Heavy rains may deoxygenate the earth, or flooding may bury it under layers of silt, but almost all the truly destructive processes affecting the soils of the world are human-mediated. The defoliation and craterization of Indochina, the ongoing destruction of the Amazonian rain forest, and the plowing of the virgin grasslands of Kazakhstan are all examples of soil destruction that will be irreversible for many human lifetimes.

The Salt Marsh

The edge of the sea presents many aspects, and perhaps the most peaceful is that of the salt marsh. Every coastal state shares in part of the approximately 6 million acres of tidal salt marsh mapped in the

Figure 12–6 A SALT MARSH ON THE MAINE COAST. The grass in the foreground is *Spartina patens*. Taller grass closer to the water's edge is *Spartina alterniflora*.

United States. But these green ribbons are at their most luxuriant along the flat sandy beaches of the southern Atlantic states and the coastal states bordering the Gulf of Mexico. In some areas, these "salt meadows" may extend a half mile out from the shore. A salt marsh is dominated by one or two species of grasses belonging to the genus *Spartina,* or cord grass (so called because a crude kind of rope was once made from their internal fibers). *Spartina* grasses grow in the intertidal zone, where they are alternately inundated by salt water and exposed by the receding tides. They survive because they have successfully solved two problems of existence in this area that other plants have not been able to solve.

The first problem involves survival in salt water, which has a higher osmotic pressure than that of the cytoplasm of most plant cells. The result for most plant cells would therefore be loss of water and death. However, the cytoplasm of *Spartina* contains more salt than ocean water and therefore has a higher osmotic pressure. It therefore does not lose water when growing in ocean water. The high internal osmotic pressure is achieved by pumping in salts, a process that requires cellular work. *Spartina* actually grows better in fresh water, but there it is outcompeted by other species.

The second problem faced by *Spartina* is oxygen supply for its roots. The mud at the bottom of a marsh is totally anaerobic, because of the heavy demand on oxygen supply made by bacteria living in the mud. The soil of a forest or prairie is porous, and oxygen from the atmosphere can diffuse down to the deepest roots. But salt marsh muck forms a tight seal to atmospheric oxygen and none can diffuse downward below the uppermost inch or so. *Spartina* avoids this difficulty by piping down its own oxygen supply through hollow stems and underground rhizomes. (A rhizome is an underground stem.) *Spartina* roots are so well supplied with oxygen that the soil immediately around them turns red instead of the normal black, indicating the oxidation of iron to its highest oxidation state.

Because of its successful adaptation to the physical environment of the salt marsh, *Spartina* rules the marsh. Stands of these grasses are so uniform that they resemble cultivated fields, and other plants are fellow travelers living on the margins of the marsh. Most of them are found on the landward side, where they tolerate the salt spray but are not faced with regular saltwater inundation. They include seaside goldenrod, sea lavender, marsh elder, and gama grass (a stout grass related to corn). Still further away are found American holly, bayberry, Eastern red cedar, and beach plum. All these were plants familiar to the first European colonists on this continent, who typically settled close to the shore.

A salt marsh may contain isolated low spots which are poorly flushed by tidal action. Evaporation can build up the salt concentration to levels that even the *Spartina* cannot tolerate. *Salicornia* or glasswort is common here. This is a low-lying succulent plant related to beets and spinach. It is edible either raw or cooked and was highly prized by the colonists as a substitute for the garden pickle.

At least for the salt marshes of the north, the testing time comes in the winter. At this time the tides scour the surface with heavy jagged splinters of ice. At the end of the winter, all free-standing vegetation has been ground away and carried off in the arms of the tides. But the *Spartina* survives by going underground. Their strong rhizomes (underground stems) remain alive and bind together the nutrient-filled substrate. Small sections at the edges of the marsh may be torn loose and then carried by the waters to distant coasts. Here *Spartina* puts down roots and begins the task of colonizing a fresh coastline.

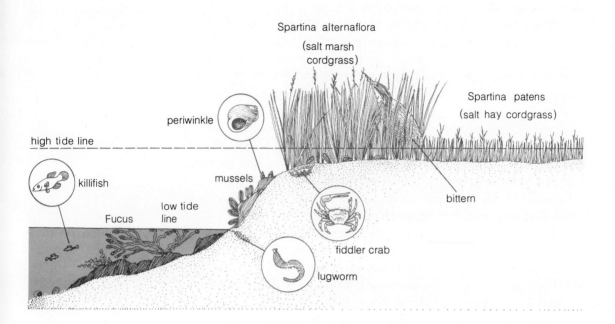

Salt Marsh Animals

The animals of a salt marsh face the same difficulties as the plants: high and changeable osmotic pressure, tidal inundation and exposure, and anoxic conditions in the mud. Most salt marsh animals burrow in the mud. While this does not solve the anoxia problem, the mud acts as an insulating blanket which buffers the drastic changes in water level, osmotic pressure, and temperature characteristic of the salt marsh.

Most salt marsh animals can alter their internal salt concentration as a means of coping with problems resulting from osmotic pressure differences. One example is offered by the lugworm, a close relative and look-alike of the earthworm, which builds U-shaped burrows in the mud and leads a carnivorous existence. If a lugworm is taken from its burrow and dropped into a glass of fresh water, it balloons out like a piece of pork sausage. This happens because water enters the worm as a result of its higher internal osmotic pressure. But then the water in the worm slowly begins to leave and, as the worm actively pumps out intracellular salts and reduces its internal osmotic pressure, it deflates back toward normal size. In nature, such a course of events

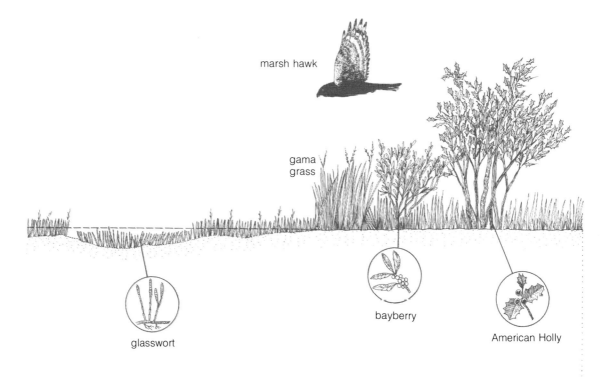

Figure 12-7 COMPOSITE DRAWING OF A SALT MARSH.

occurs after a rain or freshwater flooding. The converse experiment involves taking a normal lugworm and dropping it into a very concentrated salt solution. The worm crinkles up and collapses inward as it loses water to the external environment. But again, there is a subsequent slow return to normal size as the lugworm raises its internal osmotic pressure by pumping in extra salt. In the salt marsh world, such osmotic pressure changes are common enough that adaptive mechanisms are essential. Because the lugworm burrows in the mud, any external changes in tonicity occur slowly enough to allow the animal to remain in constant equilibrium with its environment.

Even animals that are invaders from the mainland, such as salt marsh mice, have made adaptations to life in the salt marsh. Salt marsh mice can drink salt water because they can get rid of the excess salt by excreting extremely salty urine.

Another problem common to a salt marsh is that of periodic exposure at low tide and a consequent danger of drying out. Barnacles and ribbed mussels close their shells until the next high tide. During this time, they must live without oxygen. Periwinkles are little snails

Figure 12-8 AN AMERICAN BITTERN IN A MARSH. Though fairly common in salt marshes, the bittern is more typical of fresh-water marshes. Its thin, reed-like body is built for camouflage among the vegetation.

which crawl over the surface of the marsh. At low tide they withdraw inside their shells and close a hard cover called an operculum. Thus locked up, the animal may survive without harm for as long as 6 months. Periwinkles are edible snails, and seafood stores can keep live periwinkles in open bushel baskets for months at a time, without any special measures of preservation.

A further problem that must be overcome by all burrowing salt marsh organisms is the low oxygen tension in the mud. The mud contains soil mixed with decaying organic matter, largely derived from *Spartina* above ground. Bacteria feeding on this organic material consume all available oxygen, leaving none for the animals in the mud. During high tide, mud animals such as the fiddler crab, sandworm, ribbed mussel, and soft-shelled clam circulate oxygen-rich tidal water through their burrows. But during low tide, they tolerate the anoxic conditions until the next high tide.

Perhaps the two most characteristic animals of the salt marsh are the ribbed mussel and the fiddler crab. The ribbed mussel, though edible, is less tasty than the blue mussel of rocky coasts. Ribbed mussels live embedded in the mud just below the surface. They live in enormous aggregations, anchored to each other by means of strong proteinaceous threads secreted by each mussel. During high tide, a mussel sets up a current of water which flows over its gills. From this current it filters microscopic scraps of food which are swallowed and digested (hence the mussel is called a *filter feeder*). Inedible materials are glued together by mucus and excreted as a sticky string called *pseudofeces*. Mussels are so numerous and filter so much water that they can literally bury themselves in their own pseudofeces. They must slowly move upward to stay alive.

Tracer experiments revealed a wholly unexpected contribution of the mussel to the welfare of the salt marsh. Phosphate is an essential nutrient necessary for the growth of *Spartina* grasses. As the *Spartina* dies and is decomposed by bacteria, its phosphate content is released into the water (locked up in fine organic sediment). This phosphate would be carried out to the open ocean by tides and lost to the marsh if it were not for the filter action of the mussels. They retrieve the phosphate-rich sediment, lock it up in their pseudofeces and thus return it to the marsh. The filtering of the mussels is so efficient that all the particle bound phosphate in the water is cycled through these animals in only 2½ days. This illustrates that an organism without any intrinsic commercial value may be of enormous value to the natural community of which it is a part.

The fiddler crab is likewise well integrated into salt marsh life. This little crab is hardly larger than a thumbnail. They are discovered most easily by going to a marsh at low tide and using a garden fork to dig up a small piece of *Spartina* mat. At least a few fiddler crabs will be found hiding inside in their tiny burrows. The male fiddler has one claw which is disproportionately larger than the other. This large claw is waved to any passing female who may find it enticing enough to visit the male in his burrow. Copulation follows, and the female carries the fertilized eggs on her abdomen until they are ready to hatch. At this time she deposits them at the water's edge. The young larvae are then carried by currents to a suitable landfall.

Fiddlers are active at low tide, feeding on bits of dead animal and plant life, as well as on small living animals. Most crabs breathe by means of gills, which restricts them to the water. But the fiddler has also developed a primitive lung under the shell and can survive in the air as long as it remains moist. At the same time it can remain in its burrow and resist anoxia for long periods of time. The fiddler is similarly flexible in regard to external salt concentrations. It can adjust its internal osmotic pressure both upward and downward to cope with either a briny pool or a heavy rain. This little animal, which can live almost anywhere and eat almost anything while leading an exciting love life, is perfectly suited to the changeable, colorful life of the salt marsh.

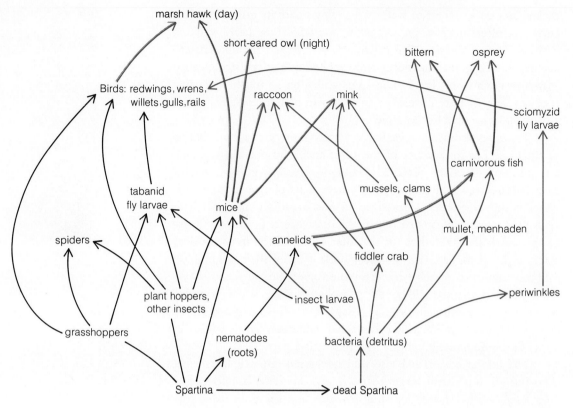

Figure 12-9 FOOD WEB IN A SALT MARSH (SIMPLIFIED). Feeding relationships based on living *Spartina* constitute the so-called grazing chain and are shown in black arrows. Feeding relationships based on decomposed *Spartina* constitute the detritus chain and are shown in colored arrows. The great bulk of food consumed in the salt marsh moves through the detritus chain.

Feeding Relationships

The feeding relationships of a salt marsh are too complex to be given in more than simplified form (Fig. 12-9). Superficially, such a diagram appears to show *Spartina* going in, and marsh hawks, bitterns, and ospreys coming out. What the diagram cannot show is the important link between salt marshes and the oceans behind them. Fully two-thirds of all commercially important ocean fish spend their larval stages in the waters of a salt marsh. They benefit from the lowered water salinity and the abundance of food and shelter. A salt marsh ranks among the most productive communities in the world, and the open ocean among the very least.

Salt marshes everywhere are today under unprecedented assault. A marsh can be filled in a matter of weeks by sand dredged from an adjacent channel. (It takes a little longer if the fill is garbage.) The site then becomes a mass grave for every living thing that called it home. The "Bayview Estates" or "Oceanside Gardens" that then rises on top of the old marsh is always a sad echo of the vibrant life it has buried.

The Freshwater Pond

In 1883, at the age of 43, the great French painter Claude Monet settled in the village of Giverny, about 50 miles from Paris. His property contained a pond shaded by huge willows and overgrown with water lilies. It was by the side of this pond that Monet was to spend most of the final 26 years of his life, painting an extraordinary series of water scenes which represents the culmination of his art. These pictures form a universe, a universe of eternal summer, light, and life.

Figure 12-10 EDGE OF A POND, WITH WATER LILIES AND RUSHES.

442 / Communities: Field Biology

It is significant that Monet sought this universe in a small pond. Along with coral reefs and salt marshes, shallow ponds are the world's most productive communities, and the ones most bursting with life.

But a pond is also nature's textbook. The distribution of life in a pond follows a beautiful logic which makes the fullest use of every segment of the pond. Strangely made plants and animals may be found in predictable locations within the pond. Such stratification of habitat is also observable in a salt marsh. But, though a salt marsh and a freshwater pond are both aquatic communities, they share almost no species in common. The additional presence of salt and tidal movements in the marsh have produced a set of limiting conditions which can be met only by a handful of organisms in the pond, and we therefore meet a striking new cast of characters.

The Deep Pond

Ponds are generally defined as bodies of water shallow enough to allow surface plants to reach the bottom with their roots. In local areas that may be too deep to permit this, the plants are restricted to free-floating types. One of the most remarkable of these is the bladderwort, a delicate, vinelike plant distributed throughout ponds rich in small animal life. The plant carries thousands of small pouches or "bladders" which open when small sensory bristles in the vestibule of the bladder are stimulated by a tiny animal. As the bladder opens, it sucks in a gulp of water, carrying the animal with it. The bladder closes and, over the course of the next 20 minutes, the animal is digested and absorbed, its remains expelled, and the bladder readied for the next victim. Suitable prey include rotifers, protozoa, small insect larvae, and crustaceans. When a light-green, starved bladderwort is placed in an aquarium containing these animals, it quickly turns dark as its bladders become engorged with prey. A single plant may hold an estimated 150,000 small animals in its traps at one time.

Another common floating plant (especially in well-fertilized waters) is duckweed, the tiniest of all flowering plants. Their roots trail in the water as rudders, not as organs of absorption, and the upper plant consists of an unbroken, leaflike body which is continually budding off new duckweed organisms.

Other photosynthesizers of the deeper pond include floating freshwater algae called *phytoplankton* ("plant swimmers"). The most common are the unicellular diatoms and desmids. Phytoplankton face the problem of remaining afloat—they generally lack organs of self-propulsion. If they sink too deep in the water, the light intensity becomes too low to permit photosynthesis. Both desmids and diatoms have elaborately sculptured, flattened cell walls which retard sinking.

Figure 12-11
A BLADDERWORT PLANT PREYING ON SMALL ANIMALS IN A POND. Inset shows a bladder with trapped *Daphnia* (a small crustacean) inside. A rotifer is using the bladder as a platform to fish for bacteria and single-celled algae.

Some species carry internal droplets of oil which reduce density. But perhaps eddy currents are the factor most important in keeping these organisms afloat. Browsing on phytoplankton are small animals—rotifers and small crustaceans such as *Daphnia* and *Cyclops*. Rotifers ("wheel bearers") carry paired circlets of cilia which under the microscope look like tiny wheels. They set up water currents which sweep the phytoplankton into the mouth, where a special grinding organ crushes them. *Daphnia* and *Cyclops* are both crustacean arthropods and swim through the water in jerky motion by means of their antennae which they use as oars. They can just barely be seen with the naked eye and are so abundant in rich pond water that the water seems to be undergoing a diffuse pulsation. They are unspecialized in their food preferences and swallow anything from phytoplankton to protozoa to decaying plant material. *Daphnia* and *Cyclops* are so abundant in ponds that they form the chief diet of young fish.

The Shallow Pond

Closer to the shore, plants such as pondweeds and water lilies anchor themselves to the bottom by means of roots. They may grow in water up to 8 feet deep and are kept erect by means of air chambers in the stems and leaves. Adult leaves of the water lily float on the surface. The upper surface of the leaf bears the stomata and is heavily waxed to prevent them from clogging.

Pondweeds and water lilies have important links to the animal life of the pond. Pondweeds are a favorite food of wild ducks and are often deliberately planted in order to encourage bird life. The stems and leaves of both plants are crawling with tiny animals. Freshwater *Hydra* hang upside down from the leaves, fishing for *Daphnia* and *Cyclops*. Filamentous green algae such as *Spirogyra* use them as anchoring points. Snails browse on the algae mat. Water insects lay their eggs on the stems and leaves. Planarians, red water mites, and bizarre protozoa such as *Stentor* and *Vorticella* hunt for unwary water animals and for each other. Down among the stems glide sunfish, flattened like vertical saucers for easy maneuvering among the tangled vegetation.

The Edge of the Pond

In the shallow water close to the shore, the plants no longer remain satisfied with merely reaching the surface. Instead they pierce the surface and then continue growing upward toward the sun. These plants form the zone of emergent vegetation, and their stems are stout and stiff, unlike the long ropelike stems of the waterlilies.

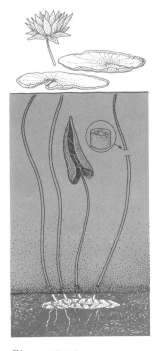

Figure 12-12
A WATER LILY STEM, SHOWING AIR CHAMBERS THAT ADD TO BUOYANCY AND FACILITATE OXYGEN TRANSPORT TO ROOTS.

Figure 12-13
EDGE OF A SMALL POND WITH CATTAILS (BACKGROUND) AND WATER LILIES.

The most common of the emergents are the cattails. Tall and striking, they form a thick jungle which is relieved by patches of bullrushes, reeds, pickerel weeds, arrowheads, and occasional woody species such as water willow and swamp loosestrife.

The air is filled with the flash of dragonflies and the sharp calls of red-winged blackbirds, which feed on the dragonflies and on other insects. Down in the water are bullfrogs and green frogs, waiting in ambush for a flying insect. But bullfrogs are adventurous feeders and have been known to swallow crayfish, other frogs, and even an unwary farm chicken that wanders too close to the shore.

Pond Insects

The real stars of the shallow pond are the water insects. They find sufficient shelter here to escape from their enemies, and an endless food supply and places to lay their eggs. The evolutionary thrust of insect development has been to adapt to a land existence. Insects have become air-breathing, internally fertilizing, cuticle-waterproofed, wing- and leg-propelled animals of air and land. The insects of the pond have in many cases had to circumvent these hard-won adaptations in returning to an aquatic existence.

Backswimmers, diving beetles, and water boatmen are all diving insects that carry a supply of air with them. Backswimmers swim on their backs while on the surface. Below the water they swim right side

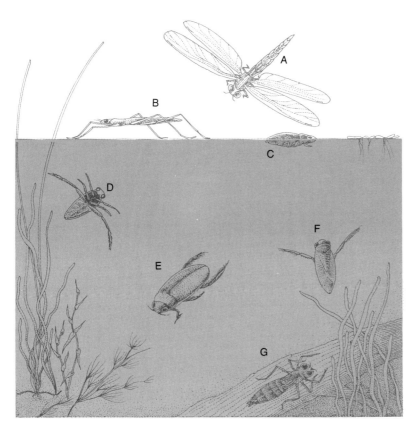

Figure 12-14
COMMON POND INSECTS.
A. dragonfly; B. water strider; C. whirligig beetle; D. backswimmer; E. diving beetle; F. water boatman; G. dragonfly nymph.

up. Their backs are keeled like a boat and are countershaded a light blue so the insect is less visible from below. The backswimmer is a carnivore, feeding on everything in its size range: tadpoles, fish fry, and other insects. It can give a bite as sharp as a bee's sting if picked up by hand. Diving beetles share the food habits of the backswimmers, but water boatmen have gentler tastes and feed on decaying organisms, plankton, and filamentous algae.

All three insects have bristled swimming legs which serve as underwater oars. And all three have evolved a scuba-tank mechanism for underwater breathing. When these insects begin their dive, they descend into the depths wrapped in a silvery bubble of air. This bubble supplies them with oxygen and absorbs the carbon dioxide exhaled from their spiracles. Since carbon dioxide is quite soluble in water, it quickly diffuses out of the bubble. The remaining air is initially 80 percent nitrogen and 20 percent oxygen. As the oxygen is used up, the bubble remains almost undiminished in size, because of its content of water-insoluble nitrogen gas. Once the concentration of oxygen in the bubble falls below that dissolved in the pond water, oxygen from the water begins to diffuse across the air-water boundary

Figure 12-15
A WATER STRIDER ON THE SURFACE OF A POND.

into the nitrogen bubble almost as fast as it is depleted by the diving insect. The surface area of the bubble is maintained intact by its content of water-insoluble nitrogen, and this surface area is large enough to allow for an adequate rate of oxygen influx. Thus these air-breathing insects can remain under water for 20 minutes or longer. In fact, they are so superbly adapted to water life that few are ever found outside a pond. Only diving beetles will spread their wings and rise into the air, especially in response to light at night. But a backswimmer would be as improbable on a prairie as would a water lily or a duckweed.

Other pond insects make their peace with water in different ways. The water strider and the whirligig beetle remain on the surface. The feet of a water strider are tipped with velvety hairs which allow it to walk on the surface without falling in. The surface tension of the water is high enough to resist penetration by the legs of this velvet-socked insect. The water strider is a carnivore, feeding on emerging midges and mosquitoes, and on insects that fall onto the water from above. Even the ferocious backswimmer is seized and eaten.

The whirligig beetle swims like a boat with a locked rudder, endlessly circling by means of short hind paddling legs. Though the beetles are highly visible on the surface, they are almost impossible to catch by hand. This is because the ripples generated by the circling beetle serve as a kind of surface sonar to warn the insect of approaching danger. The beetle must keep circling to maintain a continuous train of ripples, like a kind of water bat. The beetle has two pairs of eyes: one pair just above the water line, for vision over the surface, and the other pair just below, for vision underwater.

Still other insects spend only their larval stages in the water. These include the mosquitoes, mayflies, dragonflies, and damselflies. In addition to evolving the adaptations necessary for life under water, these insects have all evolved mechanisms for leaving the water as adults and taking to the air. Before they die, adult females must once more return to the water to lay their eggs and propagate their species.

THE FRESHWATER POND / 447

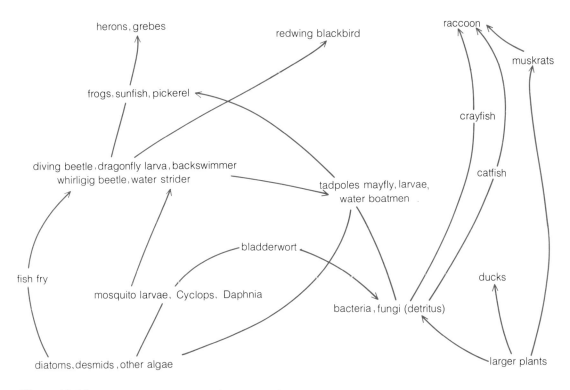

Figure 12-16 FOOD WEB OF A POND (SIMPLIFIED).

Pond Life and Pond Stability

The amount of larger animal life in a pond depends on the size and stability of the pond. Even small ponds that dry up in the summer contain frogs, salamanders, crayfish, and perhaps red-winged blackbirds and swamp sparrows. But only larger and more stable ponds can support muskrats, sunfish, pickerels, catfish, ducks, herons, and grebes. Herons and grebes (aided by an occasional raccoon) make up the top carnivores. Their hunting techniques are very dissimilar. The grebe dives after its prey (frogs or fish) and builds a floating nest on the water. Herons are stalkers. One may see them in the early morning, wading so slowly in the water that their long legs do not create a ripple. When a fish comes within range, the heron makes a lightning stab with its beak and then swallows the fish whole. The raccoon is a nocturnal animal and a hunter of opportunity. Frogs, crayfish, muskrats, and bird eggs are all welcome, as are berries and fruits. The raccoon is attracted to a pond because of its richness of animal life, not because of any intrinsic specialization on its own part. The feeding relationships of a pond are summarized in Figure 12-16.

It is perhaps a small bright spot in the regional ecological picture that the number of small ponds has been growing in the United States and Canada. While many have been filled in for residential development, even more have been deliberately created (with government subsidy) by the damming of springs and little streams. Most have been created in farming areas as an aid to water supply. Wherever they have been built or preserved, they have added greatly to the richness and diversity of local life.

The Forest

A pond in Ontario and a pond in Texas are largely similar in biological structure and species composition. But a forest in Ontario and a forest in Texas may have hardly any species in common. This is because the physical environment of water (especially deeper water) shows much smaller climatic variation than does the physical environment of air. Besides being highly variable, forests are exceedingly complex communities. A forest is in fact the most complex land community known and exists naturally wherever cold or dryness don't restrict it. Instead of surveying all the known forest types, I shall discuss here one forest with which I have been well acquainted, pointing out both its general and specific characteristics. This is a forest in the northern Catskills of New York State.

Figure 12-17 A MAPLE-ASH FOREST. The stone wall in the foreground indicates that this forest has reclaimed land once cleared for farming.

The Catskills forest forms part of the great *eastern deciduous forest* that once covered the country from New England to the midwestern prairie, and from Vermont in the North to Tennessee in the South. The forest fell to the axes of settlers in the eighteenth and nineteenth centuries, but has been making a comeback as farming has shifted to the richer soils further west. It is a common experience to walk through a New England forest and come upon a stone wall, sometimes 6 feet high, running between old trees in an uncompromising straight line. These forests were once plowed fields and meadows, and the large field stones turned up by the plow were laboriously piled up in rectangulating walls. Today the forest covers close to 50 percent of the land area of New York State, somewhat more than the national average.

It is remarkable that, after undergoing such an upheaval, the Catskills forest of today still closely resembles the forest the settlers found there 200 years ago. While some tree species (sugar maple) have increased in importance and others (hemlock) have decreased, none has entirely disappeared[1] and none are of recent origin. None of the exotic imports that thrive on city streets, such as the gingko or the tree of heaven, have been able to establish themselves among the native vegetation outside cities. This suggests that the assemblage of plant species found in the original forest was present there not by accident but as a result of a fine adaptation to locally limiting conditions.

The Forest and the Seasons

One of the important external factors to which the eastern forest has had to adapt is the drastic seasonal fluctuation in the North American climate. In Europe, fall is the "season of mists and mellow fruitfullness" as it steals gradually over the landscape. The trees only gradually lose their greens to duller browns and yellows and then equally gradually turn bare for the winter. In contrast, the great North American land mass lacks the climatic buffering of the Gulf Stream. The dramatic seasonal changes of this land mass are matched by equally dramatic responses by its plant associations. Perhaps the most spectacular of these plant responses is the mid-October color change of the trees. As the Canada geese pass high overhead, the hills and valleys below ignite into molten reds, oranges, and yellows. This autumn color change is so spectacular that early European explorers and naturalists wrote about it with wonder in their reports home.

---

1. The only exception is the American chestnut, wiped out by the chestnut blight. The American elm, attacked by Dutch elm disease, may be heading for a similar fate.

Figure 12-18 THE SEASONS IN THE FOREST. Left page, Winter: fresh snow on trees, raccoon tracks and snowshoe, frozen apples still clinging to an apple tree; Spring: melting snow, disoriented woodchuck on first day after emerging from hibernation, swollen buds on maple. Right page, Summer: white-tailed deer—doe with fawns, the forest darkens as trees leaf out; Fall: chipmunk on a log, autumn leaves in the rain.

It is this fall color spectacle and leaf drop that gives the deciduous forest its name. The term is derived from the Latin *decidere* ("to fall off"). The biome to the north, stretching from Maine across southern Canada and into the Rockies is called the northern *boreal forest* (from the Greek for "north wind"). The boreal forest is a coniferous forest, and the leaves of the trees are modified into needles. The leaves of trees in a deciduous forest are broad and distinctive, and though they may stay on a tree only from May to October, they can carry out more photosynthesis in those six months than can a comparable evergreen tree in twelve months. The deciduous leaf must be sacrificed in the autumn because of its wasteful habits with water—its water loss is so great that it cannot be supported under winter conditions.

Fall. The October leaf fall in New York State is a mighty fall. The great bulk comes down within a 2-week period, and at the end only the oaks are left clinging to their withered leaves. As one walks through the bare forest, the leaves crackle underfoot, forming a rainbow-colored blanket 2 inches deep in most parts, and up to several feet deep where they have been blown against a stone wall or a fallen tree. This warm blanket will serve both as food source and as an extra layer of insulation for the small animals of the soil preparing for the bitter winter ahead. If one stands still and listens, above the gentle sigh of the wind one soon hears the rustlings of a chipmunk, searching for fallen acorns among the leaves. It stuffs them inside its bulging cheeks and carries them down to an underground burrow built below the frost line. The chipmunk will spend most of the winter there in a state of hibernation, its body temperature poised just a few degrees above the freezing point of water. Its heart rate will drop from 300 beats per minute to 8 beats per minute, and its metabolic rate will be reduced anywhere from 20 to 100 times (Fig. 12-18).

Quiet and invisible under the soil, but still active, are other small mammals. Like the chipmunk, the star-nosed mole spends the winter in hibernation. The short-tailed shrew may weigh less than half an ounce. Its small size means that it loses relatively enormous amounts of body heat, and this must be made good by continuous feeding. Shrews are therefore the most voracious mammals known; on an ounce-for-ounce basis, they eat more than their own weight of food per day. They feed on insects, worms, snails, and even mice larger than themselves. Their small body size makes hibernation impossible, and shrews remain active all winter.

At the other end of the size spectrum lies the black bear. It remains today as the only large carnivore in the area and is restricted to rugged country and large tracts of forest. The early settlers recorded encounters with mountain lions and wolves, but today the mountain lion is extinct all over the East, and relentless hunting is threatening it in the West. The wolf was exterminated even earlier, a victim of its own gregarious habits and unremitting human hostility.

One result of the loss of these predators has been an increase in the population of white-tailed deer. The signs of deer are everywhere: narrow, well-worn paths marking their invariant routes of travel, browsed branches on young trees, footprints in the mud, the matted grass of bedding-down areas, and droppings wherever they have traveled.

Deer are most abundant in young forests. They browse on tree buds and growing branch tips, and these are seldom within reach in older forests. Along with field mice, they are important in retarding the reforestation of open areas.

Winter. The Catskills forest changes dramatically from fall to winter. At the end of October, the fallen leaves sigh and rustle with every step. At the end of November, all sound has been deadened under a thick blanket of snow. Now one can walk only with snowshoes over a surface that smoothes and hides boulders, fallen trees, and even stone walls. The same surface preserves a clear record of all resident animal life. Here are the long, hopping prints of a rabbit. The fine hatchings of a deer mouse disappear into a tunnel in the snow. A group of deer has passed in single file. Here a flock of black-capped chickadees and sparrows scrambled around for bread crumbs I have dropped for their supper. More often than not, a squirrel or a blue jay has scattered the small birds. And if one walks far enough and long enough, one can find the rarities: the fat pawprints of the bobcat, the raccoon's delicate imprint which looks like a tiny human hand, and the footprints of the gray fox which resemble those of a small dog. These little hunters are hard pressed now, and the fox and the raccoon stretch their diets by digging out frozen apples around the many wild apple trees that have escaped from the orchard at the bottom of the hill. Much of the abundant game of the summer is hibernating in burrows deep below the snow—these include woodchucks, chipmunks, and squirrels. There is no visible insect life, as these too are overwintering in the egg or pupa stage. The insect-eating birds have migrated south: robins, flycatchers, woodcocks, and orioles. Their abandoned nests dot the trees, clearly visible now, while in the summer one could stand right next to a nest and not see it through the leaves. The winter is not silent in the great forest, but there is no animal sound. There is only the whistling of the wind, the crunch of one's footsteps in the snow, and an occasional crack as a branch breaks under the snow.

Spring. But as the constellations slowly turn and the year begins to point toward spring, a new sound is heard—the quiet murmuring of water running downhill under the snow. By mid-April there are large open patches in the snow cover, and animal life begins to come back. Chipmunks and woodchucks emerge from hibernation and start to forage. Woodchucks in particular act deranged at first, wandering around in broad daylight with little concern for man. The birds are also coming back. The Canada geese again pass overhead, heading north. Local birds return and find mates for the breeding season. Orioles, juncos, tanagers, cowbirds, and flycatchers all make their appearance and make the mornings musical with song. But the greatest singer of them all is the robin, whose melodies provide the most sensuous awakening on a spring morning.

Hard behind the animals are pushing the plants. With the leaves of canopy trees still sheathed inside buds, the ground flowers rush into bloom. Trout lily, spring beauty, trillium, common blue violet, Solomon's seal, wild lily of the valley, jack in-the-pulpit, blue cohosh, and wood betony cover the ground. The forest trees are also flowering, and sunlight coming through the branches feels a little duskier as the tiny flowers put a soft down on the branches. In mid-May, just before the maple leaves erupt, there occurs one of the most magical events in the life of the forest: the flowering of the shadbush. This little tree of the rose family is so thin and inconspicuous that one hardly notices it during most of the year. But for a few days in May, all the shadbush of the forest are covered by starry, pale flowers which gleam through the woods and send out the merest suggestion of fragrance. Almost as soon as they have come, the petals flutter to the ground. The forest now rapidly grows dark as the leaf buds on the trees burst open and young leaves on every branch screen out the sun. The forest has entered its summer stage.

These dramatic changes in plant and animal life from summer to winter and back to summer again represent an essential adaptive feature of members of the deciduous forest community. Organisms that lack these abilities to adapt to the seasons are excluded from the community. The cycle of fall leaf drop, migration, and hibernation to spring snowmelt, flowering, and nest building is also perhaps the major source of excitement to the outside observer. The eastern deciduous forest is a community with a very strong sense of calendar time.

Summer. The forest of summer is best described as a working forest. Trees are busy elongating branches, adding to trunk girth, and setting fruit. Animals are busy raising their young. There is an overabundance of plant food, leaving an excess for insect, bird, and mammal. And the excess produced in the summer must be sufficient to feed all forest life during the rest of the year.

Figure 12-19 THE LAYERING OF LEAVES IN THE FOREST. (A) Leaves of second-story vegetation (witch hazel). (B) Leaves of the herb layer (wild lily of the valley, gold thread, and sessile bellwort).

The Structure of the Forest

Summer is a good time to observe the structural relationships within a forest. The working part of a plant is its photosynthetic surface, typically restricted to the leaf. Given an adequate water and mineral supply, the amount of leaf surface in a forest determines its rate of food production. The amount of leaf surface deployed is expressed as a multiple of the ground area covered and is termed the *leaf area index*. A typical, mature deciduous forest has a leaf area index of 8 to 10, which means that an acre of ground is covered by 8 to 10 acres of leaves. Of course, these leaves are stacked on top of each other.

The stacking of photosynthetic surfaces is reflected in a stratification of the plants of the forest. Rising highest above all the others are *canopy trees*—in this case sugar maples, silver maples, white ashes, beeches, red oaks, and a scattering of birches, basswoods, poplars, and hickories. Canopy trees reach more than 100 feet into the air, their crowns bathed in bright sunlight and their trunks reaching into deep shade. Their uppermost leaves alone in the forest receive a full dose of sunlight and photosynthesize at peak efficiency. Canopy trees are formidable competitors. Seedlings are able to germinate and grow in the deeep shade created by their own parents. For many years afterward the little trees must struggle to survive in the deep shade. Such saplings have their chance when an adjoining canopy tree dies,

Figure 12-20 STRATIFICATION OF THE FOREST. The leaves in each stratum are positioned so as to absorb the maximal amount of incident light.

creating an open shaft of sunlight. The saplings compete with each other in a race for the top, with essentially all growth channeled into upward extension. During this phase, a sapling may be 30 feet tall but only 3 inches in diameter at breast height. The penalty for losing the race is death. Only a rare canopy-tree sapling survives for more than a dozen years in the shade of the maple forest. While enough are always present to take advantage of opportunity, a mature forest is relatively open and easy to travel through.

However, a few species have become specialized for permanent life in the shade of canopy trees. The shadbush is one. Others in the area include striped maple, mountain maple, maple-leaf viburnum, and hobblebush. These collectively make up the *second-story vegetation*. All except the maples are insect-pollinated species. Wind speed below the canopy level is greatly reduced, and second-story trees are sufficiently scattered to make the wind an unreliable agent of pollination.

At the very bottom, hugging the forest floor, lies the *herb layer*. Here are found wildflowers, ferns, and occasional patches of moss. Conditions at ground level are drastically different from those up in the canopy. Humidity is close to 100 percent, light intensity is less than 10 percent of that at the top, wind speed is close to zero, and temperature fluctuations are much reduced. To overcome the light deficit, herbs generally follow a staggered growing season; they leaf out and flower before the canopy trees do in the spring. Mosses remain evergreen the year around, as does the Christmas fern. But as the summer progresses, most of the flowering herbs anticipate October and set fruit or provision their underground storage organs long before the first yellows and reds are seen in the canopy. The time when the herb layer is at its richest is in the spring.

The stratification of plant life in the forest is accompanied by a comparable stratification of animal life. This is perhaps most easily seen among the birds. The ovenbird and the hermit thrush are both ground-colored birds which forage and build their nests on the ground. In contrast, the bright scarlet tanager and the yellow-throated vireo are likely to be found in the canopy layer. And somewhere in between may be found birds such as the least flycatcher and the wood peewee. Similar specialization is often found among insects and, to a lesser extent, the mammals. As a result of such stratification, the assemblage of forest plants achieves the highest possible photosynthetic efficiency, and forest animals make maximal use of the forest's food resources and breeding areas.

A mature forest represents a large investment of total biomass and has a large effect on both immediate and more distant environments. As mentioned before, forest plants create a local environment that is cooler, more humid, and less changeable than that of unforested soil. A forest also acts as a giant trap for water and soil minerals, which it uses in its own life processes. Unforested soil sheds both water and soil minerals. The forest thus becomes a creating, accumulating, moderating force for life. It is apt that in so many cultures the symbol of life is a tree.

Nature in the City

Humans have lived in cities for as long as the invention of agriculture some 10,000 years ago has made urban life possible. During much of this time, the inhabitants of urban areas have taken pains to preserve natural vegetation within their cities. The following poem was written by a city dweller in Hangchow, 900 years ago.

> *Spring*
> The pear blossoms are pure
> White against the blue green willows.
> The willow cotton blows in the wind.
> The city is full of flying pear flowers.
> The petals fallen on the balcony look like snow.
> How many spring festivals are we born to see?[2]

The author of the poem is Su Tung P'o, who lived from A.D. 1036 to 1101. As a high government official of imperial China, Su was a city dweller, and his concerns were administrative and political in

---

2. *One Hundred Poems from the Chinese.* Copyright © 1971 by Kenneth Rexroth. All rights reserved. Reprinted by permission of New Directions Publishing Corporation.

Figure 12-21
NATURE IN THE CITY: THE COEXISTENCE OF STRAPHANGERS AND WILDFLOWERS.

nature. In this sense Su was a modern man. But when he talks about his city he talks about elements that are eternal: the cycling of the seasons, the wonder and transitoriness of a flowering pear tree. What Su wrote almost 1000 years ago still strikes a deep chord in the reader of today, but with one interesting difference.

In 1971 a University of Wisconsin study team made a survey of American residential preferences. They found that the largest single class of respondents preferred a rural location but within 30 miles of a central city. There seem to be two forces pulling Americans in the 1970s: a gravitation toward a large city, with its richness of human contact, wide job opportunities, and "city life," and an equivalent attraction to a green meadow, a vegetable garden, sweet air, and birdsong. Significantly, it is felt that the two are mutually incompatible. Yet they were not incompatible in the Hangchow of Su Tung P'o and perhaps they need not be incompatible in the great cities of today.

In fact, the extent to which cities are built of brick and concrete is often overestimated and may have peaked in the Unites States. For one thing, the population densities of large cities have been declining. Thus, in 1910, New York City contained 64,000 people per square mile. By 1960, the density had fallen to 13,000 people per square mile.[3] A similar though less drastic decline has occurred in all major North American urban regions. The result has been an absolute increase in open space in most such urban regions.

---

3. The population density of Manhattan remains at about 60,000 per square mile today. Part of New York City's density reduction has resulted from outmigration, and part from the annexation of adjacent boroughs.

Figure 12-22
TWO URBAN PARKS. Left, municipal park in Rowayton, Connecticut. The plants in the foreground are water dock and fescue grass. Right, Forest Park in New York. Though the visual aspect remains beautiful, topsoil in the foreground has been totally eroded by heavy traffic.

Parks

The "natural life" of the city is never entirely "natural," but one may distinguish two classes of open areas: those that have been consciously set aside as parks, forest preserves, wildlife sanctuaries, and so on, and those that have been set aside unconsciously: vacant lots, backyards, hedgerows, median strips, street plantings, and the right-of-ways of railroads and parkways. Official green spaces need not necessarily harbor the richest living communities. Thus a heavily used park such as Lincoln Park in Chicago, Central Park in Manhattan, or the central mall in Washington, D.C., all suffer from too much human disturbance to allow any but the simplest associations to develop. Perhaps the most severe kind of disturbance an urban park is subject to is the lawn mower. A close-cropped lawn offers no concealment for a small mammal or bird, and little in the way of food. There is little diversity of plant life, since all tip-growing plants (such as trees and most wildflowers) are killed as seedlings. Only plants that can grow from the base (such as grasses) and those that hug the ground (such as plantains, dandelions, and chickweeds) can survive and prosper. A well-kept lawn offers about as much biological diversity as a shag carpet, and this of course is why it is prized. The remaining plant life is exceptionally resistant to foot traffic. This is why a choice must be sometimes made between a lawn mower and bare earth. But humans can disturb a living community in many other ways as well: by flooding it with rats, dogs, and alley cats and by the use of herbicides and insecticides.

There are two factors that influence the diversity of plant and animal life possible in a city park. One has been mentioned—human disturbance. The other is the area of the park. All birds and animals require a minimum range in which to raise young and feed themselves. For a black bear it may be more than a square mile. For a horned owl, it may be 10 acres. And for a deer mouse, it may be 1/20 of an acre. This is one reason why dear mice and rabbits are certain to be found in an average 20-acre park, a horned owl is possible but doubtful, and a black bear would rate headlines. Often the map area of a city park is less important than its pattern of roads and internal obstructions. An expressway pushed through a park may eliminate a resident mink or species of salamander by cutting it off from its food source or breeding site. The same expressway may form no barrier to a scarlet tanager or a brown bat.

Often the park land of a city is simply land originally considered difficult to build on which was set aside for economic reasons. Baltimore, Washington, D.C., Cleveland, and Chicago have all made long ribbonlike parks of their floodplains, and these run like green necklaces through and around the cities. New York City created a similar string of parks along the Harbor Hills terminal moraine in the East. These parks (Alley Pond Park, Cunningham Park, Forest Park, and Prospect Park) are among the richest in the entire city parks system. Other cities have sited parks along their lakeshores (Chicago), rivers (Philadelphia and San Antonio), and oceanfronts (New York). The personalities of such large city parks may be as diverse as that of nature itself, and such parks always become a focus for the neighborhoods that surround them. It is fortunate that the acreage of city park land continues to grow. The dark ages of the 1950s seem to be lifting, when park land was the first target for interstate highways or the new stadium.

The Vacant Lot

The real genius of city wildlife lies in its ability to adapt to nooks and crannies of free space and prosper against all odds. The symbol of nature in the city must be the tree of heaven—the "tree that grows in Brooklyn." This tree of vacant lots, dark backyards, and cracks in the pavement is itself an immigrant species, as are the people among whom it makes its home. Originally a native of China, it was introduced here via Britain only 100 years ago. Its winged seeds are spread by birds and by the wind, and no human hand is needed in its dispersal. It is the most rapidly growing of our trees, sprouting as much as 8 feet in a year and shrugging off dust, smoke, and insect pests.

Figure 12-23 SCENES FROM VACANT LOTS. (A) Oats, prickly lettuce, and a tree of heaven colonizing a recently-leveled building site in Manhattan. (B) A bumblebee on red clover. The lacy-leafed plant is yarrow. (C) Thistles in seed. (D) An ambush bug waiting for its meal among goldenrod blossoms. Its muscular front limbs are used to seize and hold the victim. (E) Small ground plants: red clover, hop clover, lesser stitchwort, and basil. (F) Underparts of a female sowbug, with eggs attached to belly. Sow bugs are found under rocks and rotting wood.

Perhaps the greatest variety of life in the city is to be found in a vacant lot. Such lots may be weedy or filled with brush and trees. The soil may be deep and fertile, or it may be covered with rubble. The lot may occupy an entire block, or it may stretch for only 25 feet between two buildings. In short, ecologists do not recognize a type of community known as the vacant lot community, simply because vacant lots are too variable. But perhaps we can make a few generalizations. In the first place, a vacant lot is likely to be the least disturbed area around. In spite of constant usage by children and neighborhood dogs, it remains one of the few places in a city where a plant can flower and set fruit without being attacked and cut down as a "weed." This means that a vacant lot can harbor a diversity of plant life which in turn can give food and shelter to small animals.

However, none of the vacant lot plant life is likely to be old or established. The plants are likely to be recent colonists either because the lot has not been in existence for very long, or because human disturbance has created new sites for colonization. It is perhaps significant that so many of the plant species found on a vacant lot are of the weedy type. In nature, many of these species are found along floodplains, where the shifting work of flooding and siltation creates conditions that mimic those of a vacant lot.

Plants of the Vacant Lot. A weedy lot provides a flowering calendar of the seasons. In the spring there are the pale yellows of mustards, the deeper yellows of hawkweeds, and the blues of chicory. In the shade of shrubbery there may be found the strange blue and white dayflower, which opens in the morning and closes tightly in the afternoon. Enormous, curly leaves advertise curly dock. The leaf is edible and makes a delicate salad.

Summer may see the white and yellow of daisy and toadflax. Queen Anne's lace stands high above the field, bearing an umbrella of tiny white clustered flowers. In the very center of the cluster stand one or two dark purple flowerlets. The root is edible and resembles a tough, thin carrot. In fact, Queen Anne's lace is just the garden carrot escaped into the wild. Tangled on the ground are white and red clover, great favorites of the bumblebee.

The fall sees the yellows of goldenrod and finally the purples of fall asters. The grasses have long since finished their flowering and stand with ripe seed heads. There are the tight cylinders of timothy and bristly ones of foxtail. Crabgrass stands with outreaching thin spikes. Kentucky bluegrass holds up seed clusters arranged in pyramidal tufts. Wild oats stand as elegant as tiny chandeliers. Witch grass seeds are arranged in the most delicate of panicles. In the windblown open spaces of the West they break off and are driven around as tumbleweeds, while here they quietly fall among the weeds.

The fall is also flowering time for ragweeds. These wind-pollinated flowers produce enormous amounts of pollen which causes suffering for victims of hay fever. Ragweeds are specialized for growth in disturbed sunny soil and cannot invade either a forest or the thick root mat of an established prairie.

Plant-Feeding Insects of the Vacant Lot. The tangle of plants supports a busy animal life. The plants provide food, hiding places, moderation of temperature and humidity, and a place to mate and to lay eggs. The vigor and complexity of the plant community are directly reflected in its animal life. The animal life of a vacant lot is dominated by the insects. Insects are small enough so that all their living needs can be satisfied by a few square feet of living space. And their ability to fly makes them excellent colonizers of the little islands in an asphalt sea that vacant lots represent. The sheer variety of insect life in the city can be stunning. Frank Lutz of the American Museum of Natural History in New York City once set out to discover and catalog all of the insect species he could find in his 75 by 100 foot backyard in suburban New Jersey. On his little plot of ground he discovered 1,402 species of insects!

Among the most common insects among the weeds are the hoppers. About the size of a housefly or smaller, they may be clothed to blend in with the leaf background. When disturbed, they escape with a lightning-fast hop and then unfurl their wings in midair and fly. They feed directly on weeds and grasses by piercing the soft stems with a hollow beak and then sucking up the plant sap (phloem fluid). One of the most conspicuous of hoppers is the froghopper, or spittlebug. To find a spittlebug, walk through a weedy lot and look for gobs of spit hanging on the plants. If you pick up the spit between your fingers and gently rub it, you will discover a tiny juvenile insect in the center. A spittlebug pierces a plant stem with its beak and begins to suck the plant sap. Much of this sap is then exuded through its anus. As the fluid is exuded, the spittlebug uses its hind legs like an eggbeater to beat the fluid into a frothy mass which covers and hides the insect.

Aphids are relatives of the hoppers. Though very tiny, their sharp beaks can penetrate a plant stem and find its sweet phloem fluid. Like spittlebugs, the aphids exude much of this fluid (called *honeydew*) through their anus. But typically they do this only on demand. When an aphid's belly is gently stroked by an ant, the aphid responds by exuding its droplet of honeydew, which the ant drinks. Honeydew consists of almost unaltered phloem fluid, from which the aphid has extracted a percentage of free amino acids and sugars

Figure 12-24
A FEW COMMON GRASSES
A. Witch grass. B. Kentucky blue grass. C. Foxtail.

Ants cannot reach the phloem themselves because their mouthparts are made for biting, not piercing. Some ant species are almost totally dependent on aphids for food. In return for the feeding, the ants protect the aphids from other insects and carry them from branch to branch. Aphids overwinter in the egg stage, and some ants even carry aphid eggs down into their underground chambers for winter protection. The match between aphid and ant must be a specific one. Ant species that lack experience in aphid tending are as likely to eat the whole aphid as to drink its honeydew.

Grasshoppers, like the hoppers and aphids, feed directly on plants, but grasshoppers chew their food instead of sucking it. These large, strong-looking insects have few methods of defense other than jumping. But when caught, a grasshopper regurgitates a brown, irritating fluid which might deter a small predator. Grasshoppers and their relatives, the crickets and katydids, are noisy insects. They create most of the sound in a summer field. Typically only the male produces sound, by rubbing its hind legs together. It is the sound of love, designed to attract a female of the species.

Bumblebees are also plant feeders, but they feed specifically on the pollen and nectar of flowers. These large, furry bees with black and yellow markings live underground, sometimes in an abandoned mouse nest. The colonies are much smaller than those of honeybees, containing a queen, a few males (drones), and perhaps 50 to 200 workers. Only the queen survives the winter.

Bees share their flowers with an occasional butterfly or nocturnal moth. Butterflies and moths, if they feed at all, drink the nectar of flowers. This they suck up through a long tube which is carried coiled while in flight. The wings of butterflies and moths are covered with minute iridescent scales. The scales rub off at the slightest touch, which enables a butterfly or moth to wriggle out of many a spider web.

Much of the life of a vacant lot lies at ground level or below. Bumblebees, wasps, and mice live below the surface. The familiar click beetles feed on rotting wood. When turned on their backs, they arch themselves suddenly, flipping over right side up. The ground is also home for the secretive sowbugs, not insects at all but crustaceans living on land. Sowbugs feed on decomposing plant matter. Their breathing apparatus requires 100 percent humidity, which they seek under rocks and debris. Ants are always common; in the summer one can lift almost any large rock or board to find an alarmed colony underneath. But by the first frost, all colonies have moved deeper underground into winter quarters.

Figure 12-25
AN ANT TENDING
AN APHID.

Figure 12-26
PRAYING MANTISES COPULATING. Though the male has lost its head and most of its thorax, its abdominal section is able to finish the sex act.

Carnivorous Insects of the Vacant Lot. The life of a plant-feeding insect is never secure. Hosts of tiny assassins lurk among the stalks of grass, hide in the flowers, patrol the ground or sweep down from above. These are the insect-feeding insects. Even the large and robust bumblebee may be overpowered. Its downfall may be an ambush bug waiting for it on a goldenrod flower; well masked by its green-yellow hues and irregular outline, the ambush bug springs on its victim, impales it on a sharp beak, and literally sucks out its life.

The delicate green lacewing feeds voraciously on aphids, as does its larval form. Presumably to avoid mistakes, the lacewing lays its eggs on tiny stalks, where its own larva will not encounter and devour them in its search for aphids.

The praying mantis is the opposite of delicate. This greenish-brown insect, though as long as a human finger, is practically invisible as it waits quietly among the tangled stalks for its next meal. Its forelimbs, held outward as though in prayer, can shoot out without warning to grasp any insect within range. While the victim twists and wriggles futilely, the mantis bites off the head and then methodically devours the remaining parts of the body. The praying mantis is an unselective feeder. The male of the species may discover just how unselective when mating with the much larger female. He may literally lose his head. However, the remaining parts of his body continue with the act of copulation. The female then fashions a parchmentlike casing in which she deposits her eggs. The eggs are laid in the fall, attached to vegetation. They carry the mantid race through until the following spring.

Few insects inspire as much terror as yellow jackets. As far as humans are concerned, yellow jackets attack only when their colony is threatened and the larger the colony, the more bellicose the wasps become. The colony is built of a paperlike material fashioned by chewing up plant material and gluing it together with saliva. The paper requires a protected location, such as under a board lying on the ground or inside a hollow tree trunk. Wasps feed on carrion and on soft-bodied insects: flies, mosquitoes, caterpillars, butterflies, and even honeybees. These are pursued on the wing or caught in an exposed location. Most encounters are unsuccessful—yellow jackets have been observed to make an average of 1 catch out of 12 passes when hunting flies. This makes them clumsier than the human hand. When the insect is finally seized, the wasp chews it up in its jaws and brings the chewed mass back to its larvae in the nest. Yellow jackets inspire such respect in the animal world that a number of common and totally harmless flies lead charmed lives because they mimic the yellow and black bandings of the wasp. It takes a careful look to tell the two apart—the flies have one pair of wings and the yellow jackets two.

Many other kinds of wasps are found in a vacant lot. Most are very tiny and parasitize other insects, including other wasps and bees. They are thus important agents of population control among the insects. The vacant lot is a home or feeding ground for countless other small animals. One can find the silvery trail of the slug, the carefully built cablework of the spider, the underground entrance of the field mouse, and the droppings of the sparrow or starling. All have learned in one way or another to coexist with man, and some, such as the starling, to prosper from the association. In addition, all make modest territorial demands and have great powers of dispersion. Within these limitations, the inhabitants of the vacant lot have constructed a small world filled with color and adventure. Far from being vacant, the vacant lot pulsates with life.

We started out this chapter by asking why living communities across the world show such great differences. Part of the answer is that each association has had to adapt to a unique set of limiting factors: light, water, temperature, salt concentration, to name a few. In the case of the vacant lot, there is the additional factor of human pressure. In each case, the resultant personality of the community has been different. And yet there are invisible but powerful similarities that link all natural communities. The most basic similarity is that they all work. This is not due to chance—there is an internal logic that all real communities must follow. What keeps a community alive? We will examine this question in the last two chapters.

Bibliography

GENERAL READING

Franklin Russell, *Watchers at the Pond,* Alfred A. Knopf, 1961. A poetic recounting of the yearly cycle of life in a woodland pond. A great deal of information is woven in with the poetry.

Joseph Wood Krutch, *The Voice of the Desert,* Sloane, 1954. Joseph Wood Krutch is a great naturalist whose special love is the desert. This small collection of essays focuses on such desert life as the cactus, the kangaroo rat, the scorpion.

Rachel Carson, *The Edge of the Sea,* Houghton Mifflin, 1955. Carson's brand of natural history combines poetry with a defense of life. This excerpt is an example of the former.

> One of my own favorite approaches to a rocky seacoast is by a rough path through an evergreen forest that has its own peculiar enchantment. It is usually an early morning tide that takes me along that forest path, so that the light is still pale and fog drifts in from the sea beyond. It is almost a ghost forest, for among the living spruce and balsam are many dead trees—some still erect, some sagging earthward, some lying on the floor of the forest. All the trees, the living and the dead, are clothed with green and silver crusts of lichens. Tufts of the bearded lichen or old man's beard hang from the branches like bits of sea mist tangled there. Green woodland mosses and a yielding of reindeer moss cover the ground. In the quiet of that place even the voice of the surf is reduced to a whispered echo and the sounds of the forest are but the ghosts of sound—the faint sighing of evergreen needles in the moving air; the creaks and heavier groans of half-fallen trees resting against their neighbors and rubbing bark against bark; the light rattling fall of a dead branch broken under the feet of a squirrel and sent bouncing and ricocheting earthward.
>
> But finally the path emerges from the dimness of the deeper forest and comes to a place where the sound of surf rises above the forest sounds—the hollow boom of the sea, rhythmic and insistent, striking against the rocks, falling away, rising again.

John Teal and Mildred Teal, *The Life and Death of a Salt Marsh,* Little, Brown, 1969. The finest natural history and economic geography of the salt marsh.

TEXTBOOKS

Ralph and Mildred Buchsbaum, *Basic Ecology,* Boxwood Press, 1957. A slim volume, profusely illustrated and with a good discussion of limiting factors as shapers of community personality.

R. H. Whittaker, "Communities and Ecosystems," Macmillan, 1970. Vigorous prose and a vigorous thought process. The concept of the community is examined in a more abstract sense than by the Buchsbaums. Written at a fairly advanced level.

R. L. Smith, "Ecology and Field Biology," 2nd ed., Harper and Row, 1974. A large text of ecology whose greatest strength is perhaps its discussion of field biology.

Figure 13-1 GIRAFFE AND BUCK AT A WATER HOLE. Though both are herbivores the two species co-exist in the same area because they occupy different ecological niches.

chapter 13

There is a federal law on the books to the effect that one cannot deposit money in the bank for one's descendants 100 years or more in the future. After a certain length of time, all inactive accounts become the property of the bank. Consider what might happen otherwise. Let us imagine that, in 1776, Benjamin Franklin (who was a believer in thrift and saving) went to a bank and opened a savings account. Let us suppose that Franklin

Growth and Regulation of Populations

deposited only $1 in the account, at 5 percent compound interest. Let us further suppose that the money was to be withdrawn in 1976 and awarded as a prize for the best research in the field of electricity performed that year (Franklin was an amateur scientist). How much would the lucky scientist receive? His prize money would amount to $22,400. This is because every year the interest earned was plowed back into the principal, earning additional interest of its own. Of course, if Franklin had wanted to wait a little longer, the prize money could have reached an even more impressive figure. This is how his initial $1 would have grown through the years:

Year	Account (at 5 percent compound interest)
1776	$1 (deposited)
1876	$151
1976	$22,400
2076	$3,320,000
2176	$502,000,000
2276	$77,600,000,000
2376	$10,700,000,000,000

Money deposited at compound interest grows in a geometric fashion—that is, the rate of increase itself continually increases. If the money is left on deposit long enough, its rate of increase finally approaches infinity, and the total amounts involved likewise approach infinity. Thus in 600 years Franklin's $1 would have grown to over $10 trillion, a value 10 times larger than the present United States gross national product. By that time, the account would be earning over $500 billion per year in interest. This is why the law does not permit such a situation to develop.

The Population Growth Curve

Growth of all natural populations[1] is governed by the same laws of compound interest. The human population today shows a net growth of approximately 76 million per year. That is, the sum of all births minus the sum of all deaths equals 76 million individuals per year. This is approximately equal to the total population of Great Britain plus Scandinavia and corresponds to an annual increase of 2.0 percent. While 2 percent compound interest may sound harmlessly low, it represents in fact a situation that can only be maintained for a limited length of time. Like Franklin's dollar, the human popu-

---

1. We define a population here as a group of organisms of the same species living in a particular place. Thus we may speak of a population of fruit flies in a bottle or a population of deer in a forest or, as an extreme, of the population of humans in the world. Implicit here is the assumption that all members of a population interbreed.

THE POPULATION GROWTH CURVE / 471

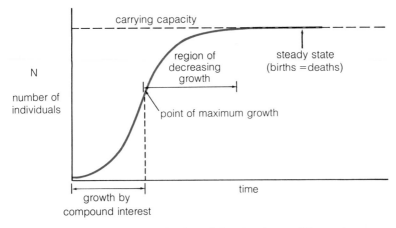

Figure 13-2
THE STANDARD GROWTH CURVE. Increase in population is measured as a function of time.

lation would eventually reach absurd levels. As an illustration, consider a world in which the human population has grown at its present rate since the time of Christ. In A.D. 0, the total population of the world was an estimated 250 million individuals. To estimate the number of descendants they would have generated by the year 1976, we use a formula similar to that used by bankers to calculate growth of compound interest:

$$N_t = N_0 e^{rt}$$

Here N_0 is the original population, N_t is the population t years later, r is the annual rate of increase, and e is a natural constant having a value of 2.718.

$N_t = 250{,}000{,}000 \times e^{0.02 \times 1976} = 250{,}000{,}000 \times e^{39.5}$

$N_t = 4 \times 10^{25}$

This is a mathematical notation for a number that begins with 4 and is followed by 25 zeros. The number is so large that it has no common name. To get a feeling for what it means, let us translate this number into population density. The total surface area of the world (including the oceans) is 4.4×10^{15} square feet. If all 4×10^{25} people were evenly distributed over this surface, there would be 5½ billion of them per square foot! Just as there exists a federal law against unmanageable compound interest in banking, there exists a natural law against unmanageable population increase in the living world. Sooner or later, all growing populations encounter environmental resistance. The maximum population a given area can support on a permanent basis is known as its *carrying capacity*. The carrying capacity is different for each species. Thus it is intuitively obvious that an acre of woods could support 20 mice on a permanent basis but not 20 mountain lions.

If one plots the rise in population as a function of time, one obtains an S-shaped curve (Fig. 13-2). Such a curve has been ob-

Figure 13-3 A FISHING BOAT OFF VALPARAISO, CHILE, RETURNING WITH A FULL CATCH. The fish are mostly hake.

served in studies with many different organisms: bacteria, protozoa, insects, and small vertebrates. The rising bottom half of the curve describes the phase of geometric growth. This highest rate of growth is reached midway up the curve (the inflection point), when population size equals 50 percent of carrying capacity. Once past this point, the population continues to grow but at an ever-diminishing rate as the environmental resistance stiffens. Finally a stable population size is reached, with the birthrate equal to the death rate. This corresponds to the carrying capacity.

While all wild populations obey the general principles illustrated in Figure 13-2, each population is unique in terms of the carrying capacity it can approach and the rate at which it approaches it. A bacterial population grows so rapidly that it can reach carrying capacity within hours, while a population of elephants may require decades. But whether a given population grows quickly or not, its final size is dictated by the environment and not by its growth rate. For instance, bacterial growth can be slowed by lowering the temperature. But the final populations achieved are the same at any temperature.

Wildlife Management

A growth table of this sort contains information that is vital in any attempt at wildlife management. How heavily should a fishery or a herd of deer be harvested in order to obtain maximum yield? Obviously, the fish or deer population should be maintained at its inflection point, where growth is most rapid. The penalty of overfishing is a depression of the population to a level where the rate of natural increase is reduced. The harvest is increasingly made up of juvenile forms which do not take part in the reproductive process.

A particularly clear example of mismanagement involves the great whales: the blue, the humpback, the gray, the right, and the finback. Of these, the blue whale is the largest animal that has ever lived. Blues average 25 feet in length at birth and reach lengths of 106 feet or more. One 83-foot male weighed 242,397 pounds. During the early 1930s, the blue whales accounted for 80 percent of all whales killed in Antarctic waters, and the yearly kill averaged about 20,000 blue whales. Such a harvest was much larger than could be made good by the normal reproductive potential of the species. The female does not reach sexual maturity until 5 years of age and has a gestation period of 2 years. This means that a 10-year-old female has produced, at the most, only two offspring. As the kill went on, all of the danger signs appeared: the number of blue whales caught declined, the percentage relative to other species declined, and the average age of the whales caught declined. The significance of these observations was clearly understood, yet the whaling went on without respite. The Japanese even insisted that the juvenile forms they were killing represented an entirely new species, the "pygmy blue whale." By 1963, blue whales accounted for only 0.06 percent of the corporate income of whaling companies.

It was not until 1963, when the blue whale had become commercially extinct, that the International Whaling Commission imposed the first quota reduction on blue whales. In 1967 a total ban was placed on the harvesting of blue and humpback whales, which thus joined right and gray whales as totally protected species. In 1975 the annual finback quota was reduced from 1,500 to 600 individuals.[2]

However, these measures may have come too late to save some of the great whales. It is estimated that at the present time no more than 8,000 blue whales (and possibly much less) remain alive in all the world's oceans. For a species that is widely dispersed and has an exceptionally low reproductive potential, this may be the end of the road. Even if no further killing takes place the species may become extinct. The blue whale is the most solitary of all animals. It meets only to mate. The few remaining individuals, separated by hundreds

---

2. Six whaling nations (with 6 percent of the world's total whale catch) do not belong to the International Whaling Commission and are not bound by its quotas. These nations are Chile, Peru, Portugal, Spain, Somalia, and South Korea. Among members of the International Whaling Commission, Japan and the USSR together account for 80 percent of the world's total whale catch. The United States and Canada, though members of the commission, have ceased all whaling operations.

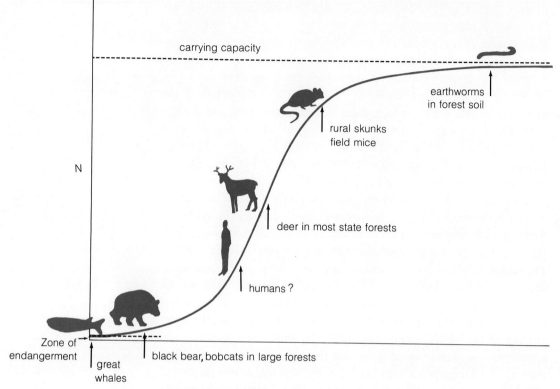

Figure 13-4 PRESENT STATUS OF SOME NATURAL POPULATIONS AS A FUNCTION OF INDIVIDUAL CARRYING CAPACITIES.

of miles of ocean, may simply be too widely separated to find each other. And when the numbers of any population drop to a critical level, simple chance becomes an important factor. Just as a tiny flame cannot survive a quick wind gust, so a small population cannot overcome reverses that would have produced only a ripple in a larger population. The blue whale offered us nothing that was essential. If it disappears, it will have disappeared before we learned little more about it other than its potential for providing pet food and transmission oil.

Several hundred other species of animals share the endangered status of the great whales. Included are many of the big cats, large birds of prey, and all the creatures most closely related to man: the gorilla, chimpanzee, orangutan, and gibbon, as well as the exotic lemurs, ayes-ayes, and the douc langur of Indochina. Not all of these have been hunted as ruthlessly as the whales and big cats. The numbers of an animal may be reduced to the danger point just by lowering the carrying capacity of the environment. The cutting of the rain forests of Madagascar has decimated the lemurs. The bombing and

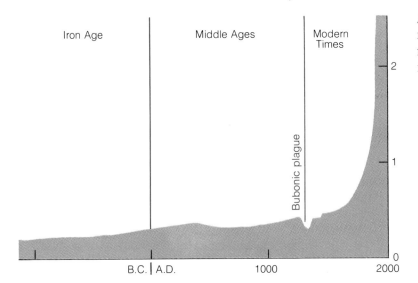

Figure 13-5
HISTORICAL RATE OF POPULATION INCREASE FOR HUMANS.

defoliation of Indochina have had the same effect on the douc langur. Human intrusiveness has pushed the California condor back to a few remote locations.

Successful organisms are those that have a large resource base and encounter little destructive interference from humans. The earthworm is a good example. It probably lives at close to its natural carrying capacity, which is enormous. Animals such as the striped skunk and the field mouse are under a modest amount of human pressure and are probably not found in numbers equivalent to carrying capacity. Deer in state forests are deliberately managed for optimum yield. The animals are maintained at levels of about 50 percent of carrying capacity by adjustment of hunting season, bag limits, and restrictions on the hunting of does. The black bear and the bobcat are under severe hunting pressure and are generally given none of the protection accorded deer. Consequently their numbers are far below carrying capacity, and it is conceivable that someday they will join the grizzly and the cougar on the endangered species list.

Carrying Capacity for Homo Sapiens

Where does the human population stand today with respect to the carrying capacity of the world? Or, to put it another way, what is the maximum human population that the world can support? A quick look at the historical rate of population increase (Fig. 13-5) shows that the human population is growing in exponential fashion. There is no hint of a slowing down of the rate of increase as natural limits are approached. There are biologists who envision a final stable world population of 50 billion (about 12 times the present population).

Figure 13-6 THE GREEN REVOLUTION. A high-yielding variety of Mexican wheat growing on an experimental farm in Tunisia.

However, other students of the problem feel that the present population trend will lead to the collapse of the agricultural and industrial base within the next century, followed by a great dying which will stabilize the human population at a level much lower than at present.

It is important to remember that the human species has itself been able to raise the carrying capacity of its environment. The earliest and perhaps most important instance of this was the change from a hunting-and-gathering to an agricultural existence. The change occurred perhaps 10,000 years ago in the Middle East, and shortly thereafter in the Far East, and laid the groundwork for the first great civilizations. The transition came much later in other parts of the world. It is estimated that an American Indian needed 2 to 3 square miles of good hunting range to survive. His farming descendants can survive much more comfortably on 2 to 3 acres of good farmland. This amounts to an enormous increase in the carrying capacity of the land, and the process continues today. In 1945, the average yield of corn in Illinois was 50 bushels per acre. Today, the yield exceeds 100 bushels per acre. Today, the world average rice yield is 1,300 pounds per acre. But on land where the new high-yield varieties have been planted to some extent, the yield is 1,900 pounds per acre. And in Taiwan, where all rice is of the high-yield type, the harvest is 3,500 pounds per acre. Comparable gains have also been realized for the wheat crop.

But before it is calmly assumed that the human species will always find a way to feed itself, it must be remembered that there is a price. A crop plant produces food out of carbon dioxide, sunlight, water, and minerals. Carbon dioxide and sunlight are free. Where natural water is insufficient, irrigation water must be supplied. This requires the drilling of expensive wells and substantial amounts of oil to drive the pumps. Oil is no longer a cheap commodity. In addition, there may be problems of salinization, as irrigation water pulls up salts from the soil.

An even greater obstacle is the supply of minerals. Illinois farmers achieve their spectacular results with corn by literally drenching the soil with fertilizer. Between 1945 and 1966, the amount of nitrate fertilizer used in the state of Illinois climbed from 10,000 to 600,000 tons. The amounts applied are so enormous that nitrate leached out by rains has contaminated the municipal water supply of Decatur, a quiet city of 120,000 lying in the middle of Illinois farm country. Nitrogen fertilizer has always been expensive. Because its manufacture requires large amounts of petroleum fuel, its price has risen to a point where all fertilizers have been priced out of the reach of many Asian farmers. Unfortunately, the high-yield varieties of rice require much greater amounts of fertilizer than their normal counterparts. We will return to this question in Chapter 14.

Beyond the simple questions of agricultural output, a rising world population puts increasing pressure on the world's energy and mineral resources. At the present time, the United States, with one-sixth of the world's total population, accounts for close to 50 percent of the world's energy and mineral consumption. Will this state of affairs persist in a vastly more crowded world? A world population of 50 billion, if it is possible at all, will be possible only with a just apportionment of the world's resources and with close cooperation among all the world's peoples. Given past human history, neither prospect seems very likely.

Population Control Factors

Let us now return to the generalized population growth curve (Fig. 13-2). The upward sweep of the lower half is due to growth at a compound interest growth rate. What accounts for the declining growth rate of the right half of the curve? That is, why does the growth rate decrease as the population approaches its carrying capacity? One should not have a simplistic conception of carrying capacity as being equivalent to food supply. While starvation is a common and very effective agent of population control, one can point to many populations that stop growing before their food supply is exhausted. The simple observation that trees are not defoliated by inch worms and prairies are not stripped to bare soil by the grazing of animals would suggest this. Mealworms will stop multiplying in a box of cereal long before all the cereal is consumed, because of toxic secretions they themselves produced. An example of midwestern farmers raising bumper crops while contaminating their own water supply presents an uncomfortably close parallel.

Biologists distinguish at least two kinds of population control factors. *Density-independent* factors are those that kill a fixed percentage of the population, no matter what its density. Thus a blizzard kills a fixed percentage of grouse in a brushland, and perhaps all those on higher ground. Or a drought kills a fixed percentage of seedlings in a field. DDT sprayed on a lake kills a fixed percentage of fish-eating birds. Factors such as these cannot be expected to control population size. If an external factor is to act selectively in limiting population growth, it must act with increasing severity as population density approaches the carrying capacity. Such factors are called *density-dependent factors*. Examples of density-dependent population control factors include predation, competition, emigration, and hormonal effects on the reproductive process.

Predation

Though everyone knows that wolves eat deer, lions hunt zebras, and the lynx runs the hare, biologists are far from agreement on the importance of predators in controlling population sizes of their prey. Many of the classical examples of supposed predator control of prey have been shown to be based on inadequate data, insufficient controls, or both. On the next several pages we look at some of these.

For example, in 1906 the Kaibab Forest, on the northern rim of the Grand Canyon, was set aside as a game preserve for mule deer that inhabited the area. To protect the deer, 6,000 major predators were shot in the forest during the next 25 years. These included wolves, mountain lions, coyotes, and bobcats. As a result, the deer population exploded from 4,000 in 1906 to 100,000 by 1923. The result was overbrowsing of the forest, mass starvation of deer, and a population collapse to about 10,000 individuals by 1939. However, the population data appear to have been based on an inadequate sampling, and only the most sensational estimates were used to buttress the argument.

Another classic study is the fluctuation in number of the Canadian lynx and the arctic hare. Records of the Hudson Bay Company dating back to 1821 show cyclic fluctuations in the populations of the lynx and the hare. Since the two cycles are out of phase with each other, with the lynx trailing the hare, it has been suggested that, as the hare population rises, it supports a higher population of lynx. The lynx then deplete their prey and fall prey to starvation, allowing the hare to recover and set up a new cycle. Unfortunately for the theory, a very similar 10-year cycle has been observed among arctic hares living on an island containing no lynx.

Still another classic study concerns the moose and wolves of Isle Royale. Studies by David Mech and others have shown that a pack of about 25 wolves controls the population of a herd of 800 moose. Because of predation by wolves, the moose herd is smaller than could be permanently supported by the 210-square-mile island. But the physical restrictions of the island have caused other biologists to question the validity of the phenomena observed. They point out that the island is little more than a giant bottle, and phenomena observed in such a restricted space may not be reproducible in the larger world outside.

The objection of smallness of scale cannot be raised against the studies of the Talbots on the Serengeti Plain. The Serengeti Plain of Tanzania and Kenya is as large as Ireland, or the states of Vermont, New Hampshire, and Delaware combined. It contains almost virgin herds of a complex fauna such as can be seen nowhere else in the world today. Lee and Martha Talbot have studied these giant herds for over 7 years. The most common large grazing animal on the plain is the wildebeest, about ½ million of which live on the Serengeti. It is hard to estimate population sizes over an area of 20,000 square miles. During the annual north-south migrations, the Talbots have stood on the top of a mountain and watched migrating herds disappear into the distance 50 miles away. The Talbots asked themselves whether or not the wildebeest numbers are regulated and how.

About 700 lions live on the plain, feeding on the wildebeests and on other grazing animals. Cheetahs, leopards, hyenas, and wild dogs together kill less than 10 percent of the number of wildebeests cropped by lions. But the total kill of the lions amounts to only 12,000 to 18,000 wildebeests per year, a number too low to control the population. Then what controls the size of the herd? The Talbots found that disease was more important than the lions, but probably the most important factor of all was the separation of young calves from their mothers in the crush of a stampeding herd (Fig. 13-7).

The young are born between December and February of the year. If the herd is stampeded at this time (and it takes only a single predator chasing a single wildebeest to start a herd of hundreds or thousands running), there are almost always lost calves wandering about afterward looking for their mothers. A cow will not accept a strange calf even if her own has been lost and her udders are burst-

Figure 13-7 A STAMPEDING GROUP OF WILDEBEESTS. A calf is running at right.

ing with milk. The doomed calf continues its searching for 2 or 3 days, its bleats growing progressively weaker. On the third day it lies down to die, if a predator has not already picked it off.

Predator Control of Small Animals and Plants. As these examples show, the role of predation as a means of population control among large vertebrates remains uncertain. The evidence among small animals and among plants appears much more conclusive. One of the earliest and most dramatic examples involved a threat to the California citrus industry. By 1880 the orange groves of California were becoming one of its most important industries. Then, in the late 1880s, the orchards were struck by *Icerya purchasi,* the cottony-cushion scale. This relative of the aphids bores into young twigs and sucks the plant sap. Slowly branches wilt and die, and no oranges are produced. Nothing could stop the tiny annihilator. In a desperate effort, the government granted $1,500 to Albert Koebele, an entomologist, to study the problem. Koebele suspected the cushion scale was an import from Australia and had arrived in California minus all its native predators. So Koebele went to Australia to search out its natural enemies. He expected to find a parasitic wasp as the most likely predator but instead found *Rodolia cardinalis,* the ladybird beetle. Both the adult beetle and its larvae eat a clean swath through a thick scale infestation. Even as a scale continues to suck the plant sap, a ladybird walks up behind it and methodically starts to eat it alive.

Koebele shipped back a carton of live beetles to California. Out of the shipment, 129 arrived alive in San Francisco in January 1889. These were carefully placed on an infected tree, and the entire tree was covered with a tent to prevent loss of the precious beetles. By June the first tree was free of scale, and many more ladybird beetles were at work (the generation time for *Rodolia* is 26 days). By July the entire orchard of 75 was liberated territory, and shipments of the miracle beetles were made all across California. The ladybird beetle became well integrated into the native fauna and kept California essentially free of scales until the 1940s. This is a clear case of a natural predator controlling natural prey.

There is a further episode in the story however. In the late 1940s, the scale erupted again. The cause was a tarnished latter-day miracle —the insecticide DDT. Well protected under a raincoat of cottonlike material, the scale insects shrug off DDT. But *Rodolia* absorbs DDT in two ways: from direct exposure and by ingestion of large doses during the course of its feeding on DDT-dusted scales. Its DDT dose then becomes additive. DDT thus causes an increase in the scale population and a decrease in the ladybug population. Not surprisingly, the chemical insecticide industry has responded by attempting to develop an antiscale insecticide.

Another clear-cut predator-prey story involves a reversal of the roles of Australia and the New World. Sometime before 1839, the prickly pear cactus was introduced into Australia as an ornamental plant. As often happens in such cases, the cactus escaped into the wild and began to spread, eventually covering 60 million acres of Queensland and New South Wales. The cactus looks small and innocuous, but it carries short, almost invisible spines which come off at the slightest touch and enter the skin of a human hand or the mouth of a grazing animal. Thus the invaded range was rendered useless for both cattle and for native herbivores. The cactus had originally come from the deserts of Central and South America, where it is found only in small patches and does not become a pest. Conse-

Figure 13-8
TWO COTTONY-CUSHION SCALE INSECTS FEEDING ON A ROSE STEM. All working parts of the insects are hidden beneath the exoskeleton and the brittle, light-colored scale produced by a process of external secretion.

Figure 13-9
PRICKLY PEAR CACTUS.

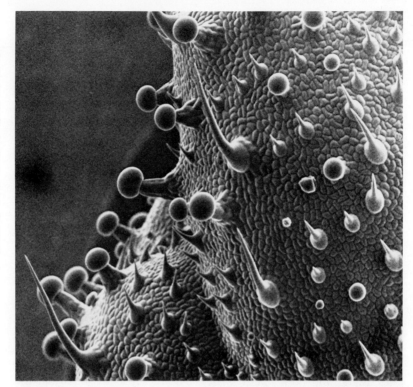

Figure 13-10 SCANNING ELECTRON MICROSCOPE VIEW OF THE SURFACE OF A MARIHUANA LEAF. The small globes sitting atop stalked extensions are packed with resin. Presumably leaf-feeding insects quickly become coated with the resin, experience mental disturbance and lose further interest in feeding.

quently an expedition was sent to Argentina to discover what kept the prickly pear in check there. Out of several insect predators, the most effective one discovered was the Argentine moth. The caterpillar stage of the moth feeds on the growing shoots of the cactus and destroys the plant. Hence the official name of the moth, *Cactoblastis cactorum. Cactoblastis* was shipped to Australia, where it discovered a promised land. The moth fed on the cactus so efficiently that the open range quickly became cactus-free, with only small, isolated pockets surviving.

But the moth was a highly specific predator of the cactus. Once the cactus population crashed, the moth starved to death, so that today both the cactus and the moth are extremely rare in Australia. The cactus flourishes and spreads briefly as a little pocket overlooked by the moth. But sooner or later the moth discovers the pocket and destroys it. Thus an unstable equilibrium is maintained between the insect predator and its plant prey.

Prey Defenses. Though plants are generally not thought of as the prey of herbivores (plant-feeding animals), a typical predator-prey relationship exists between a herbivore and its plant food. Just as animal prey have evolved defenses against their predators (such as superior speed, camouflage, mimicry, armor, and cooperative social behavior), so plants have evolved structural adaptations that protect them against their animal predators. The leaf of a hawkweed is covered with a dense mat of leaf hairs which keeps small insects away from the surface. Oak leaves and stems accumulate tannins, substances that precipitate protein and make the plant tissues unfit for animal consumption. The leaves of sweetgum, garlic, and mugwort all have pungent aromatics that repel insects. The marihuana plant secretes marihuana to drug its insect predators (Fig. 13-10). Cactus, thistle, and wild rose all have thorns that make them unattractive to grazers. Other plant adaptations include subterranean storage organs (potato), poisonous secretions (milkweed), growth areas located at ground level (grasses), and a prostrate growth form (plantain). Unfortunately, the population dynamics of plant–animal interactions have been only poorly investigated. Even a relationship as dramatic as the one of prickly pear and *Cactoblastis* may not have been discovered under the present equilibrium conditions, where both species are present in low concentrations.

Numerical and Functional Responses of Predators. In the case of a highly specialized predator such as *Cactoblastis*, the only possible response to increased prey density is a *numerical response:* as the numbers of the prey increase, the numbers of the predator also increase. But in the case of less specialized predators such as vertebrates, the predator may also show a *functional response* to rising prey density: as the numbers of prey increase, each predator may increase its take of that particular prey. This has been nicely illustrated by the small forest mammals that feed on the European pine sawfly. The sawfly is a pest which can cause heavy damage to trees grown for lumber. The larvae of the sawfly feed on the growing tips of pines. If a pine loses its topmost bud, then it cannot replace the bud. Further increase in the height of the tree is possible only if a lateral bud assumes the dominant position and continues upward growth. The result is a crooked or stunted tree which loses much of its value as lumber.

After a sawfly larva has finished its feeding program, it enters the ground to pupate. Here the heaviest predation on the sawfly takes place, and its predators are three small forest animals: the short-tailed shrew, the common shrew, and the deer mouse. Each of these opens the cocoon in a different way, hence the contribution of each can be accurately measured.

If a prey is extremely scarce in an area, a nonspecific predator does not waste time looking for it.[3] But as the prey becomes more abundant, the predator forms a specific *search image* for the abundant prey. It thus becomes specialized and takes a relatively heavier toll of its prey. This is the basis of the functional response.

---

3. An example described by Tinbergen is chicken eggs placed on a beach where normally only gull eggs are found. Crows, which feed on the gull eggs, ignore the much more conspicuous chicken eggs.

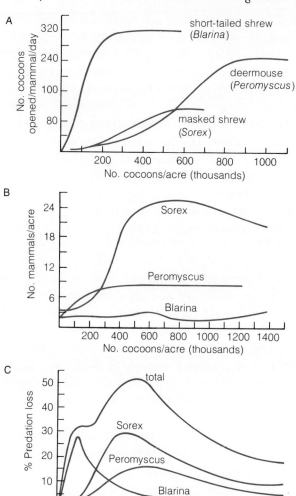

Figure 13-11
PREDATOR RESPONSES TO INCREASING DENSITIES OF PREY. The prey is the European pine sawfly, and the predators are three small mammals. (A) The functional response. As prey becomes more abundant, each predator consumes more per day. (B) The numerical response. As prey becomes more abundant, some predators significantly increase their own numbers. (C) Overall response. The intensity of predation increases with increasing density of prey, up to a maximum intensity of around 700,000 cocoons per acre. If the prey can exceed this density, it has swamped both the functional and numerical responses of its predators and will suffer proportionately less predation.

Among the sawfly predators, the short-tailed shrew (*Blarina brevicaudis*) has the sharpest functional response to increasing concentrations of sawfly cocoons in the ground (Fig. 13-11). As the cocoon concentration rises to about 300,000 per acre, the short-tailed shrew rapidly increases its toll of the sawfly until each mammal consumes 300 cocoons per day. This rate does not increase with further growth in the numbers of sawfly cocoons, because it represents the maximum that a single short-tailed shrew can eat. The functional response of the deer mouse and the common shrew is more gradual, and maximal response is achieved only with much higher sawfly concentrations.

But the small mammals of the forest also put pressure on the sawfly by a different mechanism; an increase in food supply stimulates an increase in the predator population. This is the numerical response, and here the common shrew shows a dramatic rise in population while the short-tailed shrew responds almost not at all (Fig. 13-11). This may be due to the strong territoriality of the latter. The overall response of predator to prey is given by the product of the numerical and functional responses (for example, 9 deer mice per acre $\times$ 200 cocoons eaten per day per animal $=$ 1,800 cocoons eaten per acre per day by deer mice). This may be expressed in terms of the percentage of total cocoons eaten (Fig. 13-11). As can be seen, there is a rapid overall response by the short-tailed shrew, a slower response by the common shrew, and a still slower response by the deer mouse. It is the sum total of all three interactions that dictates the effectiveness of total population control of the sawfly. Several conclusions are obvious:

1. If the sawfly population can build up to levels higher than about 750,000 per acre, it will have escaped from density-dependent predator control.

2. If the rate of natural increase of the sawfly rises to more than 50 percent per year, it will also escape from predator control.

3. Predation by a single species results in a less stable population control mechanism than predation by multiple species.

Cicadas offer a good example of how the above principles may be exploited. Annual cicadas spend many years underground as larvae and then emerge in annual classes and are attacked by cicada-killer wasps. But the periodic cicadas emerge at infrequent intervals of 13 or 17 years, coming out in such numbers that they overwhelm the wasp population and thus substantially escape predation.

The same principle was used in World War II bombing raids: it was discovered that a flight of 1,000 bombers suffered a lower percentage of casualties than a flight of 100 or 10, because the antiaircraft defenses were swamped. In fact, human warfare suggests a form of predation and exploits many of the strategies of natural predators: camouflage, mimicry, and pack hunting, to name a few.

Predation and the Generation of Organic Diversity. Before leaving the topic of predation, there is another aspect of the topic we should consider. It seems clear that in many cases a predator (whether a plant or an animal eater) can exert decisive control over a prey population. But a well-adapted predator never exterminates its prey completely—otherwise it would itself starve to death. Therefore a predator does not act to simplify the community. As Steven Stanley

Figure 13-12 INSECTS AS PREDATORS OF OTHER INSECTS. The insect predators shown here (misnamed parasites) are highly host-specific. As agents of pest control, they represent an ecologically sounder alternative than the broad-spectrum insecticides, which typically affect all animals from insects to vertebrates. (A) Gypsy moth larva. These insects feed on the foliage of trees and may totally defoliate an area. (B) The hair-like structures on the back of the gypsy moth caterpillar are designed to protect it against enemy attack. Here a tiny parasite fly, *Palexorista*, has found an opening and is preparing to deposit her eggs which will hatch into caterpillar-devouring larvae. (C) This parasitic wasp (*Coccygomimum*) attacks the pupa of the gypsy moth. The wasp penetrates the pupal case with her ovipositor and deposits an egg inside. The parasite then develops to adulthood inside the pupal case, feeding on the gypsy moth inside. (D) A gypsy moth pupa destroyed by the larva of the pupa parasite. The hole has served as the escape hatch of the adult parasite. (E) An adult female Caribbean fruit fly. The structure in the back is the ovipositor used to deposit fruit fly eggs onto citrus fruits. (F) Wasp parasite of the Caribbean fruit fly. This tiny wasp deposits her eggs inside the fruit fly pupae. (G) Pink bollworm inside a cotton boll. The bollworm destroys the growing cotton boll and causes heavy economic loss. (H) Eggs of a larval parasite attached to the pink bollworm. (I) Bollworm's end: the larval parasite eggs have hatched and the emergent larvae have devoured everything but the skin of the bollworm.

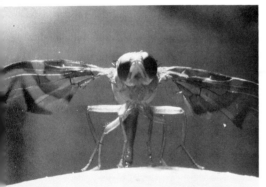

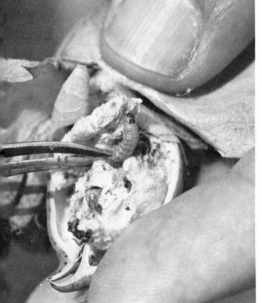

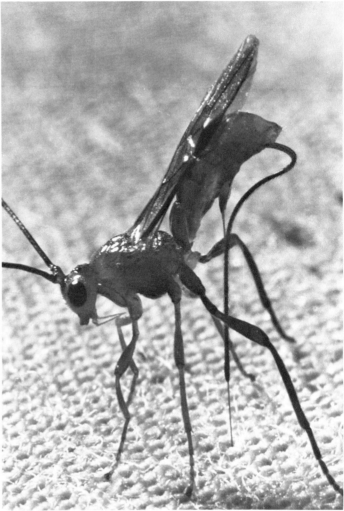

of Johns Hopkins University has pointed out, the cropping function of predators becomes an important generator of organic diversity. A community without predators is certainly possible—only photosynthesizers and decomposers need be present. But such a community would be monotonous in terms of numbers of species present. This is because the most efficient species would increase in numbers to the point where all competitors would be squeezed out. One might conceivably find forests made up of a single tree species, prairies containing a single kind of grass, and ponds with a single species of alga or waterweed. But a predator in such a community would crop the abundant species, limiting its ability to dominate and opening up space for other species. Stanley has proposed that the dramatic rise in the diversity of life during the late pre-Cambrian period occurred as a result of the evolution of the first predators.

Parasitism

If the lion is known as king of the beasts, then the louse is generally assigned to the other end of the scale. Yet the louse and the lion are both predators on other animals. And in any confrontation between the two, the louse would have at least an even chance. The louse is a parasite, and it differs from other predators in that it eats only a little bit of its prey. In addition, the parasite is always smaller than its prey (called the *host*). The parasitic relationship creates special problems and behavior patterns worth considering.

Some of the body plans and life strategies of parasites sound so bizarre that one is tempted to dismiss these organisms as biological aberrations. In fact, the parasitic way of life is a very common one. Parasites come from every kingdom and almost every phylum of the living world (including the chordates and the vascular plants). Certain phyla, such as the flatworms, the nematodes, and the arthropods are more widely distributed in terms of parasitic species, and most of the common animal parasites come from these groups.

In terms of host susceptibility, essentially all living animals contain parasites of one form or another. For example, take any whole, uncleaned fish from a fish market and examine it carefully. Even without a microscope you should be able to find wormlike forms wriggling on the surface of the gills or perhaps around the intestinal tract or the eyes. Careful slicing of the muscles and internal organs reveals more of the same, either in the free-living or encysted state. The same would be true of a wild frog, a turtle, a deer, or a human being from any nonindustrialized country. Plants suffer from a wide range of parasites; some random examples include white pine blister rust, chestnut blight, Dutch elm disease, beech bark disease, and the late blight of potatoes that devastated Ireland in the nineteenth century. In fact, all human crop plants suffer from parasites in greater or lesser degree, and breeding for disease resistance occupies much of the time of plant geneticists. Even the smallest of all living organisms, the bacteria, suffer from internal parasites. These are the bacterial viruses, made famous by the studies of molecular biologists.

What are the special problems of parasites? Because of their small size, they cannot go out whenever they are hungry and bring down a host the way a lion or weasel can. Parasites therefore live with with their hosts permanently, either on the surface (*ectoparasites*) or deep inside body tissues (*endoparasites*). But internal parasites in particular find that penetrating the host's tissues is not a simple task. A few try boring through the outer skin, but most enter through the mouth. A parasite entering through the mouth runs the grave risk of winding up as food while running across the teeth and through stomach juices and pancreatic enzymes. Only the parasite that has evolved resistance to such host defenses has a reasonable chance of winding up inside alive. And once inside, it must be able to resist the immune system of the host. The result is that endoparasites are highly host-specific. Only the parasite with an intricate "knowledge" of its host's body can enter it alive. An understanding of the adaptations involved is still very rudimentary in most cases.

Lice as Parasites. To illustrate some of the adaptive strategies of parasites, let us look at a very common parasite, the louse. Lice are small, flightless insects with poor vision and an exclusively parasitic life-style. They parasitize two classes of vertebrates: birds and mammals. Lice are ectoparasites, since they are adapted for life among feathers or hair. They cannot parasitize featherless or hairless animals. Bird lice and mammalian lice form two separate orders of insects.

Bird lice have mouthparts for chewing and feed on feathers, bits of skin, and some blood. Females lay eggs on the feathers of their host. The eggs hatch out as miniature adults which begin to feed at once alongside their parents.

A bird is typically parasitized by two separate species of lice. One species lives among the short, thin feathers of the head. Since these lice are inaccessible to the bird's beak during preening, the lice need not be flattened. Head lice have large heads and powerful mandibles for firm attachment. Some birds have a comb on their median toes which can be used to scratch the head. These birds never harbor head lice.

The other louse species parasitizing birds lives among the much larger feathers of the back and wings. These lice have flat, elongated bodies and can easily slip between the feathers to escape the preening beak. Birds with defective beaks are ineffective preeners and are much

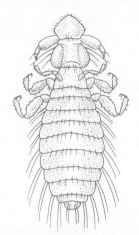 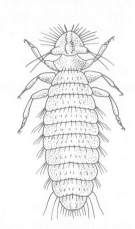

Figure 13-13
BIRD LICE. Left, the head louse of the chicken, Right, the body louse of the chicken.

more heavily parasitized than birds with normal beaks. The head and body lice of a bird may also have different coloration. A swan has a head with dark markings and a white body. The swan's head lice are almost black, while its body lice are very pale-colored.

Birds acquire their lice largely in the nest. There may be occasional transfer of lice during mating. Lice can parasitize another bird only if the bird has body contact with the infected bird. When a bird dies, its lice die with it.

Not only have bird lice become specialized for life in different regions of the body, but they have evolved a great deal of host specificity. The lice of a chicken can not infect a heron, and cuckoos raised by foster parents of other bird species do not acquire the parasites of their brood parents.

Human Lice. The lice that parasitize humans are members of the same order as other mammalian lice. Mammalian lice have mouthparts developed for sucking, hence they live on a diet of blood. No sucking lice are found on birds. Like all mammalian lice, human lice live among body hair and lay eggs at the bases of hairs. Though humans are largely hairless, this is not true of two regions of the body. One is the head region and the other the pubic area. As in the case of birds, two separate species of lice have become specialized for life in these two areas. One is *Phthirus pubis,* the crab or pubic louse. The other is *Pediculus humanus,* the head and body louse.

In people who wear no clothing, only the head and pubic areas are parasitized. But head lice respond to clothing as a substitute for body hair. Hence if individuals wear clothing, the entire body may be parasitized by *P. humanus.*

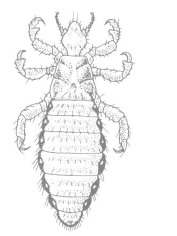
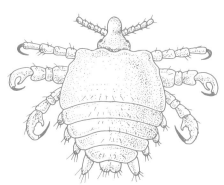

Figure 13-14
HUMAN LICE. Left, the head and body louse of the human. Right, the pubic or crab louse of the human.

As in the case of birds, dispersal is a major problem for human lice. These insects lack wings, have very reduced eyes, and depend on human blood for food. Hence direct body contact between humans again becomes the important mode of dispersal for the lice. The pubic louse travels between mates during the sex act, while the body louse has wider options. It may spread by direct physical contact, by contact with clothing, or via an adventurous agency—the domestic rat. The rat acts as a passive carrier for *P. humanus*—the louse breeds only on human skin.

To summarize, two principles can be seen at work among the life strategy of lice. First, there is a high degree of host specificity—each species of bird or mammal carries its own louse species. Specificity extends even to differences among body regions, and different louse species are present in each. Second, lice are so ill-adapted for life away from their hosts that dispersal from one host to another becomes a problem. The problem is so serious that lice can perish if their hosts die.

If dispersal is a major difficulty for the parasite, then it is to the parasite's advantage not to overtax its host to the point of killing it. A well-adapted parasite is therefore relatively benign to its host. The crab and body lice of humans are examples of well-adapted parasites. They may cause itching and discomfort, but it is doubtful that any human has ever died as a result of losing a few drops of blood to some resident lice.[4]

---

4. The human body louse spreads typhus, a disease that has killed millions. However, typhus results not from the louse itself but from a rickettsial organism (Chapter 9) infecting the louse. The louse that carries this organism is itself a sick animal and dies as a consequence. Thus the rickettsia responsible for typhus is a parasite of a parasite.

Parasites and Human Population Control. But our statement about benign parasitism is not true of parasites that are new to their hosts. A recurrent story of the European explorations from 1492 onward is the decimation of native populations after their first contact with the Europeans. This is what happened to the Hawaiians, the Eskimos, the American Indians, and the peoples of Africa and Oceania. It is a story that even today is repeated among the Indians of the Amazon. Sometimes the disease agents appear to have been viruses of the common cold. These viruses have been endemic in the populations of Europe for so long that mutual host-parasite adaptations have rendered them harmless. But no such host adaptations existed in the Eskimos or Hawaiians when they first met the Europeans, and large numbers paid with their lives. When Cortez landed in Mexico in 1519, he brought with him a malady new to the Aztec Indians: smallpox. Half of the native population died within the first year. More perished in later outbreaks of measles and typhus. In a more recent example, a worldwide epidemic of influenza killed an estimated 23 million people in 1918–1919. The viral strain involved was a new strain to which the human population had had no previous exposure.

But do parasites in fact exert any controlling influence over their host populations? Indeed they do, and some of the most compelling data come from human history. If one follows the growth of the human population over the past 1,000 years, it is obvious that major cataclysms such as World War I and World War II produced no break in the rising curve of human population. The only time the human population showed a decline during the entire period was in the fourteenth century, the century of the bubonic plague. Bubonic plague is caused by *Pasteurella pestis,* a bacterium that seems to have lived with people for as long as historical records exist. A particularly virulent strain made its first appearance in 1346 in the Crimean straits. By 1347 the Great Plague had raced to Italy and France and then through all of Europe, not disappearing until 1357. During the initial outbreak, two-thirds of the European population was afflicted. Almost all of those afflicted died. The plague returned in 1361 (killing 50 percent of the population), in 1371, and in 1382. During the last two outbreaks, only 5 to 10 percent of the population was afflicted, and the great majority of these survived. The result was a significant decrease in the human population during the fourteenth century (see Fig. 13-5). The plague has returned at irregular intervals since the fourteenth century but never with its earlier devastation. The most recent outbreak occurred during the Indochina war, when it responded easily to antibiotics.

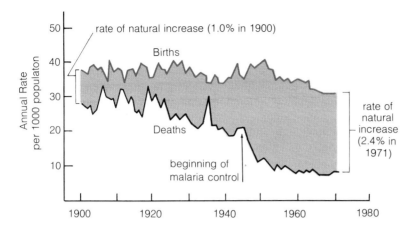

Figure 13-15
BIRTH AND DEATH RATES IN SRI LANKA (CEYLON) BEFORE AND AFTER MALARIAL CONTROL.

War and pestilence have always marched together. This is just what one would expect of a density-dependent population control factor. Humans are nowhere more closely crowded than in huge army encampments or in huddled masses of fleeing civilians. Prior to the soap and antibiotic era, every military campaign was accompanied by raging pestilence, and there is no doubt that more campaigns were decided by typhus, plague, smallpox, and cholera than by the skill of generals or the discipline of armies. In 480 B.C., the failure of the Persians under Xerxes to defeat the Greeks led to the flowering of Athens and established the basic spirit of the West. But Xerxes was defeated by disease, not by the small Greek army. The history of the Crusades is a history of disease. Time and again, armies of upward of ½ million men reached their military objective, and then melted away without a trace as pestilence struck. Some of the armies were destroyed before they could embark. History books give us the names of kings and generals, but it seems that in most military campaigns prior to the antibiotic era, the crucial decisions were made by microbes and viruses as much as by men.

Only within the last quarter-century has the battle between humans and their parasites tilted decisively against the parasites. Malaria has been controlled by the use of DDT to eliminate the *Anopheles* mosquito. Typhus has disappeared with the human body louse, likewise a casualty of DDT and improved hygiene. Cholera, plague, syphilis, and diphtheria all respond to penicillin. With the passing of these ancient enemies from the scene, their former importance as population control agents can be better measured. Thus the rate of natural population growth of Sri Lanka (Ceylon) jumped from 1.0 percent in 1900 to 2.4 percent in 1970. There was no increase in birthrate—the entire increase resulted from a decrease in the death rate. Most of this decrease occurred between 1945 and 1950, with the control of malaria. The unfortunate conclusion is that, unless the birthrate undergoes a similar decrease, the death rate will ultimately have to rise again as new and savage control mechanisms take hold.

Competition

The central biological role of competition was well appreciated by Darwin, who constructed his theory of evolution around it. When the numbers of a population exceed the carrying capacity of its environment, the members of the population must compete for the resources available. Sometimes the individuals may not realize they are competing. If a swarm of maggots (fly larvae) are let loose on a piece of meat, each larva will eat as fast as it can, and those that eat the fastest will have consumed the most meat.

This is an example of *scramble-type* competition. While this type of competition favors some larvae over others, it does not guarantee any one larva an adequate ration. It is possible that all of the larvae will end with an inadequate amount.

If a similar piece of meat is thrown to a pack of hungry dogs, a lot of fighting and snarling will ensue. Only the strongest and most aggressive dogs will receive any meat at all. This is *contest competition*. Contest competition is a more rational approach to the partitioning of a resource, since it ensures that at least some members of a population will be adequately provided for. Contest competition is often highly developed among vertebrates and some arthropods, while scramble-type competition is the rule among invertebrates (including many insects), plants (such as forest trees competing for light), bacteria, fungi, and protistans.

Territorial and Hierarchical Behavior. The fact that a vertebrate engages in contest competition with others of its kind does not mean that it spends a great deal of its time fighting. Actual fighting is greatly reduced by either *territorial* or *hierarchical behavior*. Territorial behavior in animals is observed most easily in the spring. Much of the birdsong that gives the spring its special quality results from the efforts of males in setting up and defending their breeding territories. The males may also carry brighter plumage at this time, both as an aid to territorial defense and as an attractant for the females.

A good place to observe the males at work is a marsh or the edge of a pond. The black birds with bright red or orange shoulder patches are red-winged blackbird males busy patrolling their territories. The flashing color of the shoulders serves to warn other males. The females are a dull brown color—a much more practical choice in view of the fact that they will have to sit on nests built close to the ground. Since redwings feed on marsh insects, the only suitable

Figure 13-16
SOCIAL HIERARCHY IN A BABOON FAMILY. The father is addressing a low-level threat against a contrite daughter riding on her mother's back.

nesting sites are found among marsh reeds. Areas of surrounding forest or field could not provide sufficient food for the needs of the young, and they may not even provide adequate food or shelter for the adults themselves. But since the number of adult males generally exceeds the number of desirable territories, some males wind up with no territory of their own. Such males are not able to reproduce, and are at a disadvantage in merely trying to survive. However, they make up the biological reserve from which replacements will come when any of the established males fall to predators or disease.

In general, only this sort of ultimate catastrophe can remove a male from his territory. Once established, he is imbued with courage and a "sense of right" which makes him immune to challenge by another male of his kind. The result is a greatly diminished level of conflict within the population.

Hierarchical behavior constitutes another mechanism whereby the level of conflict within a population is reduced. This kind of behavior is observed in animals that live together in larger social groupings: flocks of geese, troops of chimpanzees, packs of wild dogs, herds of gazelles. And it is not unknown in human affairs: armies, business corporations, even family groupings depend on it. The principle is well known to all: the general gives orders to the colonel, the colonel to the major, the major to the captain, the captain to the lieutenant, and so on. In a wolf pack, the chain of command is every bit as strict. The leader of the pack (the alpha male) has first choice of food and first choice for sexual intercourse with the chosen female

Figure 13-17 THE BARRED OWL (LEFT) AND THE RED-SHOULDERED HAWK (RIGHT). The two birds of prey follow almost identical life-styles. The major difference is night hunting by the owl as compared to day hunting by the hawk.

(the alpha female). In fact, only the alpha pair engage in sexual activity at all. Other members of the pack are forcibly restrained by either the alpha male or the alpha female. Because every member of the pack knows its rank with respect to the others, there is no need for fighting to settle disputes. And whatever fighting does occur is carried on with stylized bluffing and appeasement. The losing wolf bares its throat to the victor, and this gesture of submission inhibits the instinct to kill in the latter. The wolf is too deadly an animal to allow itself the possibility of a real fight. (It is unfortunate that humans, who are far more deadly than wolves, are not satisfied with a similar gesture of appeasement. But of course it is hard for a bomber pilot to see the facial gestures of his victims 30,000 feet below.) For the wolf, the result of such hierarchical behavior is that the pack reaps all the benefits of cooperative behavior (such as pack hunting), and limited resources are distributed in a rational manner.

Competition between Species. So far we have considered competition among members of the same species. Does competition ever arise among members of different species? The question is a controversial one in the ecological literature. All attempts to study interspecific competition in a laboratory setting have ended in failure. When two closely related species are thrown together in a uniform environment and allowed to compete for the same resource, one or the other species always wins and drives the other into extinction within 30 to 40 generations. This conclusion is summarized as Gause's principle: No two species can coexist if they occupy the same niche simultaneously. To an ecologist, the term "niche" does not mean the physical location of a given species, such as in trees, in the water, or under rocks. Rather, the term refers to the tolerance limits of each species. Consider two species known to coexist across most of their range: the barred owl and the red-shouldered hawk. Do they occupy the same niche? See Table 1.

TABLE 1
Tolerance limits of the red-shouldered hawk and the barred owl

TRAIT	RED-SHOULDERED HAWK	BARRED OWL
Habitat preference	Moist woodland	Moist woodland
Temperature preference	−20°F (winter) to 80°F (summer)	−20°F (winter) to 80°F (summer)
Food preference	Carnivorous; hunts on the wing	Carnivorous; hunts on the wing
Size of prey taken	Mouse-sized up to rabbit-sized	Mouse-sized up to rabbit-sized
Time of peak activity	Day	Night

As shown in Table 1, these two large birds of prey have remarkably similar preferences. Had we looked only at the first four parameters, we might have concluded that the two birds occupy the same niche and should compete with each other. But a look at the time of peak activity shows that the hawk hunts during the day and the owl at night. This immediately eliminates competition between the two species and places them in different niches.

Can the same kind of analysis be applied to other examples of potential interspecific competition? Simple intuition tells us that such clear-cut differences are not always found. For instance, the northeastern United States contains five to six species of wood warblers (small insect-eating birds) with similar body form and feeding habits. The great ornithologist Roger Tory Peterson has called these species the "troublemakers," because it is almost impossible to distinguish them in the field. It is hard to imagine that some niche overlap between these species does not exist. An even more difficult situation exists in Lake Baikal in the eastern USSR. This deep, ancient freshwater lake contains at least 239 species of gammarid crustaceans (small shrimplike animals). And Lake Malawi in southeast Africa contains over 200 species of cichlid fish (members of a family of fish that protect the young in the mouths of the adults). Even more than with birds of the American Northeast, some degree of niche overlap appears to be inevitable in these situations. Then why haven't the more successful competitors driven the less successful ones into extinction? Several possibilities suggest themselves.

Perhaps the most plausible explanation for the coexistence of so many related species is Stanley's theory of predator pressure. If predators or parasites maintain potentially competing species at levels below their individual carrying capacities, then food supply, nesting sites, and other requirements cannot be present in limiting amounts. One potential competitor species cannot put pressure on another by preempting these necessities for its own use. Another possibility exists that all such closely related species do in fact occupy separate, nonoverlapping niches, and biologists simply have not been discerning enough to discover the nature of such niche partitioning. The question remains a controversial one, and more research will be necessary to settle it.

Endocrine Control of Birthrate

Mice and elephants are two animals that seem to have very little in common. In fact, it is doubtful that each is aware of the other's existence. Yet both respond in the same way to crowding: they significantly slow down their reproductive rates.

Since mice are more often kept in laboratories than elephants are, more is known about their reproductive processes. In experiments in which mice are given unlimited food but are maintained under conditions of overcrowding, gross disturbances in reproductive function occur. Sexual maturation is delayed, ovulation is suppressed, fetuses are resorbed in the uterus, milk production is inhibited, and there is a disturbance in maternal function.

Since all reproductive functions fall ultimately under hypothalamic control, it is obvious that nervous stress could account for all the above phenomena (see Chapter 6). Recently, a *pheromone* (a hormone used in intraspecies communication) was discovered in the urine of mice subjected to crowding. As crowding increases, pheromone production also increases. The pheromone leads to a disturbance in the levels of both sex hormones and adrenal hormones. Its mode of action is under investigation. A similar crowding-induced drop in the birthrate has been observed for elephants, snowshoe hares, deer, and a variety of songbirds. Is there a possibility that similar mechanisms operate among humans? No physiological response observed in another mammalian species can be considered irrelevant to humans.

Emigration

In 1883, the greatest explosion ever known on earth destroyed the Indonesian island of Krakatoa. The island was the cone of a long-inactive volcano. Unable to find normal relief by eruption, underground pressures had built up for years and finally blew the top of the island into the stratosphere. So much dust and debris was injected into the upper atmosphere that, across the whole world, sunsets were dyed in brilliant oranges and purples for 2 years afterward. Global weather patterns were disrupted, and crop production was compromised.

As for Krakatoa, all that remained was a sterilized stub of cinders and basalt. The first human visitors found no trace of life of any sort. Yet today, less than 100 years later, the island is overgrown with a lush tropical jungle and harbors animals as large as crocodiles and pythons. But every speck of life on the island arrived there by natural agencies: floating in the wind, clinging to drift, blown by typhoons. The nearest unaffected island was 25 miles away. Biologists carefully recorded the recolonization of Krakatoa (Table 2).

TABLE 2
Return of life to Krakatoa

YEARS AFTER EXPLOSION	SPECIES FOUND
0	No life
1	1 spider species
3	Blue-green algae, 11 fern species, 15 flowering plant species
10	Coconut trees
25	Dense forest with 263 animal species
50	36 bird species, 5 lizard species, 3 bat species, 2 rat species, 1 crocodile species, 1 python species

Krakatoa is an example of the dispersive force of life; wherever an empty spot in the world exists, it is soon discovered and colonized by living organisms. But the corollary of this statement is that, whenever populations find themselves cramped in terms of food or living space, they tend to disperse to more distant areas. Such dispersion forces may work on any scale—the microscale of the family to the macroscale of regional mass migrations.

Beavers illustrate dispersion on the microscale. A beaver dam is built by a single breeding pair. In the flooded area behind the dam, the male and female build a lodge with an underwater entrance, and here they raise their young. For a whole year the young are fed

Figure 13-18 THE DESERT LOCUST OF AFRICA. (A) Female laying eggs in the ground. (B) Farmers in Morocco attacking a locust swarm with poison bait. Giant swarms like this one can strip an entire region of its field crops in a matter of days.

and sheltered, but, once they reach sexual maturity, the parents turn on them and drive them out of the lodge. They must then leave a sheltering environment and travel overland to find a new stream suitable for damming and lodge building. While many beavers perish during this difficult time, a sufficient number find new homes, with the result that the whole species prospers.

Locusts illustrate dispersion on a macroscale. The desert locust has been a major pest in Africa and part of Asia for at least 3,000 years (Fig. 13-18). The Bible mentions the plagues of locusts that afflicted Egypt. A single swarm of locusts may contain 20 billion individuals and consume 50,000 tons of green food per day. It is easy to see how an entire nation might be driven to famine by a locust plague. Locust plagues are cyclic in nature; peaks of activity occurred in 1914–1916, 1928–1933, 1941–1947, and 1952–1962. During the intervening years, the swarms seemed to disappear into thin air. The origins and disappearance of locust swarms remained a mystery for thousands of years. Then in 1921 the Russian entomologist B. P. Uvarov discovered that the common, relatively harmless grasshopper becomes transformed into the destructive locust under conditions of high population density. Structurally there is little difference between a grasshopper and a locust; the latter has longer wings, darker coloring, and a more slender body shape. But whereas the grasshopper is solitary in behavior, the locust is gregarious and travels in dense swarms. This accounts for its local destructiveness. The transformation of grasshoppers into locusts occurs within about three generations as a response to a buildup of the grasshopper population. The locust then becomes the dispersal form, traveling sometimes for hundreds of miles to new areas. If the swarm cannot find adequate food supplies, its numbers are reduced by starvation. Descendants of the survivors then become transformed back into grasshoppers which settle down quietly amid their new surroundings. However, it is only in breeding areas that grasshopper populations can build up sufficient densities to build another swarm.

The kind of migrations we have been discussing should not be confused with annual migrations, for example, of birds, whales, or caribou. These annual migrations consist of reversible, two-way traffic which follows a seasonal resource. Dispersal-type migration is a one-way process stimulated by high population density. Migrating individuals leave the mother population and gamble that somewhere else a suitable habitat will be found.

Population dispersal has played a major role in human population dynamics during the past 500 years (Table 3).

TABLE 3
World population in 1970, by continent

	MILLIONS
Africa	344
Asia	2,100
Americas	510
Europe	676
Oceania	18
Total	3,648

Most of the present population of the Americas and of Oceania, as well as part of Africa, represents a surplus exported from Europe.

In this chapter we have examined the factors that cause populations to grow and factors that bring the growth under control. Growth results when the birthrate exceeds the death rate, and the rate of growth itself increases as the population increases. Growth is checked by density-dependent control factors: predation and disease, competition, endocrine effects, and dispersal by migration. It is probable that all these factors have played a role in the past in human population control. The question of what factors will control human population growth in the future will be answered by human choice and by human intellect. One must hope that these control factors will not include the ancient scourges of war, famine, and pestilence.

In this chapter we have not examined all possible modes of interaction among different species or among members of the same species. The most significant of these is cooperation. Cooperation among different species is called *mutualism*. Examples include the mycorrhizal associations between fungi and plant roots (Chapter 9), the relationship between aphids and ants (Chapter 12), and the interdependence of flowering plants and their insect pollinators (Chapter 10). And we have touched briefly on the possible evolutionary importance of cooperation among members of the same species (Chapter 8). The sum total of all such intra- and interspecies interactions constitutes a behavioral glue which binds together the different populations within a single community. There is another way to look at the relationships between different populations in a community. This view is an economic one; it considers the flow of goods and services through a community. The goods and services of a biological community are nutrients and energy. We shall explore this vision of a community in the last chapter of this book.

Bibliography

GENERAL READING

Hans Zinsser, *Rats, Lice, and History,* Bantam Books, 1965. First published in 1934, this little book has gone through dozens of printings and has achieved a permanent place in the biological literature. Written by a Renaissance man.

Archie Carr, *So Excellent a Fishe,* Anchor Books, 1973. Archie Carr writes about the mysterious life cycle of the green turtle, as well as about the economic pressures bringing this animal to the brink of extinction. Here is a sample passage.

I have no doubt that, back in the days before turtles evolved the way they are, if somebody had suggested to their race that they ought to have fewer offspring and take better care of them—not leave them behind on land, but tend them in the sea, perhaps in the admirable way a porpoise does—the ancestral turtles would have been amenable. That it never happened doesn't necessarily mean it would have been impossible, that natural selection could not possibly have built childcare into a turtle. What it means is that except for the recent ruin humanity has brought them to, turtles are a satisfactory product of natural selection which, in spite of seeming flaws, are really doing all right. They are a surviving and thus evidently successful animal, in spite of what seems to a man the desperate plight of their young on shore. This just adds a bit more proof to the proposition that the hundred eggs a turtle lays is a package pregnant with meaning and history. In other words, ravening as the turtle predators are, they have not killed off the turtles.

D. W. Ehrenfeld, *Biological Conservation,* Holt, Rinehart, and Winston, 1970. This slim volume, dealing with factors threatening natural communities and individual species, makes use of detailed case histories. There are fine discussions of the blue whale, the large fauna of India, the fishes of the Southwest, and many others. This excerpt is about Africa.

Another example of the consequences of altering the species composition of natural communities is provided by the substitution of domestic cattle for the savanna herbivores of much of Africa south of the Sahara. The wild ungulates of the African savanna are an extremely diverse group, including scores of species from several mammalian orders, and ranging in size from the 11.5-pound dik-dik to the 11,500-pound elephant. These different species occupy a wide variety of feeding niches, and are therefore not all in competition with each other for food. Unfortunately, conventional European concepts of range management were, until recently, widely adopted in Africa; in vast areas the wild ungulates were removed and replaced with large numbers of cattle, sheep, and goats, all existing on grasses and low plants and unable to use much of the local vegetation for food. In a short time, the original carrying capacity of many areas was exceeded, plant cover disappeared, carrying capacity declined, livestock died, and dust bowls or baked clay crusts (the hallmark of ecological mismanagement) left thousands of square miles unfit for any human endeavor.

C. J. Krebs, *Ecology,* Harper and Row, 1972. A full-sized text of ecology which emphasizes factors that contribute to the growth and regulation of populations.

R. H. Wagner, *Environment and Man,* 2nd ed., Norton, 1974. A textbook of environmental issues, from water pollution to nuclear energy. There are also excellent chapters on the extinction of species and the introduction of exotic species.

Robert E. Ricklefs, *Ecology,* Chairon Press, 1973. A large text of ecology with a wider scope than most. The writing style is fresh, full of adventure and wit.

Figure 14-1 SUNLIGHT THROUGH A MICHIGAN FOREST.

chapter 14

Someone once asked the painter Marc Chagall why he painted so many flowers. Chagall replied, "Because they are evanescent. They are alive for such a brief moment—that is what makes them beautiful." A similar observation could be made about all living communities. For any community to remain in a living state, there are two requirements: a flow of energy through the community and a flow of materials. These are the same general requirements for life that must be met by a single cell (Chapter 2).

Energy Flow and Materials Cycling

The flow of materials through a community is cyclic; the same atoms or combinations of atoms are used over and over in a shuttle from the living to the nonliving world. But the flow of energy is unidirectional. The source of all energy needed to sustain a living community is the sun. This energy from the sun has a brief residence time on earth before it continues its inevitable journey out toward distant galaxies. But during its fleeting passage this energy can express itself as life, and in this sense all life is as evanescent as flowers, youth, or consciousness.

Energy Flow

In theory, any form of life could be driven directly by the sun. In fact, only one group of organisms is driven thus directly: green plants, algae, and the photosynthetic monerans. Collectively these are called the *producers*. They constitute the storage batteries of the world, which drive the motors of every other living thing, from humans to tigers to nightingales, as well as the everyday motors of washing machines and automobiles.

The sun's radiant energy is electromagnetic in nature. The energy that drives living organisms is chemical. Plants and other producers therefore act as *transducers,* that is, agents that transform one form of energy into another. The efficiency of this process is very low. The bulk of the sunlight intercepted by the earth is reflected back into space by the atmosphere, the polar ice caps, and the surface of the earth itself (Table 1). Only 43 percent of the total

TABLE 1
Uses of sunlight in terrestrial processes

Absorbed light	43%	
Absorbed light	43%	
Heating of water and land		36.0%
Absorbed by plants		7.0%
Used in photosynthesis		0.035%
Total incident radiation	100%	

sunlight is absorbed by the earth. Of the light absorbed, only a small amount strikes plants. And of the light intercepted by plants, only 0.5 percent (as a worldwide average) is used in the photosynthetic process to drive chemical bond formation.[1] The result is that photo-

---

1. The figure of 0.5 percent is the photosynthetic efficiency, obtained by dividing 0.035 percent by 7.0 percent.

synthetic production adds up to no more than 1/3000 of the solar energy striking the earth. This is the amount of energy available to drive all living processes on earth. (However, the other 2999/3000 of the energy is not "wasted." This is the energy that warms the earth to make it hospitable for life and that creates the wind and rainfall that make up the climates.)

As explained in Chapter 2, the photosynthetic machinery of plants is able to harness light energy to assemble carbon atoms of carbon dioxide into more energy-rich linkages such as sugars (carbohydrates):

$$H_2O + CO_2 + light \xrightarrow{\text{Photosynthetic apparatus}} sugar\ (carbohydrate) + O_2$$

These energy-rich linkages of sugars thus represent a form of stored sunlight. It is these sugars and their derivatives, and not the sunlight itself, that then drive all life processes, including those of the plants themselves.

Energy Flow through Communities

The questions of how much energy flows through a community and how it is distributed are fundamental to the understanding of a community. Such an energy budget tells us as much about a biological community as a financial budget tells us about a human government. Is a government rich or poor? Who produces the wealth and who

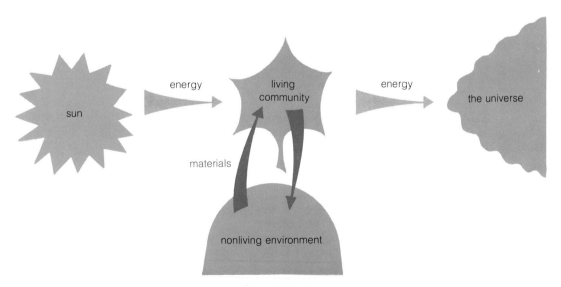

Figure 14-2 ENERGY AND MASS FLOW THROUGH A LIVING COMMUNITY.

consumes it? Is the economy simple or complex? Is the budget balanced? Such questions are fundamental to the management of a state or a city, and they are equally fundamental to the management of a biological community.

There are two energy-related factors that do the most to shape the nature of a biological community. The first factor is quantitative—it is the level of community productivity. This is a measure of the total amount of energy flowing through a community. Communities with a high energy flux are likely to be richer and more complex structurally than low-productivity communities. The other factor that shapes the nature of a biological community is qualitative—it is the manner in which the energy wealth of the producers is apportioned among other members of the community. We will discuss each of these factors in turn.

All the energy that flows through a community is of course derived from the sun[2] and is transformed into biologically usable energy by green plants and other producers. The simplest way to measure this energy input is to measure the growth of the plants. However, not all of the energy photosynthetically fixed by plants shows up as growth. This is because part of the trapped energy is used by the plants to maintain their own life processes. Plants respire just as animals do—they use oxygen to burn sugars and exhale carbon dioxide and water. But this process can be observed only in the dark. During the day, the process of plant respiration is obscured by the more vigorous process of photosynthesis, which consumes carbon dioxide and produces oxygen. The total food photosynthesized is called the *gross production*. Energy (food) spent in maintenance functions is called *respiration*. And energy directed into growth is called the *net production*. Only the net production of a plant is available for further use within the community. The three functions are related as follows:

Gross production − respiration = net production (growth)

Gross production and net production refer to total quantities of energy. If we want to speak of rates (for example, the amount of production per year), we speak of *gross productivity* and *net productivity*.

To translate these terms into the terms of human economics, the gross productivity of a biological community is analogous to the gross national product of a country. It is the sum total of all goods and services produced by the country in a year. The net production is a quantity that human economists seldom measure, perhaps because it would be so embarrassingly low. It would correspond to the real increase in the wealth of a country after maintenance costs have been

---

2. This is only indirectly true of human agriculture, in which energy input from fossil fuels may exceed the solar input. But fossil fuels are likewise products of the sun.

deducted from the gross national product (costs such as depreciation of buildings and other physical goods, costs of war and social pathology, and costs of heating, lighting, and waste disposal).

Community Productivities

The rate of energy production within a community may be expressed either in terms of dry weight (grams per square meter per year) or in terms of energy equivalents (kilocalories per square meter per year).[3] Communities across the world may differ by a factor of 100 or more in their rates of energy production (Table 2).

3. As a rough approximation, 1 gram of plant dry matter equals 4.5 kilocalories of energy. A kilocalorie of energy is the amount of energy needed to raise the temperature of 1 kilogram of water 1°C.

TABLE 2
Entimated gross productivities[a]

COMMUNITY TYPE (BIOME)	AREA (10^6 KM^2)	GROSS PRIMARY PRODUCTIVITY (KCAL/M^2/YEAR)	TOTAL GROSS PRODUCTION (10^{16} KCAL/YEAR)
Marine			
Open ocean	326.0	1,000	32.6
Coastal zones	34.0	2,000	6.8
Estuaries and reefs	2.0	20,000	4.0
Subtotal	362.4		43.4
Terrestrial			
Deserts and tundras	40.0	200	0.8
Grasslands and pastures	42.0	2,500	10.5
Scrub (dry) forests	9.4	2,500	2.4
Northern coniferous forests	10.0	3,000	3.0
Cultivated lands with little or no energy subsidy	10.0	3,000	3.0
Moist temperate forests	4.9	8,000	3.9
Fuel-subsidized (mechanized) agriculture	4.0	12,000	4.8
Wet tropical and subtropical (broad-leaved evergreen) forests	14.7	20,000	29.0
Subtotal	135.0		57.4
Total for biosphere (not including ice caps. Round figures)	500.0	2,000	100.0

[a] After Eugene P. Odum, *Fundamentals of Ecology*, 3rd ed., W. B. Saunders, Philadelphia, 1971, p. 51.

Figure 14-3 FOREST IN ROCKY MOUNTAIN NATIONAL PARK. The forest was burned in 1919 and photographed in 1971. Revegetation under the near arctic conditions is laboriously slow.

The least productive areas are deserts (because of the lack of water), tundras (because of the cold), and open oceans (because of the lack of plant nutrients). The most productive areas are reefs and estuaries (shallow waters where large rivers enter the oceans), and wet tropical forests (such as the Amazon River basin). It is ironic that precisely these are the areas now under strongest human pressure. Tropical forests are being cut down in a desperate effort to expand agriculture. It is estimated that 90 percent will have disappeared in 30 years. Estuaries are being burdened with pollutants which may not destroy their ability to fix energy but which destroy their stability and complexity. At the other end of the spectrum, deserts and tundras are equally vulnerable because their resistance to human stress is vanishingly low. The path made by a halftrack vehicle across the Alaskan tundra in 1945 is still clearly visible 30 years later. The grazing pressure put on short-grass prairies has extended the Sahara Desert southward and the western American deserts eastward across Arizona and

Figure 14-4 BURNING OF A TROPICAL FOREST IN COSTA RICA. On a worldwide basis tropical forests are being destroyed at a rate of 14 acres a minute.

Figure 14-5 AREA AROUND DUCKTOWN, TENNESSEE. A century ago these hills and valleys were covered with lush cove forests. Sulphur dioxide fumes from copper smelters have eliminated all vegetation.

New Mexico. Any strip-mining projects in these areas are almost certain to result in irreversible loss of productivity.

The productivity of a community depends not only on external factors such as temperature and rainfall, but on biological conditions created by members of the community themselves. The spectacular desert in the midst of rich forests around Ducktown, Tennessee (about 40 miles east of Chattanooga) was created because all plant life was killed by sulfur dioxide fumes from a copper smelter (Fig. 14-2). Though the sulfur dioxide fumes are long gone, the forest has not returned. The sterile soil, baked by sun and leached by rain, offers no foothold for plant seedlings which might ultimately grow and moderate the harsh environment. No one knows when the plants

Figure 14-6 A FINE FIELD OF BARLEY IN HOLLAND. An example of energy-subsidized agriculture.

will return, but the original forest has not returned for the past 70 years.

One other kind of entry in Table 2 merits special attention. This is agricultural land. In fact, two types of agriculture are practiced in the world. The agriculture of the United States, Canada, Europe (including the USSR), and Japan is a mechanized agriculture. There is a heavy energy subsidy of the growing crops: plowing, cultivation, irrigation, and application of fertilizers and pesticides. This energy input, based on fossil fuels, typically exceeds the direct photosynthetic contribution by the sun. The result is a great rise in gross productivity, when compared to agriculture with minimal fuel subsidy. The increase in net productivity may be even larger (Table 3). This is

TABLE 3
Annual agricultural yields for major food crops[a]

CROP	NET PRODUCTIVITY (KCAL/M^2/YR)	EDIBLE PORTIONS (KCAL/M^2/YR)
Wheat		
fuel-subsidized (Netherlands)	4,400	1,450
minimal fuel subsidy (India)	900	300
world average	1,300	430
Corn		
fuel-subsidized (United States)	4,500	1,510
minimal fuel subsidy (India)	1,100	350
world average	2,400	810
Rice		
fuel-subsidized (Japan)	5,500	1,840
minimal fuel subsidy (Brazil)	1,700	580
world average	2,300	760
Soybeans		
fuel-subsidized (Canada)	2,400	800
minimal fuel subsidy (Indonesia)	780	260
world average	1,400	480

[a] After Eugene P. Odum, *Fundamentals of Ecology*, 3rd ed., W. B. Saunders, Philadelphia, 1971, p. 54.

because the energy subsidy given to the crop can substitute for energy the plant would have had to use in maintenance functions. This process has been carried to its ultimate in animal husbandry, in which animals are bred as egg-laying or milk-producing machines and become so dependent on their human keepers they can barely stand.

This dependence of modern agriculture on fossil fuel subsidy must be kept in mind in all discussions of the "green revolution" and schemes for raising crop levels in underdeveloped countries. Export of miracle crop strains to the latter can have only a marginal effect unless accompanied by export of fertilizers, irrigation pumps, tractors, and fuel. And these are beyond the economic reach of all but the developed countries at present.

One more aspect of Table 2 should be pointed out. At the present time, a little over 10 percent of the total world land mass is under cultivation, yielding about 14 percent of the total gross production on land. This is a stupendous diversion of the world's energy resources by a single species! But before concluding that a great further increase in agricultural land is possible, it is worth remembering that, at the present time, domestic animals consume five times as much food as human beings do. Not all these animals feed only on agricultural land—many feed on grasslands and forests as well. And huge amounts of forest growth are harvested for use as lumber, paper, fuel, and so on. Thus very little of the world's land area escapes human utilization in some form. Channeling more of the total into direct grain production would certainly increase total food yield, but it is unlikely that more than 25 percent of the total land area is suitable for intensive farming. These limits are imposed by water supply, topography, and climate.

Finally, and most significantly, one must remember that what is important is the quality of individual human life, not the total mass of humanity that can be fed by a regimented world ecosystem. I remember vividly a friend from Europe who spent a year working in a research laboratory in the United States and then spent a month touring the western states. I saw him just before he went back to Europe and asked him what had been his most memorable experience. He answered, "Traveling a whole day through Utah with only my family and a pack burro, and not meeting another human being." Such an experience is no longer possible in much of the world today.

Energy Budgets and Community Structures

What happens to the food synthesized by the producers? As pointed out before, a considerable amount (at least 25 percent and typically much more) is used by the producers for their own maintenance. Fortunately for us, the plants as a group synthesize more carbohydrate than they need for their own life processes. The surplus then becomes available to other organisms which acquire the carbohydrate by eating the plants. These plant feeders or *herbivores* may then in turn be eaten for their content of carbohydrate and other molecules by the *carnivores*. The result may be described as a *food chain*, for example,

marsh grass ⟶ grasshopper ⟶ marsh wren ⟶ marsh hawk

or

pond algae ⟶ tadpole ⟶ yellow perch ⟶ human

Food chains seldom contain more than four or five links. This is because each act of feeding wastes perhaps 90 percent of the available energy contained in the food. Or, to put it another way, 1,000 pounds of marsh grass could be transformed into 100 pounds of grasshoppers, which could be transformed into 10 pounds of wrens which could be transformed into 1 pound of hawk.[4] As a consequence, there are fewer hawks than grass plants, and there are generally not enough hawks in a marsh to form a reliable food source for a still higher carnivore. The hawk therefore becomes the *top carnivore* in its community. While such decreases in numbers of individuals generally follow from the diminishing amounts of energy available at each step of the food chain, they need not hold if the predator is much smaller than the prey. For instance, if a marsh hawk is being eaten by lice, then 1 pound of hawk will still support no more than 1/10 pound of lice, but this may amount to many hundreds of lice.

This has important implications for human agriculture. A population feeding on grain has at least five times more food available from a given area than a population feeding on cattle. This is why meat is the first item to be eliminated from the diets of the poor. But unless special care is taken, this may also eliminate a number of amino acids essential for proper development of nervous tissue and other body parts. Though plants contain all nutrients found in animals, they do not contain them in the same ratios or in the same amounts.

Biological Magnification. However, not all materials ingested undergo such huge wastage at each step in the food chain. Let us take a common example. Marshes often support high mosquito populations, and people living in adjoining areas may demand a mosquito abatement program. On a direct-cost basis, the cheapest control measure is spraying with the insecticide DDT. DDT costs only a few cents a pound and is so longlasting that a single application may remain effec-

---

4. Such low efficiencies of transformation are perhaps surprising. The reasons for such low efficiencies must be sought in the second law of thermodynamics (Chapter 2). This is the law that states that all spontaneous natural processes produce an increase in entropy. Entropy is a form of energy that is unavailable for the performance of useful work. Work is required in order to assemble the raw materials of a food organism into structures characteristic of the feeding organism. If the efficiency of transformation involved in acts of feeding were 100 percent, then food chains could be infinitely long.

Figure 14-1
THE BALD EAGLE. This species' continued existence is in question because of the accumulation of organochlorines in the environment.

tive for months to years unless dispersed by natural processes. Though DDT is no longer applied to marshlands in the United States, it has a host of chemical cousins with similar properties (the organochlorines): aldrin, endrin, dieldrin, sevin, heptachlor, toxaphene, lindane, chlordane, and methoxychlor, to name a few. These are still used in liberal amounts.[5] When organochlorines are eaten by animals, they migrate to the fatty tissues and remain there. Because the molecules have no counterparts in the living world, the animal's body lacks enzymes to degrade them or excretory mechanisms to excrete them. Therefore substantially all the organochlorine content of an organism is passed on to the next step in the food chain. If the marsh grass is sprayed with DDT at a concentration of only 0.3 parts per million, and if a marsh hawk requires a food base of 1,000 pounds of marsh grass per pound of body weight, then it will accumulate DDT in its tissues at a concentration of 300 parts per million. This is an example of *biological magnification*.

[5] This situation may be ending. In 1974 the Environmental Protection Agency banned the use of aldrin and dieldrin on food crops. In 1975 it issued a provisional ban against chlordane and heptachlor. The reason was not the ecological havoc wrought by these compounds but the possibility that they may cause cancer in human beings.

Such a high concentration of DDT may kill a bird outright by interfering with nervous function; countless robins were killed in the 1950s after feeding on DDT-loaded earthworms. But even much lower concentrations of DDT and other organochlorines may interfere with the reproductive processes of birds. DDT inhibits an enzyme necessary to build the calcium shell of a bird's egg. As a result, eggs are laid with thin shells which break during incubation, and the parents fail to produce young. At the present time, the peregrine falcon and the bald eagle have almost disappeared from the continental United States because of interference with shell formation. The peregrine falcon and the southern bald eagle face a real threat of extinction. Even if all DDT use were stopped tomorrow, the amounts still present in the biosphere will remain lethal for a decade or longer. It is significant that both the peregrine falcon (the fastest bird in the world) and the bald eagle are at the top of their food chains.

Biological magnification is not a phenomenon restricted to insecticides. Very similar effects have been observed with polychlorinated biphenyls (PCBs), industrial compounds related to DDT. The same is true of many products of radioactive fallout such as strontium-90, iodine-131, cesium-137, and zinc-65. In the late summer of 1965, following spring hydrogen bomb tests in the Pacific, many tuna caught by Japanese fishermen were found to be strongly radioactive. One fish, caught over 1,000 miles from the test site, showed tissue radioactivity of 4,500 counts per minute. Such a radioactivity level would have been extremely dangerous to its human consumer. The radioactivity was traced to zinc-65, concentrated by oceanic food chains. The tuna stands at the top of a food chain in the ocean, just as the bald eagle does on land.

Food Webs and Trophic Levels. In the real world, the food chain is generally an oversimplification of the actual situation. Figure 12-10 shows a more reasonable (though still oversimplified) approximation of the direct feeding relationships based on marsh grass. This is called a *food web*. For simplicity, a food web of this sort may be divided into *trophic levels* ("feeding levels"). A given trophic level includes all animals that obtain their food in the same number of steps (such as periwinkles, grasshoppers, and mice). But a gull or a marsh wren can feed at several different trophic levels. Any viable community generally contains the following trophic levels.

    1. Producers (here marsh grass). This is the base on which the entire community rests.

    2. Herbivores or plant feeders. These are adapted to live on a high-cellulose diet. They may have complex stomachs and long intestines and may contain mutualistic microorganisms which help

with the digestive process. (*Mutualistic associations* are associations between two organisms that result in benefit for both organisms; see Chapter 13.)

3. Carnivores or animal eaters. They must be fast and strong enough to overcome their prey and generally have mouthparts and appendages suitable for meat feeding.

4. Decomposers. These are chiefly fungi and bacteria. They break down wastes and remains of other organisms and convert them back into simple minerals. They are essential for the recycling of minerals. Decomposers are not physically conspicuous, but they typically account for the bulk of energy flow in a community (especially in a terrestrial community). For example, decomposers consume 90 percent of the salt marsh grasses in a salt marsh and 93 percent of the plant material in a deciduous forest. When such large energy flows are involved, the decomposers give rise to their own food chains, called *detritus food chains* (as opposed to *grazing food chains*). A detritus food chain goes from dead organic matter to decomposers and then into organisms that feed on the decomposers, such as earthworms or mussels. These in turn are preyed on by carnivores.

Use of Energy by Consumers

We have seen that, in a general sense, all members of a community depend on the producers for their energy. Let us now focus more closely on the process of energy transfer at the level of the individual consumer. To illustrate, let us consider a deer (a grazer[6]) feeding on a twig of a tree (a producer). The deer is a feeder with a strong appetite, but it wastes much of its food. About 25 percent of the ingested twig is excreted without ever being digested. Only 75 percent of the plant is digested and absorbed into the circulatory system—this is called the *assimilated* fraction. The assimilated energy of an animal is analogous to the gross production of a plant. As far as the deer is concerned, food that is not assimilated is food that is lost. However, there is no loss of energy as far as the community is concerned. The excreted portion becomes an energy source for the detritus food chain.

As can be seen from Table 4, herbivores have generally lower rates of assimilation as compared to carnivores, because of the difficulty of digesting such plant materials as cellulose. Carnivores have the simpler metabolic task of converting one form of animal life into another.

[6] We use the term here to include all animals that feed on living plants or on other producers.

TABLE 4
Comparison of assimilation efficiencies

ANIMAL	FOOD	PERCENT ASSIMILATED	PERCENT EXCRETED
Millipede	Rotting plant material	15	85
Grasshopper	Grass, leaves	30	70
Mouse	Seeds & fruit	70	30
Deer	Twigs, buds	75	25
Weasel	Mice, small animals	95	5

The next question is, What does the deer do with the food it has assimilated? As with the plant originally, the deer has to use much of the food energy it takes in just to remain alive; energy is necessary for muscle movement, for replacement of worn-out body parts, and for maintenance of ion concentration differences in the cells. As with plants, such energy usage is termed respiration. All food energy left over is channeled into *growth* (equivalent to net production in plants). The deer uses 89 percent of its assimilated energy for respiration and 11 percent for growth. This compares favorably with the mouse and the weasel, each of which consumes 97 percent of its food energy in respiration. Their small body size results in higher energy loss due to radiation of body heat.

We can now summarize the fate of ingested energy as in Figure 14-8. Only that portion of ingested food that is assimilated and is not lost to respiration becomes channeled into growth. Such a diagram may be used in several ways. We may use it to describe the behavior of a single deer. More usefully, we may use it to describe the behavior of the entire deer population within a given community. Or, to simplify matters still further, we may use such a diagram to describe the behavior of all organisms at the same trophic level as the deer, that is, all herbivores. If we do the same for all trophic levels, we can begin to describe the overall pattern of energy flow within the community.

Before we consider an example of such an overall pattern of energy flow, one point must be kept in mind: the total energy flow through a given population or trophic level need have no relation to its biomass.[7] For instance, the microscopic algae of a pond can be con-

---

[7] The biomass of a population or trophic level is equal to the total weight of its living tissue. This value is usually expressed in terms of grams per square meter or kilocalories per square meter. For instance, to obtain the biomass of the producers of a forest, one would have to cut all the trees and herbs in a given area (the roots would also have to be removed), dry the materials, weigh them, and express the value in terms of grams per square meter or kilocalories per square meter equivalents.

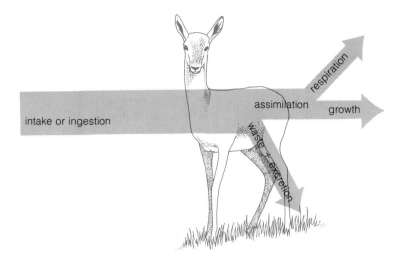

Figure 14-8 THE PARTITIONING OF FOOD ENERGY BY A DEER.

sumed by two kinds of organisms: decomposing bacteria, which feed on dead algae, and larval forms of chironomid midges (Fig. 11-9), which feed both on living and decomposing algae. In total weight, the bacteria form an insignificant fraction of living organisms of the pond, while the chironomid larvae form a substantial fraction (up to 30 percent of the total). Yet in terms of energy flow, far more of the community total flows through the bacteria than through the chironomids. This is because the bacteria have a greater respiratory loss.

Similarly, the towering trees of a forest do not necessarily show higher productivity per unit area than do the more humble grasses of the prairie. The trees have high ratios of biomass to production. Among plants, the amount of biomass needed to sustain a fixed amount of production depends largely on the amount of unproductive root and stem tissue they must support. The trees in a forest must develop deep roots and powerful stems just to bring their photosynthesizing leaves into the sunlight. But the grasses of prairies or the algae of ponds and oceans can use much more of their total biomass efficiently, therefore equivalent amounts of net production in a pond or on a prairie require a much smaller biomass than in a forest. Among forests, the production efficiency should be higher in young forests that have not yet developed the extensive root and stem systems of mature forests.

Energy Budget of a Community

Figure 14-9 shows energy relationships and the distribution of biomass in a typical young temperate forest. The overwhelming bulk of the total community biomass is tied up in tree trunks, branches, and roots. Over half of the gross production of the plants is used in respira-

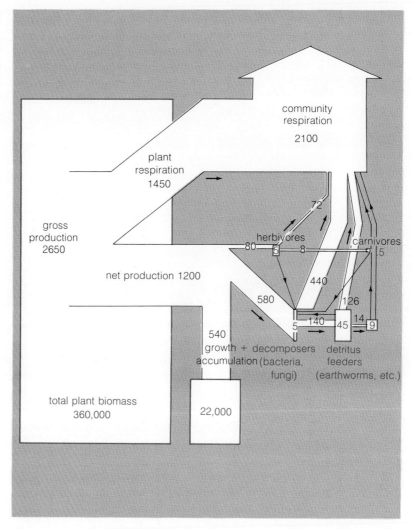

Figure 14-9
ENERGY FLOW (ARROWS) AND STANDING CROPS (BOXES) IN A YOUNG TEMPERATE FOREST (NOT DRAWN TO SCALE). Some of the values are based on educated estimates. Units for biomass are KCal/m². Units for energy flow are KCal/m²/year.

tory maintenance of the plants themselves. Of plant material consumed, the bulk passes through the detritus food chain (bacteria + fungi ⟶ earthworms ⟶ shrews, moles, and so on).

The energy budget for the forest shown is not balanced; almost half of the net plant production is neither eaten by herbivores nor decomposed by decomposers. This surplus is largely channeled into growth of the trees, and some winds up as an accumulation of litter on the forest floor. This means that every year the weight of living material in the forest increases, and so does the weight of the soil litter. Such a positive energy balance must ultimately come to a halt. This happens when community respiration comes to equal community gross production. For an oak-pine forest on Long Island, such energy equilibrium is reached in 160 to 200 years. The mature forest that results does not change from year to year and is called the *climax forest*.

Figure 14-10 RECLAIMING OF A CLEARED FIELD BY A FOREST. The saplings in the foreground are maple trees trimmed by the browsing of deer. Once growing tips of the maples exceed the reach of browsing animals, reforestation will proceed more rapidly.

Conversely, an immature forest changes in physical structure from year to year because of growth and accumulation. But this does not mean that the only difference between a young forest and an old one is that the trees are bigger. In fact, a very young forest is not recognizable as a forest at all—it may start as an area of annual herbs which is gradually invaded by herbaceous perennials and then shrubs and tree seedlings. Gradually, as the tree seedlings overtop the herbs and shrubs, these pioneer species may disappear altogether and be replaced by entirely new plant and animal species. The process is called *succession,* and it can be observed again and again as abandoned farms are reclaimed by forests, or small ponds fill in to become marshes and then wet prairies and forests. The suburban homeowner toiling with a lawn mower and plant clippers is toiling against succession.

Succession. The study of succession is the study of life power. As mentioned before, living organisms are often more important in determining their physical environment than are rainfall and temperature. Of course, humans have always taken this for granted about themselves. Humans can make the desert bloom, and they can make a blooming field into a desert. But other living organisms show the same profound powers of life enhancement (though not of desertification).

The pioneer organisms that first invade an abandoned farm must have a dual competence. They must have high powers of dispersal, and they must be able to tolerate the extremes of heat, cold, and drying of the naked earth. When we talk about pioneer organisms, we are talking about plants. Both grazers and detritus feeders follow only after a food source has been established.

An abandoned field in the North Carolina foothills is dominated for the first 4 or 5 years by only a handful of plant species that meet these requirements. During the first year, the dominants are crabgrass, horseweed, and ragweed. All these are annuals. They produce large numbers of small seeds which are widely dispersed by wind and animals and can lie dormant in the ground for many years until conditions are suitable for germination. As the plants grow up and rapidly fill the abandoned field, they shade the ground and moderate the temperature and humidity at ground level. When the plants die at the end of the growing season, their tissues are incorporated into the litter layer of the soil and gradually into the humus. Thus the moisture-retaining powers of the soil are increased. Horseweed retards its own growth by means of inhibiting chemicals produced in its roots. During the second and third years, the field is dominated by asters and broomsedge. These are biennials which flower the second year after germination and then again during the following year.

As succession progresses, the field is invaded by perennial grasses and herbs which form a thick mat on the ground. They start growth each spring from underground rhizomes or storage organs and therefore gain a competitive advantage over the annuals which must start over every year from seeds. This idea is carried one step further by shrubs and tree seedlings which also begin to invade the field. They also set aside annual stores of food in underground root systems, but they start their spring growth from buds high up on stems and branches. As the buds leaf out, they cast increasing shade which gradually becomes intolerable to the herbs of the old open field. Once the canopy above closes, the field is firmly committed to becoming a forest. The shade on the ground deepens, the litter and humus accumulation increases year by year, and forest herbs and animals replace their pioneer predecessors. The early forest of the southeastern piedmont of the United States is a pine forest, which over a span of 100 years is gradually replaced by hardwoods, primarily oaks and hickories. Ultimately, these form the climax forest for the region. The time required to reach the climax stage is at least 150 years. Table 5 lists the changes

Figure 14-11 SUCCESSION ON AN ABANDONED FIELD IN THE NORTH CAROLINA FOOTHILLS. Pioneer species are shown on bottom, the late successional species on top. A crabgrass. B. horseweed. C. aster. D. broomsedge. E. shrubs. F. pine. G. oak.

TABLE 5
Changes in dominant bird species during field-to-forest succession in the southeast piedmont

STAGE	DOMINANT BIRD SPECIES
Grass and shrub stage	Field sparrow, meadowlark, yellowthroat
Young pine forest (25 to 60 years)	Pine warbler, summer tanager, towhee, cardinal
Old pine forest (with strong hardwood understory)	Pine warbler, Carolina wren, hooded warbler, cardinal
Oak-hickory climax forest	Red-eyed vireo, wood thrush, tufted titmouse, cardinal

in bird populations that accompany the changes in vegetation and insect life.

Forests in different areas develop by different routes and have different end stages. Dunes on the New Jersey coast begin with dune grass and poison ivy, then progress to a beach plum, wild cherry, and shadbush shrub stage, and climax as a forest of American holly. An abandoned farm in upper New York State passes from annuals such as milkweed and mullein, to grassy perennials, to shrubs such as blackberry and hawthorn, to a climax forest of sugar maple and white ash. A mountain top in Pennsylvania may start with the most difficult of all substrates—bare rock. Over centuries, this may be colonized by lichens and gradually turned into soil, allowing a foothold for mosses, then grasses, and finally a scrub oak forest. The dunes at the southern end of Lake Michigan have been particularly well studied. Dune grasses give way to cottonwood seedlings, which are succeeded by pine, then oak, and finally a dense, cool climax forest of beech and sugar maple. The overall process takes about 1,000 years, as estimated from radiocarbon dating.

Though the living organisms of a community account for the driving force of succession, they do not determine the final outcome. This is a function of local climate and mineral and water supply. A beech-maple forest or a maple-ash forest requires a high average rainfall and a temperate climate. With insufficient rainfall, the succession may stop at the oak or pine stage. External disturbance may also arrest a succession. It is thought that the great prairie that once stretched from Indiana westward was maintained in the open state by frequent fires. Even southern Wisconsin was largely a prairie state until the middle of the nineteenth century, when burning was controlled. Much of Wisconsin was then bewhiskered by forests, while the rest was put under the plow.

The question of community succession is one that has caused endless disputes among ecologists. Does succession have a driving force other than habitat alteration by the resident populations? Is a late successional stage more stable than an early one? Even the question of how to measure community stability is one that lacks a simple answer. Some of the structural and functional changes of communities undergoing succession are summarized in Table 6.

TABLE 6
Structural and functional changes of communities undergoing succession

	EARLY SUCCESSIONAL STAGE	LATE SUCCESSIONAL STAGE
Community production	Low	High
Gross production/community respiration ratio	>1	1
Total number of species present	Low	High
Total community biomass	Low	High
Community accumulation	High	Low

A young community is structurally simple—it contains only a small number of species of plants and animals. This is because of the rigid demands made on pioneer organisms and because of the lack of physical complexity in the habitat later provided by larger, longer-lived species. And unlike a young married couple, the young community is in a positive energy balance, producing more than it expends in respiration.

Conversely, a mature community has achieved an overall energy equilibrium, with equal amounts of energy entering and leaving. The overall level of production is higher than it was in the young community. The community is more diverse in total numbers of species present, and the increased complexity and specialization allow for a more thorough exploitation of incoming solar radiation. The price is an increased cost of maintenance.

All human agriculture is based on the exploitation of early successional stages with their high rates of accumulation. These rates are further increased by subsidies of fossil fuels in the developed countries. Unlike much indigenous agriculture, the fossil fuel-subsidized agriculture of the West demands a drastic simplification of the agricultural community. A Kansas cornfield contains only corn, and an Iowa wheat field contains only wheat. A visitor to a native garden plot in Mexico or Guatemala would find no such monotony (Fig. 14-12).

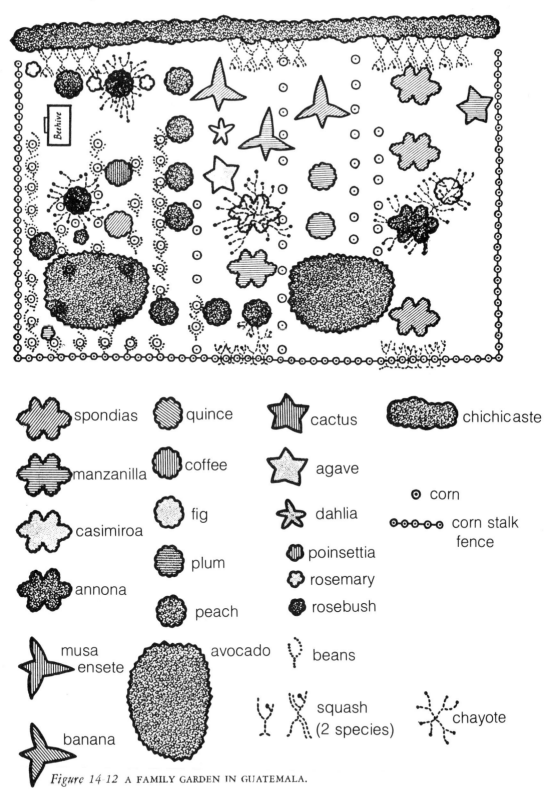

Figure 14.12 A FAMILY GARDEN IN GUATEMALA.

From Edgar Anderson, *Plants, Man, and Life,* University of California Press, Copyright © 1952. Reprinted with permission.

While the Mexican garden plot is not susceptible to mechanization, it has one great advantage that the Iowa wheat field lacks: it is much more resistant to attack by insects and other agricultural pests. Insects tend to be highly specific in their feeding habits; the corn borer has no interest in wheat, nor does the cotton boll weevil attack tomatoes. But when a pair of corn borers does find a cornfield, they can multiply almost without limit, with immense damage to the cornfield. This again is characteristic of early successional stages: while the total number of species is small, the number of individuals of some species may be astronomical. The result is greater fluctuation in population sizes than is characteristic of mature, more complex communities.

The Iowa farmer responds to an insect outbreak by using chemical insecticides which all too often spread far beyond the boundaries of the original farm and may affect an osprey or a peregrine falcon thousands of miles away. However, the more complex community represented by the Mexican garden does not offer so much food to one insect species that it can break out in plague proportions. And it does offer a stable habitat for the predatory species that silently and automatically exert the control function possible in Iowa only with repeated and expensive applications of chemical insecticides.

Summary of Energy Flow

To summarize this section, we have followed the fate of sunlight as it lingers briefly on earth in the form of chemical energy used to drive the processes of life. The transformation of sunlight into chemical energy is carried out by plants and other producers via the process of photosynthesis. Total energy trapped (gross production) is used in two ways. Roughly half is used to pay for the plant's own maintenance functions (respiration). The rest (net production) is channeled into growth or is made available as food for plant feeders.

Plant feeders in turn serve as food for other organisms, giving rise to the food chains or food webs that characterize a living community. Because of the inefficiency of the feeding process, food webs typically contain no more than four or five trophic levels. Two types of plant feeders are recognized: herbivores or grazers which feed on living plant materials, and decomposers which feed on dead plant material. Each gives rise to its own food chain, the grazing chain and the detritus chain. Typically the detritus chain carries a greater energy flux in terrestrial communities. This is true of human agriculture as well: the stalks, leaves, and other inedible portions of crop plants contain more energy than do the seeds and fruits. Part of the effort of plant breeders has focused on reversing these relationships.

Any plant material not consumed by grazers or by decomposers is channeled into accumulation, growth of plants, or an increase in litter and humus. Only young communities show this positive energy balance. The result of such accumulation is a physical change in the local environment. This environmental alteration may be more favorable for plant species other than those present at the time in the community. The result is ecological succession, as pioneer plant and animal species are replaced by a series of others, culminating in a relatively stable climax community. The climax community has a balanced energy budget; its gross community production equals community respiration, and no surplus material is created to accumulate. Western agriculture is based on the exploitation of pioneer communities via a grazing food chain. Such a strategy yields high rates of accumulation and makes the communities amenable to mechanical cultivation. However, the price paid is a lack of biological stability. This calls for continuing human intervention and creates an element of risk which may not be acceptable to a marginal economy.

The Cycling of Materials

Consider the following individuals:
    Homer (700 B.C., Greece)
    Gautama Siddhartha (500 B.C., India)
    Julius Caesar (100–44 B.C., Rome)
    Johann Sebastian Bach (1685–1750, Germany)
    Claude Monet (1840–1926, France)
    Rachel Carson (1904–1965, United States)
All the above were real people, and all are more or less familiar to you because they have changed the intellectual or political dimensions of your world. But they also have a more intimate connection with you and me: with your very next breath, you will breathe in air molecules that were once breathed by each of them. And you contain atoms in your body which at one time resided in their bodies. The amount of matter present in the world is limited, and it is used over and over again by all living things: trees, fish, and spiders as much as human beings. No life is isolated from other life. A millipede feeds on bacteria that feed on a leaf that used the carbon dioxide from your breath to build its cells. And when the millipede dies, its body will be taken apart by decomposers and there will result a little trickle of nutrients that may someday be blended into your breakfast cereal.

Each kind of molecule follows a characteristic path through the living world and travels at a characteristic rate. Thus it takes about 500 years to cycle a molecule of carbon dioxide through the living world, about 2,000 years for a molecule of oxygen, 2 million years

Figure 14-13
A DEAD DEER. Even the bones will be returned to the common pool of nutrients for use by other living organisms.

for a molecule of water, and 8 million years for a molecule of nitrogen. As an illustration, we will look at the cycling of two molecules, carbon dioxide and nitrogen.

The Carbon Cycle

Carbon forms the backbone of all large molecules that take part in the processes of life. In addition, the oxidation of carbon generates the energy that drives all maintenance functions necessary for life. The cycling of carbon revolves around two symmetrical living processes: plants convert carbon dioxide into carbohydrate, while both animals and plants burn carbohydrate to generate carbon dioxide plus energy.

$$H_2O + CO_2 \underset{\text{Animals + plants (respiration)}}{\overset{\text{Plants (photosynthesis)}}{\rightleftarrows}} CH_2O + O_2$$

This forms the basic carbon cycle. But to understand the overall cycle, we must look at the patterns of compartmentalization of carbon (Fig. 14-14):

1. A certain fraction floats in the atmosphere as gaseous carbon dioxide.

2. A much larger fraction resides in the oceans as dissolved carbon dioxide.

3. A certain fraction is tied up in the bodies of dead and living biological organisms. On land, this is largely represented by the humus of the soil. In the oceans, most biological carbon is found in the bottom ooze.

4. Some of the dead organisms are stripped of their oxygen content and converted into fossil fuels: lignite, coal, oil, and natural gas. The oxygen content of our atmosphere is derived largely from this process.

5. Finally, by far the largest fraction is buried deep under the soil surface and under ocean bottoms as limestone (calcium carbonate).

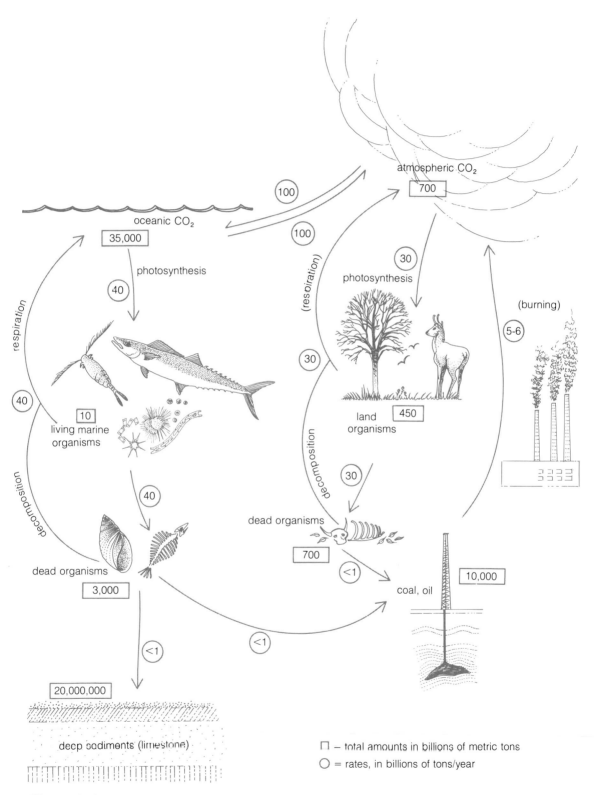

Figure 14-14 THE CARBON CYCLE.

What determines the size of the individual fractions of carbon? Each fraction is not a closed compartment but exists in a steady state with its environment. Imagine blowing air into a leaky balloon. You can blow the balloon up to a large size if you blow hard (a high input). But a very large balloon can also result from a minute input, providing the leak in the balloon is very small. This is the reason for the very large fraction of carbon tied up as limestone. Although only a very small fraction of carbon enters limestone from the living world, its rate of weathering and return to the living world is equally low.

It is important to remember that, in the natural world, the rate of input in each carbon pool equals the rate of outflow. However, Figure 14-14 shows one major imbalance. While additions to the pool of fossil fuels are extremely small, withdrawals occur at the rate of 5 to 6 billion metric tons per year, as coal and oil are burned to power human industry and agriculture. This is a rate high enough to produce a significant increase in the content of atmospheric carbon dioxide. Part of the increase is removed in the form of increased photosynthetic rates, and part dissolves in the ocean. If increased atmospheric carbon dioxide were not partially removed by photosynthesis and dissolving in the ocean, then human activities would increase the level of atmospheric carbon dioxide by about 0.8 percent per year, or a rise of 2.4 parts per million per year. Careful measurements show that the actual increase is 0.7 parts per million per year. This means that, by the year 2000, atmospheric carbon dioxide levels will rise to 375 to 400 parts per million, up from the present 320 parts per million.

A rise of this magnitude could have a global effect on the weather. Carbon dioxide acts as a blanket for infrared (heat) energy reemitted by the earth into space. An increase in atmospheric carbon dioxide levels leads to an increase in the surface temperature of the earth. It is estimated that only a 4°C increase in temperature would melt the polar ice caps and raise water levels of the oceans by 200 feet. This would drown coastal cities such as Montreal, Boston, New York, Washington, New Orleans, and San Francisco. Florida would disappear, as would the lower Mississippi River valley, giving St. Louis an ocean front.

Does this make mountain land a good real estate investment? Measurements of global temperatures over the past century have revealed no warming trend. It appears that another human activity has been countering the "greenhouse effect" of carbon dioxide. Enormous amounts of dust and soot thrown into the atmosphere by industrial effluents and agricultural activity have made the world more reflective. More of the sun's radiation is intercepted high in the stratosphere, before it can reach the surface of the earth, and reflected back into space. Which process will predominate in the end? Will the world cool and enter an ice age, or will it overheat and see its coastal areas drown? Or will the two processes continue to balance each other? No one knows, but it is sobering to realize that human activities have become a geological force on earth.

The Nitrogen Cycle

Molecular nitrogen makes up 80 percent of the air we breathe and the atmosphere we move through. Yet nitrogen deficiency is probably the most serious dietary insufficiency among the poor across the world. Both plants and animals require nitrogen to synthesize protein. But as far as they are concerned, the nitrogen of the air is an inert gas with no usefulness as a raw material. The only form of nitrogen that animal bodies can handle is "reduced nitrogen," or nitrogen in the form of ammonia (NH_3) and its derivatives. Plants can use either ammonia or nitrate (NO_3) to build proteins. In the great majority of cases, nitrogen in the form of nitrate or ammonia acts as a limiting factor in plant growth. The conversion of nitrogen gas to nitrate or ammonia can occur by biological means. The process is called *nitrogen fixation* and is carried out by soil microorganisms and by blue-green algae (Chapter 9). Many plants, such as alder and gingko, carry mutualistic bacteria on their root surfaces. Bacteria provide the trees with fixed nitrogen in exchange for organic food materials. Perhaps the most important of these mutualistic associations involve the legume crops—peas, beans, clover, soybeans, alfalfa, to name a few. These plants have root nodules housing nitrogen-fixing bacteria, and they are able to fix nitrogen at a rate 100 times greater than that of other plant communities. A small amount of nitrogen is also fixed by inorganic agents such as lightning and, more recently, automobile exhausts.

However, in 1914 an important new agent of nitrogen fixation was added to the preceding list. This was the industrial fixation of nitrogen (Table 7).

TABLE 7
Annual rates of nitrogen fixation, in millions of metric tons

Terrestrial fixation by bacteria and blue-green algae (historical)	30
Legume crops (alfalfa, and so on)	14
Fixation in oceans	10
Atmospheric fixation (lightning, and so on)	8
Industrial fixation	30
Total	92

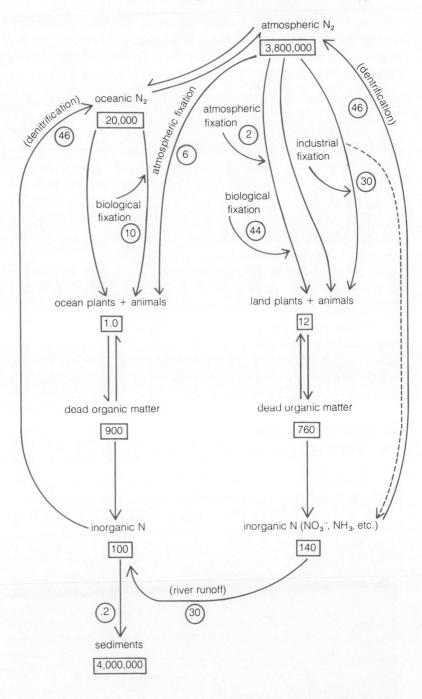

Figure 14-15 THE NITROGEN CYCLE.

In 1914, as Germany entered World War I, she found herself cut off from the Chilean guano deposits that had been the traditional source of nitrates used in the manufacture of explosives. Fritz Haber, a brilliant young chemist and later a Nobel prize winner, invented an industrial process that converted nitrogen gas from the air into ammonia. The process required high pressures and temperatures and large amounts of energy, supplied by gas or oil. Haber's process today produces as much ammonia and nitrate as had previously been produced by all land plants combined. Almost all of this finds its way into agriculture as fertilizer. There has also been a great increase in the planting of legume crops. The result is that today, half of all the nitrogen fixed in the world is fixed by human agency. In the case of the nitrogen cycle, man has become more important than natural forces.

What happens to the nitrogen that is fixed? A group of microorganisms called denitrifying bacteria convert nitrate back into atmospheric nitrogen gas to close the cycle (Fig. 14-15). Though it is assumed that on a worldwide basis the denitrifying bacteria are coping with the increased nitrate load imposed by human activity, measurements are not sensitive enough to settle the point. However, as pointed out before (Chapter 13), nitrate is becoming an increasingly common contaminant of the drinking water in agricultural areas.

If there is a single impression that emerges from a study of the world's energy and nutrient flows, it is the overwhelming impact of the human species on these processes. Few people have seen the earth in its entirety, from the perspective of outer space. One of these is James A. Lovell, Jr., an astronaut on the *Apollo 8* mission and the ill-fated *Apollo 13*. Lovell was struck by how different the earth looked from everything else in the visible universe. It was the only place that had color, warmth, and life. From outer space the earth struck him as "a grand oasis in the great vastness of space." In the years to come, it will take all of human understanding and love to keep the earth as the life-adorned oasis it is.

Bibliography

GENERAL
READING

Rachel Carson, *Silent Spring,* Houghton Mifflin, 1962. This book, perhaps more than any other, has been responsible for our present concern for the quality of our environment. In the early 1960's, Rachel Carson was bitterly attacked by the chemical industry and shamefully underdefended by the scientific community. Had she lived to the 1970's, she would have found her major contentions amply vindicated and her recommendations incorporated into the law of the land.

Silent Spring deals with the indiscriminate use of chemical pesticides, and makes an assertion that was revolutionary in 1962 but commonplace today. It is the following: one cannot judge the effects of a chemical substance by looking only at its effect in the field or forest where it was applied, or by looking only at the target organism for which it was intended. One must instead follow its path in the larger environment and judge it within the larger frame of reference.

Rachel Carson prevailed because she combined the cool, analytical logic of the scientist with the fiery moral sense of a prophet. It was a moral sense that was weighed in on the side of life. Here is a sample excerpt.

Incidents like the eastern Illinois spraying raise a question that is not only scientific but moral. The question is whether any civilization can wage relentless war on life without destroying itself, and without losing the right to be called civilized.

These insecticides are not selective poisons; they do not single out the one species of which we desire to be rid. Each of them is used for the simple reason that it is a deadly poison. It therefore poisons all life with which it comes in contact: the cat beloved of some family, the farmer's cattle, the rabbit in the field, and the horned lark out of the sky. These creatures are innocent of any harm to man. Indeed, by their very existence they and their fellows make his life more pleasant. Yet he rewards them with a death that is not only sudden but horrible. Scientific observers at Sheldon described the symptoms of a meadowlark found near death: "Although it lacked muscular coordination and could not fly or stand, it continued to beat its wings and clutch with its toes while lying on its side. Its beak was held open and breathing was labored." Even more pitiful was the mute testimony of the dead ground squirrels, which "exhibited a characteristic attitude in death. The back was bowed, and the forelegs with the toes of the feet tightly clenched were drawn close to the thorax . . . The head and neck were outstretched and the mouth often contained dirt, suggesting that the dying animal had been biting at the ground."

By acquiescing in an act that can cause such suffering to a living creature, who among us is not diminished as a human being?

Barry Commoner, *The Closing Circle,* Alfred A. Knopf, 1971. This is an original book. Commoner studies the processes of environmental degradation, but expands the frame of reference to include technological factors in addition to such obvious ones as affluence and population size. That is to say, it matters not only how many shirts a man buys per year. It matters also whether the shirts are made of cotton or of an energy-consumptive material such as Dacron. Commoner concludes that it is neither affluence nor population growth that can be blamed for the bulk of our environmental pollution. Here is a sample excerpt.

> We can sum up the possible contribution of increased affluence to the United States pollution problem as follows: per capita production of goods to meet major human needs—food, clothing, and shelter—have not increased significantly between 1946 and 1968 and have even declined in some respects. There has been an increase in the per capita utilization of electric power, fuels, and paper products, but these changes cannot fully account for the striking rise in pollution levels. If affluence is measured in terms of certain household amenities, such as television sets, radios, and electric can-openers and corn-poppers, and in leisure items such as snowmobiles and boats, then there have been certain striking increases. But again, these items are simply too small a part of the nation's over-all production to account for the observed increase in pollution level.
>
> What these figures tell us is that, in the most general terms—apart from certain items mentioned above—United States production has about kept pace with the growth of the United States population in the period from 1946 to 1968. This means that over-all production of basic items, such as food, steel, and fabrics has increased in proportion to the rise in population, let us say from 40 to 50 per cent. This over-all increase in total United States production falls far short of the concurrent rise in pollution levels, which is in the range of 200 to 2,000 per cent, to suffice as an explanation of the latter. It seems clear, then, that despite the frequent assertions that blame the environmental crisis on "overpopulation," "affluence," or both, we must seek elsewhere for an explanation.

A. L. Hammond, W. D. Metz, and T. H. Maugh II, *Energy and the Future,* American Association for the Advancement of Science, 1973. A cool, intelligent look at present patterns of energy consumption and possible alternatives for the future.

William W. Murdoch, ed., *Environment—Resources, Pollution, and Society,* Sinauer Associates, 1971. A collection of essays on such topics as world food supply, energy resources, water pollution, pesticides, and so on. There is a uniformly high quality about the essays and the selection forms a well-balanced whole.

Eugene P. Odum, *Ecology,* 2nd ed., Holt, Rinehart, and Winston, 1975. A full-size text on ecology whose major emphasis is the flow of energy through communities. A shorter edition is also available, *Fundamentals of Ecology.*

R. H. Wagner, *Environment and Man,* 2nd ed., Norton, 1974. A full-sized text dealing with environmental problems. The style is cool and lucid.

TEXTBOOKS

The Classification of Living Organisms

Kingdom Monera

NO. OF
SPECIES

    ?   *Phylum Schizomycetes:* bacteria
2,500   *Phylum Cyanophyta:* blue-green algae

Kingdom Protista

        Phylum Flagellata: flagellates
        Phylum Sarcodina: protozoa with pseudopods. Include amoebas, foraminiferans.
        Phylum Sporozoa: lack organs of locomotion. Include malarial parasites.
6,000   *Phylum Ciliophora:* ciliates
  450   *Phylum Myxomycota:* slime molds

Kingdom Fungi

NO. OF
SPECIES

500   *Phylum Zygomycota:* tube fungi. Ex. *Rhizopus* (the dark bread mold).
35,000   *Phylum Ascomycota:* sac fungi. Include morels, yeasts, some bread molds. Ex. *Neurospora* (pink bread mold).
25,000   *Phylum Basidiomycota:* club fungi. Include mushrooms, puffballs, bracket fungi.

    Fungi Imperfecti—A miscellaneous group that can't be classified because sexual stages are unknown. Most are thought to be ascomycetes or basidiomycetes.

    Lichens—Organisms formed by mutualistic association between a fungus and an alga. Most lichen fungi are members of either Ascomycota or Basiodiomycota.

Kingdom Plantae

1,100   *Phylum Pyrrophyta:* dinoflagellates
10,000   *Phylum Chrysophyta:* diatoms and golden-brown algae
1,000   *Phylum Phaeophyta:* brown algae
3,000   *Phylum Rhodophyta:* red algae
6,000   *Phylum Chlorophyta:* green algae
25,000   *Phylum Bryophyta:* mosses, liverworts, hornworts
300,000   *Phylum Tracheophyta:* vascular plants
    Subphylum Psilopsida: leafless, extinct vascular plants
    Subphylum Lycopsida: lycopods
    Subphylum Sphenopsida: horsetails (*Equisetum*)
    Subphylum Pteropsida: large-leaved plants
        Class Filicineae: ferns
        Class Gymnospermae: plants with naked seeds—pines, firs, ginkgo
        Class Angiospermae: flowering plants
            Order Monocotyledonae: plants with one seed leaf—grasses, palms, lilies, orchids
            Order Dicotyledonae: Plants with two seed leaves

Kingdom Animalia

10,000   *Phylum Porifera:* sponges
50   *Phylum Mesozoa:* small, extremely simplified parasites of marine invertebrates
9,000   *Phylum Coelenterata:* hydras, jelly fish, corals
    Class Hydrozoa: hydroids
    Class Scyphozoa: jellyfish
    Class Anthozoa: corals, sea anemones

NO. OF SPECIES	
90	*Phylum Ctenophora:* comb jellies
13,000	*Phylum Platyhelminthes:* flatworms

        Class Turbellaria: free-living forms. Ex. *Planaria*
        Class Trematoda: flukes (all parasitic forms). Ex. *Schistosoma*
        Class Cestoda: tapeworms (all parasitic)

Protostomes

750	*Phylum Nemertinea or Rhynchocoela:* nemerteans or ribbon worms. Related to flatworms but have tubular digestive tract
1,500	*Phylum Rotifera:* rotifers. Microscopic, most live in fresh water
175	*Phylum Gastrotricha*
64	*Phylum Kinorhyncha*
10,000	*Phylum Nematoda:* roundworms or nematodes. In terms of total numbers, second only to arthropods. Probably many more species remain to be discovered. Most nematods are microscopic and are found in all habitats. Tubular digestive tract without true coelom.
230	*Phylum Nematomorpha:* horsehair worms
500	*Phylum Acanthocephala:* parasitic worms
60	*Phylum Entoprocta:* moss animals
8	*Phylum Priapulida:* penis-shaped marine animals
9	*Phylum Gnathostomulida:* marine worms. First described in 1969
275	*Phylum Sipuncula:* marine worms
128,000	*Phylum Mollusca:* mollusks. Include chitons, snails, clams, oysters, squids, octopods
150	*Phylum Echiura:* marine worms
8,700	*Phylum Annelida:* segmented worms

        Class Polychaeta: Marine worms. Ex. *Nereis*
        Class Oligochaeta: Ex. earthworms
        Class Hirudinea: leeches

180	*Phylum Tardigrada:* water bears. Terrestrial, in damp moss
65	*Phylum Pentastomida:* tongue worms. Endoparasitic in mammals
73	*Phylum Onychophora:* have traits intermediate between annelids and arthropods
900,000	*Phylum Arthropoda*

        Class Trilobita: extinct trilobites
        Class Chelicerata: spiders, ticks, scorpions, horseshoe crabs
        Class Crustacea: crustaceans. Include lobsters, crabs, shrimps, sow bugs, copepods, sand fleas.
        Class Insecta: insects
            Order Odonata: dragonflies
            Order Isoptera: termites
            Order Orthoptera: grasshoppers, mantids

NO. OF SPECIES	

          Order Hemiptera: true bugs. Include stink bugs, bedbugs, water boatmen
          Order Homoptera: cicadas, aphids
          Order Coleoptera: beetles. Largest insect order
          Order Anoplura: sucking lice. Parasites of mammals only
          Order Mallophaga: chewing lice—parasites of birds and mammals
          Order Lepidoptera: butterflies, moths
          Order Diptera: flies, mosquitoes
          Order Hymenoptera: wasps, ants
          Order Siphonaptera: fleas
     Class Diplopoda: millipedes
     Class Chilopoda: centipedes
     Class Pauropoda: pauropods
     Class Symphyla: symphylans

15   *Phylum Phoronida:* marine animals
4,000   *Phylum Bryozoa:* moss animals. Most less than ½ mm long
260   *Phylum Brachiopoda:* lamp shells. Superficially resemble clams. All marine. May be connecting link between protostomes and deuterostomes

The Deuterostomes

50   *Phylum Chaetognatha:* represent most primitive deuterostomes
5,300   *Phylum Echinodermata:* echinoderms—starfish, sea urchins, brittle stars
80   *Phylum Pogonophora:* beard worms
80   *Phylum Hemichordata:* acorn worms. Related to chordates but lack notochord
39,000   *Phylum Chordata:* chordates
     Subphylum Urochordata: tunicates
     Subphylum Cephalochordata: lancelets
     Subphylum Vertebrata: vertebrates
          Class Agnatha: jawless fish. Ex. lamprey. Gave rise to Placoderms
          Class Placodermi: armored fish with jaws. All extinct. Gave rise to 2 classes below
          Class Chondrichthyes: cartilaginous fish—sharks, rays
          Class Osteichthyes: bony fish. Most modern fish
          Class Amphibia: frogs, salamanders, cecilians
          Class Reptilia: reptiles
               Order Chelonia: turtles
               Order Crocodilia: crocodiles, alligators
               Order Squamata: snakes, lizards
               Order Rhynchocephalia: tuatara
          Class Aves: birds
          Class Mammalia: mammals

Subclass Prototheria: monotremes. Egg-laying mammals. Ex. duckbill platypus, spiny anteaters

Subclass Metatheria: marsupials. Pouched mammals. Ex. kangaroo, oppossum

Subclass Eutheria: placental mammals

    Order Insectivora: include shrews, moles

    Order Chiroptera: bats

    Order Carnivora: include dogs, cats, bears, weasels, seals

    Order Rodentia: include rats, mice, squirrels, beavers, porcupines

    Order Edentata: sloths, anteaters, armadillos

    Order Lagomorpha: rabbits, hares

    Order Primates: include lemurs, monkeys, apes, humans

    Order Artiodactyla: even-toed ungulates. Include cows, deer, sheep, camel, pigs, hippopotamuses

    Order Perissodactyla: odd-toed ungulates. Include horses, zebras, rhinoceri

    Order Proboscidea: elephants

    Order Cetaceae: whales, dolphins

    Order Sirenia: manatees

Glossary

a-: combination form meaning "without."
A horizon: the uppermost layer of soil, consisting of decomposing organic material mixed with the mineral substrate, and representing the normal limits of rainwater penetration.
ab-: combination form meaning "opposite."
abscisic acid: a plant hormone that inhibits growth and maintains seed dormancy.
abscission layer: corky tissue separating the leaf petiole from the branch. Its growth causes a deciduous leaf to separate from the stem in the fall.
acetate: a salt of acetic acid, a 2-carbon carboxylic acid found in vinegar and playing a key role in cellular energy metabolism.
acetylcholine: a hormone secreted by many nerve terminals which typically induces an action potential in a postsynaptic nerve or muscle by opening Na^+ and K^+ gates.
acetylcholinesterase: an enzyme released by the postsynaptic membrane which hydrolyzes acetylcholine.
acid: any substance that acts as a hydrogen ion donor.
acirculatory: without a circulatory system.
ACTH (adrenocorticotropin; adrenocorticotrophic hormone): a hormone released by the anterior pituitary that stimulates secretion by the adrenal cortex.

action potential: a sharp and transient change in the electrical potential across a nerve or muscle membrane that is propagated along the cell. In a nerve, it corresponds to the nerve impulse.

actin: a contractile protein found in muscle and other cells. A filamentous form (F-actin) interacts with myosin to produce muscle contraction. Actin may dissociate into globular subunits (G-actin).

active site: the specific surface region of an enzyme that combines with the substrate.

active transport: an energy-dependent process that allows a cell to transport ions and molecules against a concentration gradient.

adaptation: any genetically controlled characteristic that allows an organism to survive and reproduce in a specific environment or way of life.

adaptive radiation: the evolutionary proliferation of organisms derived from a single basal stock so as to fill all available ecological niches. For example, the evolution of flying, swimming, climbing, burrowing, carnivorous, and herbivorous forms among the mammals.

adaptor RNA. See **transfer RNA.**

adenine: one of 4 nitrogenous bases commonly found in nucleic acids. It is a purine and specifically base-pairs with thymine or uracil.

ADP (adenosine diphosphate): a molecule containing adenine, the sugar ribose, and two phosphate groups. A "high-energy" bond links the two phosphates.

adrenocorticotropin. See **ACTH.**

aerobic: with oxygen.

alcohol: an organic molecule with one or more —OH groups attached to a carbon-hydrogen backbone.

all-, allo-: combination form meaning "other" or "different."

allantois: one of the four fetal membranes of reptiles, birds, and mammals. In birds and reptiles, it forms a receptacle for waste disposal. In mammals it contributes blood vessels to the placenta.

allele: any of several alternate forms of a gene which occur at the same locus on a chromosome.

allopolyploid: a polyploid in which the replicated chromosomes are contributed by two different species. For example, a polyploid species with 14 chromosomes formed from parental species with 8 and 6 chromosomes respectively.

all-or-none law: states that a nerve or muscle cell either conducts an impulse maximally or not at all.

allosteric: literally, "alternate shape." The allosteric site of an enzyme binds a small inhibitor so as to make it catalytically inactive.

alveolus: a small sac, such as those found in lungs for oxygen exchange or in breasts for milk production.

amino acid: a small molecule containing an amino group ($-NH_2$), a carboxyl group ($-COOH$), and a functional group which comes in 20 different varieties.

amniocentesis: withdrawal of a small sample of amniotic fluid from a pregnant woman for purposes of testing the fetus for chromosomal or metabolic abnormalities.

amnion: in birds, reptiles, and mammals, the fetal membrane that encloses a volume of plasmalike fluid (the amniotic fluid) in which the fetus is suspended.

an-: combination form meaning "without."

anaerobic: without oxygen.

anaphase: during cell division, the stage when centromeres have divided and chromatids begin moving to opposite poles of the cell.

angstrom (Å): a unit of length equal to 10^{-10}m. A water molecule is about 1.5 Å long.

angio-, -angium: combination forms meaning "vessel" or "container."

anion: a negatively charged ion.

annual: a plant that lives for one year, sets seed, and then dies. New growth starts next year from the seeds.

antheridium: that portion of a gametophyte plant which produces sperm.

antibody: a protein produced by a cell in response to challenge by a foreign substance (antigen), which specifically reacts with the antigen.

anticodon: three nucleotides in transfer RNA which are complementary to three nucleotides in m-RNA forming the codon.

antigen: a protein, polysaccharide, or other large molecule which elicits the formation of a specific antibody in a competent cell.

archegonium: the structure in the gametophyte plant which produces the egg.

archenteron: during embryonic development, the cavity in the gastrula that becomes the future gut.

ascus: a sausage-like sac housing the eight spores of an ascomycete fungus.

association areas: areas of the human cortex in which neurons make connections with other areas and centers. These cortical areas are neither motor nor sensory in nature and are thought to be involved in higher or more complex behaviors.

atom: the smallest unit of an element, consisting of a nucleus made up of protons and neutrons, and surrounded by electrons.

ATP (adenosine triphosphate): a molecule that plays a central role in all energy metabolism. It contains the nitrogenous base adenine, linked to the sugar ribose, which in turn is linked to three phosphate groups in series. The two terminal phosphate groups are linked to their neighbors by "high-energy" bonds. This bond energy may be used to drive energy-requiring cellular processes.

auto-: combination form meaning "self" or "same."

autonomic nervous system: the portion of the peripheral nervous system that innervates internal glands and smooth muscle, and is concerned with vegetative functions, such as breathing, heartbeat.

autotroph: an organism capable of manufacturing organic nutrients from inorganic raw materials. This includes all photosynthetic organisms.

auxin: a plant hormone that regulates many aspects of plant growth and development: it stimulates elongation of stem cells in phototropism and negative geotropism; inhibits elongation of root cells in geotropism; inhibits lateral bud formation, inhibits leaf and fruit-drop; etc.

axon: a long process in a nerve cell, normally conducting impulses away from the cell body and releasing neurotransmitters at its terminus.

B horizon: the layer of soil below the A horizon. It accumulates substances leached out from above by rainfall.

bacteria: moneran organisms lacking distinct nuclei or internal cell organelles of the type found in eukaryotes. They are typically heterotrophic and include the smallest living organisms known.

bacteriophage: a virus that attacks bacteria.

base: a substance that acts as a hydrogen ion acceptor. When added to an aqueous solution, it increases the (OH^-) ion concentration.

basidium: the spore-enclosing structure of a basidiomycete fungus such as the common mushroom.

bilateral symmetry: symmetry about a central plane. Examples of bilaterally symmetrical objects include a hammer, a book, and the external human being.

biofeedback: a set of techniques that provide an animal or human with additional information about ongoing events inside the body or brain, as for example visual or auditory cues about heart rate or brain waves.

biomass: the weight of all the organisms forming a given population or trophic level, or inhabiting a defined region. Biomass is usually expressed in terms of dry weight.

biome: a very large, easily-recognized community type, classified according to its dominant vegetation. Examples include the deciduous forest, the prairie, the tundra.

blasto-: combination form meaning "embryo."

blastocoel: the large central cavity of an embryo at the blastula stage.

blastopore: in the gastrula stage of an embryo, the opening of the archenteron to the outside.

blastula: an early embryonic stage in which the walls are arranged in the form of a hollow sphere.

blastulation: the formation of the blastula during embryonic development.

bond, covalent: a strong chemical bond between two atoms formed by the sharing of a pair of electrons.

bond, hydrogen: a weak chemical bond in which a component hydrogen atom serves as the bonding group between two electronegative atoms such as oxygen or nitrogen.

bond, ionic: a weak chemical bond produced by electrostatic attraction between oppositely charged ions.

bond, Van der Waals: a weak chemical bond, acting over very short distances, resulting from the attraction of mutually induced electrical charges of opposite sign.

C horizon: the layer of soil lying below the B horizon, and consisting of relatively unaltered parent material.

caecum: a blind pouch in the digestive tract. In mammals it represents a terminal portion of the large intestine.

cambium: a layer of embryonic, rapidly dividing cells in a plant.

capillary: the smallest class of blood vessels, across whose walls the exchange of foods, gases, and wastes takes place. The capillary wall is an endothelium a single cell layer thick.

carbohydrate: a chemical compound consisting of carbon, oxygen, and hydrogen in the approximate ratio of 1:1:2. Examples include starch, sugar, and cellulose.

carboxyl group: an acidic function of a molecule, with the chemical structure —COOH.

cardiac: pertaining to the heart.

carnivore: an organism that feeds on animals.

carotenoid: an accessory pigment typically found in plants. The colors range from reds to yellows, and the pigment molecules consist of long polyunsaturated chains of carbon and hydrogen.

carrying capacity: the number of individuals of a particular species that a particular area can support on a permanent basis.

caryo- or karyo-: combination form meaning "nucleus."

catalyst: a substance that accelerates a chemical reaction without itself being consumed in the reaction.

cation: a positively charged ion.

cellulose: a very high-molecular weight carbohydrate consisting of polymerized glucose. Cellulose is the major constituent of plant cell walls, and hence of wood and paper.

centriole: a microtubular organelle consisting of paired rodlike structures which form the mitotic spindle in animal cells.

centromere: the region of the chromosome which holds two chromatids together and attaches to the spindle fibers prior to mitosis.

cerebellum: a portion of the hindbrain, concerned largely with muscle coordination.

cerebrum: paired outgrowths of the vertebrate forebrain. The cerebrum makes up the great bulk of the human brain and is the highest center of coordination.

character displacement: the divergence of characters in closely related species occupying the same area. Character displacement serves to reduce competition between two species.

chitin: a nitrogen-containing polysaccharide that forms the exoskeleton of insects and crustaceans.

chlorophyll: a flat, complex, magnesium-containing molecule that functions in photosynthesis to convert solar energy into the energy of excited electrons.

chloroplast: a cell organelle of plants that contains the photosynthetic machinery.

chorion: the outermost fetal membrane of reptiles, birds, and mammals. In higher mammals, it forms part of the placenta.

chrom-: combination form meaning "colored."

chromatid: one of two identical strands of a chromosome still held together by a single centromere. After division of the centromere, a chromatid is called a daughter chromosome.

chromosome: in viruses and prokaryotes, a single naked strand of nucleic acid. In eukaryotes, chromosomes consist of DNA, histones, and other proteins. A single chromosome contains many genes.

cilium: a microtubular organelle capable of whiplash movements. Cilia have the same structure as flagella, but are shorter and more numerous.

citric acid cycle (Krebs cycle): a cyclic series of metabolic reactions during which an acetate group is oxidized to CO_2, with energy-rich electrons trapped by NAD and conveyed to the electron transport system. During the process, acetate is chemically bound to other molecular fragments.

climax community: a community that reproduces itself more or less permanently if external conditions do not change. It represents the final stage in a process of ecological succession.

cloning: the production of one or more offspring from a single parent by inserting diploid cell nuclei of the parent into enucleated eggs and allowing the eggs to develop. The offspring are genetically identical to the parent.

closed circulatory system: a circulatory system in which blood flow is restricted to well-defined vessels.

co-: combination form meaning "with" or "together."

cochlea: the coiled portion of the inner ear concerned with hearing.

coelom: a large body cavity lined with a membrane of mesodermal origin, and housing the digestive tract and associated structures.

coenocyte: an organism with many nuclei but a single cell membrane.

coenzyme: a nonprotein component of an enzyme which is required for enzyme activity. Coenzymes are usually vitamins or metal ions.

commensalism: a type of association between two organisms whereby one organism benefits without harming the other. An example is Spanish moss hanging on a tree branch.

community: a relatively self-sufficient assembly of organisms adapted to a common set of limiting conditions such as water supply, temperature, nature of substrate.

compound: a substance composed of more than one element and possessing a unique composition and chemical properties.

cone: in the retina, a light-receptor cell that is responsible for color vision and functions in bright light.

conifer: a cone-bearing gymnosperm. Examples include firs, spruces, and pines.

continental drift: the theory stating that the surface features of the earth are in constant motion, carried on the backs of free-floating plates.

contractile vacuole: a cell organelle found in many fresh-water protistans. It functions to expel water that enters because of osmotic pressure differences between the inside and outside of the cell.

convergence: the independent evolution of a superficially similar body form by unrelated organisms as a response to similar selection pressures.

corpus luteum: an ovarian structure formed from a ruptured follicle in response to LH stimulation. It produces progesterone.

cotyledon: a seed leaf borne by the embryo of a seed. It acts as a food store or as an organ for digestion of food stored in the endosperm.

Cowper's glands: two small accessory sexual glands in the male that secrete a small amount of lubricating mucus prior to ejaculation.

crossing over: the exchange of comparable chromosomal segments by two homologous chromosomes during meiosis.

cuticle: a waxy waterproofing layer covering the outer surface of a leaf, insect, etc.

cyclic AMP: a form of adenosine monophosphate in which the phosphate function forms a ring with the ribose. Cyclic AMP functions as an intracellular "second messenger" in target cells of hormones derived from amino acids.

cytochrome: a protein that functions in electron transport. It contains a heme group bearing a central atom of iron which can accept or donate electrons.

cytology: the study of cell structure.

cytoplasm: cellular material located inside the cell membrane but outside the nucleus.

cytosine: one of the four nitrogenous bases found in DNA or RNA. It specifically base-pairs to guanine.

deciduous: referring to the seasonal loss of leaves.

decomposer: an organism that feeds on ready-made organic materials by the release of extracellular enzymes. Most decomposers are bacteria or fungi.

dehydrogenation: the removal of hydrogen atoms from a molecule.

deletion: a type of mutation in which a portion of genetic material is lost.

dendrite: that portion of a neuron which conducts impulses toward the cell body. It can respond to neurotransmitters but cannot release them.

depolarization: the abolition of the electric potential across a cell membrane by chemical or electric means. In muscle and nerve cells, this initiates an action potential.

-derm: combination form meaning "skin" or "tissue layer."

deuterostomes: animals such as chordates and echinoderms whose mouths are formed secondarily by creating a new opening opposite the blastopore.

dicot: a flowering plant whose seeds contain embryos with two cotyledons.

differentiation: the embryological process whereby a cell or tissue develops into a form specialized for a specific task.

diffusion: the movement of molecules or ions from regions of high local concentration toward regions of lower concentration as a result of thermal agitation.

digestion: the chemical breakdown of large polymerized molecules into smaller units by the process of hydrolysis.

diploid ($2N$): containing two copies of each kind of chromosome per cell.

divergence: the evolution of two related organisms in such a way as to increase mutual differences.

DNA (deoxyribonucleic acid): a molecule capable of self-replication and capable of storing genetic information in the form of nucleotide sequences. It is a polymer of deoxyribonucleotides.

dominance: the phenotypic expression of an allele in the presence of a second allele specifying a different character.

Down's syndrome: a birth defect characterized by gross mental retardation and assorted physical changes, resulting from the presence of three copies of chromosome 21 in body cells.

ecology: the study of the relationships between living organisms and their biological and physical environments.

ecto-: combination form meaning "outside."

ectoderm: the outermost of the three fundamental body tissue layers. The ectoderm gives rise to a portion of the skin and to the nervous system.

edema: a local swelling due to the leakage of water out of capillaries and into the interstitial space.

effector: the target organ of a motor neuron, typically a muscle or gland.

egg: the female gamete. It has a haploid chromosome number and typically a large amount of cytoplasm serving as a food source for the developing embryo.

electron: the smallest unit of negative charge. It is found in all atoms, where it orbits around the nucleus.

electron transport chain: a series of molecules able to undergo gain or loss of electrons, and arranged in a sequence of increasing electron affinity (decreasing energy content). The energy lost by electrons is trapped to synthesize ATP.

element: one of the approximately 105 chemically different atoms.

embryo: an immature organism, typically still contained in the seed or uterus.

embryo sac: the female gametophyte of a higher plant.

endo-: combination form meaning "inside."

endocrine: internally secreting. An endocrine gland is a ductless gland which releases its hormone directly into the blood stream.

endoderm: the innermost of the three fundamental tissue layers. It gives rise to the inner lining of the digestive tract and to associated areas such as lungs, liver, etc.

endodermis: a sheath of cells surrounding the vascular bundle of the root of a plant. The pumping of salts by the endodermis is essential for the passage of water into root xylem.

endoplasmic reticulum: a system of intracellular membranes arranged as stacked sheets or tubules and found in eukaryotic cells. In combination with ribosomes, this system is responsible for protein synthesis.

endoskeleton: a skeleton located inside the body. The vertebrate skeleton is an endoskeleton.

endosperm: in flowering plants, a $3N$ tissue found in the seed and used as a food source by the embryo.

entropy: the amount of energy which is unavailable for doing work. It constitutes a measure of disorder in a system.

enzyme: a protein catalyst, found in every living cell.

epi-: combination form meaning "upon" or "outermost."

epidermis: the outermost cells of an organism.

epididymis: a long, narrow, coiled tubule lying on top of the testis. It conducts sperm from the seminiferous tubules to the sperm duct and is used for sperm storage.

erythrocyte: a red blood cell (RBC).

estrogen: one of a group of steroids which acts as a female sex hormone. It induces female secondary sexual characteristics.

eu-: combination form meaning "true" or "good."

eugenic: refers to attempting to improve the human race by the interbreeding of individuals with desirable genes.

eukaryotic cell: a cell with a distinct nuclear membrane and intracellular organelles such as mitochondria, endoplasmic reticulum, etc.

euphenic: attempting to improve human development by physiological or embryological means.

eversible: capable of being pushed out.

evolution: descent with modification; the change with time of the gene pool of a population.

ex-, exo-: combination form meaning "outside."

exoskeleton: a skeleton located outside the body, such as that found in mollusks and arthropods.

fat: a molecule formed by the reaction of glycerol with three molecules of fatty acid. Examples include butter, margarine, olive oil.

fatty acid: a molecule with a hydrophobic tail consisting of a hydrocarbon function and a hydrophilic head consisting of a carboxyl group (—COOH).

fermentation: the anaerobic breakdown of an organic compound via the glycolytic pathway or some variant of it. The products may include ethanol, lactic acid, etc.

fixation: the conversion of a raw material into a biologically useful form. Examples include the conversion of CO_2 into carbohydrate, or the conversion of nitrogen gas into ammonia.

flower: the generative organ of angiosperms. The complete flower gives rise to both the male and female gametophytes and provides a method of internal fertilization. The fertilized egg develops into a seed enclosed inside a fruit.

follicle stimulating hormone: See **FSH**.

fossil: physically surviving evidence of ancient life.

fruit: the matured ovary of an angiosperm, enclosing one or more seeds.

FSH (follicle stimulating hormone): a gonadotropic hormone released by the anterior pituitary. In males, it stimulates sperm production. In females, it stimulates the growth and secretion of the Graafian follicle.

fungus: an organism that feeds by decomposition, typically has a body form made up of hyphae, and reproduces by means of spores.

β-galactosidase (also called lactase): an enzyme that hydrolyzes lactose and other beta-galactosides (molecules containing galactose in a beta linkage).

gamete: a haploid cell specialized for fusion with another haploid cell of opposite sex or mating type. Gametes of advanced organisms are called sperm or eggs.

gametophyte: the haploid generation of a plant, which produces sperm and/or eggs.

ganglion: a mass of neuronal cell bodies which typically make synapses with each other.

gastrovascular cavity: the large central cavity of primitive animals such as coelenterates. It functions both as a digestive and vascular system.

gastrula: an early embryonic stage in animals consisting of two tissue layers and a single external opening leading to the future digestive tract.

gastrulation: process of forming a gastrula.

Gause's principle: no two species can coexist in the same geographical area if they occupy the same ecological niche.

gene, control (promoter, regulator, operator): a gene that controls a structural gene. The regulator gene codes for a repressor protein which attaches to the operator gene. The promoter gene is the site of attachment of RNA polymerase.

gene, structural: a gene that codes for a protein other than the repressor protein. Most of the proteins coded for are enzymes.

gene pool: the sum total of all genes carried by a population.

genetic code: the specification of amino acid sequence in proteins by the sequence of nucleotide triplets in messenger RNA.

genetic drift: the shift in time of gene frequencies of small populations as a result of statistical fluctuations.

genotype: the genetic makeup of an organism, as opposed to its expressed characteristics (phenotype).

gibberellins: a class of plant hormones concerned with stem elongation, breaking of seed dormancy, and in some cases induction of flowering.

gill: a feathery structure projecting outward from the body of an animal and used in gas exchange, typically in a watery medium.

gill slit: an opening in the side of the throat of primitive chordates which houses gills and allows for the passage of water.

glucose: a common six-carbon sugar which plays a key role in the energy metabolism of most living organisms.

glycerol: a three-carbon molecule bearing three alcohol groups (CH_2OH–$CHOH$–CH_2OH). It is a raw material in the biological synthesis of fats, and plays an important role in energy metabolism.

glycogen: animal starch, consisting of glucose molecules linked in large, branching arrays.

Golgi body: a cell organelle of eukaryotes, concerned with secretion processes.

gonad (testis or ovary): an animal structure that produces gametes.

gonadotropin: a hormone that stimulates the gonads. Examples include FSH, LH, and chorionic gonadotropin.

granum: a stacked array of photosynthetic membranes (lamellae) in a chloroplast.

grazing chain: a food chain based on decomposer organisms (fungi and bacteria).

greenhouse effect: the warming of an area by means of trapped infrared radiation (IR). For example, visible light from the sun can penetrate the windows of a greenhouse. The soil re-radiates the bulk of the absorbed light as IR. IR is unable to penetrate the window glass and remains trapped inside, raising the internal temperature. Atmospheric CO_2 functions in a way similar to window glass on a global scale.

guanine: one of the four nitrogenous bases found in DNA and RNA. It specifically base pairs with cytosine.

guard cell: one of a pair of cells which acts to open or close the stomatal apertures of plants.

habitat: the type of environment inhabited by an organism. Examples are prairie, rich woods, roadsides.

haploid (1N): containing a single copy per cell of each kind of chromosome.

Hardy-Weinberg equation: an equation that predicts the incidence of all possible genotypes in a large, randomly interbreeding population. For example, in a 2-allele system, if dominant allele A has a gene frequency of p in the population and recessive allele a has a frequency of q, then the predicted ratios of the genotypes $AA:Aa:aa$ will be given by the equation $p^2 + 2pq + q^2 = 1$, with $AA = p^2$, $Aa = 2pq$, $aa = q^2$.

haustorium: an absorptive organ of a fungus or parasitic vascular plant that penetrates a host cell and absorbs nutrients from it.

herbicide: a chemical used to kill plants. Common examples are 2,4-D, 2,4,5-T, cacodylic acid.

herbivore: an animal that feeds on plants.

hermaphrodite: an organism, such as the earthworm, that contains both male and female reproductive organs.

heterotrophic: dependent upon ready-made organic compounds for food.

heterozygous: containing two dissimilar alleles of a given gene.

hibernation: a state of reduced metabolic activity in which some animals pass the winter. Examples include snakes, bats, chipmunks.

histo-: combination form meaning "tissue."

histone: one of a group of basic proteins associated with the chromosomes of eukaryotes.

homeostasis: a series of physiological adjustments made by an organism to maintain a constant and optimal internal environment despite changing external conditions.

homo-: combination form meaning "self" or "same."

homologous chromosomes: two chromosomes present in the same cell which carry genes for the same cellular functions.

homologous structures: structures which have been evolutionarily derived from a single structure in a common ancestor.

homozygous: carrying two identical alleles for a given gene function.

hormone: a chemical released by one cell in an organism which elicits a response in another cell. Some hormones (e.g., cyclic AMP) are intracellular, being produced by one cell organelle and eliciting a response from another organelle.

host: an organism being preyed upon by a parasite.

humus: finely divided, partly decomposed organic material found in the soil.

hybrid: a plant or animal produced by parents that are genetically unlike. The parents may belong to different species or to well-defined varieties within the same species.

hydrocarbon: a molecule containing only carbon and hydrogen. Examples include gasoline, oil, and benzene.

hydrolysis: the splitting of a covalent bond by adding the elements of water, or in general, R—X—R' + HOH ⟶ R—X—H + R'—OH. Examples include the splitting of starch to produce glucose, the splitting of protein to produce amino acids, etc.

hydrophilic: dissolves readily in water. Hydrophilic functions include acid groups, alcohol groups, amino groups, and all ionized groups.

hydrophobic: dissolves poorly in water. Examples includes fats and all hydrocarbons.

hydroxyl group: a functional group in a molecule consisting of a hydrogen atom linked to an oxygen atom (—OH).

hydroxylase: an enzyme which introduces a hydroxyl group into a molecule.

hyper-: combination form meaning "more than" or "above."

hypha: the working part of a fungus, consisting of long filaments with or without internal divisions into distinct cells.

hypo-: combination form meaning "less than" or "below."

hypothalamus: a structure in the forebrain concerned with regulation of vegetative functions and the expression of a number of basic drives and emotions.

incus: one of the three small bones of the middle ear, lying between the malleus and the stapes.

insulin: a protein hormone produced by the islets of Langerhans in the pancreas. It has widespread effects on the metabolism of sugars and amino acids. It is necessary for the entry of glucose into body cells from the circulatory system.

inter-: combination form meaning "within."

interphase: in the cell division cycle, the stage between mitoses. Chromosomes are not visible in the light microscope, the nuclear membrane is intact, and nucleoli are visible in the nucleus.

intra-: combination form meaning "within."

invaginated: folded inward.

ion: an electrically charged atom or molecular fragment. The charge is due to an excess or deficiency of electrons.

ionizing radiation: radiation (either particulate or electromagnetic in nature) which is sufficiently energetic to expel bonding electrons from molecules, producing molecular breakdown and generating ions. X-rays are the most common form of ionizing radiation.

iso-: combination form meaning "same."

isolating mechanisms: behavioral, physiological, or environmental barriers to interbreeding between organisms belonging to different species.

isotope: an atom that occupies the same place as another atom in the periodic table and hence has the same number of protons in the nucleus, but differs in the number of nuclear neutrons.

kine-: combination form meaning "movement."

Krebs cycle. see **citric acid cycle.**

lactic acid: a three-carbon carboxylic acid bearing a hydroxyl group, with the structure (CH_3—CHOH—COOH). It is formed as a result of the anaerobic breakdown of glucose in the human body.

lamella: a structure resembling a small plate; for example, the photosynthetic membranes found inside chloroplasts.

larva: an immature form of an animal, which does not resemble the adult form.

leaf area index: ratio obtained by dividing the total surface area of all leaves by the area of the ground covered.

leuko-: combination form meaning "white" or "colorless."

leukocyte: a white blood cell. In vertebrates, the leukocytes include lymphocytes, granulocytes, and monocytes. They function in defense of the body against foreign organisms or substances.

LH (luteinizing hormone): a gonadotropin produced by the anterior pituitary. In males, it stimulates the testes to produce testosterone. In females, it stimulates the corpus luteum to produce progesterone.

ligament: cartilaginous connective tissue of joints which binds bone to bone.

lipid: a chemically unrelated group of compounds all of which have a low solubility in water. Examples include fats, steroids, oils, and waxes.

locus: location. A gene locus is the location on a chromosome of a particular gene.

luteinizing hormone. See **LH**.

lymph: the fluid that fills lymphatic vessels and surrounds body cells. Its composition is that of a protein-deficient plasma.

-lysis, lyso-: combination form meaning "to dissolve."

lysogenic: a type of viral life cycle in which the viral nucleic acid is incorporated into host DNA and reproduces in synchrony with it. As long as the virus remains integrated, the host is not harmed.

lysosome: an organelle of eukaryotic cells consisting of an array of hydrolytic enzymes enclosed by a membrane. Lysosomes are particularly prominent in phagocytic cells such as granulocytes.

malleus: one of the three little bones of the middle ear, lying between the eardrum and the incus.

medulla: in general, the inner portion of an organ (such as the adrenal medulla). The medulla oblongata is a portion of the hindbrain which is continuous with the spinal cord and controls a number of vegetative functions and nerve reflexes.

medusa: the dispersive and sexually reproducing stage of a coelenterate life cycle. A jellyfish has the medusa body form.

mega-: combination form meaning "large" or "million."

meiosis: a type of cell division which halves the chromosome number and leads to crossing-over of chromosomes.

membrane, semipermeable: a membrane with pores so small that some molecules are unable to cross, while smaller molecules such as water can cross with ease. All biological membranes are semipermeable.

messenger RNA (m-RNA): a complementary RNA copy of a structural gene. It conveys genetic information from the nucleus to ribosomes, where it is translated into protein.

meso-: combination form meaning "middle."

mesoderm: one of the three fundamental tissue layers, lying between the ectoderm and the endoderm and giving rise to tissues such as muscle, gonads, kidneys, and blood.

metabolism: the sum total of all chemical reactions taking place in a cell. Metabolism represents the chemical work necessary for an organism to stay alive, and to perform external work on the environment.

metamorphosis: a type of development accompanied by a dramatic change in body form and function; common among the advanced insects.

metaphase: a stage of cell division during which chromosomes have attached to the center of the spindle by means of their centromeres.

micro-: combination form meaning "small" or "one millionth."

micron: a millionth of a meter, or 10^{-6}m.

microtubule: one of a number of extremely fine tubules found in all eukaryotic cells. Microtubules are assembled from subunits of the protein tubulin. They form the spindle during cell division, and function in cellular movement and changes of shape.

mitochondrion: a rod-shaped organelle of eukaryotic cells made up of two membranes (with the inner membrane folded into cristae). Mitochondria contain enzymes needed for aerobic energy metabolism (Krebs cycle plus electron transport system).

mitosis: a type of nuclear division which produces daughter nuclei carrying the same chromosome number and chromosome types as present in the parent nucleus.

molecule: the smallest unit of a chemical compound which still retains the chemical properties of the compound.

molting: the shedding of the outer layer, such as the exoskeleton of insects, the plumage of birds, or the skin of snakes.

Monera: a kingdom made up of prokaryotic organisms; the bacteria and the blue-green algae.

monocot: a flowering plant whose seeds contain embryos with a single seed leaf. Examples include the grasses, lilies, orchids, palms.

morph-, morpho-: combination form meaning "form."

motor neuron: a neuron leading away from the central nervous system and toward an effector.

mutation: a chemical or physical change in the makeup of the genetic material of a cell.

mutualism: a type of association between two organisms that leads to a benefit for both.

mycorrhizae: a close association between fungi and tree roots which facilitates the uptake of nutrients by trees.

myo-: combination form meaning "muscle."

NAD (nicotinamide adenine dinucleotide): a complex molecule which functions as an electron carrier, transferring electrons from organic substrates to other electron carriers such as flavins. It plays a key role in all energy metabolism.

NADP: a phosphorylated derivative of NAD. It functions as an electron carrier and plays a key role in photosynthesis and other ATP-generating pathways.

neo-: combination form meaning "new."
nematocyst: an intracellular structure found in the cnidoblast (a specialized cell of the coelenterates). It consists of a rigid oval capsule containing an explosively eversible thread.
nephron: the basic functional unit of the vertebrate kidney, consisting of a Bowman's capsule (which absorbs the filtrate from the glomerulus), proximal tubule, loop of Henle, distal tubule, and collecting duct.
nerve cell body: the portion of a neuron which contains the cell nucleus.
neuron: a nerve cell.
neurosecretion: the release of chemical substances by the axon terminals of a neuron.
neutron: a fundamental subatomic particle. It is electrically neutral and has a mass comparable to that of a proton (hydrogen nucleus).
niche: the functional role of an organism in a community, including its tolerance limits, mode of food procurement, and associations with other organisms.
nitrogen fixation: the conversion of atmospheric nitrogen into a biologically usable form such as ammonia or nitrate.
notochord: a stiff yet flexible cartilaginous rod present beneath the spinal cord of primitive chordates.
nucleic acid: DNA or RNA—a complex molecule composed of polymerized nucleotides.
nucleolus: a dense structure within the nucleus where ribosomes are manufactured.
nucleotide: a complex molecule consisting of a nitrogenous base, a sugar, and a phosphate group. Nucleotides can be polymerized into nucleic acids.
nucleus: the central area of a eukaryote cell, bounded by a double membrane and housing the great bulk of the genetic material of the cell.
oo-: combination form meaning "egg."
open circulatory system: a type of circulatory system in which the blood does not remain inside fixed vessels but moves through irregular spaces between body cells. It is present in insects and other arthropods.
operator: a region of DNA within the operon where repressor protein attaches to block RNA polymerase transcription of the DNA.
operon: a cluster of juxtaposed control genes and structural genes described in bacteria which can be turned on or off as a unit. The structural genes code for proteins used in a single metabolic pathway.
organelle: a subcellular structure with a specific function and organization.
organizer: a cluster of cells in an embryo that influences the development of an adjacent area.
osmosis: in biological systems, the net flow of water through a semipermeable membrane. Osmotic phenomena depend on the diffusion of water from regions of high water concentration (low solute concentration) to regions of low water concentration (high solute concentration).
ov-, ovi-: combination form meaning "egg."
ovary: in flowering plants, the basal portion of the pistil, which encloses the future seed. In animals, the female gonad which produces eggs and female sex hormones.

ovule: in higher plants, the structure which encloses the female gametophyte. After fertilization, it develops into the seed.

ovulation: the release of the ovum by the Graafian follicle.

ovum: the egg.

oxidation: the loss of electrons by a molecule or atom.

pampas: the South American equivalent of the steppe, or dry prairie.

pancreas: a vertebrate digestive organ which secretes a number of important digestive enzymes into the duodenum. It also produces the hormones insulin and glucagon.

parasite: an organism that feeds upon a larger organism (the host) and generally inflicts only a moderate amount of damage. However, insect parasites usually kill their host and behave more like typical predators.

parasympathetic nervous system: the region of the autonomic nervous system which connects to the central nervous system in the cranial and sacral regions of the spinal cord, and whose activity predominates during periods of digestive activity and low stress.

parathyroids: small endocrine glands of vertebrates located near or inside the thyroid and concerned with regulation of calcium and phosphate levels in the blood.

parthenogenesis: virgin birth. The development of an unfertilized egg into an adult.

pathogenic: causing damage to the host. The term is normally used in reference to bacteria.

peptide bond: the chemical bond between two amino acids, or in general, between an amino group and a carboxyl group (–CO–NH–).

peripheral nervous system: the portion of the nervous system located outside the brain and spinal cord.

peristalsis: alternate waves of constriction and relaxation moving down a muscular tube such as the digestive tract.

pH: a measure of acid-base concentration. The pH is expressed as the negative log of the hydrogen ion concentration, i.e., a hydrogen ion concentration of $10^{-1}M$ has a pH of 1, and a hydrogen ion concentration of $10^{-7}M$ has a pH of 7. Values of pH less than 7 are characteristic of acids, and values higher than 7 are characteristic of bases.

phagocytosis: the ingestion and digestion of particulate food by a cell.

phenotype: the physical appearance of an organism (as opposed to its genotype, or genetic makeup). The phenotype results from the expression of a portion of the genetic potential.

phenylketonuria (PKU): a birth defect involving phyenylalanine metabolism leading to high blood levels of phenylalanine and the excretion of phenylpyruvic acid in the urine. If uncontrolled, it leads to severe mental retardation.

phloem: conductive tissue in vascular plants used to transport sugars and other organic materials.

-phore: combination form meaning "carrier."

phospholipid: a modified fat, containing a hydrophobic tail and a hydrophilic head that contains a phosphate group.

photo-: combination form meaning "light."

photosynthesis: the light-energized reduction of CO_2 to carbohydrate in the presence of chlorophyll.

-phyll: combination form meaning "leaf."

phylogeny: the evolutionary ancestry of an organism.

phyto-, -phyte: combination form meaning "plant."

phytoplankton: microscopic floating algae found in oceans and fresh waters.

pinocytosis: the cellular ingestion and digestion of liquids.

pistil: the female reproductive structure of the flower, consisting of a stigma, style, and ovary. The ovary produces a totally enclosed female gametophyte inside an ovule. After internal fertilization, the ovule develops into a seed which remains enclosed by the mature ovary wall.

pituitary: a small endocrine gland lying just below and controlled by the hypothalamus. It secretes a number of hormones which control other endocrine glands or which have a direct effect on target organs throughout the body.

placenta: an extraembryonic structure in placental mammals which supplies food and oxygen to the fetus and removes its wastes. In addition, it synthesizes progesterone needed to maintain pregnancy. The placenta is formed in part from embryonic tissue (the chorion and allantois) and in part from maternal tissue (the uterus).

plankton: microscopic, passively floating organisms of fresh and salt waters.

plasma: the noncellular fraction of blood, consisting of water, proteins, and salts as the major constituents.

platelet: a cell fragment found in the blood that participates in fibrin clot formation and stops small breaks in blood vessels by means of the platelet plug.

pleio-: combination form meaning "more."

pleiotropic: referring to a gene which affects more than one body character.

pollen: the male gametophyte of flowering plants.

poly-: combination form meaning "many."

polygenic: a trait controlled by more than one gene.

polymer: a molecule made up of repeating subunits.

polyploid: having a chromosome number greater than 2N.

polysaccharide: a polymer formed from many sugar molecules. Examples include starch, cellulose, and chitin.

population: all the members of a given species which live in a specified area, and interbreed with each other.

portal vein: a vein that begins and ends with a capillary bed. Examples include the hepatic portal vein and the portal veins between the hypothalamus and the pituitary.

post-: a combination form meaning "after."

predator: an organism that feeds on another living organism.

prey: an organism hunted and eaten by another organism.

primary structure: in proteins, the specific sequence of amino acids making up the protein.

primitive: an organism or biological structure which is unspecialized and resembles the presumed ancestral state.

pro-: combination form meaning "before."

procaryotic. See **prokaryotic.**

producer: an autotrophic organism.

production, gross: total amount of food photosynthesized.

production, net: amount of energy channeled into growth. It is equivalent to gross production minus respiration.

progesterone: a steroid female sex hormone which stimulates thickening of the uterine wall, prepares breasts for milk production, inhibits ovulation, and relaxes the uterus during pregnancy.

prokaryotic: lacking a membrane-bounded nucleus or intracellular organelles such as mitochondria, Golgi body, etc. Pertaining to the monerans.

promoter: the adenine-thymine-rich region of an operon to which RNA polymerase attaches to initiate the transcription process.

prostaglandin: one of a number of widely distributed body chemicals which stimulate uterine contraction and may have future potential as birth control agents. Chemically they are related to fatty acids.

protein: a long polymer of amino acids.

Protista: a heterogeneous kingdom, all of whose members have a eukaryotic cell structure but show a very low degree of differentiation into tissues and organs. The prostitans have evolutionary affinities with all three of the higher kingdoms—the plants, fungi, and animals.

proto-: combination form meaning "first."

proton: a positively charged subatomic particle with mass equivalent to one atomic mass unit. It forms the nucleus of the hydrogen atom.

protostome: an animal whose mouth is embryologically derived from the blastopore. Examples include arthropods, annelids, and mollusks.

purine: a nitrogenous base formed from a fused aromatic ring system made up of 5 carbon atoms and 4 nitrogen atoms. Examples include adenine, guanine, and uric acid.

pyrimidine: a nitrogenous base formed from a single aromatic ring containing 4 carbon atoms and 2 nitrogen atoms. Examples include cytosine, thymine, and uracil.

pyruvic acid: a three-carbon carboxylic acid with the structure CH_3–CO–COOH. It is a key intermediate in the metabolic breakdown of glucose.

race: a population with gene frequencies significantly different from those of other populations in the same species; a subspecies.

radial symmetry: symmetry around a central axis. Examples of radially symmetrical objects include a bell, a lampshade, a cylinder.

RAS. See **reticular activating system.**

RBC. See **red blood cell.**

reaction center: the chlorophyll molecule which actually emits the reducing electron during photosynthesis.

receptor: a cell which translates an environmental stimulus into a nerve signal.

recessive: an allele whose genetic information is not phenotypically expressed in the heterozygous state. Recessive alleles are expressed only in the homozygous state.

recombination: the result of the exchange of equal fragments between homologous chromosomes during the process of crossing over.

red blood cell: a blood cell concerned with oxygen and CO_2 transport.

reduction: the gain of electrons by an atom or molecule.

reflex arc: a basic functional unit of the nervous system, consisting of a receptor, sensory neuron, central nervous system correlation, motor neuron, and effector.

regulator gene: a control gene which codes for a protein repressor.

repression: the shut down of DNA transcription for an enzyme or group of enzymes as a result of the intracellular accumulation of certain metabolites. These are called corepressors. In some systems, repression is a spontaneous process which does not require the buildup of metabolites.

repressor: a protein coded for by the regulator gene which combines with the operator gene to prevent transcription of all structural genes in the operon.

respiration: the expenditure of food energy for maintenance functions by a living cell, organism, or community.

reticular activating system (RAS): a mixture of nerve cells and fibers in the lower brain stem, extending from spinal cord to thalamus and playing a crucial role in regulation of sleep and waking and in behavioral arousal.

ribosome: a small intracellular organelle made up of protein and RNA. It is the site of protein synthesis.

rickettsias: small bacteria incapable of survival outside a host cell. They cause a number of serious diseases in humans, including typhus and Rocky Mountain spotted fever.

RNA: (ribonucleic acid) a polymer of ribonucleotides (nucleotides with ribose as the sugar component). Three classes of RNA exist in eukaryotes, and all are concerned with protein synthesis: messenger RNA, transfer RNA, and ribosomal RNA.

saprophyte: an organism which feeds on dead material by a process of decomposition. Most saprophytes are fungi and bacteria.

sarcomere: the basic unit of contraction in a muscle fibril, extending from one Z line to the next.

savannah: open country with scattered trees, typical of dry regions of Africa.

secondary structure: regions of protein that show a regular repeating pattern. Common examples include the alpha helix and the pleated sheet. Such structures are maintained by weak-bond interactions.

secretion: the external release of useful cellular products by a cell. Secretion differs from excretion in that the latter refers to the release of an unwanted product.

seed: an embryo plant plus a food source (endosperm) enclosed in a protective coat. Some seeds, as beans, contain little or no endosperm.

selection: enhanced reproductive success of individuals best adapted to a local environment.

seminal vesicle: in vertebrates, a male accessory sex organ that adds to the fluid making up the semen. In humans, this fluid contains the sugar fructose as well as prostaglandins.

semipermeable membrane. See **membrane, semipermeable.**

sensory neuron: a neuron conducting impulses from a receptor toward the central nervous system.

sessile: immobile, usually attached to the substrate.

sex-linked inheritance: in humans, the inheritance of genes carried on the X chromosome. Examples include hemophilia and red-green color blindness.

sieve tube: a cell with perforated end-plates that functions in phloem conduction.

soma-, -some: combinaton form meaning "body."

speciation: the formation of new species.

species: the basic unit of evolution, consisting of one or more populations which may be separated in time and space, and are potentially capable of successful gene exchange with each other but not with other populations.

spindle: the football-shaped assembly of microtubules which appears during cell division in eukaryotes and pulls daughter chromosomes to opposite poles of the dividing cell.

spore: in bacteria, a dehydrated, thick-walled, highly resistant cell which carries all the DNA and a small amount of cytoplasm. Bacterial spores are produced during times of environmental stress. In plants and fungi, the spore is a $1N$ cell produced by meiosis and often serving as the dispersal stage (though not in tracheophytes). The spore germinates into a $1N$ organism.

sporophyte: in plants, the $2N$ generation which produces spores.

sporulation: the production of spores.

stapes: one of the three small bones of the middle ear, lying between the incus and the oval window.

starch: a polysaccharide formed by the polymerization of many glucose molecules.

steady state: a dynamic metabolic state characteristic of living organisms. While materials from the external environment are being continually incorporated and other materials are being released, the internal composition and metabolic relationships remain stable.

steroid: a very complex lipid molecule, formed from a fused 5-ring nucleus composed of carbon and hydrogen. Many steroids function as hormones, and participate in cell membrane formation.

stigma: the region of the pistil on which pollen is deposited.

stoma (pl. **stomata**): an opening in the epidermis of a leaf which allows for exchange of gases and water vapor between cells inside the leaf and the atmosphere.

substrate: the molecule on which an enzyme acts.

succession: the replacement of one kind of community by another in an orderly and predictable pattern.

sucrose: table sugar—a molecule formed by the chemical linkage of the simple sugars glucose and fructose.

sugar: a small carbohydrate molecule containing numerous –CHOH linkages plus an aldehyde or ketone function (–CHO or –CO–).

sym-: combination form meaning "together."

symbiosis: the coexistence of two organisms belonging to different species. Some symbiotic associations are mutually beneficial, such as those of lichens. Others are one-sided, such as those between parasite and host. Still others are intermediate between the two extremes.

sympathetic nervous system: that portion of the autonomic nervous system originating from the thoracic and lumbar portions of the spinal cord. It is concerned with regulation of the body during states of emergency, and antagonizes the activity of the parasympathetic system.

syn: combination form meaning "together."

synapse: the juncture between two neurons.

synergistic: working together in such a way that the total effect is more than the sum of the individual effects. An example is the synergism between alcohol and barbiturates in depressing the nervous system.

taxonomy: the science of classification of organisms on the basis of their evolutionary affinities.

telophase: the final stage of cell division, just before the cell enters interphase. The nuclear membranes are re-forming around daughter nuclei but full cytoplasmic division has not yet occurred.

tendon: cartilaginous connective tissue linking muscle to bone.

territory: a fixed geographical area defended by an individual or group against others of the same species for purposes of reproduction or feeding.

tertiary structure: in proteins, the specific three-dimensional folding of the molecule. Such structure is maintained by weak-bond interactions.

thalamus: a mass of gray matter in the forebrain concerned with the functions of discrimination and attention.

thymine: a nitrogenous base found in DNA but not RNA. It specifically base-pairs with adenine.

thyroxin: a hormone produced by the thyroid gland which speeds up the basal metabolic rate.

tissue: a group of cells with similar form and function. Examples include muscle, cartilage, bone.

transfer RNA: (t-RNA) a small-molecular weight RNA which binds a specific amino acid at one end and binds to a specific m-RNA codon at the other end. It is also called adaptor RNA.

transformation: the introduction of new genetic material into a cell by exposing it to a suspension of DNA.

transpiration: the loss of water by a plant through its open stomata.

triplet code: the coding for specific amino acids used in protein synthesis by nucleotide bases taken in groups of three.

-trophic: combination form meaning "nourishing."

-tropic: combination form meaning "to turn."

tropism: a growth response by plants or sessile animals to an environmental stimulus. Positive tropism means growth toward the stimulus (ex. light, gravity); negative tropism means growth away from the stimulus (ex. gravity).

uracil: a nitrogenous base found in RNA but not DNA. It specifically base-pairs with adenine.

urea: a small molecule formed from two molecules of ammonia and one molecule of carbon dioxide, with the structure $H_2N-CO-NH_2$. It represents the chief vehicle for excretion of nitrogenous wastes in amphibians and mammals.

uric acid: a nitrogenous base with extremely low water solubility. It represents the chief vehicle for excretion of nitrogenous wastes in birds and reptiles.

vacuole: a fluid-filled, membrane-enclosed structure in cells. Vacuoles serve various functions, from food digestion to water expulsion.

vascular tissue: conductive and supportive tissue in higher plants. The two forms of vascular tissue present are xylem and phloem.

villus (Pl. villi.): a finger-like projection from the inside wall of the intestine, serving to increase the internal surface area.

virus: an encapsulated piece of nucleic acid. The capsule consists typically of protein. The nucleic acid may be DNA or RNA but not both, and it contains information for diverting the metabolic machinery of a host cell into synthesis of more virus particles.

vitamin: an essential molecule which can not be synthesized by the body and must therefore be obtained in the diet. Vitamins serve as coenzymes.

-vorous: combination form meaning "feeding on."

weak bond: a chemical bond which is easily broken. It is based on electrostatic attraction between two molecules or molecule fragments. Examples of weak bonds include hydrogen bonds, ionic bonds, and Van der Waals bonds.

X chromosome: the larger of the two sex chromosomes. In vertebrates, the presence of two X chromosomes in the cell makes the individual a female.

X rays: an energetic form of electromagnetic radiation, having wavelengths shorter than ultraviolet radiation.

xylem: a conductive tissue present in tracheophyte plants which consists of stacked, open-ended tubular cell wall and conducts water and minerals from roots to leaves and growing regions.

Y chromosome: the smaller of the two sex chromosomes. In vertebrates, the presence of a single Y chromosome in a diploid cell makes the individual a male.

zoo-: combination form meaning "pertaining to animals" or "motile."

zooplankton: small, floating marine or fresh-water animals which feed mainly on the phytoplankton.

zygote: the fertilized egg.

Photo Acknowledgments

Opening Part I—R.O. Kelley

Chapter 1

- 1-1   Yerkes Observatory
- 1-8   L. L. Millecchia and M. A. Rudzinska, *J. Protozoology 19,* 473-483 (1972). © Society of Protozoologists
- 1-9   Courtesy Dr. J. Rhodin, from Rhodin: *Histology, a Text and Atlas,* Oxford University Press, New York, 1974.
- 1-13 A. S. Breathnach, M. Gross, B. Martin, "Freeze-fracture Replication of Melanocytes and Melanosomes," *J. Anat. 116,* 303-320 (1973). © Cambridge University Press.
- 1-14 Peter Wellauer, R. Weber, T. Wyler, *J. Ultrastructure Research, 42,* 377-393 (1973). © Academic Press
- 1-15 Keith R. Porter
- 1-16 L. L. Millecchia and M. A. Rudzinska, *J. Protozoology 17,* 574-583 (1970). © Society of Protozoologists
- 1-17 Myron C. Ledbetter
- 1-18 Paul Burton, R. E. Hinkley, G. B. Pierson, *J. Cell Biol. 65,* 227-233 (1975). © Rockefeller University Press
- 1-22 L. L. Millecchia and M. A. Rudzinska, J. *Protozoology 19,* 473-483 (1972). © Society of Protozoologists

1-23   L. L. Millecchia and M. A. Rudzinska, *J. Protozoology 17*, 574-583 (1970). © Society of Protozoologists
1-24   Albert S. Klainer, *American J. Medicine 58*, 674-683 (1975)

Chapter 2

2-1   Keith R. Porter
2-2   George Gardner (below)
2-3   Leonard Lee Rue III
2-18   Naval Research Laboratory
2-19   Fisher Scientific Company

Chapter 3

3-1   H. Delius, A. Worcel, *J. Mol. Biol. 82*, 107-109 (1974). © Academic Press
3-2   Courtesy of the IBM Corporation and R. T. Miller
3-15   O. L. Miller, B. A. Hamkalo, C. A. Thomas, *Science 169*, 392-395 (1970). © AAAS

Opening Part II— Courtesy of Raymond Schwarz

Chapter 4

4-1   Emil Bernstein and Eila Kairinen, Gillette Research Institute, cover photo *Science 173*, Aug. 27, 1971. © AAAS
4-8   S. Donald Greenberg, *Amer. J. Medicine 58*, 441 (1975)
4-9   David E. Birk, NYU Laboratory of Cellular Biology

Chapter 5

5-1   E. R. Lewis, Y. Y. Zeevi, and T. E. Everhart, *Science (165)*, 1140-1143 (1969). © AAAS
5-19   Private collection
5-22   E. R. Lewis
5-24   L. M. Beidler
5-27   Brenda R. Eisenberg, A. M. Kuda, *J. Ultrastructure Research 51*, 176-187 (1975). © Academic Press.
5-33   David Meszler

Chapter 6

6-1   Top and bottom right: FAO   Bottom left: La Leche League International
6-3   New York Zoological Society Photo
6-4   Uldis Roze
6-21   Florinda Minutoli for the material; Jared Rifkin for the photo.
6-24   R. O. Kelley, R. A. F. Dekler, J. G. Bleumink, *J. Ultrastructure Research 45*, 254-258 (1973). © Academic Press
6-28   Raven Lang, *Birth Book*, Genesis Press, Palo Alto, Calif. (1972)

PHOTO ACKNOWLEDGMENTS / 567

Opening Part III—Michael Gochfeld

Chapter 7

7-1   Nancy Hays for Monkmeyer Press Photo Service
7-2   Ada and Don Olins, *Science 188*, 1097-1099 (1975). © AAAS
7-4   National Institutes of Health
7-7   Akira Morishima

Chapter 8

8-1   American Museum of Natural History
8-3   Uldis Roze
8-4   New York Zoological Society Photo
8-10 Raccoon—Leonard Lee Rue III; Vicuna—FAO; Hedgehog and gibbon—New York Zoological Society Photos; Rhino—SATOUR; Echidna—Gordon Gilbert.
8-17 Wombat and Tasmanian wolf—H. Burrell, Australian Museum; Sugar glider and marsupial mouse—H. Hughes, Australian Museum; Hyena—SATOUR; Flying squirrel—New York Zoological Society Photo; Placental mouse—Uldis Roze

Chapter 9

9-1   Emil Bernstein, Gillette Research Institute
9-4   *Eobacterium isolatum*—E. S. Barghoorn, J. W. Schopf, *Science 152*, 758-763 (1966). © AAAS; *P. aeruginosa* and *S. pyogenes*—David Greenwood
9-6   Left—Donald Tipper; right—Stanley C. Holt
9-7   Charles E. Herron, USDA
9-8   *Oscillatoria*—Milton E. Nathanson, *Spirulina*—Jared Rifkin
9-9   Left—S. D. Lin and M. K. Lamvik, *J. of Microscopy 103*, 249-257 (1975). © Blackwell Scientific Publications, Ltd.
      Right—R. W. Horne, I. P. Ronchetti, J. M. Hobart, *J. Ultrastructure Research, 51*, 233-252 (1975). © Academic Press
9-13 (A) Maria A. Rudzinska, *J. Gerontology 16*, 213-224 (1961). © Society of Gerontologists; (B) Maria B. Rudzinska, *Transactions of the New York Academy of Sciences 29*, 512-525 (1967). © New York Academy of Science
9-14 (A) M. A. Rudzinska et al, *J. Protozoology 12*, 563-576 (1965). © Society of Protozoologists; (B) M. A. Rudzinska and W. Trager, *J. Protozool. 15*, 72-88 (1968). © Society of Protozoologists
9-16 Elm Research Institute
9-18 Uldis Roze

Chapter 10

10-1 Uldis Roze

10-2   (A) Daniel Habib; (B and C) Exxon Production Research Co.; (D) Uldis Roze; (E) Carl Strüwe for Monkmeyer Press Photo Service; (F) Jared Rifkin; (G) Fisher Scientific Company
10-4   Annemarie C. Reimschuessel, Allied Chemical Corp.
10-8   (B) Andrew Greller; (C) Uldis Roze
10-10 Uldis Roze
10-13 Annemarie C. Reimschuessel, Allied Chemical Corp.
10-18 FAO
10-20 Uldis Roze
10-21 UPI

Chapter 11

11-1   Courtesy of Eastman Kodak Company
11-5   Russ Kinne, Photo Researchers, Inc.
11-7   Courtesy of Eastman Kodak Company
11-8   Courtesy of Eastman Kodak Company
11-9   (A) Courtesy of Eastman Kodak Company; (B and C) James E. Sublette and Mary F. Sublette; (D,E,F) Lawrence M. Beidler
11-11 Annemarie C. Reimschuessel and John M. Kolzer, Allied Chemical Co.
11-17 (A) Jacalyn Madden; (B,C,D) David Ewert; (E) Robert Madden
11-19 (A,D,E) David Ewert; (B) Michael Gochfeld; (C) Andrew Greller
11-20 (A) FAO; (B,F) Michael Gochfeld; (C,D,) SATOUR; (E) New York State Department of Environmental Conservation
11-21 Leonard Lee Rue III
11-24 New York Zoological Society Photo
11-25 SATOUR
11-27 (A) American Museum of Natural History; (B,C) Lee Boltin and Time-Life, Inc.

Opening Part IV—Uldis Roze

Chapter 12

12-1   Clyde W. Hare
12-3   USDA-SCS Photo by B. C. McLean
12-6   Andrew Greller
12-8   Jacalyn Madden
12-10 Uldis Roze
12-13 Uldis Roze
12-15 Clyde W. Hare
12-17 Uldis Roze
12-18 winter: snow on branches, apple tree—Clyde W. Hare; tracks—Uldis Roze. Spring: Uldis Roze. Summer: doe—Leonard Lee

PHOTO ACKNOWLEDGMENTS / 569

        Rue III, leaf cover—Uldis Roze. Fall: chipmunk—Leonard Lee Rue III, fallen leaves—Uldis Roze
12-19 (A) Uldis Roze; (B) Andrew Greller
12-21 Andrew Greller
12-22 Left: Gerald F. Laird; Right: Kim Estes
12-23 (A) Susan Munger; (B) Gerald F. Laird; (C) David Ewert; (D,F) Robert Madden; (E) Uldis Roze
12-26 David Ewert

Chapter 13

13-1 SATOUR
13-3 FAO
13-6 FAO
13-7 Gordon Gilbert
13-8 USDA
13-9 David Ewert
13-10 Myron C. Ledbetter and A. D. Krikorian, *Phytomorphology 25*, 166-176 (1975). © International Society of Plant Morphologists
13-12 All photos USDA. (B) Murray Lemmon. (C-I) Charles E. Herron
13-16 Joe Popp / Anthro-Photo
13-18 FAO

Chapter 14

14-1 David Ewert
14-3 Andrew Greller
14-4 David Ewert
14-5 USDA Photo
14-6 FAO
14-7 Michigan Dept. of Natural Resources
14-10 Uldis Roze
14-13 Andrew Greller
p. 533 Monkmeyer Press Photo Service

Index

abortion, 225, 230
abscisic acid, 371
abscission layer, 368
Acetabularia, 345
acetylcholine, 141, *141*, 142, 175, 178, 181
Achilles tendon, 416
acrosome, 194
actin, 25, 28–29, *31*, 175
active transport, 11–12
adaptive radiation, 289, *290*, 291–296, *293*, *295*, *296*
adenine, *72*, 73–74, 78, 79
adenyl cyclase, 183
ADP (adenosine diphosphate), 52
adrenal cortex, 181
adrenal hormones, 181, 498
adrenal medulla, 181
aggressiveness, 199
agriculture, 476–477, 512–513, 514, 524, *525*, 526

albumin, 101, 102
algae, 342, *343*, 344–345
  blue-green, 326–327, *327*, 333, 342
  freshwater, 442–443
allantois, 219, 220
alleles, 240–242, 273
  See also gene frequencies
allosterism, 47, 48, *48*, 49, 88–89, *90*
alpha chains, 240
alternation of generations, 345–348, *346*, 349
ambush bug, 465
ameboid motion, 110
amino acids:
  evolutionary studies of, 300–301, *302–303*, *304*, 304–309, *305*, *308*
  in protein formation, 37, *44*, 44–45, 81–82, *84*
aminopeptidase, 122–123
ammonia, 326–327, 531
Ammophila, 391–392

amniocentesis, 220–221, *221*
amnion, 220
amniote egg, 218–219, *219,* 400
amniotic fluid, 219, 220
Amoeba, 15, 25
amphibians, 201, 390, 396, *397,* 398, *401*
analogous structures, 298
angiosperms, 347–372
  anchoring function, 366–367
  carbohydrate transport, 363–366
  flowers, 349–351, *350, 351, 352,* 353–356, *355, 356, 357*
  hormonal system of, 178, 367–371, *367, 369, 370*
  leaves, *358,* 358–359, *359,* 368
  phloem, 349, *363,* 363–366, *364, 365, 366*
  roots, 349, *361,* 361–363, 366–367, *369, 369*
  supportive function, 367
  water transport, 360–363, *361, 363*
  xylem, 349, 360–361, 367
angiotensin, 128
angiotensinogen, 128
animals, 375–423
  annelids, 380–382, *381, 382, 383,* 393
  arthropods, *383,* 383–384, *385, 386,* 387, 393, 443
    See also insects
  carrying capacity for, 471–475
  chordates, *see* chordates
  in city parks, 459–460
  coelenterates, 376–378, *377,* 393
  flatworms, *379,* 379–380, 393
  forest, *450–451,* 452–454, 457
  salt marsh, 436–441, *438, 440*
  in soil, 430–433, *431, 433*
  in vacant lots, 463–466, *464, 465*
  See also population control factors, *and specific kinds*
annelids, 380–382, *381, 382, 383,* 393
Anopheles, 493
antibodies, 70, 102, 111, *113*
anticodon, 82
antigen, 111, *113*
ants, *385, 386,* 431, 432, 463–464, *464*
apes, 408–411
aphids, 463–464, *464*
arachnids, 387
arboreal life, 409
Argentine moth, 482
arms, 414–415
arteries, 115–116, 121
arterioles, 116
arthropods, *383,* 383–384, *385, 386,* 387, 393, 443
  See also insects
ascomycetes, 338, 339
Ascophyllum, 344
assimilated energy, 517–518
atherosclerosis, 104, 108
ATP (adenosine triphosphate), 51, 74*n*
  action of, 51–52
  muscle contraction and, 175–176
  structure of, *51*
  synthesis of, 52–56
atrium, 398
Australopithecus, 417–418
autonomic nervous system, 146, 147–148, 155, 156
auxin, *367,* 367–370, *369, 370*
axon, 135

B cells, 111, *112, 113*
baboons, 415, *415, 495*
backswimmers, 444–446
bacteria, 315–326, *321, 322, 324, 325*
  biological warfare and, 325–326
  carrying capacity of, 472
  cell division in, 322
  cell wall, 7, 321–322, 323
  food habits of, 323–324, *325*
  in food web, 517
  in large intestine, 124
  nitrogen-fixing, 531, *532*
  parasites, 489
  pathogenic, 324–326
  photosynthetic, 58
  protein synthesis in, 86–92
  size of, 322, *322*
  in soil, 432
  spores, 323, *324*
  structure of, 321–322
barnacles, 437
Bartholin's gland, *209,* 210
basal bodies, 25, 28, *30*
basal metabolism, 183–186
basidiomycetes, 338, *338, 339*
basilar membrane, 170, *171*
bats, 406
bear, 453–454, 460, 475
beavers, 499–500
bees, 390–391, 461, 464, *465*
behavioral isolation mechanisms, 276–277
beta chain of hemoglobin, 240, 307
bile, 124
bile salts, 124
biofeedback, 148–149
biological environment, 289
biological magnification, 514–516

biological oxidation, 52–54, 57
biome, 428
bird lice, *489,* 489–490
birds, 402–403, *403, 405*
  beak forms, *404*
  in city parks, 460
  DDT and, *515,* 515–516
  eggs of, 201, *219*
  endangered, 474
  forest, 453, 454, 457
birth, *222,* 223–225
birth defects, 49–50, *50,* 70
bitterns, *438, 440*
bladder, 125
bladderwort, 442, *442*
blastula, **216**
blood, 100–112
  distribution in body, 120
  plasma, 100–105
  red blood cells, 19–21, *96,* 105–108
  white blood cells, 19–21, 108–112, *109, 110, 111, 112*
blood clotting, 103–104, *104,* 117, 124
blood group frequencies, 260, *261*
blood vessels, 112–118
  arteries, 115–116, 121
  capillaries, 103, 116–117
  closed system, 113, 114, *114*
  kidney and, 128
  lymphatics, 117
  open system, 113–114, *114*
  regulation of, 120–121
  vascular spasm, 103
  veins, 115, 117–118, *118, 119*
blue-green algae, 326–327, *327,* 333, 342
blue whales, 473–474
BMR (basal metabolic rate), 183–186
bobcat, 475
bollworm, *486–487*
bonds, chemical, 46, 62–65, *63, 64, 65*
bone growth, 198, *198,* 203
bone marrow, 109
boreal forest, 452
Botulinus toxin, 142
bradykinin, 121
brain, 133–134, 149–167
  bird, 403, *405*
  caudate nucleus, 157
  cerebellum, 151, 152, *152*
  cerebral cortex, 154, 155, 160–166, *161, 162, 163*
  development of, 135, *136,* 150–151, *151*
  forebrain, 153–160
  hindbrain, 151–152
  hypothalamus, *see* hypothalamus
  mammal, 406
  medulla, 151–152
  midbrain, 153
  pituitary, *156,* 159–160, 180
  primate, 411
  reptilian, 400
  reticular activating system, *153,* 154–155
  size, 149–150
  thalamus, *153,* 154
breasts, 186, 203, *204*
brown algae, 342, *343,* 344
bryophytes, 345–347, *346*
Bryopsis, 345
bubonic plague, 492
bud development, 368
bullfrogs, 444
bumblebees, 390–391, 461, 464, 465
butterflies, 464

Cactoblastis cactorum, 482
calcitonin, 180
calcium retention, 198
cambium, *363,* 363–364
cancer, 112, 244
canines, 417
canopy trees, 455
CAP protein, 90, *91*
capillaries, 103, 116–117
carbohydrate transport in plants, 363–366
carbon cycle, 528, *529,* 530–531
carbon monoxide, 108
carboxypeptidase, 122–123, 124
cardiac muscle, 174, *175*
carnivores, 514, 517–518
carpel, *357*
carrying capacity, *see* population growth curve
catalysis, 47
cattails, 444, *444*
caudate nucleus, 157
cecum, 124
cell body, 135
cell differentiation, 72, 92
cell division, *see* meiosis; mitosis
cell membranes, 7–17
  bacterial, 322
  flow of water and, 13–15, *14, 15*
  functions of, 11–13
  phospholipids in, 8–12, *9, 10, 12*
  transport across, 10–13
cell metabolism, 33–50
  See also enzymes; respiratory energy metabolism
cell walls, 7

cell walls (*continued*)
  algal, 342, *343*
  bacterial, 7, 321–322, *323*
cells:
  architecture of, 3–32
  complexity of, 6–7
  cytoplasm, 7, 215
  daughter, 247, *248,* 249
  destruction of, 19–21
  diploid, 239–242
  external membranes, *see* cell membranes
  haploid, 249
  internal membranes, *17,* 18, 21–25, *22*
  intracellular movement, 28–29, *30, 31*
  meiosis in, *see* meiosis
  mitosis in, 28, 29, *191,* 239, 346, 349
  muscle, 175
  nucleus of, 17, 21–22, *22*
  sizes of, 5–6
  See also DNA; neurons; protein synthesis
cellulase, 323
cellulose, 7, 37–38, 41
Cenozoic era, 288
centipedes, 384, *387,* 431, 432
central nervous system, *see* brain
centrioles, 25, 28, *28,* 248
Cepaea nemoralis, 252–253
Ceratocystis ulmi, 337
cerebellum, 151, 152, *152*
cerebral cortex, 154, 155, 160–166, *161, 162, 163*
cerebrospinal fluid, 151
cervix, *209,* 210–211, 224
Cetaceae, 406
chameleon, *399, 401*
chance, laws of, 253–256, *255, 256*
character displacement, *290,* 291
Chelonia, 400
chemical bonds, 46, 62–65, *63, 64, 65*
chimpanzees, 474
chipmunks, 452, 454
Chiroptera, 406
chitin, 388
Chlamydomonas, 344–345
Chlorella, 344
chlorophyll, 25, 58–61, 326, *327,* 342
Chlorophyta, 342
chloroplasts, 25, 57, *58,* 333
cholinesterase, 142
chordates, 393–411
  embryological development of, 215–216
  evolution of, 393–396
  See also amphibians; birds; mammals; reptiles

chorion, 219–220
chorionic gonadotropin, 181, 197, 212, *213*
chromosomes, 190
  composition of, 238–239
  defined, 238
  homologous, 239–240, *248,* 252
  independent assortment of, 253–256, *255, 256*
  in meiosis, 247–256, *248*
  in mitosis, *29,* 239
  in polyploid organisms, 280, *281,* 282
  sex, 250–251, *251*
  structure of, *238,* 239
Chrysophyta, 342
chymotrypsin, 122, 124, 300, 301
cicadas, 485
cilia, 25, 28, *30*
circular canals, 171
circulatory system, 97–121
  amphibian, 396, *397,* 398
  bird, 402
  closed, 113, 114, *114*
  component parts of, *see* blood vessels; heart
  evolution of, *98,* 98–100
  fetal, 224
  fish, 396, *397,* 398
  function of, 101
  open, 113–114, *114*
  regulatory mechanisms in, 118–121
  reptilian, 399
city wildlife, *457,* 458–466
  parks, *459,* 459–460
  vacant lots, 460, *461,* 462–466, *463, 464, 465*
class, in classification system, 317, *318*
classification of organisms, 315–320
climatographs, *428*
climax forest, 520
clitoris, *209,* 209–210
cloaca, 405
club fungi, 338
cnidoblast, 378
cochlea, *170,* 170–171
codons, 81–82
coelenterates, 376–378, *377,* 393
coelom, 381
collecting duct, 125
community, defined, 428
companion cell, 364, *364,* 365
competition, 494–498, *495, 496*
condom, 225, 226
cones, 167, *168*
contest competition, 494
continental drift, 287, *288,* 289
contraception, 225–230

contraceptive isolating, 276
contractile vacuoles, 15, *15*
control genes, 238
convergent evolution, *293*, 293–296, *295*, 393
copepods, 387
copulation, 207
corals, 377–378
cord grass, 434–435
corona radiata, 211, *211*
corpus luteum, 202, 205, 212, 213
cortex, 361, *361*, 362
cottony-cushion scale, 480–481, *481*
coupled reactions, 51–53, *53*
covalent bonds, 37*n*, 46
Cowper's gland, *207*, 209
cranial nerves, 151
crayfish, 447
Crick, Francis, 72–75
crickets, 464
cristae, 23, *32*
crocodiles, 400
crossing over, *248*, 252–253
crustaceans, 443
cyclic AMP, 90, 91; 178, 181, 183–184, *184*
cyclic photophosphorylation, 58, *59*
cycling of materials, 527–533, *528*, *532*
Cyclops, 443
cytochrome c, 301, *302–303*, *304*, 304–307, *305*, 309
    cytochrome c oxidase, 309
    cytochrome c reductase, 309
cytokinins, 370
cytoplasm, 7, 215
cytosine, *72*, 73–74, 78, 79

damselflies, 446
Daphnia, 443
Darwin, Charles, 268–269, 291–292, 367
daughter cells, 247, *248*, 249
DDT, 481, 493, 514–516
decomposers, 517
deer, 453, 475, 478, 517–518, 519
deer mouse, 453, 460
delivery of infants, *222*, 224–225
dendrites, 135, *135*
dendritic spine synapses, *166*
DES (diethyl stilbestrol), 182, 203*n*
desmids, 442–443
detritus food chains, 517
deutorostomes, 394
Devonian period, 267, 396
diapedesis, *109*, 109–110
diaphragm, 225, 227, *228*
diatoms, 342, *343*, 344, 442–443

diffusion, 10–11, 12
digestive system:
    annelid, 380–382
    bird, 403
    coelenterate, 378
    flatworm, 380
    human, 121–124
digger wasps, 391–392
dinoflagellates, 342, *343*, 344
dipeptidases, 122–123, 124
diploid cells, 239–242
distal tubule, 125
diving beetles, 444, 445–446
DNA (deoxyribonucleic acid), 37, 39*n*
    composition of, 70–71
    control mechanisms, 86–92
    difference between RNA and, 78–79, *79*
    hydrogen bonds, 63–65, *64*
    mitochondria, 24, *24*
    protein synthesis and, *68*, 71, 78–82
    replication, 71–78, *75*, *77*, 214–215
    structure of, *72*, 72–74, *73*, *74*
    viral, 328–331, *329*, *330*
    See also genes
dolphins, 406
Down's syndrome, 221, 244
dragonflies, 446
Drosophila, 271–273, *272*
duck-billed platypus, 405
duodenum, 123
Dutch elm disease, *336*, 336–337

eagle, *515*
ear bones, 298
eardrum, 170
ears, 154, 160, 161, 169–171, *170*
earthworm, 379–382, *381*, 430–431, *432*, 475
eastern deciduous forest, 449
echidna, 405
echinoderms, 215–216
ecological niche, 273–276
ectoderm, 216, *377*
ectoparasites, 489
edema, 118
egg yolk cell, 6
eggs, *192*, 193, 211
    amniote, 218–219, *219*, 400
    bryophyte, 347
    production, 200–203
    sex chromosomes in, 250–251, *251*
    See also embryonic development; fertilization
ejaculation, 208
electron transport chain, 54, *54*

elementary particles, 23
embryonic development, 71, *212*, 213–221
  fetal membranes, 218–221, *219*
  permanent structures, 214–218, *217, 218*
emigration, 499–502, *500*
emission, 208
emotions, hypothalamus and, 155–158, *157*
endocrine control of birthrate, 498
endocrine glands, *178, 179*, 180–181
endoderm, 216, 377
endodermis, *361*, 362–363
endoparasites, 489
endoplasmic reticulum, *17, 18*
endosperm, 351
endothelium, 115–116
energy flow, 506–507
energy flow through communities, *507*, 507–527
  biological magnification, 514–516
  community productivities, 509–513, *510, 511*
  community structures, 513–517
  consumer use of energy, 517–519, *519*
  energy budgets, 513–517, 519–524, *520, 521, 525, 526*
  food webs, *433*, 516–517
  succession, 521–524, *525, 526*
enzymes, 40–50
  active site of, 47–48
  allosteric inhibition of, 48, *48*
  birth defects and, 49–50, *50*
  as catalysts, 40–42
  control valves, 48
  in digestion, 122–124
  instability of, 46–47
  as proteins, 44–46
  specificity, 42–43
epidermis, 361, *361*, 362
epididymis, 194, 207, *207*
erectile tissues, 207
erection, 207
Escherichia coli, 87, 87–91, *89, 90, 91*, 253, 320n
esophagus, 123
estrogen, 181, 202, 203–206, *206*
ethylene, 371
eugenics, 236
eukaryotes, 331–333, 338
eukaryotic cells, 304–305
eustachian tube, 169–170, *170*
evolution, 267–312
  adaptive radiation, 289, *290*, 291–296, *293, 295, 296*
  chordates, 393–396
  circulatory systems, *98*, 98–100
  continental drift, 287, *288*, 289
  convergent, *293*, 293–296, *295*, 393
  Darwin's theory of, 268–269
  evolutionary changes in species, 273–278, *274, 275, 277, 279*
  evolutionary trees, 296–309
  faunal regions, 283, *284, 285*, 286, 287
  formation of new species, 279–280, *281*, 282–283, *283*
  homology, 297, 297–299, *299*
  human, 418–423
  at molecular level, 300–301, *302–303, 304*, 304–309, *305, 308*
  phyletic, 279, *283*
  process of, 269–296
  reproductive isolation, 275–278, 280, *281, 282, 283*
  species definition, 269–273
exoskeleton, 384, *385*, 388, 390
eyes, 154, 160–161, *167*, 167–169, *218*, 409–410, *410*

facilitated diffusion, 12–13
family, in classification system, 317, 318
fats, 37
faunal regions, 283, *284, 285*, 286, 287
feedback inhibition, 143, *143*
feed-forward inhibition, *143*, 143–144
feet, 416
fermentation, 336
fertilization, 190–191, *192*, 193
  amphibian, 390
  angiosperm, 349–351, *352*
  bryophyte, 347
  external, *192*, 193
  internal, *192, 193*, 207–211
  reptilian, 400
fetal membranes, 218–221, *219*
fibrin, 103–104, *104*
fibrinogen, 101–104
fibroblasts, 111
fiddler crab, 438, *439*
field biology, *see* city wildlife; forest; freshwater pond; salt marsh; soil
filter feeder, 439
finches, *290*, 291–292
fireflies, 276–277, *277*
fish, 201, *395*, 395–396, *397*, 398
flagella, 25, 28, *30*
flatworms, *379*, 379–380, 393
floating plants, *442*, 442–443
florigens, 371
flowering hormones, 371
flowering plants, *see* angiosperms

flowers, 349–351, *350, 351,* 352, 353-356, *355, 356, 357*
food chains, 514, 517
food vacuole, 19, *19*
food webs, *433, 447,* 516–517
foramen magnum, 412, *412*
forebrain, 153–160
forest, *448,* 448–457, 510, *510,* 513
  energy budget, 519–521, *520, 521*
  seasons in, 449, *450–451,* 452–454
  structure of, *454,* 454–457, *456*
  succession and, 522–523, *523*
founder effect, 264
fox, 453
freshwater pond, 441–447
  deep, 442–443
  edge of, *441,* 443–444, *444*
  feeding relationships, 447, *447*
  insects in, 444-446, *445, 446, 447*
  shallow, 443
  stability, 447–448
frog egg, *214,* 215, 216
froghopper, 463
frogs, *401,* 444, 447
fructose, 208
fruit, 354–355, *355,* 368
fruit flies, 271–273, *272*
FSH (follicle stimulating hormone), 180, 183, 194, 196, 200, 202, 206
Fucus, 343, 344
fungi, 333, *334, 335, 336,* 336–338, *337, 338, 339,* 517

GABA (γ-aminobutyric acid), 141, 178, 181
galactose, 11, *11*
galactosidase synthesis, *87,* 87–91, *89, 90, 91*
Galapagos Islands, *290,* 291–292
gall bladder, 123, 124
gamete incompatibility, 277
gametes, 191, 193, 200, 249
gametophyte, 346, 349, 351, *352*
gamma globulins, 102, 111
gastrovascular cavity, 378
gastrula, *214,* 216
Gause's principle, 497
gene expression, 238–247
  alleles, 240–242
  environment and, 242–244
  pleiotropic effects, 245–247
  polygenic effects, 245, *246*
gene frequencies, 257–265
  alleles, 273
  genetic drift, 263–265, *264,* 278

  in human populations, 260–262
  mutation, 262, 273, 278, *308,* 308–309
  probability laws, *257,* 257–259, *258*
  selection, 262–263
gene transmission, 247–253
  crossing over, *248,* 252–253
  independent assortment, 253–256, *255, 256*
  maternal and paternal contributions, 247, *248, 249*
  sex chromosomes, 250–251, *251*
genes, 49, 71, 190
  control, 238
  recombination of, 252–253, *253*
  structural, 88–91, *89,* 238
  See also gene expression; gene frequencies; gene transmission
genetic code, 81–82
genetic drift, 263–265, *264,* 278
genetics, 235–237
  See also gene expression; gene frequencies; gene transmission
genitalia:
  female, *199,* 200–205, *209,* 209–210
  male, *193, 194*–199, *195, 207,* 207–209
genotype, 242
genus, in classification system, 317, 318
geographical isolation, *279,* 280, 284
GH (growth hormone), 180, 182–183
gibberellins, 370–371
gibbons, 414, *414,* 474
gills, 382
glasswort, 435
globulin, 101, 102
glomerular filtrate, 125
glomerulus, 125, 128
glucagon, 124, 181
glucose:
  active transport, 11, 12
  difference between galactose and, 11, *11*
  oxidation of, 52, 53, 54, 57
  in photosynthesis, 365
  polymerization, 38, *38,* 40
  resorption, 125
  respiratory breakdown of, 54–56, *55*
glucose-6-phosphate, 42–43
glutamic acid, 141
glycine, 141
glycolysis, 54–55, *55,* 56
Golgi body, *17,* 18
gonads, *see* ovaries; testes
gorillas, 414, *415,* 417, *474*
Graafian follicle, *201,* 201–202, 211, *211*
grana, 25
granulocytes, 110, *110*

grasses:
  prairie, 519
  salt marsh, 434–435, 439, *440*
  in vacant lots, 462–463, *463*
grasshoppers, *388,* 464, *500,* 500–501, 518
grease, 8–9, *9*
grebes, 447
green algae, 342, *343,* 344–345, 443
green plants, *see* plants
GRF (growth hormone releasing factor), 180, 182
gross production, 508–509
growth inhibitors, 371
guanine, *72,* 73–74, 78
gymnosperms, 353
gypsy moth larva, 486–487

habitat, as isolating mechanism, 275–276
hair cells, 170–171
hair growth, 198
hallucinogenic drugs, 144
hands, 416–417
haploid cells, 249
Hardy-Weinberg equation, 259
hares, 478–479
haustorium, 337
hawkweed, 482–483
hearing, sense of, 154, 160, 161, 169–171
heart, 97–98, 112–115
  amphibian, *397,* 398
  bird, 402
  contraction of, 114–115
  energy expenditure of, 114–115
  regulation of, 119, 120–121
heme groups, 307
hemlock, 449
hemoglobin, 105–108, *107,* 307–309, *308*
hepatic portal vein, 124
herb layer, *455,* 456
herbivores, 513–514, 516–518
herons, 447
heterozygotes, 240–242
hierarchical behavior, 494–496, *495*
hindbrain, 151–152
histamine, 121
histone, 238
hominids, 418*n*
Homo erectus, 421
Homo sapiens, 411–421
  body form, 411–416
  carrying capacity for, *475,* 475–477
  evolution of, 416–421
  genetic frequencies in, 260–262
  parasitism, 490–494, *491, 493*

  See also circulatory system; digestive system; kidney; nervous system; reproduction
Homo sapiens neanderthalensis, 419–420
homologous chromosomes, 239–240, *248,* 252
homology, *297,* 297–299, *299*
homozygotes, 239
honeydew, 463
hoppers, 463
horizons, soil, *429,* 429–430
hormones:
  adrenal, 181, 498
  defined, 158*n,* 179
  hypothalamic, 155, *156,* 158, 159–160, 180, 182–186, *183, 184, 185,* 194, *194,* 200, 202, 206
  major, table of, 180–181
  plant, 178, 367–371, *367, 369, 370*
  regulation of circulatory system, 120–121
  sex, 181, 182, 186, 196–199, *197,* 202, *202,* 203–206, 211–212, 224, 498
horse, 273–276, *274, 278*
human lice, 490–491, *491*
humans, *see Homo sapiens*
humus, *429,* 430
Hydra, 377, 443
hydrochloric acid, 123, 124
hydrogen bonds, 63–65, *64*
hydroids, 377, 378
hydrolysis, 122
hydrophilia, 8–9, *9*
hydrophobia, 8–9, *9*
hydrostatic pressure, 15
hyperpolarization, 143
hypha, 333, *335,* 336
hypothalamus, 155–160, *157,* 194
  blood supply, 158, 159
  hormones secreted by, 155, *156,* 158, 159–160, 180, 182–186, *183, 184, 185,* 194, *194,* 200, 202, 206

Icerya purchasi, 480
ICSH (interstitial cell stimulating hormone), 198
iguana, *401*
ileum, 415
immune cells, 111–112, *112*
implantation, 220
independent assortment, 253–256, *255, 256*
induction, 89, *90,* 218
inferior ovary, 354, *355*
inguinal canal, 194
inhibitory neurons, 142–144, *143*
inner ear, 169, *170,* 171

insects, *383*, 383–384, *385, 386,* 387–392, *388, 389*
  abundance of, 387
  bacterial control of, 324–325
  behavior, 390–392
  metamorphosis, 388–389
  nervous system, 390
  pollination by, 354
  pond, 444–446, *445, 446, 447*
  respiration, 388
  size of, 390
  structure of, 384, *385, 386,* 387
  in vacant lots, 463–466, *464, 465*
  wings of, 389
  See also population control factors
insulin, 124, 181
intercourse, 210
interstitial cells, *195*, 196
interstitial fluid, 117
intrauterine device (IUD), 225, 227–228, *229*
ionic bonds, 62, *63*
ionic pump, 128
iris, 169
isotopes, 75–76

Japanese beetle, 324–325, *325*
jellyfish, 377, 378

karyogram, *250*
katydids, 464
Kelvin, Lord, 268–269
kidney, 125–128, *126, 127,* 181
  artificial, 128–129, *129*
  function of, 125
  transplanted, 129–130
kidney tubule, 125
killer whale, 293–295
kingdom, in classification system, 317–320, *319*
Krebs cycle, *55,* 55–56

labia majora, *199,* 203
labia minora, *199,* 203
lacewing, 465
lactose metabolism, 87, 87–91, *89, 90, 91*
lactation, 225, 227
ladybird beetle, 480–481
land plants, *see* angiosperms
large intestine, 124
larva, 389, 398, 494
leaf area index, 455
learning, 164–166
leaves, *358,* 358–359, *359,* 368
leeches, 380
lemurs, 474–475

lens, 168, 169, 218, *218*
Lesch-Nyhan syndrome, 247, 251
LH (luteinizing hormone), 180, 183, 194, 197–198, 202, 206
lice, 488, *489,* 489–491, *491*
limiting factors, 429
lions, 479
lipase, 124
lipids, 8–12, *9, 10, 12,* 102
liver, 123, 124
liver fluke, 379
liverworts, 345
lizards, 399, 400
lobe-finned fishes, 395, *395,* 396
locusts, *500,* 500–501
loop of Henle, 125, 126–128, *127*
LRF (luteinizing hormone releasing factor), 180
lugworm, 436–437
lungfish, 279, 289, 396
lungs, 224
lymph, 117
lymph nodes, 117
lymphatics, 117
lymphocytes, 70, 110–112, *111, 112*
lynx, 478–479
lysosomes, *17,* 19–21, *19,* 110

macromolecules, 36–39, *39*
  See also fats; nucleic acids; polysaccharides; protein
macrophages, 110
magnolia flower, *357*
malaria, *493,* 493–494
malleus, 170
maltase, 124
mammals, 405–406, *407*
  brains of, 406
  convergent evolution, 294–296, *295*
  eggs of, 201
  metabolic rates of, 405
  primates, 408–411, *410*
  reproduction, 193
man, *see Homo sapiens*
manatees, 406
Marfan's syndrome, 246–247
marihuana plant, *482,* 483
marine worms, 379, 380, 382, *382*
marsh hawks, 440
matrix substance, 23
mayflies, 446
medulla, 151–152
medusa, 377, *377*
megaspore, 351

meiosis, 190, 191, *191,* 193, *200,* 247–256, *248, 352*
   crossing over, *248,* 252–253
   egg formation and, *200,* 200–201
   in fungi, 338, *339*
   independent assortment, 253–256, *255, 256*
   maternal and paternal contributions, 247, *248,* 249
   sex chromosomes, 250–251, *251*
   sperm production and, 194, *195*
membrane depolarization, 138
memory, 164–166
Mendel, Gregor, 236–237
menopause, 206
menstrual cycle, *201,* 205, 206
Meselson-Stahl density gradient experiment, 75–76, *77*
mesoderm, 216, 217, 379, *379*
mesoglea, 377
Mesozoic era, 400
metabolism, 33–50
   See also enzymes; respiratory energy metabolism
metamorphosis, 388–389
mice, *35, 217,* 437, 453, 460, 464, 498, 518
microfilaments, 25–26, *26*
microtubules, 25, *26,* 26–29, *27, 28, 29, 30, 31*
microvilli, *17, 27, 27,* 171
midbrain, 153
middle ear, 169–170, *170*
migration, 278
millipedes, 431, 432, 518
mites, 432
mitochondria, *17,* 23–24, *24, 32, 184,* 185, 332–333
mitosis, 28, *29, 191,* 239, 346, 349
molecules, defined, 10, 37*n*
moles, 452
molting, 388
monerans, 320–327, 338
   See also bacteria; blue-green algae
monkeys, 408–411
mons pubis, *199,* 203
mosquitoes, 446, 493, 514
mosses, 345–347, *346,* 456
moths, 388, 464, 482
motor cortex, 161–163, *163*
motor neurons, 161
mouth, 123
MSH (melanocyte-stimulating hormone), 180, 182
multicellularity, evolution of, *98, 99*
muscle fiber, 174–176, *176,* 177
muscles, 174–177, 413–414
   cerebellum and, 152, *152*
   contractile process, 174–176, *176,* 177
   motor cortex and, 161–163
   types of, 174, *175*
mussels, 437, 438–439
mutation, 262, 273, 278, *308,* 308–309
mutualism, 502, 517
mycorrhizae, 337
myelin sheath, *139,* 140
myoglobin, 307–309, *308*
myosin, 25, 28–29, *31,* 175–176, *176,* 177
Myxotricha paradoxa, 331–332

NAD, 54, 61
NADP, 61, *61*
Neanderthal, 419–420
nematocysts, 378
nematodes, 432
nephrons, 125, 126
nerve impulse, 135–140
nerve reflex, *146,* 146–147
nervous system, 133–173
   central, *see* brain
   classification of, 134
   development of, 135, *136*
   formation of, *214,* 216–217
   functions of, 134
   hearing, 154, 160, 161, 169–171
   insect, 390
   muscle movement and, *146,* 146–147, 152, *152,* 175–176
   neurons, *see* neurons
   pain, 154, 172–173
   peripheral, 145–149
   proprioceptive sense, 172
   sight, 154, 160–161, *167,* 167–169, 218, *218,* 409–410, *410*
   smell, sense of, 154, 172, 412
   taste, sense of, 154, 171
net production, 508–509
neurohumors, 178, 181
neuromuscular junction, 175
neurons, 135–145, *166*
   fast transport, 144–145
   inhibitory, 142–144, *143*
   motor, 161
   nerve impulse, 135–140
   synapse, 140–144, *141, 143,* 166, *166*
nitrogen cycle, 531, *532,* 533
nitrogen fixation, 326–327, 531, *532,* 533
norepinephrine, 141, 178, 181
notochord, 217, 394
nuclear membrane, 21–23, *22*
nuclear pores, *17*

nucleic acids, 37, 81–82
    See also DNA; RNA
nucleolus, *17,* 83
nucleotides, 37, 70–73, *72, 73, 75,* 81–82
nucleus of cell, *17,* 21–22, *22*
nursing, 186, *188*

olfactory receptors, 171, *172*
operator protein, 88–91
operon, 90–91
optic vesicle, 218, *219*
order, in classification system, 317, *318*
organochlorines, 515
orgasm, 207
osmotic pressure, 13–15, *14, 15,* 126–128, *127,* 360, 362, 365
Osteichthyes, 395
osteoporosis, 198
ova, *see* eggs
oval window, 170
ovaries, *199,* 200–206, *211,* 368
oviducts, *199,* 202–203, 211
ovulation, 200–203, 227, *227*
ovule, 351, *352*
oxidation, 52–54, 57
oxygen, transportation of, 106–108, *107*
oxytocin, 180, 186, 224

pain, sense of, 154, 172–173
Palexorista, 486–487
pancreas, 123, *124,* 181
pancreatic amylase, 124
pancreozymin, 181
Paramecium, 28
parapodia, 382
parasitism, 488–494
    bacterial, 324–326
    human population control and, 492–494, *493*
    lice, 488, *489,* 489–491, *491*
parasympathetic nerves, 120–121
parks, wildlife in, *459,* 459–460
Pasteurella pestis, 492
pathogenic bacteria, 324–326
Pediculus humanus, 490–491, *491*
pelvic girdle, 415
pelvis, 203
penis, 207, *207*
pepsin, 122, 123, 124
peptides, 83, *84,* 122, 123, 124, 182
peripheral nervous system, 145–149
peristalsis, 123, *123*
periwinkles, 437–438
permease, 12–13
petals, 350, *350*

petiole, 368
Phaeophyta, 342
phagocytosis, 19, *19,* 110, 219
phenotype, 242
phenylalanine, 49–50, 242
phenylketonuria (PKU), 49–50, *50,* 70, 242
phenylpyruvate, 49–50
pheromone, 498
phloem, 349, *363,* 363–366, *364, 365, 366*
phosphate, 439
phospholipids, 8–12, *9, 10, 12*
Photobacterium fisherii, 320
photosynthesis, 56–61, 344, 508
Phthirus pubis, 490–491, *491*
phyletic evolution, 279, *283*
phylum, in classification system, 317, *318*
Phytophthora infestans, 336
phytoplankton, 442–443
pistil, 350, *350*
pituitary, *156,* 159–160, 180
placenta, 181, 213
placental hormone, 197
planarians, 443
plant cells, 15, 28, *28*
plant starch, *38,* 40
plants, 341–372
    bryophytes, 345–347, *346*
    in city parks, 459–460
    flowering, *see* angiosperms
    forest, 454–457
    freshwater pond, *442,* 442–444, *444*
    hormonal system of, 178, 367–371, *367, 369, 371*
    parasites, 488–489
    polyploidy in, 280, *281,* 282, 283, *283*
    predation, 480–482, *481, 482*
    reproduction, 345–346, 349–356
    respiration, 508
    succession and, 521–524, *522, 525, 526*
    in vacant lots, 462–463, *463*
    See also algae; bacteria; fungi
plasma, 100–105
plasma membrane, *17*
platelets, 103, 104
Platyhelminthes, *379,* 379–380
pleiotropic effects, 245–247
polar bodies, 201
polar nuclei, 351
pollen, *350,* 350–351, *351,* 353–354, 368
polygenic effects, 245, *246*
polymers, 37
polyp, 377, *377*
polypeptides, 158
polyploidy, 280, *281,* 282, 283, *283*

polysaccharides, 37
    See also glucose
pond, *see* freshwater pond
pondweeds, 443
population, defined, 470*n*
population control factors, 477–502
    competition, 494–498, *495, 496*
    emigration, 499–502, *500*
    endocrine control of birthrate, 498
    See also parasitism; predation
population genetics, *see* gene frequencies
population growth curve, 470–477, *471*
portal veins, 159
praying mantis, 465, *465*
predation, 478–485, *486–487*, 488
    large animals, 478–480, *480*
    numerical and functional responses, 483–485, *484*
    organic diversity and, 485, 488
    prey defenses, 482–483
    small animals and plants, 480–482, *481, 482*
pregnancy, 211–221
    fetal membrane development, 218–221, *219*
    permanent structure development, 214–218, *217, 218*
    progesterone and, 211–212
prehensile limbs, 411
PRF (prolactin releasing factor), 180
prickly pear cactus, *481*, 481–482
primary speciation, 279–280, *283*
primates, 410–413
probability, laws of, *257*, 257–259, *258*
progesterone, 181, 186, *202*, 203, 204–206, 211–212, 224
prokaryotes, 322
prolactin, 180, 182
promoter protein, 88–91
proprioceptive sense, 172
prostaglandins, 181, 186, 208, 230
prostate, 207, 208
protein synthesis, *17*, 18, 78–86
    control of, 86–92
    DNA in, *68*, 71, 78–82
    genetic code, 81–82
    memory trace and, 166
    messenger RNA in, 79–81, *80, 84, 85*, 86, 88–92, *89, 91*
    ribosomal RNA in, 83, *84, 85*, 86
    transfer RNA in, 82, *83, 84*, 86
proteins, 37
    in cell membrane, 11–12, *12*
    digestion of, 122–123
    enzymes as, 44–46
    function in cells, 37

plasma, 101–104
    sequence analyses, 300–301, *302–303*, 304, 304–309, *305, 308*
    structure of, *44*, 44–46, *45*
prothrombin, 104
protistans, 331–333, *332*, 338
protostomes, 394
protozoa, *28*, 443
proximal tubule, 125
pseudofeces, 439
psilopsids, 347
pupa, 392
pupil, 169
Purkinje cells, 143
pus, 21
Pyrrophyta, 342
pyruvate, *55*, 55–56

rabbit, 410, *410*, 453
raccoon, 447, 453
radius, 408
ragweeds, 463
Ramapithecus, 417
red algae, 342, *343*, 344
red blood cells, 19–21, *96*, 105–108
red-shouldered hawk, *496*, 497
red water mites, 443
reduction, 54
redwood tree, 367, *367*
releasing factors, 182
renin, 128, 181
repressor protein, 88–91, *89*
reproduction, 189–230
    amphibian, 390
    angiosperm, 349–356
    bacterial, 322
    birth, *222*, 223–225
    bryophyte, 345–347, *346*
    cellular, *see* meiosis; mitosis
    embryonic development, 213–221
    estrogen, 202, 203–206, *206*
    fertilization, *see* fertilization
    fungal, 338, *339*
    ovaries, *199*, 200–206, *211*, 368
    progesterone, 181, 186, *202*, 203, 204–206, 211–212, 224
    reptilian, 193, 400
    testes, 181, *193*, 194–199, *195*
    testosterone, 181, 196–199, *202*, 203
    viral, 328–330, *329, 330*
reproductive isolation, 275–278, 280, *281*, 282, *283*
reptiles, 398–400, *399, 401*
    brain of, 400

convergent evolution, 293–294
  eggs of, 201, 218–219, *219*
  reproduction, 193, 400
respiration, 382, 388, 399, 508, 518
respiratory energy metabolism, 38, 50–56
  biological oxidation, 52–54, 57
  coupled reactions, 51–53, *53*
  electron transport, 54, *54*
  glycolysis, 54–55, *55*, *56*
  Krebs cycle, *55*, 55–56
reticular activating system, *153*, 154–155
retina, 167–169, *168*, *218*
rhodopsin, 168–169
Rhynchocephalia, 400
rhythm method, 225, 226–227
ribose, 18, 79, *79*
ribosomes, *17*, 18, 83, *84*, *85*, 86
rickettsiae, 326
ripeners, 371
RNA (ribonucleic acid), 37
  composition of, 78–79, *79*
  differences between DNA and, 78–79, *79*
  messenger, 79–81, *80*, *84*, *85*, 86, 88–92, *89*, *91*, 239
  ribosomal, 83, *84*, *85*, 86
  structure of, 79
  transfer, 82, *83*, *84*, 86
  viral, 331
robin, 453, 454
Rodolia cardinalis, 480–481
rods, 167, *168*
root cells, 368
root hairs, 361, 362
roots, 349, *361*, 361–363, 366–367, 369, *369*
rotifers, 443
round window, 170–171

sac fungi, 338
sarcoplasmic reticulum, 176, *177*
salamanders, 396, *396*, 447
Salicornia, 435
salt marsh, 433–441, *434*, *437*
  animals, 436–441, *438*
  feeding relationships of, *440*, 440–441
  grasses, 434–435, *439*, *440*
salts, 101, 105, 125, 135–140
sarcomeres, 175, 176
sawfly, 483–485, *484*
schistosomiasis, 379
Schwann cells, 140
scramble-type competition, 494
scrotum, 194
sea urchin eggs, 216
search image, 483

sebaceous glands, 198
second-story vegetation, *455*, 456
secondary sexual characteristics, 196, 198, 203
secretin, 181
secreting cells, *17*, 18
seeds, 351, *352*, 353, 354–355, *355*, *367*, 367–368, 371
selection, 262–263
semen, 207, 208
seminal vesicles, 181, 207, 208
seminiferous tubules, 194, *195*, 196
sensory association area, *161*, 164
sensory cortex, 161, *162*, 163
sepals, 350, *350*
serotonin, 178
sex, *see* reproduction
sex chromosomes, 250–251, *251*
sex drive, 199
sex hormones, 181, 182, 186, 196–199, *197*, *202*, *202*, 203–206, 211–212, 224, 498
short-tailed shrew, 484, 485
shrew, 484, *484*, 485
sickle cell genes, 240–242, *241*, 263
side nectar, 354, *355*
sieve tube cell, 364, *364*
sight, sense of, 154, 160–161, *167*, 167–169, 218, *218*, 409–410, *410*
Sirenia, 406
sister species, 280
skull, 412, *412*
slugs, 432, 465
small intestine, 123, 124, *124*, 181
smell, sense of, 154, 172, 410
smooth muscles, 174, *175*, 186
snakes, 400
soap, 8–9, *9*
social behavior, 411
sodium, 101, 105, 125, 135–140
sodium bicarbonate, 124
sodium pump, 128
soil, 429–433
  animals in, 430–433, *431*
  feeding relationships in, 432–433, *433*
  horizons, *429*, 429–430
sowbugs, 464
Spartina, 434–435, 439
speciation, 279–280, *283*
species:
  in classification system, 317, *318*
  evolution of, *see* evolution
sperm, 193
  path of, 207, *207*, 210–211
  production of, 194, *195*, 196
  sex chromosomes in, 250–251, *251*

sperm ducts, 194, 207, *207,* 226
spermatids, 194
spermicides, 225, 227
Sphagnum, 345
sphincter muscle, 116–117
spiders, 466
spinal column, 412–413, *413*
spinal cord, 145, *150*
spirochetes, 331–332
Spirogyra, 343, 344, 443
spores, 323, *324,* 346–347
sporophyte, 346, 349
Squamata, 400
stamen, 350, *350*
stapes, 170
starch, 37–40, *38*
stems, 360–361
Stentor, 443
steroid pill, 225, 228–229, 230
stigma, *350,* 351
stomach, 123
stomata, 359, *359*
striated muscle, 174, *175, 176*
stroke, 104
strontium-90, 109
structural genes, 88–91, *89,* 238
    See also gene expression; gene frequencies; gene transmission
suberin, 362
subspecies, 270
succession, 521–524, *522, 525,* 526
sugars, 37
synapses, 140–144, *141, 143,* 166, *166*
syndrome, 246–247

T cells, 112, *112*
T system, 176, *177*
tadpole, *20,* 20–21
tapirs, *275,* 275–276
taste, sense of, 154, 171
taste buds, 171
taxonomy, 315–320
teeth, *408, 415,* 415–416
temperate viruses, 328, *329,* 330
territorial behavior, 494–496, *495*
testes, 181, *193,* 194–199, *195*
testosterone, 181, 196–199, *197, 202,* 203
thalamus, *153,* 154
thermodynamics, second law of, 35–36
thrombin, 103
thymine, *72,* 73–74, 79
thymus, 112
thyrocalcitonin, 180
thyroid, 180, 183

thyrotropin, 183–184, *184*
thyroxin, 180, 183–186, *184, 185*
toads, 276, *401*
Tokophrya, 15, *15, 334*
tongue, 171
topsoil, *429,* 430
tracheophytes, 341–342, 347–349
    See also angiosperms
transaminase, 50
transduction, 329–330, *330*
transpiration, 358–359, 360
trees, 519, *520*
    canopy, 455
    See also forest
TRF (thyrotropin releasing factor), 158, 180, 183–186, *183*
trophic levels, 516–517
trypsin, 122, 124, 300, 301
tse-tse fly, *374, 385*
tubal ligation, 225, 229–230
tubulin, 27, *28,* 28–29
turtles, 400
twins, 242–244, 270
typhus, 493

ulna, 410
Ulva, 343, 344, 345
uracil, 78–79, *79*
urethra, 208
urine, 125–128, *127*
uterus, 186, *199,* 203, 204–205, *205, 209,* 210, 211, 224

vacant lots, wildlife in, 460, *461,* 462–466
    insects, 463–466, *464, 465*
    plants, 462–463, *463*
vagina, *199,* 203, *209,* 210, 224
vaginal epithelium, 203
Valonia, 345
Van der Waals bond, 64, *65*
vas deferens, 194, 196
vascular spasm, 103
vasectomy, 196
vasoconstriction, 117
vasodilation, 117, 121
veins, 115, 117–118, *118, 119*
ventricle, 398
vertebrae, 415, *418*
vertebrates, *see* chordates
vesicle, *17,* 18
villi, 124, *124*
viruses, *328,* 328–331, *329, 330*
vitamin K, 124
voluntary nervous system, 146–147

Volvox, 343, 344
Vorticella, 443

wasps, 391–392, 464, 465–466, 485, 486–487
water:
    osmotic pressure, 13–15, *14, 15*
    phospholipids and, 8–10, *9, 10*
    resorption, 125–128
    solvent property of, 8, 100–101
    transport in plants, 360–363, *361, 363*
water boatmen, 444, *445,* 445–446
water lilies, 443, *443, 444*
water striders, 446, *446*
Watson, James, 72–75
weak bonds, 46, 62–65, *63, 64, 65*
whales, 406, 473–474

white blood cells, 19–21, 108–112, *109, 110, 111, 112*
wildebeest, 479–480, *480*
withdrawal, 225, 226
wolves, 478, 479, 495–496

X chromosomes, 250–251, *251*
xylem, 349, 360–361, 367

Y chromosomes, 250–251, *251*
yellow jackets, 465–466
yolk, 6
yolk sac, 219, 220

zebras, 310–311, *312*
zygote, 191, 211, 249